AF517673

SALINITY TOLERANCE IN PLANTS :
Methods, Mechanisms and Management

Dr. B.K. Garg (b. 1947), obtained M.Sc. and Ph.D. degrees from CCS, H.A.U. Hisar in the years 1972 and 1977 respectively. He served at Central Arid Zone Research Institute (CAZRI), Jodhpur from 1977 to 2009 in various capacities from Scientists S-1 to Principal Scientist (Plant Physiology). Dr. Garg has specilization in stress physiology and worked on Drought and Salinity problems of the arid zone in more than 20 research projects. He has published over 100 research papers, 5 books, 50 chapters in books and many review articles. Dr. Garg was FAO consultant at Republic of Yemen in 1998 for optimum use of saline water in agriculture. He has participated in several International Projects like BIOSAFOR (2006-2009) and Indo-US project on Drought Tolerance (1990-1998). He has attended numerous national and international conferences and project meetings. Dr. Garg has special expertise in research techniques and methodologies related to plant metabolism, photosynthesis and water relations. Since his retirement in 2009, he has published several research papers and two books namely Plant Analysis and Salinity Tolerance in Plants.

Dr. I.C. Gupta (b. 1941), was awarded Ph.D. on a subject thesis of environmental soil science by CAZRI/University of Jodhpur in 1968. He worked at Central Soil Salinity Research Institute from 1970 to 1986 (Principal Scientist). Dr. I.C. Gupta served Central Arid Zone Research Institute from 1986 to 2001 where he worked as Head of Regional Research Station, Bikaner and Head, Division of Natural Resources and Environment. A Fellow of Indian Water Works Association, United Writers Association of India & Indian Society of Salinity Research Scientists, Dr. Gupta was elected Vice-Chairman of Arid Zone Research Association of India. He has been working as Secretary of Indian Society of Salinity Research Scientists & Chief Editor Current Agriculture since 1977. Dr. Gupta had been conferred several awards notable being Glaxo Scientific Honours Club, Dr. Rajendra Prasad Puraskar, First Prize, Outstanding Book Award by ISAE, Dr. Gorakh Prasad Puraskar First Prize, S.P. Unvala Memorial Prize by IWWA and Life Achievement Award by IARS. He was nominated as member of the 'Environment Conservation Council of Rajasthan' and of 'Special Committee on Environment Impacts' of INCID, Ministry of Water Resources, Govt. of India. He has about 250 publications including 30 books and two inventions (Jaltripti and Jalshuddhi). Dr. Gupta was appointed Ph.D. examiner by ten universities. He was also nominated for one of the most INTRIGUING PEOPLE of 2003 by American Biographical Institute, North Carolina, USA, after close scrutiny of biographical archives of over one million individuals.

SALINITY TOLERANCE IN PLANTS : Methods, Mechanisms and Management

2nd Revised and Enlarged Edition

B.K. Garg
Formerly Principal Scientist (Plant Physiology),
Central Arid Zone Research Institute, (CAZRI)
Jodhpur

I.C. Gupta
Formerly Principal Scientist,
CSSRI, Karnal
Head Regional Research Station, Bikaner
Head, Division of Natural Resources and Environment
Central Arid Zone Research Institute, Jodhpur

Published by:

Scientific Publishers (India)
5-A, New Pali Road, P.O. Box 91,
Jodhpur – 342 001 (India)
E-mail: info@scientificpub.com
Website: www.scientificpub.com

Print: 2019

ISBN: 978-93-88449-34-2 (Hardbound)
978-93-88449-36-6 (E-book)

Printed in India

PREFACE

Population pressure, shortage of good arable land and good quality waters are forcing crop production into more marginal environments facing abiotic stresses particularly drought and salinity and thereby limiting the production and productivity of staple food crops. The situation is assuming serious proportion, as almost half of the existing irrigation system of the world is under the influence of secondary salinity, sodicity or waterlogging. Therefore, to maintain productivity of existing arable land, the sustainability of the agricultural and irrigation systems that has been generated at a huge cost, is more important than immediate increase in yield. Much work done in the last century in several countries has increased our understanding of the genetics and physiology of salt tolerance of plants. Crop responses to salt stress are governed by a number of complex and interrelated, morphological, physiological, and biochemical processes. However, we still do not have a complete understanding of the underlying mechanisms of salt tolerance.

The salt affected soils and saline irrigation waters widely spread in the arid and semi-arid regions of world pose a serious problem for crop production. Reduction in plant growth and crop yields under saline environment has been reported as a result of adverse changes in the physiological processes including the salinity induced effects on the nutrient uptake and metabolic processes. Intensive studies have been undertaken on the responses of several crops and their genotypes to the diverse salinity conditions. A critical evaluation and thorough understanding of the principles governing the plant behavior under adverse saline environment is of paramount significance for developing salt tolerant genotypes with high yield potential through modern breeding techniques and selection procedure, as such genotypes have a key role in reclamation and utilization programme of the salty wastelands.

A vast body of scientific information generated on this topical subject is scattered in numerous publications. We have made an attempt in putting together this valuable information in the form of the present book. The book comprises twelve chapters, of which the first one attempts to trace the origin and history of development of this undesirable menace and lucidly describes the various research and development measures adopted to combat the problem. The chapter two deals with the nature and properties of the salt affected soils and waters including the latest concepts regarding their physical, chemical and biological characteristics, which form the basis for reclamation technology development. Various parameters for quality of irrigation water have been also discussed. In chapter three the principles of amelioration and management of saline soils and saline waters have been discussed at length. These include land development, leaching and drainage methods, selection of salt tolerant crops and cropping sequences and optimum use of saline waters..

Chapter four deals extensively with the plant responses to salinity, crop tolerance at various growth stages and latest criteria followed for evaluating the same. In addition, the crop responses to soil sodicity have also been discussed which are particularly relevant under the Indian conditions. Mechanisms of salt injury in plants are often attributed to the osmotic and ionic effects, while the effect of nutrient imbalance is often ignored. However, we have adopted a balanced approach integrating all these mechanisms in chapter five. The chapter six deals comprehensively with the salinity effects on the metabolic processes as the salt induced growth inhibition is generally accompanied by many metabolic disturbances and changes in enzymatic activities. The recent information in respect of photosynthesis, respiration, carbohydrate metabolism, nitrogen and protein metabolism, lipid metabolism, enzyme dynamics and growth hormones has been elaborated systematically. The process of nodulation and symbiotic nitrogen fixation in legumes that are more sensitive to salinity has also been discussed in detail.

In chapter seven the interaction of essential nutrients with salinity has been presented in great details. The alleviation of adverse effects of salt stress through adequate nutrient management has received special attention. The role of calcium in alleviation of salinity effects has also been given due attention. We have presented systematic information on the relative effects of salinity as a function of environmental conditions in chapter eight, and have stressed the need of focussing future research in this area. The physiological and biochemical mechanisms of salt resistance forms the subject matter of chapter nine. Despite the vital significance of salt resistance in crop improvement, much less attention is given because of multitude of factors involved in the complex phenomena including both avoidance and tolerance. Another chapter (chapter 10) has been incorporated in the revised edition dealing with molecular mechanisms of salt tolerance. Recent information on antioxidant defence ion homeostatis osmolytes and transcription factors is included with emphases on molecular mechanisms related to these important factors. Updated knowledge on bioengineering and signaling proteins has also been added in this chapter.

The chapter eleven deals with the various methods of introducing salt tolerance in plants such as breeding, genetic variations, physiological approaches, tissue culture, somaclonal variation, somatic hybridization and recombinant DNA technology. Recent literature on role of genetic markers and QTLs in salt tolerance has been incorporated and discussed in this chapter. Another addition is upto date knowledge on marker assisted selection (MAS) for salt tolerance in selected crops. Chapter twelve describes Afforestation and Agroforestry techniques in salt affected soils with emphasis on salt-tolerant species and suitable tree-crop combinations for different situations. Role of biodrainage has also been discussed at length in the revised version of the book.

The entire subject matter of the revised edition of this book is supported extensively with Indian data along with appropriate illustrations and tables. In all the chapters recent information has been incorporated to present the latest knowledge on salinity tolerance of crops. We hope that the book will be of tremendous benefit to the research workers, teachers, students, planners and all those who are interested in utilizing saline soils and saline ground waters for crop production and agroforestry. We understand that present book will also be a very good addition to the scientific literature in the libraries of the research institutes and universities located in such areas of the world as are afflicted with the problem of salinity.

B.K. Garg

I.C. Gupta

CONTENTS

Chapter - 1

HISTORICAL DEVELOPMENTS

INTRODUCTION

Although the knowledge of fertile and infertile lands called as `Urvar' and `Anurvar' is as old as farming but the scientific knowledge of the factors causing infertility came into being only in the beginning of twentieth century. `Anurvar' soils were also called as `Usar soils'. `Usar' word in real sense is derived from `Ushtra' a Sanskrit word. In the past, all such lands which used to be unfit for agriculture were called as `Ushtra'. Apart from such lands as were adversely affected with and climate or scarcity of water, lands adversely affected with excess of salts, neutral (NaCl) or alkali ($NaHCO_3$, $Na_2CO_3$), were also called as `Reh' by geologists in mid-nineteenth century to characterise the appearance of salt efforescence on the surface of lands, as a result of the introduction of canal irrigation. The word 'Saline, meaning salt like was used for the surface soil covered with white salt efforescence. There may be several factors for the development of saline wastelands but since their rise is timed with the birth of canal irrigation, the latter has been held more often than not as one of the most potential factors responsible for their formation. In this context, it is desirable to trace out the development of canal irrigation in India.

Canal Irrigation

Archaeologists have found a canal constructed dating back to the Pre-Harappan period at Kunal (Hisar, Haryana). It is reported that this canal was linked to the Saraswati river about 5000 years ago. The excavation and remnants of Kunal reveal that its civilization dates back to 3000 B.C. and is related with Pre-Harappan period.

Irrigation is an age old practice in India. During the Vedic period (C 200-600 B.C.) the peasants dug wells and constructed canals to apply water to the crop lands. Till this period there is no record of the rise of salinity in irrigated tracts.

Western Jamuna Canal

The development of canal irrigation began in the second half of the 14th century at the initiative of Sultan Feroz Shah Tughlak (1351-1388), a pioneer in the history of canal irrigation. During his reign extensive irrigation works were planned and executed to stabilise the agricultural economy. Five big canals were dug to distribute the waters of Jhelum, Jamuna and Sutlej for irrigation. The most important was the Western Jamuna Canal, the oldest canal of north India, constructed in 1355, which carried the waters of Jamuna river to the newly constructed city of Hisar Firiza (Hisar), hunting grounds at Safidon, orchards of Hansi and additionally provided irrigation to the limited cultivated areas in what are today known as Karnal, Rohtak, Hisar and Jind districts. In aligning it, advantage was taken of natural depressions or drainage channels whose slope and direction were found suitable. The channel thus took the form of a linked series of drainages and depressions rather than a canal as the word is understood today. Perceiving that there was a great scarcity of water, two streams were conducted into the city of Hisar from two rivers viz. Jamuna and Sutlej. Both these streams were conducted through the vicinity of Kernal, and, after a length of about 80 kos (1 kos = 2.8 km), discharged their waters by one channel into the city of Hisar. Sultan Feroz was occupied for two years and a half in building the city. When it was built he laid out many trees, including all sorts of fruit trees. Prior to this time, spring harvest failed because wheat could not grow without water.

The canal that Sultan Firoz Shah had made contained only a scanty supply of water, became ruinous after the Sultan's death. The use of canal water for irrigation got an impetus during the time of Sher Shah Suri (1540-1545) and later during the reign of Akbar (1555-1605) in areas near Delhi but the proportion of area it may have irrigated relative to that served by the wells can safely be regarded as minimal. He put it in repair and set it flowing again to fertilise the places in his *jagir*, and hence it was called Nahr - i - Shahab; but for want of repairs, however, it again stopped flowing.

Shah Jahan (1628-1658) restored the canal and brought a channel from it to Shahjahanbad. At the time when the sublime attention was turned to the building of Tajmahal fort and palace, it was, commanded that the canal from Khirzabad to Safidon should be repaired, and a new channel excavated from the latter spot to the regal residence, a distance of 30 kos.

The salty patches in the soil to have developed under canal irrigation, disappeared when the canal went into disuse in the year

1750, after the fall of the Mughal empire. In the primitive system of well irrigation there was no significant area affected with reh (salt efflorescence).

In 1813, the second Earl of Moira, later called Marquis of Hastings was appointed Governor-General of India. In January, 1815, he saw the neglected bed of the Western Jamuna Canal and the surrounding country side. He observed for a considerable distance the vestiges of the canal. It began where the Jamuna bursts through the hills into the plain and it took a direction nearly parallel to that of the river quite close to Delhi. Its object was to fertilise the long tract of country from its source to its termination; in which extant no tolerable water was to be procured but by sinking wells to such as enormous depth as was beyond the compass of ordinary funds. All the water found in the higher strata was brackish and deleterious to vegetables as well as unwholesome for man. During the wars, the banks of the canal were broken in many places, its course stopped. The country had then in consequence, an air of desolation. Ruins of villages met the eye everywhere. The possibility of re-establishing the canal had early struck Hastings. On a rough estimate the engineers computed expenditure of three lacs of rupees to put the whole canal into perfect condition, but the estimates were seemingly on higher side. In 1817, Blane was appointed by the Government of India to restore the Western Jamuna Canal. It took three years to do the job and due to paucity of funds, alignment of the old Mughal canal consisting of natural channels and depressions was adhered, which resulted in the formation of large swamps and extensive waterlogging. The canal took off from the River Jamuna near Hathnikund, flowed down an old creek of the river until it joined the Pathral and Somb torrents. There were no permanent headwords built and the supplies rose and fell with the seasonal changes in the river level.

The famine of 1832-33 led to the enlargement and multiplication of branch and distributory channels. The financial aspects of the project proved satisfactory; for, in 1947, it was yielding a net return of 13 per cent on the capital invested, which amounted to Rs. 13,81,000.

Eastern Jamuna Canal

Little is known about early history but it seems that Eastern Jamuna Canal taking off from the river Jamuna on its eastern bank near Naushehra in Uttar-Pradesh, reported to have been originally designed by Ali Mardan Khan in the days of Shahjahan was constructed probably during the reign of Mohammed Shah. It was abandoned soon after construction due to declining power of the

Mughals. Rohilla chief attempted in 1780 to restore water supply into the canal but it proved failure. The alignment was faulty, and no great stream of water could have been carried down the channel for any prolonged period and the works were abandoned in the time of Ahmad Shah. The canal, 250 km in length, for irrigation was opened in January, 1830. It irrigated tracts in Saharanpur, Muzaffarnagar and Meerut districts. Bentinck Colonel Baird Smith mentioned ``Most beautiful in all parts it truly is, with its broad road, smooth as an English lawn, its double rows of trees dropping over the stream, its long graceful sweeps, its bordering of the most luxuriant crops, its neat station houses, and the peculiar case with which all its works are maintained. It is certainly one of the most interesting and attractive of Indian sights."

When first constructed, irrigation was affected directly from the canal, water being supplied to the village water courses through openings in the bank. This led to much waste, and the area commanded was small. An attempt at extension was made by utilising the natural drainage lines, across which earthen dams were thrown, but the channels quickly silted up and it became necessary to excavage regular distributaries.

The Eastern Jamuna Canal runs throughout its length almost parallel to, but a few kilometers distant from river.

Ganga Canal

The history of the Ganga Canal belongs to the last years of the East India Company's rule. The great work was commenced by Lord Auckland when he had to deal with the famine which ravaged northern India in 1837-38. In this famine, about 80,00,000 people died. His successor, Lord Ellenborough, however, suspended the work on the Ganga canal. In December 1839, Cautley, a military engineer, undertook, a close examination of the country near Haridwar. He proposed to adopt a direct line from Haridwar to Roorkee. The project was sanctioned in 1814 by Lord hardinge. The canal was opened in April 1854. Irrigation in the grand manner began, however, with the opening of Ganges canal in 1878. Large areas in Uttar-Pradesh (Muzzaffarnagar, Meerut, Bulandshahar, Aligarh, Mathura, Etah, Mainpuri, Farrukhabad, Etawah, Kanpur, Fatehpur, Allahabad, Hardoi, Unnao, Lucknow, Rai Bareilly, Sultanpur, Pratapgarh, Jaunpur etc.) and the then Punjab (Rohtak, Gurgaon, Karnal) were covered under canal irrigation.

DEVELOPMENT OF SALT AFFECTED WASTELANDS AND EARLY INVESTIGATIONS

First Report from Haryana

Alarm at the progressive deterioration of land in the command of the Western Jamuna Canal promoted the Government of Punjab for the first time in 1858 to draw the attention of the then Governor-General to the seriousness of the problem of reh and usar tracts. The Chief Commissioner of Punjab reported. ``For some time past, it has been known that many villages on this canal in the Panipat, Delhi and Rohtak districts, have been suffering from a destructive saline efforescence. The accounts of poverty in some of these villages have been quite distressing. At present, about 60 villages, or one-tenth of the total number of canal villages, are affected. The mischief is increasing yearly; that it would soon attain to very considerable proportion, and would entail fiscal loss to Government and suffering to the people. On consultation. Colonel Baird Smith, Superintendent General of Irrigation in the NWP, attributed appearance of the efflorescence to the percolation under pressure of the canal water through the soils in various directions, the controlling factor being the volume of the canal and the difference between the surface level of the water within the channel and the surface of the surrounding country (Whitcomb, 1971).

Early Investigations

Medlicott, the then Professor of Geology at Thomason College, Roorkee, was probably the first scientist who made scientific investigation of the problem of reh in 1863. His findings were the following :

(1) Because of the changes in the rate of movement of sub soil water through percolation, capillarity, and by the poor quality of irrigation water itself, progressive concentration of soluble and partially soluble salts, originally distributed more widely throughout the profile, takes place in the upper strata of the soil to a degree toxic to plant growth.

(2) Samples from the Ravi and Ganges rivers, which fed the Western and Eastern Jamuna Canals, contained high concentrations of sodium sulphate and sodium chloride, the predominant salts characterising reh and usar lands.

(3) The canal water was an exacerbating but not a prime cause and the natural salt lands were made increasingly so by inefficient drainage, and insufficient application of water.

(4) Soil in which salts accumulated through high rates of evaporation and of capillarity were perched, never thorough soaked; there was no natural sub-soil drainage and no free connection" therefore could be made with ground water stratum.

(5) Canal water could well be used to provide the requisite soaking for the soil, to wash down the salts and establish a ground water connection.

Meanwhile, samples of *reh* had been sent for analysis to Dr. Thomas Anderson, Professor of Chemistry at Glasgow. He corroborated in general the reports of his predecessors that *reh*, may be described as mixture of common salt (sodium chloride), sulphates, soda, potash, lime and small quantity of alumina. The appearance of *reh* was ascribed to the presence in the soil of some alkaline minerals, the decomposition of which is promoted by the irrigation water and a large quantity of these substances is converted into a soluble form, and gradually accumulated until they become so abundant as to become noxious to plants. Drainage was suggested as the cure of the problem.

By 1864, *usar* was well known in Faizabad, Behraich, Lucknow, Sitapur and Hardoi and its reports were received by the Government of India. *Usar* was in general a combination of *reh*, a mixture of highly soluble salts, and *kankar*, a residual insoluble precipitation, or hardpan; *kankar* is impure calcium carbonate. Amongst soluble salts, carbonates rather than sulphates and chlorides predominated. Occurrence of *usar* was ascribed to the capillary action.

This correspondence on redeterioration published in 1864 represents the first collected documentation on the problem and its scientific analysis. Copies were sent to the Secretary of State in England in 1865, with specimens of reh soil collected from village Munak and Asan Kalan, both now in district of Karnal, on the Western Jamuna Canal, and also samples of canal and spring water. The specimens were analysed by W.J. Ward in the metallurgical laboratory of the Royal School of Mines. Ward observed that canal waters contained notably low amounts of chlorides and sulphates, a finding contrary of Medlicott. He also concluded that reh was a mixture of highly soluble salts, that water was the vehicle by which it was brought into prominence in the soil and that the cure was the combination of sufficient irrigation with adequate drainage. Subsequent to the receipt of the reports and accompanying papers, the Lieutenant Governor, NWP, recommended by way of immediate action that experiments on leaching and drainage should be begun, for which the central portion of the Eastern Jamuna Canal provided an opportunity. The progress of the modest experiments in reclamation was reported on annually, and annually they were shown to be of little practical value in that a return on investment was not immediately forthcoming. For example, canal silt enabled crops to be grown on *usar*, but the operation was not directly remunerative. By 1875, the most practical operation seemed to raise non-agricultural plantations. Meanwhile, coordination between the analysis and experiments on reh and, in particular, its connection with canal irrigation, and the implementation of plans to extend the canal systems was non-existent.

Second Report from Uttar Pradesh

And in 1876, another alarm was raised about the disasterous spread of *reh* in parts of Aligarh. Meerut and throughout Kali Nadi Valley of Uttar Pradesh, through a petition from David Robarts, a substantial Zamidar of pargana Sikandra Rao, Aligarh district, to the Board of Revenue for relief on the ground that his land had deteriorated fast due to the appearance of saline efflorescence as a result of the introduction of canal irrigation. The condition had become far more obvious, sufficient to cause serious, if sowewhat academic, concern on the part of government officers and this led to the formation of the historical Indian Reh Committee in 1877 (Agarwal and Gupta, 1968).

Report of `Reh' Committee

The report of the Reh Committee appeared in 1886 which focussed attention on many important points related to canal irrigation, drainage and spread of alkali in the soil. The enquiry was far from adequate. No account, for instance, was given of the extent of usar tracts in the provinces; they were said to cover ``immense areas", without details as to acerage. No Agricultural Chemist was appointed to the committee nor even consulted during the investigations. The committee reported, ``The canals were built for the purpose of affording abundance of irrigation water; and in order that it might be easily available, they were laid high so as to allow their waters when used to flow naturally over or through the agricultural lands. No provision for drainage was made. It was suggested that indiscriminate use of water through the flush system of irrigation, should be stopped by greatly economising the distribution of water, which could be done by the raising of rates charged by government on flush irrigation. Since flush rates were already high, an increase would deter farmers altogether with disasterous results for the revenue accruing from canal charges. Alternatively, lift irrigation was suggested. Provision of deep drainage, straightening and opening of natural drainage ways and preparation of groundwater maps were also suggested and it was also recommended that experiments in reclaiming *usar* tracts which had begun in 1874 under the supervision of the newly created Department of Agriculture should be continued. These consistently showed that usar could be brought back into cultivation only by careful watering accompanied by intensive manuring. Nothing, however, was done on any significant scale to increase the local supply of manure near these tracts in order to keep pace with the increase in irrigation from the canal. Obviously the outcome of the modest series of reclamation experiments was as before. Whilst it was stated, for example, in 1888 that Government of India attached great importance to the experiments, Colonel J.G. Forbes, Irrigation Secretary to Govt. NWP and Oudh, admitted that, so far, all attempts at reclamation had proved unsuccessful.

Dr. J.A. Voelcker was the first Agricultural Scientist appointed by Government to report on Indian agarian conditions. He recorded that it is hardly surprising that enormous tracts, specially in the plains of Northern India", are affected by reh. In the NWP alone, it was estimated to cover between 10,000 and 12,000 km^2 (Voelcker, 1897). In the midst of this desolated usar land, patches of "valuable crops such as opium, sugarcane, wheat, castor oil plant, and cotton stood out 'like oasis' in the salt covered desert around them". For the first time, he emphatically indicated the necessity for scientific soil study to tackle the problem of saline-alkali soils in this country. He also recommended that a permanent post of an Agricultural Chemist should be created for the provinces.

SCIENTIFIC INVESTIGATIONS

Uttar Pradesh

Subsequently, with the implementation of Dr. Voelcker's recommendation, Dr. J.W. Leather was appointed as first Agricultural Chemist, Dr. Leather deserves to be called the first scientific worker on soil salinity in this country. He made a thorough systematic and scientific multipronged attack on reh (alkali) problem. Beginning his studies in 1893, Leather published his several research findings on the "Reclamation of *reh* or *usar* lands" in Agricultural Ledger of 1893 and 1897. The first and the foremost scientific treatise on the subject "Investigations on Usar Land in the United Provinces" by Leather, was published, in the form of a report, by Government Press, Allahabad, in 1914. Some of the salient conclusions which emerged from his studies of *usar* (saline alkali) patches at many locations in the districts of Etah and Mainpuri, in the vicinity of Lower Ganges canal in 1893 are as follows :

(1) Usar soils contained practically sodium carbonate and bicarbonate which amounted to 0.06 to 0.09% respectively, in the surface 60 cm depth. Total soluble salt content seldom exceeded 2%. A good arable soil contained less that 0.1% soluble salts.

(2) Infertile (Usar) patches were impervious, the water penetration being 1/10 to 1/100 of normal soil and always less than 0.1 cm per hour and associated with waterlogging at about 7 to 9 feet (2 to 3m) from the surface.

(3) Sub soil water was considered to rise and fall freely below the impervious patches of usar soil which were 7 to 8 feet (2 to 3 m) thick.

(4) In respect of frequency, infertility, nature and amount of salts, or physical condition, there was substantially no difference between usar patch irrigated or unirrigated, with canal water. The alkali in the soil was presumed to be a product of the soil itself. Canal water could possibly not create such patches but, there could be possible extension of reh due to waterlogging on the introduction of canal irrigation.

(5) Lowering of sub-soil water or opening of drains was though necessary to reclaim these soils but experiments showed that surface and sub-surface drainage were ineffective.

(6) Gypsum and canal silt were found to reclaim usar land but prohibitive because of the high cost.

(7) Deep and good cultivation coupled with heavy manuring reclaimed only surface 1 ft (30 cm) of the soil.

(8) Scrapping of the salts was thought useless. A typical soil surface scrap collected from Karnal consisted of 48.3% Na_2CO_3, 33.4% $NaHCO_3$, 13.3% Na_2SO_4 and 3.9% NaCl. Total salts were 22.8%.

(9) Digging out the impervious soil, if not more than 4 ft (120 cm) thick was suggested as also one of the measures to reclaim the soil.

(10) Allowing 2 ft (60 cm) of water to percolate through the soil, but for a prolonged period, in well maintained bunds, coupled with small amounts of gypsum of fine size was also supposed to remove salts to a lower stratum.

(11) Reclamation by tree (babul or shisham) planting was very encouraging. Growing of grasses after enclosing land from grazing was also found encouraging.

(12) Pot culture experiments revealed that sodium carbonate upto 0.4% and chloride and sulphate upto 1.0% were not generally immediately fatal to maize, wheat, barley, cotton etc. but germination was severely affected by 0.7% sodium carbonate or sodium sulphate. The germination of leguminous plants, gram, peas and arhar was interfered with 0.2–0.4% sodium carbonate or sodium chloride and 0.7% sodium sulphate.

(13) Investigation on the quality of waters of Mathura district showed that well waters were considerably saline as compared to canal waters which were the purest form of irrigation water available. Some of the well waters contained 1,250-10,000 ppm salts. The upper limit of TSS of waters suitable for irrigation was fixed up as 1000 ppm (EC 1.5 dSm^{-1}). This limit was considered safe only when the irrigation water was used in excess of what was required for maturing the crops.

(14) The origin of the salts was ascribed to four possible ways : (a) Sub soil bed of salt; (b) brought at least in part by the rivers when the alluvium of the plains was deposited; (c those portions of the soil which have been rendered soluble by water, air and plants and rejected by the latter and which have not drained away; (d) brought in part by the canal water used for irrigation. Imperfect drainage, surface washing, wind drift and canal were considered as causes for the accumulation of reh.

Duthie (1896) reported on the *usar* reserves in Chherat and Gursikran of Aligarh district in 1893-94 as compared to 1888. The grass called as bat (*Diplachne fusca*) was observed to be more useful than *usar* grass *(Sporobolus pallidus*) on the low lying regions, where water was liable to accumulate for any length of time. Planting of *Dhak (Butea frondosa*) tree was recommended. The fresh tree (*Tamarix articulata*) also thrived on saline soil.

Although, Leather recognised as early as 1897 that calcium salts coagulated usar soils and made them permeable, he calculated gypsum requirement of the soil as simply the quantity which was equivalent to soluble carbonates present. The formation, characterisation and reclamation of saline alkali soils, in fact, remained obscure until the implications of the principles of base exchange were fully known.

Deccan

This awareness to the problem of soil salinity was extending throughout India with the spread of irrigated farming since land was deteriorating due to canal irrigation. In the Deccan tract, the Nira valley irrigation canal was opened in 1884 to irrigate for the first time on a large scale the deep black cotton soil area and soon after, the land began to be affected by salt and waterlogging. In 1886, Dr. Ozanne, the then Director of Agriculture, was directed to make a survey of the problems. He reported that damage was due to seepage and percolation from canals. In 1893, Leather visited the tract and sampled some of the affected soils. The problem in this tract also became so serious that in 1903, the Government of Bombay appointed a committee to consider the measures to be adopted to check the spread of the damage. the committee recommended that all nalas (natural drainage lines) should be opened without delay, salt affected lands be surveyed and reclamation trials carried out. Mann and Tamhane (1910) estimated that 6-7 per cent of the area was being damaged annually, and even those lands on which sugarcane was being raised had accumulated soluble salts, whose adverse effect was being suppressed by the amounts of manure and irrigation water being used. Drainage trials laid out remained unsuccessful.

Talati (1941) reported that Deccan soils had four types of damaged soils excluding waterlogged areas : (i) Mixed saline soils, (ii) Saline soils, (iii) Alkali soils and (iv) Strong alkali soils. These required different treatments for their improvement. The first type of soils improved merely by leaching while the saline soils become alkaline either under natural rainfall or artificial conditions of leaching and required fertilizer for their improvement. Gypsum and sulphur in combination with FYM had been found the best. Sugarcane variety CO 290 was observed to be alkali resistant and was tried after *dhaincha (Sesbania aculeata)* green manuring with a basal dose of gypsum and FYM. Talati (1947) conducted field experiments on the reclamation of salt lands in Baramati of Bombay Deccan. The water table was lowered to a depth of more than 1.2 m by drainage followed after ploughing and levelling. Rainfall of 400 cm reduced salt content from 3 to less than 1% and reduced pH in top 15 cm but increased in the lower horizon. For reclaimed soils under sugarcane, tilth was good upto pH 8.6 above which yields decreased. $CaSO_4$ with FYM or sheep manure gave best results.

Punjab

Most of the canals in the Punjab were constructed in the last quarter of the nineteenth century. The survey of Lower Chenab canal area was conducted by Bernee in 1908 and field reclamation experiments started at Narwala in district Layallpur in 1914. By 1918 the problem of waterlogging and salinity was so much of concern that Government set up a Drainage Board. After the abnormal excessive rains in 1925, a Waterlogging Enquiry Committee was constituted to carry out statistical investigation and experimental work pertaining to waterlogging and *thur* (saline soil). The Chakanwali Reclamation Farm in Gujranwala district was also opened with the simultaneous establishment of the Punjab Irrigation Research Laboratory at Lahore to investigate all the problems connected with irrigation, drainage and salinity in 1924. This research laboratory rose to the status of a full-fledged Research Institute by 1931-32. Scientists like Mackenzie, Taylor, Puri, Asghar and Mehta devoted themselves to research on the problems of soil salinity and water use. Standards of soil and water quality were established and the means of reclamation of deteriorated soils found. The implementation of these research findings in the field started in the late thirties. Since the problem continued to increase beyond all proportions, Waterlogging Enquiry Committee was expanded and converted into Waterlogging Board. Likewise Chakanwali Reclamation Farm, was established at Baramati, in Deccan black cotton soil tract. In 1929, Punjab Government agreed to construct main and branch drains but the responsibility for construction of field drains was left with the land owners. The position was reviewed annually in Waterlogging Conference presided over by the Governor.

Punjab Action Committee

The problem created by the abnormal floods of 1947, 1950 and 1952 resulted in the formation of Waterlogging Board for Punjab. The floods of 1954, 1955 and 1956 further aggravated the waterlogging and thur (usar) and problem of drainage became more serious.

The Punjab Government set up an Action Committee in 1957, in order to tackle this problem effectively on a top priority basis. The Action Committee recommended the following remedial measures :

(1) Minimisation, if not total prevention of flooding of the affected area from river, big streams and creeks.

(2) Improvement of land surface drainage.

(3) Installation of an adequate number of tubewells and revival of percolation wells fallen into disuse to stabilise the subsoil water at the proper level.

(4) Lining of water courses, main canals and distributaries if economically possible.

(5) Increase in the water allowance in thur affected areas so as to attain an intensity of 200 per cent.

(6) Construction of seepage drains parallel to the existing channels.

Saline-Alkali Soils Seminar

The Indian Council of Agricultural Research organised a well planned Seminar on the Recent Advances in Agronomy and Soil Science at Shimla in 1959. The programme included a subject on saline and alkali soils. While discussing many aspects of this subject, it was felt that the existing information on soils in the country is voluminous but the same has not been brought together and published. The Council desired that Dr. R.R. Agarwal and Dr. R.N. Gupta of Uttar Pradesh collected this information from the available sources and write a bulletin on the subject. In consequence, a bulletin entitled ``Saline-Alkali Soils in India" was prepared which covered research work on the subject since the time of inception to 1962. This was published by ICAR in 1968. (It was revised subsequently by Agarwal *et al.* (1979) and published again by ICAR).

The Seminar on ``Salinity and Alkali Soil Problems", held at Indian Agricultural Research Institute, New Delhi in 1962, recomme-nded reclamation of saline and alkali soils on a national basis and setting up of suitably manned and equipped central salinity and alkali laboratories, already approved in the Third Five Year Plan, to be given the highest priority.

Planning Commission Report on Waste Lands

While more or less systematic research work continued going on in some parts of the country, increasing need was being felt seriously for the intensification of the studies on the problem soils. Planning Commission, Committee on Natural Resources (1963) prepared a study report on waste lands, including saline alkali and waterlogged lands and their reclamation measures, and recommended setting up of three Regional Salinity Laboratories in the country. This recommendation was not accepted immediately and concerted efforts to solve the problem on national basis remained lacking, though the salinity continued to induce creeping death to the soil.

NEDECO Investigation in Gujarat

NEDECO (1968) a foreign consultant agency operated a pilot polder project (1500 ha) in Gujarat, in order to suggest measures for the reclamation of the Bhal area, representing a former tidal flat, then abandoned by the sea, consisting of vertisols, in the highest part and Halaquepts in the lower part, lying as waste land or in use as poor pasture. The results obtained in the project had been poor due to various reasons including the presence of a layer of very low permeability in the upper metre of the soil profile, swelling nature of the soil, low amount of fresh water carried by the rivers during the monsoon, irregular and short duration floods and absence of adequate sources of fresh groundwater. Discharge of water from the pilot polder could not be affected due to severe siltation of the outlet channel. Desalinisation of the soil did not respond to improved drainage. The recommendations emerged from the studies were not to proceed with plans for reclaiming the entire area of saline soils in Bhal, unless after extensive research under the given circumstances more was known about desalinisation, whether or not in combination with flooding and drainage, possibilities for storage of water (dams) for irrigation in the area west of the Bhavnagar - Ahmedabad road, availability of fresh ground water in the same area and possibilities for storages of water in the Mithapur Depression. Measures to check erosion especially in the valley of the Khalkhalia and encroachment of the sea in the Sonrai creek area by low levels around the threatened arable lands, were also suggested for investigation.

USAID-INDIAN Team Recommendations

In 1967, Indian Council of Agricultural Research constituted a USAID-INDIAN Water Use and Soil Management Research Consultant Team consisting of 3 American and 3 Indian scientists to suggest the need of and plan for research on water use and soil management towards meeting India's food shortages. The team made a thorough study of the problem and made several recommendations.

The team recommended that for increasing research effectiveness on immediate food production problems, putting research on an applied or adaptive nature immediately at a level of high concern to the technical people should be considered. For improving interdepartmental and inter-agency relation and coordination, in principle, the Ministry providing the leadership for the design and construction aspects of land and water development should also be responsible for the specialised research necessary to implement their programs. It is essential that these research activities be coordinated so that the efforts interlock but do not overlap.

With the Ministry, ICAR should coordinate water use and soil management research and training activities and in states, the relevant research pertaining to regional and national significance should be carried out by Agricultural Universities. To provide essential technological information for the water and soil management component of the "Package of Practices", several research projects were recommended. Great emphasis was laid on pilot projects.

In addition, the team recommended for early initiation of the following measures:

1. Establishment of coordinated drainage-salinity pilot research project at one national centre and three regional centres (Black soils - Maharashtra or Madhya Pradesh; Mixed Red and Black soils - Karnataka or Orissa; Coastal alluvial soils - Orissa or West Bengal), to the management of salinity and alkalinity in cultivated soils of India.
2. In states other than those selected for the centres and where salt problems are acute, appointment of an Indian Scientist having a depth of knowledge, to aid cultivators.
3. A research programme on basic soil properties in relation to soil and water management practices and production of crops would be initiated at four centres in India.
4. Establishment of a Central Water Management Research Centre at the Indian Agricultural Research Institute to study water management problems of national significance.
5. Research on problems of well design, construction and maintenance, including materials and equipment and evaluation of pumping costs.
6. Research on tanks at catchment and storage structures for water supplies.
7. Studies on the problems of recharge of underground aquifers.
8. Examination of existing approaches to research on rainfed lands and a substantial strengthening of programs at the soil conservation research, demonstration and training centres throughout India.

As a consequence of these recommendations, ICAR launched in 1968 an All India Coordinated Scheme for "Studies on Soil Salinity and Water Management" at different locations, and set up Central Soil Salinity Research Institute at Karnal and Water Technology

Centre at New Delhi (Indian Agricultural Research Institute) in 1969. Another Coordinated Scheme on "Use of Saline Water in Agriculture" came in operation in 1972, at 5 centres in the country in the states of Uttar Pradesh, Rajasthan, Karnataka, Andhra Pradesh and Maharashtra. With the establishment of two Research Institute/Centres and two Coordinated Schemes, and with sufficient budget grants, the lacunae on the part of the Government had almost been removed. The outcome of the findings on the problem and its solution was expected to pave the path of irrigated farming with success in future.

UNDP Investigation in Haryana

Introduction of canal irrigation in 1.2 mha in Haryana has proceeded with the rise of water-table and consequential development of waterlogging and salinity. It has been estimated that state of Haryana is suffering annually a loss of Rs. 420 million due to twin problems of waterlogging and salinity. Further, this loss is likely to increase to Rs. 1000 million by 2000 A.D., if the problem is not contained. Out of the total geographical area of 4.4 mha in Haryana, 3.5 mha is under cultivation. About 2.8 mha is underlain with saline ground waters. The cost of reclamation of the entire area in Haryana for potential agricultural production through drainage and other measures has been estimated to be Rs. 7200 million (HSMITC, 1984).

Salinity Problem in Rajasthan

One of the biggest irrigation project of the world, the Indira Gandhi Canal Project, in respect of which work is now going on a fast pace under stage II, has touched the figure of 1326 crore by the end of 1995, in the construction of 649 km long canal from Harike barrage to Jaisalmer. The total agricultural production is stated currently to be worth Rs. 800 crore, in addition to dense tree plantation changing the environment.

Loonkaransagar terrain comprising of a sub soil hard pan, traditionally known as the no-man's land (Kalapani of Bikaner) has been transformed into a prosperous land of plenty with the arrival of water by lift irrigation from Indira Gandhi Canal. In about a decade now, the scarcity-prosperity-scarcity cycle seems to be coming a full circle. Introduction of canal irrigation in desert of western Rajasthan through canal has completed journey from one wasteland (water starved) to another wasteland (water soaked).

There are two different sources from which soluble salts may be derived and be caused to accumulate in irrigated soils namely, (i) the irrigation water itself, and (ii) the sub-soil or the deep underground where salts were present before irrigation was begun. The ground water in arid and semi-arid zones usually contains considerable amount of dissolved salts. Seepage from canals and irrigation laterals and the application of excessive amount of irrigation water may cause the water level to rise, sometimes almost to the surface of the soil. As the ground water rises, the soluble salts native to the substrata dissolve and this increases the concentration of the ground water. When the ground water comes sufficiently near the surface, evaporation and transpiration take place, leaving the salts in the soil. The development of salinity in the newly canal irrigated areas has arisen chiefly from pre-existing deposits in the substratum rather than from the irrigation water. Sub-soil conditions resulting in waterlogging, a rising water table (43 to 83 cm a year) and the inevitable salinity has already rendered major part of the land useless for agriculture. It is anticipated that 20 to 60 per cent of the canal irrigated area may develop drainage and salinity problems within 30 to 100 years.

Salinisation problems are more severe if the salinity of ground water is high such as prevailing generally in arid regions whereas alkalisation problems are more prominent when the ground water contains residual sodium carbonate. The latter situation is generally encountered in semi-arid regions. However, in both the situations, evaporation far exceeds precipitation, the magnitude of difference being more in arid regions. The intensity of the problem magnifies with time in close basins where there is no outflow of water.

IRRIGATION COST AND BENEFIT - AN OVERVIEW

The situation is not grim only in Haryana and Rajasthan but Punjab, Uttar Pradesh, Gujarat, Andhra Pradesh, Madhya Pradesh and Maharashtra states are equally affected with soil salinity problems, accelerated through waterlogging, arisen after the introduction of canal irrigation. No doubt that first ten to twenty years of the inception of canal irrigation shows spectacular results in terms of bumper production. The Sarda Sahayak canal for irrigation in U.P. has created such waterlogging problems that a drain had to be dug all along the canal and tubewells sunk along its banks to drain off the water. Tungabhadra, Nagarjunsagar and Sriramsagar projects in Andhra Pradesh have created grave waterlogging problems. In the Tungabhadra canal area alone 83,000 ha of irrigated land has been affected and crop yields have been reduced by about 30 per cent. The Tawa Command area in Madhya Pradesh, Purna Command area in Maharashtra and Chambal Command area in Madhya Pradesh and Rajasthan have become the farmer's ruin.

Although India has the world's largest irrigation system (83.0 mha), it has not been able to benefit from it as the system has become highly inefficient (Table 1.1). The gap between the potential created and actual utilisation has been estimated to be 9.1 mha in 1991-92. The cost of creating irrigation potential through major and medium schemes has increased from Rs. 1.520 per ha during First Plan to Rs. 60,000 per ha in Eighth Plan.

Table 1.1. State / Source-Wise Net Area Irrigated during 2000-2001.

(Thousand hectares)

S.No.		Canals	Tanks	Wells	Other Sources	Total Area (All Sources)
1	Andhra Pradesh	1649	727	1954	197	4528
2	Arunachal Pradesh	–	–	–	42	42
3	Assam	148	–	2	20	170
4	Bihar	1136	155	2093	241	3625
5	Chhatisgarh	678	55	170	81	984
6	Goa	4	–	19	–	23
7	Gujarat	492	18	2452	18	2980
8	Haryana	1476	1	1467	14	2958
9	Himachal Pradesh	3	(a)	14	107	124
10	Jammu & Kashmir	284	3	2	22	311
11	Karnataka	966	261	1018	397	2642
12	Kerala	105	50	116	110	381
13	Madhya Pradesh	808	85	2651	591	4135
14	Maharashtra	1047	–	1912	–	2959
15	Manipur	–	–	–	65	65
16	Meghalaya	54	–	–	–	54
17	Mizoram	9	–	–	–	9
18	Nagaland	–	–	–	72	72
19	Orissa	878	282	774	–	1934
20	Punjab	676	(a)	2880	46	3602
21	Rajasthan	1354	38	3473	42	4907
22	Sikkim	–	–	–	17	17
23	Tamilnadu	833	589	1449	16	2887
24	Tripura	22	5	4	5	36
25	Uttar Pradesh	3091	82	9384	259	12816
26	West Bengal	261	173	1397	523	2354
27	A & N Islands	–	–	–	–	–
28	Chandigarh	1	–	1	–	2
29	Dadra & Nagar Haveli	2	–	2	3	7
30	Daman & Diu	–	–	–	–	1

S.No.		Canals	Tanks	Wells	Other Sources	Total Area (All Sources)
31	Delhi	2	–	30	2	34
32	Lakshadweep	–	(a)	1	–	1
33	Pondicherry	8	–	13	(a)	21
	All India	15989	2524	33277	2892	54682

Source: Ministry of Agriculture, Directorate of Economics & Statistics

Note: (a) Below 500 hectares Total may not tally due to rounding off.

The performance has not been commensurate with large scale public investments and subsidy given to farmers due to poor crop productivity, uneven water distribution and rising problems of water-logging and salinity (Table 1.2). Despite huge subsidy on surface irrigation systems, the actual yield levels are far below the projected targets. The deprivation of tails enders is serious and an area of 6-10 million ha is losing productivity because of uneven water distribution (Joshi, 1990). While increasing waterlogging and salinity have led to wastage of resources, the time and cost overrun in the irrigation works have resulted in an expenditure of nearly Rs. 160 million by Central and State Government between 1950 and 1984 to create an additional potential of 30 million hectares.

Table 1.2. Extent of waterlogging and soil salinity in selected irrigation projects

Irrigation project	State	Extent (Thousand hectares)	
		Waterlogging	Soil salinity
Sriramsagar	Andhra Pradesh	60.00 (47.62)	1.00 (0.79)
Tungbhadra	Andhra Pradesh and Karnataka	4.65 (1.27)	24.48 (6.69)
Gandak	Bihar and Gujarat	211.01 (21.11)	400.00 (40.03)
Ukai Karapar	Gujarat	16.25 (4.32)	8.29 (2.20)
Mahi kadana	Gujarat and Rajasthan	82.00* (16.81)	35.76 (7.33)
Malprabha	Karnataka	1.05* (0.99)	–
Chambal	Madhya Pradesh and Rajasthan	98.70 (20.31)	40.00 (8.23)
Tawa	Madhya Pradesh	–	6.64 (3.79)
Indira Gandhi	Rajasthan	43.10 (7.98)	29.11 (5.39)
Sarda Sahayak	Uttar Pradesh	303.00** (7.98)	50.00 (4.68)
Ramganga	Uttar Pradesh	195.00 (32.99)	352.42 (59.62)
Total		1,104,76	947.70

Bowonder and Ravi (1984).

Nevertheless, 60 per cent of total food production came from 32 percent of irrigated land while farm produce from remaining 68 per cent unirrigated land accounted for 40 per cent of the total food grain production in the country. Irrigated area in the world has increased from 90 million hectares in 1950 to 250 million hectares in 1980. India continues to lead as the country with the largest irrigated area of 75 million hectares in the world, followed by China with 47 million hectares, USA with 27 million hectares and the Soviet Union with 21 million hectares.

Global estimates according to FAO and UNESCO show that as much half the world's irrigation schemes suffer in some degree with waterlogging, salinisation and alkalisation of soils. These estimates indicate that some 10 mha of irrigated land are being abandoned each year. It is distressing to note that the world spends on armaments 1200 billion dollars a year. Just 0.5 percent of the world annual military expenditure can pay for all the agricultural machinery required to make poor countries self sufficient in food according to FAO. About 500 million people in world are suffering from hunger and chronic malnutrition. The three pronged strategy in relation to irrigation and development consists of measures for the alleviation of poverty and famine, sustainability of agricultural production whether by irrigation or drainage or both and steps for the environmental protection. Increasing use of chemical fertilisers and pesticides and relative neglect of organic manure in agriculture has led to a concentration of nitrates in the soil and aquifers which form a major source of drinking water. In fact there is need for a shift from the narrow concepts and practices of irrigation to the larger scenario of water resources development taking into account all water use projects whether it be for irrigation strategy in India, thus, would have to take into consideration entire river basins rather than being project specific. Such an approach would take care of not only water but all other vital elements of the life support system like land, flora and fauna etc. which become part of an ecologically sound water use policy. The per hectare cost of irrigation at present varies from Rs. 40,000 to Rs. 60,000, whereas watershed method of water harvesting in Karnataka costs only Rs. 6,000 per hectare and in Maharashtra between Rs. 6000 and Rs. 10,000 per hectare. The present strategy on irrigation needs to be reviewed to formulate a new national policy on water use stressing the traditional water harvesting systems.

Interestingly, Eighth Plan period in India have now no new irrigation project. The emphasis will be on the completion of the on going works in the medium and major projects and on the improvement of quality of irrigation in the command area of the existing projects. However, greater emphasis will be given to minor irrigation works both surface and groundwater. Better utilisation of completed projects, speedy completion of ongoing projects, and efficient utilisation of available water are expected to be substantially more in Eighth Plan than previous plans.

History has witnessed rise and fall of the civilizations associated with the development of irrigation and waterlogging respectively. In the ancient period, in Mesopotamia, people had taken recourse to wheat cultivation which required more water than the area could digest in an ecologically non-destructive way. The result was that the area became infertile for millennia.

The exact cost due to waterlogging and salinity to agricultural production in India is not known but has been conservatively estimated at Rs. 10,000 million per annum. The cost to society accounted for more than Rs. 6000 million alone in the eleven selected irrigation projects (Table 1.2 and 1.3). In view of the huge cost involved in the development of canal irrigation and ruin of long-term economic potential, source of social destabilisation and environmental destruction, it is highly desirable that well irrigation is given greater emphasis and priority over canal irrigation. The development of well irrigation does not cause deforestation and seriously disturb the eco-system. It does not lead to over irrigation, seepage and waterlogging in the nearby areas leading to salinisation and long term infertility.

Table 1.3. Projective expenditure on canal irrigation development, production losses of rice and cost due to waterlogging and salinity

Irrigation project	Expenditure per ha (thousand Rs.)	Production loss of rice		Total cost to society (million Rs.)
		Quantity (Thousand tons)	Value (Million Rs)	
Sriramsagar	15.48	26.26	49.61	944.04
Tungabhadra	2.72	27.65	53.77	79.34
Gandak	2.95	296.75	582.61	1804.00
Ukai-Kakrapar	3.87	13.07	26.69	95.05
Mahi-Kadana	4.02	60.97	149.77	473.80
Malaprabha	9.84	0.75	1.50	10.29
Chambal	2.62	41.58	97.35	362.70
Tawa	4.70	3.75	6.27	31.20
Indira Gandhi	4.66	18.54	53.22	336.70
Sarda Sahayak	2.84	182.97	317.09	1002.00
Ramganga	2.20	297.77	516.04	1215.00

Joshi (1990).

NEED FOR RECLAMATION

The present population of the country is 915 million and the maximum annual food grain production has been to the tune of 190.36 million tonnes in 1995-96, though it was of the order of only 50 million tonnes in the 1950's. By the end of the 20th century, the population of the country is expected to increase to more than 1000 million and the food grain requirement to more than 240 million tonnes. In order to cope up with the food requirement of the increasing population in the near future, the agricultural production has to be stepped up substantially. This can be done either by multi-cropping on the existing cultivated lands or by bringing the additional good area under cultivation. In view of the projected increase in food grain requirement and remote possibility for the increase in the net cultivated area, it is obvious that our country must produce a greater yield per unit land area than ever before and limited resources of land and water must be used more efficiently. If irrigation targets are kept and land and water are managed properly, India has the capacity to feed 2.621 million people by the turn of the century, 2.5 times the estimated population, according to a recent study by the Food and Agriculture Organization-Agriculture Towards 2000. Reversing the degradation of land which has been continuing for so long is very important task. Our precious crop land and soils have to be protected particularly keeping in view the needs of the next few generations.

Apart from the reclamation and utilisation of salt affected soils potentially suitable for food grain production, the second priority should be given to afforestation of such lands as are otherwise difficulty utilisable or economically unviable. This assumes all the more significance because nearly 50 per cent of the country's land (175 mha) is degraded and more forest and non-forest lands are getting degraded every year, according to the 30th report of the Parliamentary Standing Committee on Environment and Forests (1996). Social forestry, agro-forestry, individual forestry, canal bank and road side afforestation programmes need to be given added impetus to combat desertification, salinisation, waterlogging, and soil and water erosion.

DEGRADATION OF ENVIRONMENT

The occurrence of salt affected lands is generally linked with the canal system. This linkage, tends to increase opportunities for human/water/vector contact in addition to their creation of habitats well suited to vector production. There is always a risk to health from vector borne diseases such as malaria, lymphatic filariasis, encephalitis, schistosmiasis etc. Adverse soil conditions in alkali lands characterised with low rates of infiltration result in longer periods of stagnant water during rainy season. Where the intensity of land use under irrigation is relatively low due to exposed water surface being greater than the irrigated area, the period of standing water is sufficient for a number of breeding cycles of vector and pest insects. The development of dense aquatic weeds and algal growth in canals is also responsible for the rise of some insect vectors, pest insects and snails. During last one decade and half, sharp rise in malaria has been observed in Indira Gandhi Canal Command Area in Western Rajasthan, Greening of desert has proceeded with a line of despair and gloom, alongwith the so called as life line. Malaria was insignificant in this region upto 1980. About 350 persons died of lethal disease alone in 1995. According to the information given by National Malaria Eradication Programme, increase in malarial cases in 1994 was 26 times compared to 1961, in Rajasthan. The number of cases touched the figure of 2, 29, 700 in 1994. This grim situation can be observed almost in all canal irrigated areas, which is very serious problem. The use of domestic waste water in irrigation, if any also serves the cause for increase in breeding of mosquitoes. The association between alkali lands and vector borne diseases is both complex and specific to the site and human population.

Appraisal of existing as well as potential health hazards in canal systems is therefore an essential first requirement in controlling vector borne diseases. For successful appraisal, a multidisciplinary approach should be adopted. This appraisal should also identify the specific environmental parameters that require monitoring. Canal managers should have adequate understanding of vector biology and related subjects so that they are able to identify existing health hazards and potential health hazards, and trends, and decide what specific action then should be taken for minimizing or eliminating vector production. The relationships between vector-borne diseases and water regimes, or agricultural production systems, are not well understood. To enforce effective measures for control of diseases, a better understanding of these relationships is essential.

Lining of canals, in addition to improving irrigation efficiencies and performance is a effective measure in reducing vector breeding and snail survival through minimising water seepage and puddle formation, increasing water velocities in canals and reducing weed growth. However, problems associated with the quality of construction, sedimentation and maintenance of canals, have to be considered.

Modification to water regimes at the farm level through intermittent irrigation and diversified cropping offers considerable scope of reducing vector breeding. When canals are closed either for repairs or due to water storages, stagnant water in canal beds serves as breeding ground for vectors. In such situations, desilting and weekly flushing wherever possible should be carried out to prevent vector breeding.

Monitoring, surveillance and diagnosis of vector borne disease in existing irrigation systems require different criteria at local and national levels. At a local level, simple indicators are required which enable project managers to respond to health problems as they occur during project operation.

Some municipal corporation pour MLO (mosquito larvicidal oil) and other larvicides into pools and tanks of water as well as large canals during mosquito breeding season until February. Between May and September, adult mosquitoes are dealt with BHC (benzene hexachloride) spray and the newer sps (synthetic pyrethroids).

Pyrethrins, an extract from the chrysanthemum plant, mixed with kerosene is the sp most commonly used for spraying in the city. The spray of malathion to control mosquito population has been banned recently as the insecticide caused lung and heart related illness as well as skin allergies and perhaps even high blood pressure. Malathion has also been reported to be ineffective since the local mosquitoes seem to have built up resistance to it. The use of DDT was stopped in 1977 due to build up of resistance to it by the mosquitoes.

The saying `Prevention is better than cure' holds good in this respect. As the cure is much more difficult, it is better to check the rise of mosquitoes which would not breed so freely if water and sewage lines are not allowed to leak and stagnate water and also soils are not allowed to become saline, alkali and/or waterlogged.

REFERENCES

Agarwal, R.R. and Gupta, R.N. 1968. Saline Alkali Soils in India. Tech. Bull. (Agric. Series). ICAR, New Delhi, 221 p.

Agarwal, R.R., Yadav, J.S.P. and Gupta, R.N. 1979. Saline and Alkali Soils of India, ICAR, New Delhi, 286 p.

Bowander, B. and Ravi, C. 1984. Waterlogging from irrigation projects : an environmental management problem. Centre for Energy and Technology, Administrative Staff College of India, Hyderabad, 2-10.

Duthie, J.F. 1996. Reclamation of reh or usar land. Agric. Ledger. 3 : 339-46.

HSMITC 1984. Studies for the use of Saline Water in the Command Areas of Irrigation Projects Haryana, India - Appraisal Document (Summary). HSMITC, Karnal, 115 p.

Joshi, P.K. 1990. Influence of surface irrigation on crop productivity. Indian J. Agric. Sci. 60 : 91-6.

Leather, J.W. 1914. Investigations on *Usar* Land in the United Provinces. Govt. Press. U.P. 87 p.

Mann, H.H. and Tamhane, V.A. 1910. The Salt Lands of Nira Valley. Bull. Deptt. Agric. Bombay, 35 p.

NEDECO 1988. Bhal Reclamation Scheme (Govt. of Netherlands and State of Gujarat, India). Delft Hydraulics Laboratory, Delft, The Netherlands. Vol. 1-6.

Planning Commission India 1963. Study on Wastelands including Saline Alkali and Waterlogging Lands and their Reclamation Measures. Rep. Committee of Natural Resources. New Delhi. 237 pp.

Talati, R.P. 1941. Damaged lands in Deccan and their classification. Indian J. Agric. Sci. 11 : 959-77.

Talati, R.P. 1947. Field experiments on the reclamation of salt lands in Baramati of Bombay Deccan. Indian J. Agric. Sci. 17 : 153-74.

Voelcker, J.A. 1897. Report on the improvement of Indian Agriculture. Govt. Press. United Provinces.

Whitcomb, E. 1971. Agrarian Conditions in Northern India. Vol. 1 (1880-1900). Thomson Press, New Delhi.

■■■■

Chapter - 2

NATURE AND PROPERTIES OF SALT AFFECTED SOILS AND GROUND WATERS

INTRODUCTION

Though `salt-affected soils' broadly cover saline soils, saline-alkali soils but occurrence of pure saline soils without alkaline salts and pure alkali soils without neutral salts is very rare. Salt affected soils are therefore, generally classified as saline or alkali, depending upon the dominance of neutral salts or alkaline salts. More conservatively, salt affected soil represents saline soil or salty soil, unless otherwise specified and the other category in which alkaline salts dominate is called alkali soil.

Development and management of salt affected soils are decisively governed by the water which acts as a reactant, solvent and transporting agent. In fact, it is the cause as well as the ameliorating agent for the salt affected soils. Whereas, the concentration and composition of soil water governs the ionic composition of the soil adsorption complex and naturally therefore the properties of solid soil constituents, the irrigation water controls the concentration and composition of soil water itself. Generally the movement, retention and transport capability of the irrigation water is governed by its own composition *vis-a-vis* that of the soil water. Chemical environment of the soil water not only influences the crop growth, but decisively determines the physical environment of the soil and thus controls the water regime and its availability to plant roots. It is natural, therefore, that the salt affected soils have been classified on the basis of their chemical characteristics. These groups, however, can be easily identified from each other as they differ considerably in their physical, chemical and biological characteristics. The studies on soil properties are necessary because of the following factors.

(1) Diagnosis of salt affected soils.

(2) Identification and characterisation of processes in the formation and evolution of salt affected soils.

(3) To analyse rationally, the techniques for amelioration and management.

(4) To evaluate the efficiency of treatments and to know the stage of reclamation.

(5) To estimate post-reclamation changes in the watersheds of which these soils are a constituent.

Visual Characters

Salt affected soils can be easily identified from the normal lands in the vicinity by some visual observations. It may be possible even to identify the nature of the problem, e.g. whether the problem is of salinity or alkalinity. The relevant characters listed in Table 2.1 are important in carrying out any such investigation.

NATURE AND PROPERTIES

Chemical Properties

1. Soil Reaction (pH)

The pH of the saturated soil paste (pHs) has been widely recognised as an index, distinguishing salt affected soils relative to normal soils, though pH of the 1:2 soil water suspension (pH_2) has also been reliably used for the same purpose in specific cases. Due to hydrolysis upon dilution, pH_2 values tend to increase as compared to pHs, but the differences are small in normal or saline soils as compared to alkali or saline-alkali soils. Whereas in normal and saline soils pH_2 values are higher than pHs by 0.2 to 0.5 units, in saline-alkali and alkali soils, the difference varies from 0.5 to 1.0 unit. Higher values of pH_2 in alkali or saline alkali soils are due to the nature of electrolytes. The saline soils contain predominantly NaCl, which is a strong electrolyte. The alkali soils contain predominantly Na_2CO_3, weak electrolyte. On hydrolysis, the pH_2 of weak electrolyte increases sharply with dilution, whereas changes in pH with dilution are less pronounced in strong electrolytes (Gupta, 1972). The pH_2 values greater than 8.2, usually correspond to ESP values greater than 15 in alkali soils or saline alkali soils (Agarwal and Yadav, 1956). The pH values of salt affected and normal soils are shown in Table 2.2. The most productive normal soils have pH in the range of 6 to 8. Whereas inland saline soils have pH_2, generally higher than 7, coastal saline soils may have pH_2, as low as 5. Whereas normal soils may have pH_2 values around 8, the alkali soils show values usually higher than 8.2.

Table 2.1. Visual characters in saline, alkali and normal soils

Visual character	Identifying character for the soil		
	Normal	Saline	Alkali
Climatic zone	Every zone	Arid and semi - arid zone	Semi-arid and sub-humid zone
Rain or irrigation water	Absorbed quickly	Absorbed quickly	Stagnates for long
Standing water quality	Clear	Clear	Muddy and soapy
Soil surface	No salt efflorescence	White salt efflorescence	Ash coloured/dark brown salt efflorescence
Ground water quality	Good and poor both	Generally poor	Generally good
Crust	May or may not be there, usually thin and not very hard	May or may not be there, usually thin and not very hard	Usually present, thick and hard
Natural vegetation	Cropped in good condition	*Cressa cretica, Cyperus rotundus, Chloris pallida, Sporobolus pallidus, Sesuvion portulacastrum. Haloxylon salicornicum, Aeluropus lagopoides, Zygophyllum simplex, Dichanthium annulatum, Sueda fruticosa, Butea monosperma*	*Sporobolus marginatus, Desmostachya bipinnate. Sueda martima, Diplachne fusca, Dichanthium annulatum. Cynodon dactylon, Chloris barbata Brachiaria mutica, Kochia indica, Panicum antidotale*

Table 2.2. The pH values of soils

Nature of soil	pH_5	pH_2
Normal	< 7.5	<8.0
Saline	7.5 – 8.5	8.0 – 9.0
Alkali (Sodic)	8.2 – 10.0	9.0 – 10.8
Saline-Alkali	8.5 – 10.0	9.0 – 10.5

In addition to dilution, several factors affects the pH values of the salt affected soils, e.g. total concentration of soluble salts, nature of salts, composition of exchangeable cations, nature of cation exchange materials and the presence or absence of gypsum and alkaline earth carbonates. The excessive soluble salt concentration, particularly those of neutral salts, tends to depress pH values of the soils. The pH values of the normal soils, therefore, decrease when the soil is salinised particularly due to NaCl. The highly saline soils having an EC_e of about 100 dSm^{-1}, usually have pH value close to 7. It is for this reason that pH cannot be depended upon for distinguishing saline soils relative to normal soils.

In soil testing laboratories, the soil is evaluated as normal to most of common crops if pH_2 is 8.5; tending to become alkali soils if pH_2 is between 8.5 and 9.0 and alkali soil, harmful for growing crops if pH_2 is >9.0 (Muhr *et al.,* 1963). Gupta (1972) has suggested that last two limits of pH_2 should be changed to 8.5-9.5 and >9.5.

The appraisal of the soil reaction in the salt affected soils on the basis of pH_2 is rapid and convenient. It is a very useful determination, particularly where a large number of soil samples are to be analysed and where repeated samplings are to be made in the same soil to determine the change in pH with time or treatment. However, determination of pH is less meaningful in saline soils than alkali or saline-alkali soils, because appreciable changes do not take place with time or treatment in the former soils.

2. Salinity (Soluble Salts)

Salinity, which indicates the presence or excess of soluble salts in the soil is not desirable beyond a certain limit. Salinity of the soil is measured through electrolyte conductivity of the soil solution extracted from the saturated soil paste, denoted as EC_e. Conventionally, a soil having EC_e more than 4 dSm^{-1} has been called as saline and less than 4 dSm^{-1} as non-saline. The most productive normal soils have EC_e less than 1.0 dS/m. The primary adverse effect of excessive salinity is to raise the concentration of soil solution, as a result of which the flow of water into the plant by osmosis is reduced or reversed and the plant is starved of water even though the soil is moist (physiological drought condition).

The saline soils may be associated with high pH and/or ESP but their physical properties are not affected adversely because of the high electrolyte concentration, particularly due to neutral salts. The kind of soluble salts, the textural properties of the soil and nature of plants being grown are some of the important factors which determine the plant growth. On a soil having similar salinity, chloride salts have been recognised to be more harmful to the plants than sulphate salts (Table 2.3). As the EC_e of the soil increases from 4 to 20 dSm^{-1} in the case of chloridic soils and 6 to 40 dSm^{-1} in the case of sulphatic soils, the relative productivity of the soil in most cases decreases from almost 100 to nearly zero.

Table 2.3. Relationship between nature of salts in the soil and plant growth

EC (dSm^{-1})		Relative plant growth (%)
Chloride	Sulphate	
4	6	100
4-8	6-12	80-90
8-12	12-20	50-70
12-20	20-40	20-50
20	40	0

Source : Kovda *et al.* (1973).

Compared to magnesium sulphate and calcium chloride the adverse effect due to sodium sulphate is two times, sodium chloride and sodium bicarbonate three times and sodium carbonate ten times, at equivalent concentrations.

As a rule, coarse textured soils have low moisture retention and high permeability, whereas, fine textured soils have high moisture retention and generally lower permeability. Therefore, at a given EC_e the adverse effect to plant growth is more in fine textured soils than coarse-textured. However, owing to a high degree of aggregation of the particles and the presence of neutral salts, there are notable examples of fine-textured soils that are moderately permeable.

The tolerance of plants also differs from one to another. The horticultural plants except date palm are sensitive to salinity (EC_e 1.5-3.0 dSm^{-1}). The sunflower, groundnut, maize, sugarcane and rice comprise the semi-tolerant crops (EC_e 3-5 dSm^{-1}). Cotton, sorghum, pearl millet, wheat, barley, mustard, safflower and sugarbeet fall in the tolerant group (EC_e 5-10 dSm^{-1}).

The soluble salts in saline soil mainly comprise the chlorides and sulphates of sodium, magnesium and calcium. Increasing from low to high salinity, the proportion of sodium salts over magnesium and calcium salts increases. Likewise at higher salinity levels, the proportion of magnesium salts over calcium increases. Higher concentrations of sodium and magnesium salts cause adverse effect to the physical properties of the soils and plant growth. Bicarbonates occur in very low amounts and carbonates are usually absent, or occur only in traces. Although NaCl or Na_2SO_4 usually predominate in saline soil, instances are not lacking where $CaCl_2$, $MgCl_2$ and $MgSO_4$ also dominate e.g. saline soils occurring in Iraq are called as *Sabakh*. The later group of salts being highly deliquescent absorb moisture from the atmosphere, thus causing the soil to appear moist. Such soils often have a greasy appearance and are almost black in colour, and are likely to be mistaken for alkali soils.

In soil testing laboratories, EC is usually determined in 1:2 soil water extracts. Singhla *et al.* (1974) examined the relationships between EC_e and EC_2. It has been observed that EC_2 is highly and significantly correlated with EC_e. The ratio of EC_e/EC_2 decreases with the increase in the saturation percentage of the soil. On the basis of EC_e, it has been reported that salinity effects are mostly negligible if EC_e is below 2 dSm^{-1} and that yields of many crops are restricted at EC_e of 4 through 8 dSm^{-1} and that only tolerant crops may yield satisfactorily at EC_e between 8 and 16 dSm^{-1}. Above 16 dSm^{-1} only a few very tolerant crops yield satisfactorily (Richards *et al.,* 1954). On the basis of EC_2, the soils are rated as normal, critical for germination, critical for the growth of salt sensitive crops and severely injurious to crops if EC_2 is 1, 1-2, 2-3 and > 3 dSm^{-1} (Muhr *et al.,* 1963). The salinity appraisal on the basis sample is limited or a large number of soil samples are to be analysed and where repeated sampling are to be made in the same soil to determine the change in salinity with time or treatment. In view of the latest available information, following scale of conductivity is suggested (Table 2.4).

Table 2.4. Conductivity scale of evaluating plant growth restriction

ECe (dSm^{-1})	Plant growth
< 1.5 (normal)	Normal for all crops
1.5 – 3.0 (low salinity)	Yields of very sensitive crops restricted
3.0 – 5.0 (medium salinity	Yields of many crops restricted
5.0 – 10.0 (high salinity)	Only tolerant crop yield satisfactory
> 10.0 (very high salinity)	Only a few very tolerant crops yield satisfactory

3. Exchangeable Sodium Percentage (ESP)

This is defined as the degree of saturation of the soil exchange complex with sodium. It may be calculated by the following formula :

$$ESP = \frac{Exchangeable\ sodium\ (meq\ /100\ g\ soil)}{Cation\ exchange\ capacity\ (meq/100\ g\ soil)} \times 100 \qquad (2.1)$$

Sodium adsorption ratio (SAR) is employed for describing the equilibrium relation between soluble and exchangeable cations. This is defined as follows :

$$SAR = \frac{Na^+}{\sqrt{\frac{Ca^{2+} + Mg^{2+}}{2}}} \tag{2.2}$$

where, Na^+, Ca^{2+} and Mg^{2+} refer to the concentrations of the designated soluble cations expressed in milliequivalents per litre.

The relationship between SAR and ESR (exchangeable sodium ratio = exchangeable sodium/[(cation exchange capacity-exchangeable sodium)] at the saturation moisture percentage is linear. The correlation coefficient between SAR and ESR is highly significant and permitted practical use of the relation. However, this did not include those soils for which ESR was greater than one, which corresponded to ESP of more than 50. Using the relationship between ESR and SAR, the relation between ESP and SAR is given by the following equation :

$$ESP = \frac{100\,(-\,0.0126 + 0.01475\ SAR)}{1 + (-\,0.0126 + 0.1475\ SAR)} \tag{2.3}$$

ESP is a very important chemical property which defines the physical character of the salt affected soils. This parameter is usually a good indicator of the structural stability of soil and of the physical response that may be anticipated when water is applied.

The alkali soils containing expanding type 2 :1 clay minerals (illite, vermiculite, montmorillonite) exhibit unfavourable physical properties at levels of ESP greater than 15. This limit is somewhat arbitrary and tentative. In some cases, for example 2 or 3 meq of exchangeable sodium per 100 g has equal or even greater usefulness as a critical limit (Richards *et al.*, 1954). In general the physical properties become increasingly unfavourable with increasing ESP but at a given level of ESP physical properties are usually poorer in soils with expanding type 2 : 1 clay minerals, containing a low electrolyte concentration, more so of salts capable of alkaline hydrolysis, viz. sodium carbonate, sodium bicarbonate, silica, etc., than in soils with clay minerals of the non-swelling 1:1 type (kaolinite) or mixed type, containing a high electrolyte concentration, more so of neutral salts, viz., sodium chloride or sodium sulphate. It is for this reason that physical properties of saline-alkali soils are better than alkali soils. Also, therefore, ESP values are of less significance for saline soils.

The soil texture and the tolerance of plants to ESP are two more factors which are of practical importance. The coarse textured soils exhibit less deterioration in physical properties than the fine textured at a given ESP. The adverse effect due to higher values of ESP on plant growth is induced because of the lower concentration of soluble calcium (<2 $meql^{-1}$). Since saline soils are seldom deficient in soluble calcium, crop growth on saline soils is primarily a function of salinity of the soils and ESP is not much meaningful.

4. Sodium Adsorption Ratio (SAR)

Conventionally, salt affected soils have been defined on the basis of pH, EC and ESP. Due to the difficulties in the determination of ESP and because ESP is directly related with the SAR, increasing shift is now in favour of SAR as a substitute for ESP. Accordingly, it is recognised that the alkali soil containing expanding type 2:1 clay minerals exhibit unfavourable physical properties at levels of SAR greater than 15. However, SAR does not mean much for saline soils. According to Soil Science Society of America (McNeal, 1976), saline soil is defined as one having EC_e >2 dSm^{-1}, irrespective of any values of SAR and pH whereas sodic soil is one which has SAR >15, irrespective of any values of EC and pH. A soil having EC_e > 2 dSm^{-1}, SAR > 15 and pHs > 8.5 is called saline-sodic.

In general, ESP increases with increase in SAR. There are occasional deviations, but generally low SAR values are associated with low ESP values in the soil, and high SAR values denote high ESP values. The important deviations are observed in soils which have much higher proportion of magnesium over calcium ions in the saturation extract. In the calculation of SAR, both Ca and Mg have been placed together. It is well established now that the chemical effect of Mg is different from that of Ca and further that increasing Mg/Ca ratio in the soil solution enhances soil sodicity and causes adverse effect on the physical properties of the soil and plant growth. High values of Mg lower SAR values although ESP values determined actually remain high. Therefore, this is one of the drawbacks limiting the usefulness of SAR for all practical purposes. Gupta (1984) suggested that for such situations where Mg is very high as compared to Ca, SAR should simply be calculated as $Na/\sqrt{Ca}$, and called as Sodium to Calcium Activity Ratio (SCAR).

5. Macronutrients

The testing of the soil for its major nutrients, viz. nitrogen, phosphorus and potassium, is proven, practical method for evaluating the fertility status of the soils and for providing a sound basis for making recommendations in respect of fertiliser applications for increased crop production. The most productive normal soil is high in organic carbon (above 0.75 per cent) which is used as a measure of available nitrogen, high in available P_2O_5 (above 25 kg ha^{-1}) and high in available K_2O (above 280 kg ha^{-1}). The soil is rated as low in organic carbon, in the available P_2O_5 and in the available K_2O, if the respective values are below 0.5 per cent, 10 kg ha^{-1} and 110 kg ha^{-1}, respectively. The corresponding values for soil to be rated medium are 0.5-0.7 per cent, 10-25 kg ha^{-1} and 110-250 kg ha^{-1}, respectively. The salt affected soil is usually low in nitrogen, medium to high in phosphorus and high in potassium.

The indiscriminate use of fertilisers should be avoided in saline soils. The fertiliser requirement also depends upon the crops and their varieties. The kind of fertilisers, their method and time of application as related to the growth stage of the crops, are some of the important considerations which play a vital role in crop response. The degree of salinity also influences the response of fertilisers. In general, whereas, ammonium sulphate or liquid ammonia is more beneficial to alkali soils, urea is efficacious in saline soils. If the degree of salinity is initially high, the response of fertilisers is limited.

6. Micronutrients

Extensive research carried out in plant nutrient has established that 16 elements are essentially required for optimum plant growth and development. Nine of these elements (C, H, O_2, Ca, Mg, S, N, P and K) are required in relatively larger amounts and are called macronutrients. The remaining seven, *viz.* B, Fe, Mn, Zn, Mo, Cl and Cu, are needed by plants in relatively smaller amounts and therefore called micronutrients.

Table 2.5. Micronutrients in saline and alkali soils (mgl^{-1})

Soil	Zn	Cu	Fe	Mn
Saline	0.1	0.1	67.6	13.2
Alkali	0.5	1.0	2.9	8.0

Source : Gupta *et al.* (1970).

The different crops have different requirements of micronutrients. The deficiency can be identified on the basis of soil or plant analysis, the visual symptoms on plants, etc. The availability of micronutrients is affected by several factors, the most important being the pH of the soil. Boron is usually not deficient in saline soils. Likewise, Mo, Fe and

Mn deficiency may not be there in many saline soils. However, deficiency of Zn and Cu has been observed to be very common problem in saline soils (Table 2.5).

7. Toxic Elements

Almost all the elements which are otherwise recognised as micronutrients, if present in soil beyond specific limits, act as toxic elements and adversely affect the growth of certain plants. In addition, some more elements, viz. Al, As, Be, Cd, Cr, F, Pb, Ni, Se and V, are also known to cause a toxic effect to plant growth, if present beyond permissible limits. The elements which have been recog-nised to be of practical importance from the toxicity point of view in salt affected soils are B, Li, F, Se and Mo. Out of these five elements, three viz. B, Li, and F are phyto-toxic, whereas Se and Mo many not adversely affect plant growth but the crops, particularly the forages grown on them, may contain an excess of these elements which cause adverse effects on the health of animals who graze on them. The salt affected soils may contain one or more of these elements in toxic levels under specific situations.

(a) Boron

The permissible limits for soluble boron (available in the saturation extract) were given in the past on a tentative basis (Richards *et al.,* 1954). Concentrations below 0.7 mgl^{-1} was considered as safe for sensitive plants; from 0.7 to 1.5 mgl^{-1} as marginal and 1.5 mgl^{-1} boron as unsafe. Recent work shows that permissible limits for sensitive, semi-tolerant and tolerant plants can be considered as 1-3, 3-5 and 5-10 mgl^{-1}, as the soil texture changes from light to heavy, because heavy textured soils adsorb higher amounts of boron. Boron tends to increase with increase in the salinity of the soils. The presence of $CaCO_a$ and gypsum, however, depresses the solubility of boron (Gupta, 1979).

High levels of boron in saline soils can be easily reduced by leaching alone, and in saline alkali soils by leaching after treatment with gypsum. During the normal leaching process, boron may be removed almost in the same proportion and additional depth of leaching water may not be necessary to reduce boron to a safe value for good growth. The boron can also be leached efficaciously by rainfall, alone, if appreciable and effective (>500 mm).

Under non-saline conditions, the growth of wheat is not affected adversely even with 20 mgl^{-1} of boron in irrigation water (Gupta,

1980). Total fruit weight of tomato is not affected significantly by 10 mgl^{-1} of B in the soil solution (Francois, 1984).

(b) Selenium

Selenium is not an essential element for plant growth but traces of selenium (0.05 –0.1 mgl^{-1}) are required by the animal for normal metabolism. Vitamin E and selenium, are interrelated in the metabolic functions but vitamin E can not completely replace the need for selenium. Selenium is the only mineral elements known to be absorbed by food and forage plants in sufficient amounts to make them lethal when consumed by animals. Tentative limits of the toxicity of selenium in soils to affect adversely the growth and quality of plants are considered to be equal to 1 mgl^{-1} and plants grown on such soils might contain 20 to 500 mgl^{-1} of selenium which could prove lethal to the animals.

(c) Lithium

Lithium is an element which is neither essential for human nor for plants. On the contrary, even relatively small amounts of lithium are toxic to plants. The permissible limits for lithium in the soil solution can be given tentatively with respect to the tolerance of plants for lithium. The sensitive, semi-tolerant and tolerant plants can tolerate lithium in concentrations of 5, 10 and 20 mgl^{-1}, respectively. The tolerance by plants in general seems to be higher than that of B. Under non-saline conditions, the growth of wheat was not affected significantly with even 40 mgl^{-1} of Li in irrigation water (Gupta, 1982). Saline soils tend to contain higher amounts of Li than normal soils (Gupta *et al.,* 1974; Lavado *et al.,* 1978).

(d) Fluorine

Like selenium, fluorine is also essential for animals but non-essential for plants. Till now no deficiency symptoms have been recorded in plants and animals subsisting on foods containing even the minutest amount of the element. The importance of fluorine is due to the harmful action in animal nutrition even when ingested in minor doses. The fluorine content in saline waters or soils has been reported to be up to 15 mgl^{-1}. At this level of fluorine, plant growth has not been reported to be affected adversely.

(e) Molybdenum

The soils with high amounts of available molybdenum result in the accumulation of the molybdenum in fodder crops, in concentrations which are toxic to the livestock. Waterlogged soils tend to contain excessive amounts of molybdenum. The plants may accumulate molybdenum to a level which is toxic to cattle, however, exhibiting any toxicity to the plant itself.

PHYSICAL PROPERTIES

1. Bulk Density

Bulk density is defined as the dry weight of the soil per unit volume and is commonly expressed in g/cm^3. The bulk density considerably depends upon the texture of the soil; usually coarser textured soils have higher bulk densities compared to medium or fine textured soils. Bulk densities of salt affected soils, whether saline or alkaline are generally higher. It is because these soils over a long period of time have been lying barren and uncared for. The surface gets hardened because of human and animal movement and due to impact of raindrops over the uncropped barren surface. Surface layer exhibiting bulk density as high as 1.8 g/cm^3 is not uncommon. High bulk density of these soils results in low infiltration and hydraulic conductivity (due to decreased bulk volume and total pore space, impeded drainage and less root proliferation and penetration). Instances of increasing sodium saturation of exchange complex leading to decreased bulk density are also reported (Varallyay, 1977).

2. Depth of Soil

Plant growth is greatly influenced by the depth of the soil as in deep soils roots can penetrate to greater depths to meet their moisture and nutrient requirements. Except for the shallow rooted crops, a root zone depth of 45-90 cm is considered necessary for optimum yields of most crops. The depth of the root zone in salt affected soils is restricted due to formation of *kankar* (impure $CaCO_3$) layer arising from precipitation of the salts or from a layer of soil with high exchangeable sodium or high in soluble salts. Gypsum accumulation at a given depth in saline soils also restricts the depth of the root zone. Crop characteristics usually decide the depth of reclamation in salt affected soils. The operation of levelling which is an essential part of any reclamation technology is influenced by depth of the soils as in shallow soils, depth of cut will be limited. When *Kankar* layer in salty soils is near the surface, levelling is often a problem.

3. Water Stable Aggregates

Water stable aggregates is one of the important aspects of soil structure problems of salt affected soils. By and large cation exchange capacity and exchangeable Ca^{++} play an important role in enhancing the water stability of the aggregates. It is particularly true in alluvial and acidic soils. In alkali soils it is because of high sodium on the exchange complex. Water stable aggregates decrease with increase in ESP (Table 2.6).

Table 2.6. Cumulative percentage of water stable aggregates of various sizes as affected by ESP

ESP	Size greater than (mm)					
	2.0	1.0	0.40	0.20	0.075	0.045
6.2	40.0	61.0	77.8	88.2	92.0	92.7
10.0	35.8	54.8	74.0	83.2	86.6	88.3
15.6	20.6	44.6	70.7	83.9	90.1	92.0
21.8	18.2	36.0	63.6	79.1	84.7	86.2
37.5	18.0	38.3	49.7	64.6	71.1	74.2
58.0	0.8	4.0	19.4	34.3	45.9	51.4
CD at 5%	6.58	4.51	6.23	5.44	10.20	7.85

Gupta and Verma (1983)

4. Pore-Size Distribution

It is well known that geometry of the soil pores and pore size distribution, rather than total pore volume govern both saturated and unsaturated hydraulic conductivity. Soil permeability of water is known to be a function of square of the mean pore radius and thus any process which results in reduction in the size of the pores due to swelling or clogging will have a significant influence on soil permeability. The non-capillary porosity of alkali soils is reported to be as low as 2 per cent compared to normal soils where it is in the range of 10-30 per cent. Clay-Na interaction causes internal swelling in soils, thereby resulting in increase in volume of solids at the expense of volume of macropores. Moreover, dispersion of clay and its downward movement results in clogging of micropores reducing pore size and macropore volume. Even in saline sodic soils where due to high electrolyte concentration of the soil solution a reversible flocculation of the primary particles takes place due to low zeta potential, the floccules are only stable in certain concentration range and once the concentration drops below a particular threshold value, soils again get dispersed. Some evidence on increase in total pore volume with Na^+ saturation indicates that the clogging of micropores, radical or significant changes in pore-size distribution and pore-size and macropore volume decrease can only explain the reduction in hydraulic conductivity in salt affected soils.

5. Crack and Crust Formation

Alluvial, heavy texture alkali soils may develop a hard surface crust on drying after rainfall or irrigation because of low stability of structure, high dispersion and slaking nature. This crust reduces the infiltration rate causing over saturation and aeration problem and retards the root penetration. Drought sensitivity of alkali soils may as well arise from the crust formation in these soils. Swelling clay in alkali soils on the other hand results in cracking which poses another problem of water management in these soils. The water enetering these cracks is lost directly to lower layers, diminishing the water storage in the soil. Shrinkage during drying results in deep and wide cracks through which high evaporation losses dry out the soil early and to deeper depths.

The problem of crust formation and cracking is also common in saline soils with low organic matter content. Crust thickness, crack width and crust strength for saline water irrigated soils at Pali in Rajasthan are given in Table 2.7. The data clearly indicate increasing crust thickness, crust strength and crack width with increasing salinity of the irrigation water. the crust formation in these soils with saline irrigation water is traced to stagnation of water containing high sediments which accumulate in surface and subsurface layers. Cementing action of silica present in saline water imparts hardness to the crust (Goyal and Jain, 1982). Emergence of seedlings can be improved by frequent irrigations, mulching and by adding chemical amendments such as gypsum to the soils.

Table 2.7. Effect of saline water irrigation on crust properties and crack width at Pali in Rajasthan

Salinity of irrigation water (dSm^{-1})	Crust thickness (mm)	Crust strength ($kg\ cm^{-2}$)		Crack width (mm)
		7.7.1979	3.11.1979	
2.7	1.03	1.70	4.55	4.00
6.0	1.56	4.84	20.20	4.66
9.0	1.90	8.11	26.41	6.30
12.0	2.26	8.27	19.85	6.36
Unirrigated	0.94	1.27	–	3.12

WATER RETENTION AND TRANSMISSION CHARACTERISTICS

1. Water Retention and Availability

Low availability of moisture to the plants in soils is either because of low available moisture capacity (AMC) of the soil, poor transmission characteristics or the high osmotic potential. The availability of water to the plants is controlled by the total water potential which is the sum of gravitational potential, osmotic potential, and the tensiometer-pressure potential which itself consists of the pneumatic potential and matric potential. Whereas, gravity and pneumatic potential are independent of the salt status of the soil, osmotic potential is a function of the concentration of the soil solution. Matric potential on the other hand is affected by the change in soil physical environment for which no simple relationship can be established. Abrol (1977) reported that amount of water released from saturation to 0.5 atm. increased from 27.0 to 33.4 per cent as the ESP decreased from 70 to 5. It was observed that with increasing ESP, soil water storage and availability is drastically reduced. Contrary to it, however, there is evidence that available moisture capacity in the available moisture ranges from field capacity (pF 2.3 - 2.7) and wilting percentage (pF 4.2) increases with increasing ESP and the reduced availability to plants is because of very low flux of water to the roots through the relatively dry moisture depletion zone in which transport coefficients are extremely low (Varallyay, 1977).

The limitations of low transport coefficients were well documented by the soil moisture depletion of two soils at a ESP of 4 and 38. The moisture depletion in a period of 10 days in soil with ESP 38 was negligible at a depth of 16 cm inspite of the high suction gradients, mainly due to low transport coefficients. Quality of irrigation water influences the moisture retention in soils (Yadav and Girdhar, 1981). Moisture retention at 0.30 and 15 bar suction, increases with increasing SAR and Mg : Ca ratio of the water. It decreases with increased electrolyte concentration of 80 mel^{-1} compared to 20 mel^{-1}. The results were similar in alluvial as well as black clay soil but more marked in the latter because of heavier texture dominated by montmorillonitic clay.

2. Infiltration Rate

Infiltration rate of water through the soil surface is extremely important in soils, because it determines the total water storage in the profile. From the standpoint of soil salinity it determines how much and how fast the leaching of a given soil can be accomplished. From hydrological point of view, it is probably the most important hydrological process in a watershed as it determines the runoff and ground water recharge. In a saline environment it will determine whether salts will get leached (infiltrate), accumulate or flushed with runoff.

Infiltration rate is dynamic and changes with elapsed time. When excess water is applied continuously to the soil surface, not only infiltration rate decreases with time, the rate of this decrease also reduces with increasing time. The asymptotically decreasing curve has a tendency to approach a final minimum value commonly referred to as the basic infiltration rate. This basic infiltration rate is important in the design of drainage system and to evaluate leachability of soils and the fastness with which a soil can be leached. Another parameter influencing the infiltration rate is initial moisture content which has a great bearing on the infiltration rate curve of a soil. The infiltration rate curve of a wet soil will always lie below that of an initially dry soil though the two curves may meet at a time when the infiltration rate approaches the basic infiltration rate. The decrease in infiltration rate of an initially wet soil is due to the reduced hydraulic gradient between the surface layer and soil just ahead of the wetting front. Wetting front, however, will move faster in a wet soil than in a dry soil.

The effect of physico-chemical properties of soils on the infiltration rate of irrigation water is similar to that on the hydraulic conductivity. Infiltration rate of a light textured soil is usually greater than a fine textured soil, unless aggregation in the latter because of electro-static or other forces causes improvements in the infiltrability. The stability of aggregates to water then plays a major role. Leaching of divalent ion (Ca) and their replacement by monovalent ones (Na) may decrease the infiltration because of poor flocculating efficiency of sodium ions, compared to that of calcium.

The cumulative infiltration curves for a highly saline soil irrigated with fresh canal water show that infiltration rate of the initially barren saline soils is low but it increases after cultivation of the fields. Initial infiltration rate is increased to about three times because of shallow tillage only. The saline soils which have been lying barren has a periodic hardening of the surface due to impact of raindrops. Breaking of this hardened surface and decrease in bulk density appear to be the most relevant reasons for this increase in infiltration rate. the continuous cultivation further increases the initial infiltration rate but consequent reduction in the value of basic infiltration rate shows that the decay of the infiltration rate with time is faster during the period of test. Increase in organic matter content, improved structure and reduced soil pore fluid concentrations can be possible reasons for these changes.

3. Hydraulic Conductivity

The proportionality coefficient, K of Darcy's Law which is designated as the hydraulic conductivity specifies the quality of a soil horizon that enables water or air to move through it. It expresses the interaction between the fluid and the media and has the units of velocity, cm sec. Smith and Browing (1946) grouped the soils on the basis of hydraulic conductivity (Table 2.8). Pertinent comments have also been made in respect of leaching and drainage of these soil groups to serve as a useful guide for characterising the soil. Hydraulic conductivity of 0.1 cm hr^{-1} is considered reasonable, below which the leaching and irrigation may present serious difficulties.

Table 2.8. Classes of permeability (percolation rates) for saturated subsoils.

Class	Hydraulic conductivity or percolation rate (cm hr^{-1})	Comments
Extremely slow	< 0.0025	Nearly impervious, leaching is insignificant
Very slow	0.0025 – 0.025	Poor drainage. Too slow for artificial drainage
Slow	0.025 – 0.254	Too slow for favourable air-water relations and for deep root development
Moderate	0.254 – 2.540	Adequate permeability
Rapid	2.54 – 25.4	Excellent water holding relations as well as excellent permeability
Very rapid	> 25.4	Associated with poor water holding

It is a well known fact that geometry of soil pores and pore-size distribution influence the hydraulic conductivity than the total pore space. Investigations have confirmed that soil permeability or hydraulic conductivity in a salt affected soil is affected by solution salt concentration and its composition. The ionic composition on the exchange complex more particularly the cationic composition given by Na^{+} to Ca^{++} ratio or Ca^{++} to Mg^{++} ratio have a great bearing on the hydraulic conductivity. The salt concentration of irrigation water in relation to soil solution concentration also governs the hydraulic conductivity. The hydraulic conductivity changes with ionic composition of the exchange complex have been studied by several authors with almost similar findings that the hydraulic conductivity reduces in order of sodium > magnesium > calcium ions on the exchange complex (Kanwar *et al.*, 1969, Patel and Shah, 1979; Girdhar and Yadav, 1981). Quantitative reduction in the hydraulic conductivity of various mono-ionic soils generally shows this trend (Table 2.9).

Table 2.9. Fractional change in hydraulic conductivity (K/K_{max})

Soil type	Baroda	Khed-brahma	Himmat-nagar	Bardoli	Miyagam	Ahwa	Piplod	Mean
Normal soil	1.000	1.000	1.000	1.000	1.000	1.000	1.000	1.000
Ca-soil	0.785	0.429	0.631	0.604	0.379	0.116	0.673	0.510
Mg-soil	0.682	0.141	0.065	0.220	0.000	0.067	0.467	0.234
Na-soil	0.000	0.000	0.000	0.000	0.000	0.000	0.000	0.000

Patel and Shah (1979).

The lower conductivity values in case of Ca-soil are attributed to removal of organic matter during soil preparation and continuous leaching over large periods of time. The magnitude of decrease in Mg and Na-soil might also have increased because of the reasons stated above. Decrease in size of large pores due to swelling of clays and migration of dispersed Na-clay along with the percolating solution and blockage of interstices is mainly responsible for the highly reduced hydraulic conductivity of sodium soils. Higher dissociation which causes greater dispersion of Mg-soil has been suggested as the reason of reduced values in Mg-soils. The increased dispersion as a consequence of higher Mg : Ca ratio of leaching water in both calcareous and non-calcareous soils justifies the observations and findings reported earlier (Table 2.10).

Table 2.10. Effect of leaching with waters of different Mg : Ca ratio and SAR on the degree of soil dispersion percentage.

Mg : Ca ratio	Calcareous soil			Non-calcareous soil		
	10*	25	50	10	25	50
2	5.0	15.5	56.0	8.1	29.4	79.0
4	5.3	16.0	57.0	8.7	30.3	80.4
8	5.9	16.7	58.5	9.6	31.4	81.6
16	6.7	18.0	61.5	11.0	33.7	85.3

*SAR of water. Yadav and Girdhar (1981).

Soil water diffusivity and unsaturated hydraulic conductivity decrease with increasing ESP. It is reported that in a sandy load alluvial soil at high soil-water contents, hydraulic conductivity decreases more sharply at low ESP values compared decreases more sharply at low ESP values compared to soils of high ESP, although in the latter case hydraulic conductivity is several folds less in magnitude. Dispersion and movement of clay particles and their blocking action of large conducting pores are considered as major factors for reduced hydraulic conductivity in these soils, which is probably apparent as water dispersible clay (illite type) has been found to increase from about 10 per cent at a ESP of 5 to around 85 per cent at a ESP of 50.

The electrolyte concentration and ionic composition of the irrigation water profoundly influence the water transmission characteristics. When electrolyte concentration of water is increased from 0 to 40 mel^{-1}, weighted mean diffusivity increased nearly 2.8, 3.1, 20, 40 and 164 times for ESP 5, 14, 22, 30 and 70, respectively. In fact for a given ESP value there is a threshold concentration of leaching water, below which the soil permeability decreases markedly due to the swelling of clay domain and subsequent clay deflocculation. Lower the electrolyte concentration of the soil solution, lower the ESP at which the structure becomes unstable. In saline soils, a typical situation arises during rains which is distinctly non-saline compared to the waters which are being used for irrigation. It is reported (Sharma *et al.*, 1982) that hydraulic conductivity is largely reduced under these conditions (Table 2.11).

Table 2.11. Effect of water quality on saturated hydraulic conductivity of sandy-loam and clay-loam soils

Salt concentration (mel^{-1})	Hydraulic conductivity (mm hr^{-1}) for water of SAR							
	Sandy loam				Clay loam			
	5	15	30	45	5	15	30	45
Synthetic water								
15	3.8	3.5	2.5	2.3	0.44	0.19	0.03	0.05
30	5.6	5.1	2.9	2.1	0.67	0.10	0.03	0.04
60	6.2	5.5	3.8	2.8	1.05	0.12	0.04	0.03
90	5.7	5.5	3.8	3.6	0.96	0.12	0.06	0.04
Distilled water								
15	0.35	0.19	0.19	0.13	0.08	0.03	0.02	0.02
30	0.46	0.27	0.17	0.11	0.08	0.03	0.01	0.01
60	0.44	0.31	0.11	0.10	0.09	0.03	0.02	0.01
90	0.38	0.38	0.12	0.12	0.09	0.03	0.01	0.01

This emphasises that in saline alkali soils, the leaching through rainfall may not be as efficient as is usually considered because of such permeability hazards. Chloride salt systems produce higher hydraulic conductivity than its corresponding bicarbonate salts in red, laterite, alluvial and black soils and an increase in bulk density to 1.7 gcc^{-1} reduced the hydraulic conductivity to 26.5, 33.3, 44.9 and 15.1 per cent of its value at a bulk density of 1.5 gcc^{-1} in the four soil groups respectively (Shivakumaraswamy and Satyanarayana, 1982).

During management of saline or sodic soils, hydraulic conductivity modifications can be attempted by tillage practices including deep ploughing and subsoiling. These modifications will not be permanent and the processes may have to be repeated say once in a few years.

BIOLOGICAL PROPERTIES

1. Organic Matter

Soil organic matter, generally referred to as humus found in small amounts in productive agricultural soils, generally representing less than 5 percent of the dry weight, imparts unique and remarkable properties to the inorganic matrix of sand, silt and clay. Soil organic matter can conveniently be classified into two main groups of compounds, non-humic and humic substances. The non-humic substances are the unaltered remains of plant and animal tissues and also organic compounds. The humic substances are chemically and biologically modified substances and represent the most active fraction of the organic matter in most soils. Humic substances are resistant to microbial degradation and based on their solubilities, these can be separated into four fractions viz. (a) fulvic acid, which is soluble in both alkali and acid, (b) humic acid, soluble in alkali but insoluble in acid and alcohol, (c) Hymatomelanic acid, soluble in alkali and alcohol, and (d) humin, insoluble in both alkali and acid.

In saline soils, organic matter is not dispersed as in alkali soils and therefore, the content is higher, generally varying from 0.2 to 1.0 per cent. The saline soils which occur in arid regions, where the transport of salts and of sodium ions (chloride, etc.) occurs by capillary movement as a precursor of evapotranspiration, these processes retard plant growth and therefore the humus dynamics.

The decomposition of *dhaincha* plus paddy straw, in equal proportions in saline sodic soils builds up maximum organic matter under low temperature conditions of winter (Bajpai *et al.,* 1980).

2. Micro-Organisms

The physical and chemical properties of a soil determine the nature of the environment in which micro-organisms are found. These environmental characteristics in turn affect the composition of the microscopic population, both qualitatively and quantitatively. The soil reaction plays a vital role in the growth of micro-organisms. The total number of bacteria, actinomycetes and fungi are usually the highest in productive normal soil, followed by saline and alkali soils (Neelkanthan and Pahwa, 1977). In an incubation experiment on the mineralisation of gulmohur (*Delonix vegia* L.) leaves added to the soil, it was observed that CO_2 evolution and total carbon mineralisation decreased with increased concentrations of salt mixture (NaCl + $CaCl_2$) from 0.1 to 5.1 per cent and the process of nitrification was completely inhibited between 0.6 and 0.9 per cent salt (Laura, 1974). Bajpai and Gupta (1979) reported that ammonifying power which ranged from 43 to 54.7 per cent was the highest in normal (50.8 per cent) and lowest in saline-alkali soil (43.4 per cent). The nitrifying power was considerably low in salt affected soils as compared to normal soils. Carbon mineralising power determined as the capacity to oxidise carbon of FYM (semi-rotten) was observed to be maximum in normal soil (8.5 per cent) and minimum in saline-alkali soils (1.7 per cent).

BENCH MARK SOIL SERIES

National Bureau of Soil Survey and Land Use Planning, Nagpur (Lal *et al.,* 1994) have identified broadly 180 soil series in India out of which two series representing saline soils viz. Fatehgarh series (Kachchh, Gujarat) and Ghabdan series (Patiala, Punjab) are described below.

1. Fatehgarh Series

The Fatehgarh series is a member of the coarse-loamy over clayey, mixed, hyperthermic family of Typic Salorthids. Typically, Fatehgarh soils have light yellowish brown to brown, moderately alkaline, loamy sandy loam A horizons; dark yellowish brown to dark greyish brown, moderately alkaline, sandy loam to silty clay Bz horizons and dark greyish brown, moderately alkaline, sandy loam C horizons.

Typifying pedon : Fatehgarh loamy sand-uncultivated

A1 0-8 cm--- Light yellowish brown (10 YR 6/4 D) and dark yellowish brown (10 YR 4/4 M) loamy sand : weak medium subangular blocky structure; slightly hard, friable, nonsticky and nonplastic; few fine grass roots; few fine disseminated indurated lime nodules; violently effervescent; few fine gravels; moderately alkaline (pH 8.0); clear smooth boundary.

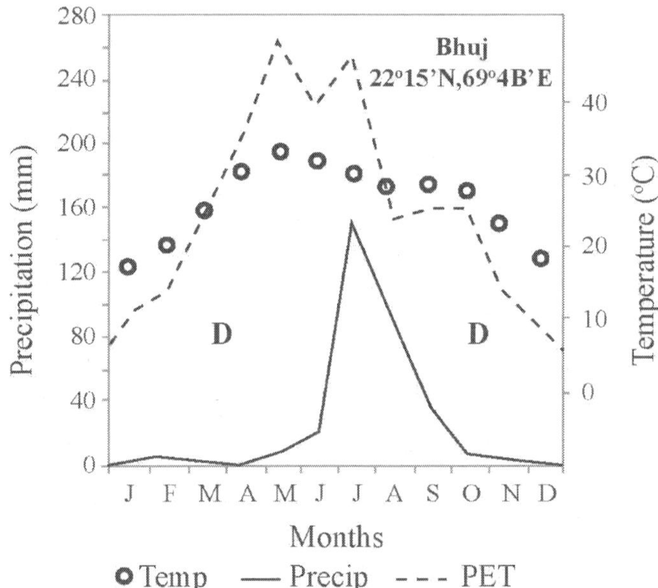

Fig. 2.1. Precipitation, temperature and water balance at Bhuj

A2 8-31 cm --- Yellowish brown (10 YR 5/4 D) and brown (10 YR 4/3 M) sandy loam; weak medium subangular blocky structure; slightly hard, friable, slightly sticky and nonplastic; few coarse and fine roots; few disseminated indurated lime nodules; violently effervescent; few gravels; moderately alkaline (pH 8.3); clear smooth boundary.

AB 31-44 cm --- Dark yellowish brown (10 YR 4/4 M) sandy loam; few scattered greyish patches; moderate medium subangular blocky structure; slightly hard, friable, slightly sticky and nonplastic; few fine and coarse roots; few indurated lime noduls; violently effervescent; few ferruginous gravels; moderately alkaline (pH 8.1); abrupt smooth boundary.

2Bzl 44-74 cm--- Dark yellowish brown (10 YR 4/4 M) silty clay; many grey mottles; moderate coarse prismatic structure breaking into strong coarse angular blocky peds exhibiting white coating of salts on the surface when dry; hard, firm, very sticky and very plastic; violently effervescent; moderately alkaline (pH 7.9); diffuse smooth boundary.

2Bz2 74-101 cm --- Dark greyish brown (10 YR 4/2 M) silty clay; patches of dark yellowish brown (10 YR 4/4) mottles; strong coarse angular blocky structure coated with whitish salt material when dry; very hard, very firm, very sticky and very plastic; violently effervescent; water oozes out of the matrix; moderately alkaline (pH 7.9); abrupt smooth boundary.

3C 101-113 cm --- Dark yellowish brown (10 YR 4/4 M) sandy loam; many dark grey mottles massive coated with whitish salt when dry; nonsticky and nonplastic; violently effervescent; water collects when dug; moderately alkaline (pH 8.3).

Type location : 23°45′ N, 70°51′ E; about 6.5 km north of village Fatehgarh, tehsil Rapar, district Kachchh, Gujarat.

Range in characteristics : The thickness of the solum ranges from 90 to 110 cm. The A horizon is 40 to 50 cm thick with the colour in hue 10 YR, dry value 5 or 6 and moist value 4 and chroma 4 both for dry and moist soils. Its texture varies from sand to sandy loam. The B horizon is 50 to 70 cm thick. The colour of the B and C horizons is in hue 10 YR, moist value 4 and chroma 4 or less. The effective rooting depth extends upto 50 cm of the surface. The water table is within 1 m of the surface.

Competing series and their differentiae : The competing soils are those of Mayawandh series which are deep, sandy loam to sandy clay loam in the surface with dominantly brown to yellowish brown colours. Watertable in Mayawandh soils is found to be relatively deeper and it is about 2 metres of the surface. The Mayawandh soils belong to Typic Salorthids.

Geographic setting : Fatehgarh soils are formed in coastal alluvium and occur on nearly level beach ridge slough and marine salt waste at an elevation of about 13 m above MSL. The climate is arid subtropical with mean annual temperature of 26.7°C, and mean annual rainfall of 350 mm (Fig. 2.1). The estimated MAST is 27.7°C, MSST 29.0°C and MWST 22.1°C.

Geographically associated soils : These are Balasar, Desalpar and Mayawandh series which are Typic Torripsamment. Typic Paleargid and Typic Salorthid, respectively.

Drainage and permeability : Imperfectly drained with slow to very slow permeability.

Use and vegetation : Mostly dry salt waste in Rann, devoid of any vegetative growth of mesophytic species; sporadic groves of halophytes found.

Table 2.12. Physico-chemical characteristics of Fatehgarh soil series

Horizon	Depth (cm)	Size class and particle diameter (mm)				Coarse frag-ments >2 mm % of whole soil	Organic carbon %	$CaCO_3$ < 2 mm%	$pH_{2.5}$	$EC_{2.5}$ (dSm^{-1})
		Coarse sand (2-0.2)	Fine sand (0.2 - 0.02)	Silt (0.02 - 0.002)	Clay (<0.002)					
		% of < 2 mm								
A1	0-8	36.5	52.1	3.5	7.9	8	0.63	1.0	8.0	5.5
A2	8-31	29.9	51.6	4.4	14.1	6	0.61	1.2	8.3	5.8
AB	31-44	35.7	50.9	3.4	10.1	3	0.16	Tr	8.1	4.8
2 Bz1	44-74	8.2	10.8	27.6	53.4	–	0.16	9.8	7.9	40.1
2Bz2	74-101	5.8	10.3	30.1	53.8	–	0.19	7.8	7.9	42.1
3C	101-113	28.0	53.8	5.9	12.3	1	0.35	2.1	8.3	7.0

Distribution and extent : Extensive (about 1,55,000 ha) in Kachchh district of Gujarat.

Series proposed : National Bureau of soil Survey and Land Use Planning Regional Centre, Nagpur 1977.

Interpretation : Fatehgarh soils are highly saline and have drainage problems. The water table is very high and soil remains saturated upto 40 cm depth from the surface. The soils are coarse textured upto 40 cm depth underlain by finer texture. Mostly they are dry salt waste ad devoid of vegetation and not suitable for growing mesophytic plants. Aridity is the main limitation.

Interpretative grouping :

(i) Land capability subclass V/s

(ii) Irrigability subclass 6s

(iii) Productivity potential Low

Note : The soil texture is based on international particle size classification.

2. Ghabdan Series

The Ghabdan series is a member of the fine-loamy, mixed, hyperthermic family of Aquic Naturstalfs. Typically, Ghabdan soils have light olive brown, very strongly alkaline, loam A horizons, and light olive brown to olive brown, very strongly alkaline, loam to clay loam B horizons.

Typifying pedon : Ghabdan loam-barren and salt affected land

A 0-6 cm --- Light olive brown (2.5 Y 5/6 M) loam; weak thin platy structure; friable; many fine to medium vesicular pores; very strongly alkaline (pH 10.3); clear smooth boundary.

BA 6-24 cm --- Olive brown (2.5 Y 4/4 M) loam; moderate medium subangular blocky structure; firm; common fine pores; few iron - manganesde concretions of 2 to 3 mm size; very strongly alkaline (pH 9.8); clear smooth boundary.

Bt1 24-48 cm --- Olive brown (2.5 Y 4/4 M) clay loam; few fine faint mottles of higher chroma; moderate medium subangular blocky structure; firm; few fine pores; patchy thin brown (10 YR 5/3) clay coatings; few iron-manganese concretions of 1 to 2 mm size; strongly effervescent; very strongly alkaline (pH 9.6) gradual smooth boundary.

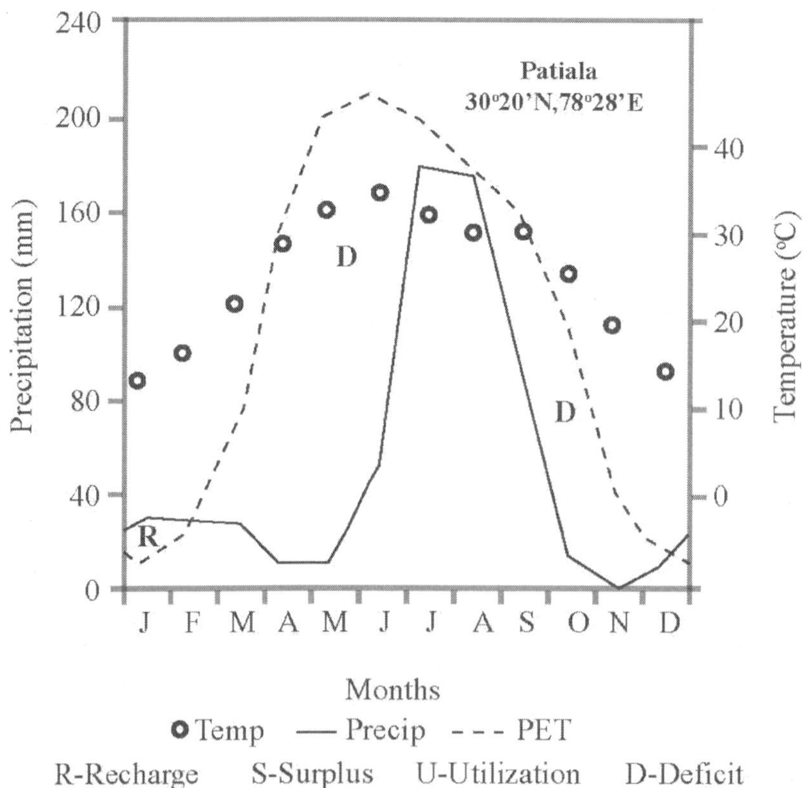

Fig. 2.2. Precipitation, temperature and water balance at Patiala

Table 2.13. Physico-chemical characteristics of Ghabdan soil series

Horizon	Depth (cm)	Size class and particle diameter (mm)			Organic carbon %	$CaCO_3$		$pH_{2.5}$	$EC_{2.5}$ (dSm^{-1})
		Sand (2–0.05)	Silt (0.05 – 0.002)	Clay (<0.002)		>2 mm	< 2 mm		
		% of < 2 mm				%			
A	0–6	37.0	42.1	20.9	0.22	–	7.6	10.3	6.6
BA	6–24	34.7	41.6	23.7	0.17	–	9.4	9.8	4.2
Bt1	24–48	33.7	35.3	31.0	0.14	–	6.2	9.6	3.1
Bt2	48–73	36.3	35.5	28.2	0.14	–	4.4	9.6	0.9
Bt3	73–97	37.0	31.6	31.4	0.10	1.8	2.3	9.4	0.8
BC	97–124	36.8	38.6	24.6	0.10	5.3	4.0	9.4	0.9
Ck	124–145	32.9	41.5	25.6	0.09	7.9	19.4	9.3	1.0

Depth (cm)	Extractable bases				CEC NH_4 OAc	ESP	Satur -ation %	pH	Water extract from saturation paste							EC_e (dS m^{-1})
	Ca	Mg	Na	K					Ca+ Mg	Na	K	CO_3	HCO_3	Cl	SO_4	
	cmol (p+) kg^{-1}								mel^{-1}							
0-6	2.0	0.7	6.8	0.3	8.2	83	35.4	10.1	5.6	296.0	0.6	125.0	32.4	148.0	2.8	32.4
6-24	1.4	1.0	8.9	0.3	11.8	75	37.8	9.7	5.0	91.6	0.8	20.0	10.5	48.0	2.3	10.5
24-48	2.0	1.1	10.2	0.2	13.6	75	44.2	9.4	3.7	65.1	0.6	–	8.0	46.0	0.6	8.0
48-73	2.8	0.9	10.8	0.4	14.2	76	45.0	9.3	4.0	46.0	0.6	–	5.6	32.0	0.4	5.6
73-97	3.5	0.8	9.4	0.2	13.8	68	45.4	9.2	6.2	31.6	0.4	–	3.7	26.0	0.3	3.7
97-124	3.3	0.9	6.2	0.2	9.8	63	46.0	9.2	10.0	27.2	0.6	–	4.1	25.2	0.2	4.1
124-145	2.1	0.7	6.0	0.1	9.0	67	47.3	9.1	7.5	38.0	0.3	–	4.6	24.2	0.2	4.6

Bt2 48-73 cm --- Light olive brown (2.5 Y 5/4 M) clay loam; few fine faint mottles; moderate medium subangular blocky structure; firm; few fine pores; patchy thin greyish brown (2.5 Y 5/2) clay coatings; few iron-manganese concretions of 1 to 3 mm size; strongly effervescent; very strongly alkaline (pH 9.6); clear smooth boundary.

Bt3 73-97 cm --- Light olive brown (2.5 Y 5/4 M) clay loam; few fine faint mottles; moderate medium subangular blocky structure; firm; few fine pores; patchy thin greyish brown (2.5 Y 5./2) clay coatings; few lime nodules of 5 to 7 mm size; few iron-manganese concretions of 1 to 3 mm size; strongly effervescent; very strongly alkaline (pH 9.4); clear smooth boundary.

BC 97-124 cm --- Light olive brown (2.5 Y 5/4 M) loam; common fine distinct yellowish brown (10 YR 5/5) mottles; moderate medium subangular blocky structure; firm; few fine pores; common lime nodules of 5 to 10 mm size; few iron-manganese concretions of 2 to 3 mm size; violently effervescent; very strongly alkaline (pH 9.4) : clear smooth boundary.

Ck 124-145 cm --- Light olive brown (2.5 Y 5/4 M) loam; few fine faint mottles; massive; firm; common small to large lime nodules; few iron manganese concretions of 1 to 3 mm size; violently effervescent; very strongly alkaline (pH 9.3).

(N.B. A thin layer of salt crust on the surface was present).

Type location : 30°15′N, 75°58′E; 300 m NE of a hermitage in the village of Ghabdan, situated 12.5 km from Sangrur on Sangrur to Patiala road, district Sangrur, Punjab.

Range in characteristics : The A horizon is 6 to 15 cm thick and ranges in colour in hue 10 YR and 2.5 Y, value 4 to 5 and chroma 4 to 6 in texture from loam to clay loam and in structure from weak fine subangular to weak thin platy. The B horizon is more than 100 cm thick and ranges in colour in hue 2.5 Y, value 4 to 5 and chroma 4, in texture from loam to clay loam and in structure from weak strong medium to coarse subangular blocky. These soils are moderately to strongly calcareous. A few small lime nodules are usually observed scattered in lower part of the B horizon and the calcic horizon is observed below 1 m of the surface. Few iron-manganese concretions are present.

Competing series and their differentiate : The Ghabdan soils are competing with Kaheru series which are also highly saline-sodic and show the development of argillic horizon. The Kaheru soils are comparatively coarse in texture and imperfectly drained and belong to Aquic Naturstalfs.

Geographic setting : Ghabdan soils are formed in alluvium and occur on relatively lower topographic position with slopes of less than 1 per cent at an elevation of about 240 m above MSL. The climate is semiarid subtropical with mean annual temperature of 24.0°C and mean annual rainfall of 435 mm (Fig. 2.2). The estimated MAST is 25.0°C, MSST 31.3°C and MWST 15.5°C.

Geographically associated soils : Principal associated soils are Kaheru soils which belong to Aquic Natustalfs.

Drainage and permeability : Poorly drained with slow to very slow permeability.

Use and vegetation : Recently reclaimed areas cultivated for paddy, berseem and wheat but the yields much lower than the associated normal soils. These soils support patchy growth of *Desmostachya bipinnata* (dab) and *Sporobolus marginatus* (chirriya ka dana) when uncultivated.

Distribution and extent : Occur on nearly level, highly saline sodic alluvial terraces : in Sangrur district found in a compact area starting from Amargarh and extending to Ghabdan along with Choa Nadi.

Series proposed : Punjab Agricultural University, Ludhiana and Department of Soil Conservation and Engineering, Punjab, 1972-73.

Interpretation : Ghabdan soils are poorly drained and are saline-sodic. These soils need to be reclaimed before bringing under cultivation.

QUALITY OF IRRIGATION WATERS

Because of accelerating population and rising living standard, water requirement is increasing with time. By 2025, 48 countries will be experiencing water stress and the number of people adversely affected will exceed 1.4 billion. In India, water requirement in the year 2025 is assessed at 1027 Billion Cubic Meter (BCM) as against the requirement of 525 BCM in the year 1990 showing an increase of 86%. Shrinking of fresh water resources and more than half of the world's ground water resources being saline, alternate sources of water need to be tapped to meet the increasing demand. According to the estimates, waste water generated through irrigation, water supply, industry and energy sectors is expected likely to be 556 BCM by 2025 in India. Treatment of wastewaters will not only protect the environment but their reuse will also meet the persistent shortage. However, suitability of such waters to be used for irrigation needs to be ascertained carefully lest the irrigated agriculture should be doomed due to development of salinity and/or sodicity.

Parameters for Suitability

Salinity Hazard

The total concentration of soluble salts is the single most important criterion which has been used conventionally to determine the quality of irrigation water. It is measured quantitatively in terms of Electrical conductivity (EC), because it is very closely related with the sum of major cations (or anions) determined by chemical analysis, and because this correlates well with the value of total dissolved solids, as well as osmotic potential. The EC of a water sample is determined by measuring the electrical resistance between two parallel electrodes immersed in the solution. The EC, also called 'specific conductivity' or the conductance per unit cross-sectional area and across unit distance, is obtained from the resistance recorded across a conductivity cell from the following relationship:

$$EC = K / R \tag{2.1}$$

where, K is the cell constant, and R is the resistance

Expressed in the past as micromhos per centimeter (μmhos cm^{-1}) for low salinity waters and millimhos per centimeter (mmhos cm^{-1}) for high salinity waters, it is expressed now according to SI Units (from the System International d' Unite's). as deciSiemens per meter (dSm^{-1}). Accordingly, EC equal to 1 $mmhoscm^{-1}$ at 25°C is equal to 1 dSm^{-1}. Also, 1 mmhos cm^{-1} and 1 $\mu mhoscm^{-1}$ is equal to 1 $mScm^{-1}$ and 1 μScm^{-1}, respectively. EC equal to 1 dSm^{-1} is nearly equal to 10.0 $meqL^{-1}$ of total cation concentration, TCC, (8 to 11 $meqL^{-1}$ for bicarbonate/sulphate waters in which calcium and magnesium are high to chloride waters that are high in sodium), or 640 mgL^{-1} (550-750 mgL^{-1}) of total dissolved solids (TDS). Both the relationships hold good in EC range from 1 to 5 dSm^{-1}. With further increase in EC from 5 to 10 dSm^{-1}, the range of TCC and TDS tend to increase from 12 to 13 $meqL^{-1}$ and 800 to 900 mgL^{-1}, respectively. The reporting unit used in the past as ppm, has now been replaced with mgL^{-1} (milligrams per litre). The two units are essentially identical for irrigation waters having EC below 5 dSm^{-1} but numerical values in mgL^{-1} are somewhat higher than corresponding values in ppm, because of solution densities appreciably greater than one for brines, sea waters and high salinity ground water or drainage waters.

Total salt concentration used to be measured gravimetrically in the past as TDS by evaporation of a known amount of irrigation water. Determination of TDS has become obsolete because different salts after drying retained differed amounts of moisture, depending upon the drying temperature and the degree of rehydration, between drying and weighing. About half of the bicarbonates are lost in the process of heating resulting in lower values of TDS. This method therefore, is less accurate for waters which contain bicarbonates and salts of hygroscopic nature ($CaCl_2$, $MgCl_2$ etc.). Because of this ambiguity and time consuming process, determination of EC has been widely preferred to evaluate water quality in terms of salinity hazard. The standard temperature for reporting EC is 25°C as the EC of aqueous salt solution increases at the rate of approximately 2% per degree centigrade rise in temperature.

The plant growth is governed by the salinity of the soil solution in the field-moisture range. However, due to difficulty of obtaining extracts of soil solution, and because field-moisture range is directly related with the saturation percentage, EC of the soil extracts at saturation (EC_e) has been widely recognized as an index to evaluate the plant growth. In the field, the moisture content of the soil fluctuates between a lower limit represented by the permanent-wilting percentage and the upper wet end of the available range, which is approximately two times the wilting percentage. The saturation percentage (SP) is approximately equal to four times the 15-atmosphere percentage, which in turn, closely approximates the wilting percentage. The soluble-salt concentration in the saturation extract, therefore, tends to be about one-half of the concentration of soil solution at the upper end of the field moisture range and about one-fourth the concentration that the soil solution would have at the lower dry end of the field moisture range. The salt-dilution effect that occurs in fine-textured soils, because of their higher moisture retention, is thus automatically taken into account.

Although, the general plant response is related with EC_e, actual plant responses to salinity are related to the sum of osmotic pressure (OP), and matric suction in the root zone. The OP values corresponding to some common salts, namely, NaCl, $CaCl_2$ and Na_2SO_4 are 24, 32 and 36 $meqL^{-1}$ per atmosophere respectively. The salts which produce large number of ions per molecule and which remain completely dissociated with individual ionic components produce the highest osmotic effects. However, the differences evident for single salt solutions become less pronounced for a mixed salt solution. OP dSm^{-1} in atmosphere (atm) is equal to 0.36 times the EC of irrigation water in. With a negative sign fixed, OP values in atm become nearly identical with osmotic potential values expressed in bars.

Salts normally found in natural waters or irrigated soils have a wide range of solubility (Table 2.14). A concentration of 80 $meqL^{-1}$ in the soil solution at field capacity or 40 to 50 $meqL^{-1}$ at saturation is usually considered to be indicative of incipient salinity. Therefore, the salts of low solubility such as $CaCO_3$, Ca-$MgCO_3$ or $CaSO_4$ precipitate before they reach a saline concentration in the soil solution. These soluble salts are not sufficiently soluble to produce a saline soil condition, whereas all Cl salts and Mg-Na_2SO_4 are highly soluble. It is obvious that fresh waters or sweet waters are dominant in calcium bicarbonate and high salinity waters are charged with sodium chloride. More recently, a concentration of about 15 to 20 $meqL^{-1}$ at saturation (EC 1.5 to 2.0 dSm^{-1}) has been recognized to be indicative of incipient salinity.

Table 2.14. Solubility ($meqL^{-1}$) of salts in water

Salt	Solubility
Low solubility	
Calcium carbonate, $CaCO_3$	0.5
Calcium bicarbonate, $CaHCO_3$	3-12
Calcium sulphate, $CaSO_4$, $.2H_2O$	30
Magnesium carbonate, $MgCO_3$	2.5
Magnesium bicarbonate, $MgHCO_3$	15-20
High solubility	
Calcium chloride, $CaCl_2.H_2O$	25470
Magnesium sulphate, $MgSO_4.7H_2O$	5760
Magnesium chloride, $MgCl_2.\ 6H_2O$	14955
Sodium bicarbonate, $NaHCO_3$	1642
Sodium sulphate, $Na_2SO_4.10H_2O$	683
Sodium chloride, NaCl	6108

Source: Doneen (1975)

A salinity problem related to water quality occurs if the total concentration of salts as measured by EC is high enough for salts to accumulate in the crop root zone to the extent that yield are adversely affected. If excessive quantities of soluble salts accumulate in the root zone, the crop has extra difficulty in extracting enough water from the upper part of the root zone and respond more critically to the salinity in this part than to the salinity level in its lower depths. When using normal irrigation practices, management of this critical upper root zone is quite important.

Ideally, it would be inferred that EC of irrigation water should be as low as possible but the water which is completely free of the soluble salts is never the best for irrigation. The waters having EC less than 0.2 dSm^{-1} have no fertility value and are well known to create permeability problem in the soil. The irrigation water should, however, have EC preferably less than 1.5 dSm^{-1} so that the irrigated soils do not ever become saline and there is full choice to grow the crops. Unfortunately, ground waters occurring in arid and semi-arid zones are highly saline. The percentage of low salinity waters decreases with increasing aridity (Table 2.15). In the state of Rajasthan, whereas 90% of the waters have EC less than 2.25 dSm^{-1} in humid districts, percentage of such waters decreases to 70 and 40 for semi-arid and arid districts respectively.

Table 2.15. Percent distribution of ground waters of Rajasthan in different EC classes

EC range ($dS\ m^{-1}$)	Arid districts (2817)*	Semi-arid districts (4000)	Humid districts (2614)
<0.75	10	23	41
0.75-2.25	29	48	49
2.25-5.00	27	19	9
5.00-10.0	21	8	2
10.0-15.0	9	2	–
>15.0	5	–	–

Source: Gupta (1979a); * No. of samples

To evaluate the quality of irrigation waters only on the basis of EC, it is presumed that low salinity waters have high RSC and high salinity waters have high SAR. Therefore, it is necessary that low salinity waters are evaluated for RSC, and high salinity waters for SAR. The medium salinity waters may be subjected to both.

Sodium Hazard

Sodium Adsorption Ratio (SAR)

Apart from the total salinity, the next important consideration is whether the use of irrigation water of given quality will cause sodicity in the soil. Water which might be considered suitable for irrigation on the basis of EC may not be suitable if sodium predominates. In the past sodic (alkali) soil was defined as the one, exchangeable sodium percentage of which was greater than 15 and pH_s (pH of the saturated soil paste) values usually ranged between 8.5 and 10. It implies that the quality of irrigation water should be such that Exchangeable Sodium Percentage (ESP) of the irrigated soil does not exceed 15 and pH_s value 8.5. However, now the sodic soil is defined as one the SAR of which is greater than 15. Obviously, the irrigation water which does not increase SAR of the soil beyond 15 is considered to be of good quality.

Gupta (1987) reported that SAR of the irrigated soils is highly correlated with the product of SAR_{iw} and salt concentration factor according to the following equation.

$$Y = X1 \, . \, X2^{1/2} \quad (2.2)$$

where Y is the SAR of the soil, X1 is the SAR of irrigation water, and X2 is the salt concentration factor, such that

$$X2 = \sqrt{(EC_e / EC_{iw})} \; (=1/\sqrt{LF}) \quad (2.3)$$

The high value of the coefficient of correlation (r=0.99) shows that processes such as precipitation, dissolution and mineral weathering etc. do not play a significant role under the prevailing conditions in western Rajasthan where saline ground waters have been used for irrigation for decades. This shows that as irrigation water after entering the soil becomes more concentrated, the SAR value increase in proportion to the square root of the concentration factor (Table 2.16). SAR of the irrigation water is therefore, useful index for evaluating the quality of irrigation water in so far as sodicity in irrigated soil is to be appraised on the basis of SAR. SAR of the soil is related with SAR of irrigation water through LF according to the following relationship:

$$SAR_e = SAR_{iw} / \sqrt{LF} = SAR_{iw} \sqrt{(EC_e / EC_{iw})} \quad (2.4)$$

Rhoades (1972) pointed out that there are several difficulties associated with equating SAR of irrigation water with soil water. These include:

Table 2.16. Observed and calculated values of SAR of the soils

Site	EC (dSm⁻¹)		X2	SAR_{iw}	SAR of soil	
	IW	Soil		X1	Observed	Calculated
Dantiwada	5.7	10.9	1.4	18	26	25
Kaparda	9.8	11.8	1.1	25	31	28
Jelwa	5.3	6.7	1.1	21	21	23
Shikarpura	2.3	22.2	3.1	9	32	28
Surayta	8.7	11.1	1.1	7	6	8
Dhakri	7.5	14.4	1.4	20	28	28

IW= Irrigation water

- The increase in soil water concentration over that of the applied irrigation water by evapotranspiration does not change the relative proportions of sodium and calcium + magnesium ions, but will change the SAR value because of the square root relationship. Thus, the process of evapotranspiration will produce higher ESP values in soils than predicted from SAR_{iw}.
- The effect of variation of soil water concentration with time between irrigations and with depth in the soil profile on plant growth is largely unknown.
- The loss in calcium and magnesium contents due to precipitation as the carbonates is a major factor affecting the

final soil water SAR value. Weathering of certain soil minerals may also affect the SAR of the soil water by introducing additional amounts of calcium, magnesium and bicarbonate ions into the soil solutions.

High SAR waters cause permeability problem in the soil wherein the rate of water infiltration into and through the soil is reduced to such an extent that the crop is not adequately supplied with water and consequently the yield is reduced. The adverse effect due to SAR depends upon total electrolyte concentration, bicarbonate, carbonate and silica of irrigation water and clay mineral of the soil. Higher the total electrolyte concentration, lower the bicarbonate, carbonate and silica content, lesser the adverse effect on the permeability. For stable permeability, the irrigation waters having SAR 10, 20, 30 should not have total electrolyte concentration less than 10, 20 and 30 meqL^{-1} (EC 1, 2, and 3 dSm^{-1}) respectively. Montmorillonite and illite-vermiculite (2:1 crystal lattice) minerals are affected more adversely than kaolinite-sesquioxides (1:1 crystal lattice). At similar concentration, adverse effect due to sodium adsorption ratio in the soil is highest when the irrigation water contains Na_2CO_3 followed by $NaHCO_3$, Na_2SO_4 and NaCl. SAR and ESP of soils, both increase with increase in the proportion of sodium ions in relation to calcium + magnesium, total cation concentration remaining the same (Table 2.17).

Table 2.17. Effect of soluble sodium percentage on SAR and ESP of soils

TCC (meqL$_{-1}$)	Na (%)	Ions (meqL^{-1})		SAR	ESP
		Na	Mg+Ca		
15	50	7.5	7.5	3.9	4.2
15	66	10.0	5.0	6.3	7.4
15	80	20.0	3.0	9.8	11.7
15	93	30.0	1.0	20.0	22.0

SAR of the waters normally increases with an increase in EC of the water. For example, if EC is less than 1.5 dSm^{-1}, 96% waters have SAR less than 10. When EC is more than 10 dSm^{-1}, only 15% waters have SAR less than 10 and 64% waters have SAR from 10 to 20 (Table 2.18). Although, SAR values of highly saline ground water may be as high as 100, the percentage of such waters as having SAR higher than 30 is extremely low.

Table 2.18. Per cent distribution of ground waters of Mathura district Uttar Pradesh, on the basis of EC and SAR

SAR	EC (dSm^{-1})				
	<1.5	1.5-3.0	3.0-5.0	5.0-10.0	>10.0
<10	96	76	64	34	15
10-20	4	22	31	59	64
20-30	-	1	4	6	17
30-40	-	-	-	1	1
>40	-	-	-	-	1

An unusual aspect of the SAR is that Ca^{++} and Mg^{++} have been lumped together. According to one group of scientists, Mg acts like Ca in influencing the physico-chemical properties of the soil, owing to divalent nature of the cation. Others, however, believe that Mg should not be coupled with Ca in view of the varying affinity for adsorption on the exchange complex of the soil. There seems to be no theoretical basis for combining the concentrations of these two ions, except for the fact that ion valency is considerably most sensitive parameter for predicting ion-exchange relations that is the ions size. Garrels and Christ (1965) have shown the Debye Huckel distance of the closest approach for magnesium to be approximately one-third larger than for calcium. Hence, over the range of solution salt concentration common to irrigation agriculture (2 to 200 $meqL^{-1}$ or 0.001 - 0.1 molar) the activity coefficient for magnesium should be 2 to 20% higher than for calcium. Many soils and clays also tend to prefer Ca^{++} over Mg^{++} in ion-exchange, with a relative increase in preference of about 20% being common. These two effects have opposing trends and may account for at least a part of the widespread success when using the SAR for ion-exchange predictions. Another important factor is that many waters and soil solutions contain two to five times more Ca^{++} than Mg^{++}. Mg is only a minor constituent of the vast majority of irrigation waters in the U.S.A. More important and adequate, justifications for the widespread usage of the SAR are its relative simplicity and the fact that it generally works (Bresler *et al.*, 1982). It has been estimated that approximately 80% of the river irrigation waters throughout the western U.S.A. have EC less than 1.5 dSm^{-1} and less than 10% have values higher than 2.25 dSm^{-1}. Ground waters used for irrigation have EC commonly between 3 and 5 dSm^{-1}. The examination of chemical composition reveals that Mg/Ca ratio is generally less than one and mostly it varies from 0.2 to 0.5. The fact remains that it was for this typical group of waters that SAR worked as a reliable index for predicting the sodium hazard of irrigation waters, in the western U.S.A.

Table 2.19. Effect of irrigation with saline water (TCC = 112 meqL^{-1}; SAR = 8) having Mg/Ca ratio as 5.4 on properties of irrigated soils

Depth of soil (cm)	TCC (meqL^{-1})		SAR		ESP	
	A	B	A	B	A	B
0-20	16	131	3	9	8	4
20-40	11	72	4	7	6	18
40-60	11	54	3	5	13	22
60-100	49	64	5	8	14	27
100-150	91	98	8	9	24	40
150-200	60	97	6	8	29	42

A = Before irrigation, B = After irrigation

Sodium to Calcium Activity Ratio (SCAR)

The application of SAR to the group of waters which have EC higher than 5 dSm^{-1} and Mg/Ca ratio more than 1 is obviously questionable. About 35% ground waters occurring in 11 arid districts of Rajasthan have EC greater than 5 dSm^{-1} and nearly all waters are dominated by magnesium over calcium. Although some waters have Mg/Ca ratio as high as 16, the majority falls between 2 and 4. Gupta and Abichandani (1970b) examined the effect of irrigation with high salinity water having Mg/Ca ratio as 5.4 and SAR 8, on a deep sandy to sandy loam soil, containing $CaCO_3$ from 1 to 2%. Although, SAR of the irrigated soil was low and in line with SAR of the irrigation water, ESP values were much higher than expected from the SAR of irrigation water or irrigated soil (Table 2.19). The high salinity irrigation waters therefore, seem to promote an increase in exchangeable sodium. But, when the SAR of the irrigation water is calculated simply as Na/√Ca, the value comes as 15 in the month of October (before the beginning of irrigation) and 17 in the month of next March (after the irrigation is over). These values correlate highly with the observed values of ESP, namely, 14 and 18 in 0 to 20 and 20 to 40 cm depth of the irrigated soils. Likewise, for other locations also, SAR values calculated as Na/√Ca correlate much better with observed values of ESP (Table 2.20). This suggests therefore, that SAR of saline irrigation waters which have high Mg/Ca ratio, should be calculated simply as Na /√Ca and called as Sodium to Calcium Activity Ratio (Gupta, 1984a). Karen (2000) also advocated likewise. Yadav and Girdhar (1981) observed that increasing Mg/Ca ratio in the leaching water at a given SAR increased ESP of the soil more conspicuously in non-calcareous soil. The adsorption of Na on the exchange complex increased with the increase in the Mg/Ca ratio in the leaching water at a given SAR, due to lower binding energy of Mg, as compared to that of Ca. The degree of dispersion increased and hydraulic conductivity decreased with increase in Mg/Ca ratio and SAR. At a given SAR of the applied water, a higher ESP than normal has been reported to result in the soil when using a water of Mg/Ca ratio greater than 1 (Rahman and Rowell, 1979).

Table 2.20. Comparison of SAR and SCAR values of irrigation waters in relation to observed ESP of soils

Location	Irrigation water				Irrigated soils	
	EC(dSm^{-1})	Mg/Ca	SAR	SCAR	SAR	ESP
Kaparda	10.8	4.4	28	59	31	60
Jelwa	5.9	8.1	23	36	28	38
Shikarpura	4.5	16.0	8	26	32	35

Alkalinity Hazard

Residual Sodium Carbonate

Eaton (1950) gave the concept of residual sodium carbonate (RSC) and observed that development of alkali soils (Saline or non-saline) may be expected when water containing CO_3+HCO_3 higher than Ca+Mg is used for irrigation provided that it is used so sparingly that little leaching occurs. In water containing high concentrations of bicarbonate ions, there is tendency for calcium and to some extent for magnesium to precipitate as carbonates as the soil solution becomes more concentrated. On the other hand, if sodium is not too greatly in excess of calcium + magnesium and enough of such water is used such that the drainage water continues to contain a substantial proportion of Ca and Mg, there may be no ill effects. The following reaction takes place:

$$Ca^{++} + 2HCO_3^- = CaCO_3 + H_2O + CO_2 \uparrow \quad (2.5)$$

This reaction does not go to completion under ordinary circumstances, but in so far as it does proceed, the concentrations of calcium and magnesium are reduced and the relative proportion of sodium is increased causing thereby a decrease in soil salinity and increase in ESP.

The residual bicarbonate reacts with sodium and forms sodium bicarbonate which upon evaporation is converted into sodium carbonate according to the following reactions:

$$Na^{+}+HCO_3^{-} = NaHCO_3 \quad (2.6)$$

$$2NaHCO_3 = Na_2CO_3+H_2O+CO_2\uparrow \quad (2.7)$$

The sodium carbonate, thus formed, reacts with normal soil and replaces exchangeable calcium with sodium.

$$\text{(exchange complex) } Ca^{++}+Na_2CO_3= \text{(exchange complex) } 2Na^{+} + CaCO_3 \quad (2.8)$$

In the presence of gypsum, sodium carbonate is converted into sodium sulphate as follows:

$$Na_2CO_3 + CaSO_4 = CaCO_3 + Na_2SO_4 \quad (2.9)$$

In the presence of carbonic acid, calcium carbonate is converted into calcium bicarbonate:

$$CaCO_3+H_2CO_3 = Ca(HCO_3)_2 \quad (2.10)$$

The solubility of calcium carbonate increases from 0.01 to 0.06-0.14 gL^{-1} and pH of the equilibrium solution drops from 10-10.2 to 7.5-8.5.

The above conclusions were largely based on the field observations of the effects of use of water for centuries in two of the world's oldest irrigation developments, one in Egypt along the Nile and the other in Iraq along the Tigris and Euphrates. Alkali soils have been found prevalent in the Nile valley whereas saline soils have developed in areas associated with Tigris and Euphrates. The Nile water contained RSC from 0.23 to 1.44 $meqL^{-1}$ whereas in Tigris and Euphrates waters, RSC was absent (Table 2.21). Since the river waters did not contain carbonates, RSC was actually determined as (HCO_3) - (Ca+Mg).

Table 2.21. Composition of the river irrigation waters

Source	Ions ($meqL^{-1}$)						RSC
	Ca	Mg	Na	HCO_3	SO_4	Cl	($meqL^{-1}$)
Flood							
Nile	0.64	0.57	0.50	1.44	0.16	0.14	0.23
Tigris	2.39	0.85	0.38	2.75	0.49	0.39	0
Low							
Nile	0.87	0.72	2.04	3.03	0.23	0.46	1.44
Tigris	2.78	1.64	0.71	3.56	0.94	0.63	0
Normal							
Nile	3.93	1.99	0.94	3.00	1.81	1.55	0
Tigris	3.42	2.84	0.66	2.92	1.79	2.21	0

Source: Eaton (1950)

It was further stated that the formation of alkali soils along the Nile by use of a water containing RSC is directly contingent upon heavy soil texture, poor internal drainage, low rainfall (<150 mm) and high evaporation. The Nile water contained 40% HCO_3 among anions and 60 mgL^{-1} silica. Interestingly, the richest agriculture of the world is found even today on many of the lands served by the Nile, implying thereby that it is the drainage and leaching which govern the successful utilization of waters containing RSC. The alkali soils in Egypt are not as widespread as saline soils. The saline soils have developed due to excessive water application, under the flood irrigation system causing a rise in the water table in consequence of water logging.

Wilcox *et al.* (1954) conducted a pot experiment in which Rhodes grass was cropped intensively in loamy soil using irrigation waters containing different amounts of RSC. The soils were irrigated with the water under test and then allowed to dry to a soil moisture tension of about 700 to 800 cm of water between irrigations. In a period of less than two years, 86 irrigations were applied which corresponded to its simulated use for more than 20 years in the field. The rainfall was completely excluded. There were low and high-leaching regimes in some treatments. The high leaching treatment (LF=0.25) provided for the application of excess water so that about 25% of the applied irrigation water was collected as percolate after each irrigation. The low-leaching regime (LF=0.06) provided for the same proportion of leachate every fourth irrigation. The composition of irrigation waters that were tested and analysis of soil samples collected after the forty-second and the eighty-sixth irrigation are reported in Table 2.22.

Table 2.22. Composition of the irrigation waters and properties of irrigated soils (low leaching non-calcareous treatments)

Irrigation water							Irrigated soil					
Na (%)	Ca (%)	HCO_3 (%)	Cl (%)	TCC (meq L^{-1})	SAR	RSC (meq L^{-1})	After 42 irrigations			After 86 irrigations		
							pH_s	ESP	EC (dSm^{-1})	pH_s	ESP	EC (dSm^{-1})
75	25	0	100	20	9.5	0	6.8	12	3.7	6.8	16	5.3
75	25	50	50	20	9.5	5	8.6	52	5.9	9.4	72	16.0
75	25	0	100	10	6.7	0	7.2	8	2.2	7.3	15	3.7
75	25	50	50	10	6.7	2.5	8.6	20	2.0	9.0	42	7.3
75	25	0	100	5	4.8	0	6.8	9	1.3	6.7	11	1.8
75	25	50	50	5	4.8	1.25	8.4	10	1.2	7.7	20	2.4

The use of the two bicarbonate waters, containing RSC 5 and 2.5 meqL^{-1}, gave rise to substantially higher values than the corresponding chloride waters. It was found that the percentage of applied HCO_3 which precipitated as $CaCO_3$ varied from 0 to 54 and the extent of exchangeable sodium accumulation from the waters was related to RSC of the irrigation waters (depending upon leaching fraction (LF), composition of irrigation water and partial pressure of CO_2. High leaching regime and presence of calcium carbonate resulted in lower ESP values than low leaching regimes and absence of calcium carbonate. The effect of $CaCO_3$ was, however, insignificant. Based upon these observations, it was concluded that waters containing RSC, to be calculated, (CO_3+HCO_3)-(Ca+Mg), more than 2.5 meqL^{-1} were not suitable for irrigation, between 1.25 and 2.5 meq L^{-1} as marginal and only those containing less than 1.25 meqL^{-1} were probably safe. Interestingly the waters used for irrigation in this experiment did not contain either carbonates or magnesium ions. As such RSC was actually calculated as HCO_3-Ca.

It is surprising that although suggested critical limits of RSC based upon a pot experiment in which rainfall was completely excluded were stated to be tentative and subject to modification by the degree of leaching and scrutiny of the additional data, the quality of irrigation waters continues to be subjectively evaluated in many parts of the world on the basis of the above limits, even in areas where rainfall is appreciable and effective (> 400 mm).

Thorne and Thorne (1954) observed that irrigation with 12 different waters at 14 sites in Utah, containing bicarbonates from 1.25 to 6.92 meq L^{-1}, RSC from 0 to 3 meqL^{-1} and SAR from less than 1 to 26, applied under average farm conditions for relatively long periods at 15 different soil locations, did not lead to either the accumulation of carbonates in soils or increase in ESP. They concluded that there was not sufficient evidence to justify any special classification of waters based on bicarbonates. Moderate to high leaching, CO_2 from soils and plant roots, presence of NaCl and Na_2SO_4 in the soil solution were considered as some of the factors that can check the rise of ESP in soils irrigated with such waters (Thorne and Peterson, 1954).

Kelley (1962) explained that significant amounts of Na_2CO_3 cannot be formed in the soils by the use of irrigation water containing bicarbonate. He argued that as $CaCO_3$ is precipitated from a bicarbonate solution (eq. 2.5), one molecule of HCO_3 is set free per molecule of $CaCO_3$ precipitated. In addition CO_2 is given off by root respiration and micro-biological processes. Hence, the soil solution practically always contains dissolved CO_2. The result, the so-called residual sodium carbonate of irrigation water, will be $NaHCO_3$, rather than Na_2CO_3. As a matter of fact, the mere removal of H_2O from bicarbonate irrigation water, the soil temperature will not produce Na_2CO_3, regardless of its content of dissolved CO_2 or its ratio of HCO_3 to divalent cations. The fact that $NaHCO_3$ rather than Na_2CO_3 will be formed is an important distinction, for bicarbonate solution is able to retain Ca in solution, while Ca is practically insoluble in Na_2CO_3 solution. Na_2CO_3 could be formed indirectly as a result of applying bicarbonate water that is through hydrolysis of adsorbed Na^+ that may be formed, but the probability is remote that agriculturally significant amounts will be formed in this way. In California, and probably in many other semi-arid areas, leaching takes place to some extent practically every time irrigation is applied. Consequently, the percentage of adsorbed Na^+ usually remains low and this is true even where the irrigation water contains HCO_3 considerably in excess of Ca+Mg. In San Joaquin valley of California, irrigation with water containing RSC has not resulted in injurious amounts of adsorbed Na^+ or high concentrations of soluble salts except in localities where leaching is made ineffective by a high water table. Thousands of hectares have been irrigated for almost a century with water containing HCO_3 in excess of Ca+Mg. Many of the crops are highly sensitive to alkali in various forms but generally high yields have been obtained. The same is true in the Nile valley of Egypt, where large areas have long been irrigated with water from the Nile which also contains so-called 'residual Na_2CO_3'. As in California, crop yields have been high without evidence of alkali soil development except where the water table is high.

Formation of indurated layers of $CaCO_3$ has often been associated with the use of irrigation waters containing bicarbonate ions higher than calcium ions. Many layers of carbonate deposition become indurated or hardened, and essentially are impermeable, except for cracks and other discontinuities. Indurated layers normally are considered to be a product of hundreds or even thousands of years of carbonate accumulation, so that their formation during short-term agricultural operations is probably unimportant. The quantities of calcium and bicarbonate present in representative irrigation waters are so low that 10 to 20 years of typical water usage would be required to change the lime content of a 15 cm layer of soil by 1% even if all the lime was being precipitated within that single layer (McNeal, 1974).

Irrigation waters having RSBC as high as 10 $meqL^{-1}$ can be used continually without significant problem on sandy loam calcareous soil, containing illite as the principal mineral provided the SAR is less than

10 and EC is low, in areas where annual rainfall is appreciable and effective (650 to 750 mm) and evaporation is less than 2000 mm (Table 2.23). Nevertheless, if the winter rainfall is also appreciable and effective (150 to 250 mm), irrigation waters with RSC up to 10 $meqL^{-1}$ may be used successfully even if SAR is as high as 20 irrespective of the source (Gupta, 1983c).

Table 2.23. Effect of irrigation with waters (TCC 20 $meqL^{-1}$) containing RSBC on soil properties and crop growth of wheat (average of three years)

Irrigation water		Soil properties						Wheat yield (g/lysimeter)			
RSBC ($meqL^{-1}$)	SAR	pH		ECe (dSm^{-1})		SAR		Grain		Straw	
		NG	G	NG	G	NG	G	NG	G	NG	G
2	5.2	8.2	-	2.0	-	5.2	-	489	-	895	-
6	7.0	8.4	8.1	1.7	2.5	8.0	4.4	408	442	849	952
10	9.5	8.8	8.2	1.3	2.2	10.3	5.3	417	443	786	822

*G = Gypsum treated; NG = No gypsum

Lysimeter experiments conducted during four years revealed that compared to control (NaCl waters), the average grain and straw yields of wheat were higher in soils irrigated with waters containing residual sodium bicarbonate and lower with waters containing residual sodium carbonate (Table 2.24).

Table 2.24. Average (four years) grain and straw yields of wheat (g/lysimeter) as affected by different quality waters

Treatment (EC 3 dSm^{-1})	Grain	Straw
Control (NaCl)	448.5	734.5
RSBC ($NaHCO_3$)	473.5	766.5
RSC (Na_2CO_3)	443.5	728.7

Gupta (1983d) reviewed the concept of residual sodium carbonate and concluded that under conditions, with absolutely no provision for leaching, continued irrigation with high RSC water causes adverse effect on soil properties and crop growth. The normal soil is converted into alkali or saline-alkali depending upon the salinity and RSC of irrigation water. High water table linked with ground water containing RSC, low rainfall and high evaporation accentuate the formation of alkali soils. Fortunately, the irrigation waters containing RSC occur only in semi-arid zones, where rainfall is appreciable and effective (500 mm). In arid zones, the rainfall is generally very low but then there do not occur irrigation waters which contain high RSC. However, under field conditions, in well-drained soil with satisfactory provision for leaching, high RSC waters (10 $meqL^{-1}$) have not been observed to cause any adverse effect to soil properties or crop growth where rainfall is appreciable and effective and evaporation is not high. RSC higher than 10.0 $meqL^{-1}$ has been observed to be critical, particularly if SAR is also high. An interesting example is illustrated with the detailed composition of the water of a tube well located at Tatesar village, Nangloi block, Delhi (Table 2.25). This water caused severe sodicity in the irrigated soil and therefore, was abandoned soon after the first season.

Table 2.25. Chemical composition of highly sodic water, unsuitable for irrigation

Property	Value ($meqL^{-1}$)	Property	Value
Na	21.5	EC	2.1 dSm^{-1}
Ca	0.5	RSC	14.5 $meqL^{-1}$
Mg	0.7	RSBC	15.2 $meqL^{-1}$
HCO_3	15.7	SAR	27.7 $(mmolL^{-1})^{1/2}$
$CO_3^=$	0.0	SCAR	30.7 $(mmolL^{-1})^{1/2}$
Cl^-	4.4		

Residual Sodium Bi-carbonate (RSBC)

Though, Eaton (1950) suggested that RSC should be determined as (CO_3 + HCO_3) – (Ca + Mg) yet RSC was actually calculated as (HCO_3) - (Ca +Mg). In subsequent pot culture experiment conducted by Wilcox *et al.* (1954) to substantiate RSC concept, RSC was determined as (HCO_3-Ca). Similarly, in green house and lysimeter experiments, Bower (1961) and Pratt *et al.* (1960), considered only HCO_3 and Ca. Likewise, in many experiments conducted in India only $NaHCO_3$ salt was added in very low salinity waters to prepare waters of different RSC (Gupta, 1980c; Anonymous, 1983; Bajwa *et al.*, 1983; Manchanda *et al.*, 1985a).

A condition not provided for by the RSC concept has been encountered in recent years. If the concentrations of both calcium and bicarbonates are about equal and high i.e. of the order of 10 $meqL^{-1}$ or greater, the RSC will be low or possibly zero. Such waters will precipitate some calcium carbonate and should be considered at least marginal. Relative Ca^{++} precipitation for LF 0.1 was compared with RSC for waters containing more HCO_3 than Ca^{++}. Over half of the Ca^{++}

was predicted to precipitate in the majority of cases even when RSC had a negative value (Oster and Rhoades, 1977).

Specific Ion Toxicity Hazard

Major Constituents

Sodium: The sodium is the most abundant cation in all natural waters whereas its presence is least desirable. Excess of sodium ions characterizes that the water is saline or alkaline depending upon its occurrence in association with chloride/sulphate or carbonate/bicarbonate ions, and accordingly the irrigated soil may develop salinity or sodicity problem. For some time in the past, the quality of irrigation water used to be evaluated with respect to sodium on the basis of soluble sodium percentage (SSP) calculated as below:

$$SSP = (\text{Soluble sodium concentration} / \text{Total cation concentration}) \times 100 \quad (2.11)$$

Where, ion concentration is expressed as $meqL^{-1}$.

It has been useful in characterizing water since a high value indicates a soft water and low value a hard water. When a water containing excess of sodium is used for irrigation, a part of it is adsorbed by the soil. The adsorbed sodium is an index of the sodicity hazard in the soil. If SSP in the water or in the resulting soil solution is 66 or more, cation exchange will inevitably take place and cause an increase in adsorbed sodium. Qualitatively, SSP indicated sodicity hazard alright but this has not been found a satisfactory index to predict the sodicity hazard on quantitative basis. A perusal of data of Gupta and Abichandani (1967) revealed that after rainfall there is significant improvement in quality of waters on the basis of EC and SAR but this fact is not borne by SSP values which remain nearly the same. Even when EC and SAR/ESP of the soils increase, SSP values remain nearly constant. The use of SSP as one of the criteria in evaluating the quality of irrigation water has therefore, become obsolete. A better index called "Sodium Adsorption Ratio (SAR)" was developed by United States Salinity Laboratory (USSL) (Richards, 1954).

Both, soils and plants are adversely affected by irrigation waters containing high amounts of sodium. Sodium soils are relatively impermeable to air and water. They are hard when dry, difficult to till, plastic and sticky when wet. These adverse physical conditions retard or prevent germination and are generally unfavourable for plant growth. Unfortunately, sodium is taken up freely by many plants but is probably not essential in the same sense, as the other nutrients and it may even be specifically toxic to some plants. Most tree crops and other woody type perennial plants are particularly sensitive to sodium. Most annual crops are not so sensitive, but may be affected by higher concentrations. Sodium sensitive crops include deciduous fruits, nuts, citrus and beans. These plants suffer injury as a result of sodium accumulation in the leaves. Although, all sodium salts are harmful, the relative effects depend upon the associated anion. Sodium carbonate, sodium bicarbonate and sodium chlorides are 5, 2 and 1.5 times more harmful than sodium sulphate. In general, concentration of sodium ions increases with increase in EC. Whereas low salinity waters may be found dominant in sodium bicarbonate, high salinity waters will be dominant in sodium chloride.

Magnesium: The magnesium is the second most abundant cation usually found in high salinity waters. However, in low salinity waters, calcium dominates over magnesium ions. Obviously, it can be stated that with an increase in EC of waters, Mg/Ca ratio tends to increase. It was believed in the past that if the proportion of magnesium plus calcium ions is high, the sodicity hazard is low: conversely, if sodium predominates, the hazard is high. However, this held good so long as EC of the waters was not very high (<5 dSm^{-1}) and Mg/Ca ratio was <1. Szabolc and Darab (1968) considered that one of the most important criteria in determining quality of water for irrigation is the Mg content of the irrigation water. A harmful effect on soils appears when the ratio Mg:Ca+Mg exceeds 0.5. Occurrence of magnesium ions in higher proportion than calcium ions tends to increase the adverse effect due to sodicity. Therefore, if the proportion of magnesium amongst divalent cations is high, the sodicity hazard is high, and conversely, if calcium predominates, the hazard is low. An index called "Sodium to Calcium Activity Ratio (SCAR)" has been suggested by Gupta (1984 a) to account for magnesium hazard.

Magnesium is known to affect plant growth mainly by reducing calcium uptake and causing calcium deficiency. However, when high concentrations of magnesium occur together with high concentrations of calcium there is no specific ion toxicity effect due to magnesium (Meiri and Shalhavet, 1973).

Chlorides: The occurrence of chloride ions in natural irrigation waters increases with an increase in EC and sodium ions. Therefore, these ions are most dominant in very high salinity waters. Unlike sodium ions, neither the chloride ions have any effect on the physical properties of the soil nor are they adsorbed by the soil. For this reason, absolute or relative concentration of chloride ions has not found any importance as general criteria in the evaluation of quality of irrigation water. However, it has been recognized that it causes specific ion toxicity to some tree crops and woody type perennial plants. For example, chloride hazard for citrus in the coastal area of Israel has been evaluated (Table 2.26). The critical tolerance limit of chloride for sensitive citrus root stock is considered to be 10.0 meq L^{-1}.

Table 2.26. Chloride hazard for citrus

EC_{iw} (dSm^{-1})	Cl_{iw} (meq L^{-1})	Soil texture*		
		Sandy	Loamy	Clay
<1.2	6	-	-	-
1.2-1.5	5-7.5	-	-	+
1.5-1.75	7.5-9.0	-	-	+
1.75-2.25	9.0-15.0	-	+	+++

Source: FAO/UNESCO (1967).*-, no risk; +, low risk; ++, medium risk; +++, dangerous

Chloride salts are more harmful than sulphates. The adverse effect due to salinity of 20 dSm^{-1} caused in the presence of chlorides is the same as that of 40 dSm^{-1} in the presence of sulphates (Table 2.27). This is because when both the ions occur in high concentrations, only half of the sulphate ions contribute to salinity due to the fact that approximately half of the sulphates get precipitated as $CaSO_4$ while the other half remain in soluble form as Na-$MgSO_4$ in the soil. Doneen (1963) introduced the term 'Potential salinity' of irrigation water and suggested its determination as shown below:

Potential salinity = Cl^- + 1/2 $SO_4^{=}$, all ions are expressed as meq L^{-1}, and recommended permissible limits as 5-20, 3-15 and 3-7 meqL^{-1} for soils of good, medium and low permeability, respectively.

Table 2.27. Relative plant growth as affected by chlorides and sulphates

EC_e (dSm^{-1})		Relative plant growth (%)
Chlorides	Sulphates	
4	6	100
4-8	6-12	80-90
8-12	12-20	50-70
12-20	20-40	20-50
20	40	0

Bicarbonates and Carbonates: Bicarbonates occur in low salinity waters and its concentration usually decreases with an increase in EC. The proportion of bicarbonate ions is higher than calcium ions and has been considered to be undesirable because after evaporation of irrigation water, bicarbonate ions tend to precipitate calcium ions. The precipitation of calcium carbonate as a scale in distribution lines, boilers, water heaters etc. is familiar phenomenon. The same reaction may take place when such a water is applied to the soil. The effect of bicarbonates together with carbonates used to be evaluated in the past through an index called "Residual Sodium Carbonate (RSC)" as follows:

$$RSC= (CO_3 + HCO_3) - (Ca+Mg) \quad (2.12)$$

All ions in this equation are expressed as $meqL^{-1}$.

Carbonates generally occur in trace amounts and seldom more than 1.0 $meqL^{-1}$ in natural waters. Their presence is not very significant unless pH of the water is greater than 8.5. An examination of about 900 water samples collected from four districts of western Rajasthan showed that the percentage of waters having pH greater than 8.5 varies from less than one in Barmer and Jaisalmer districts to about two in Jodhpur and Bikaner districts (Bhandari *et al.*, 1970). Since carbonate ions do not occur very frequently in appreciable concentrations and because bicarbonate ions do not precipitate magnesium ions, Gupta (1984a) suggested that the alkalinity hazard should be determined through the index called "Residual Sodium Bi-carbonate (RSBC)" to be calculated as below, all ions expressed as meq L^{-1}.

$$RSBC = HCO_3\text{-}Ca \quad (2.13)$$

Absolute concentration of bicarbonate ions in amounts greater than 10.0 $meqL^{-1}$ is likely to affect plant growth in a number of ways such as: (i) may affect mineral nutrition, (ii) tends to reduce the availability of iron in many plants thereby causing chlorosis, and (iii) may also create severe problems if used for irrigation through overhead sprinklers. High concentration of bicarbonates (12.5 - 14.0 $meqL^{-1}$) associated with low content of calcium are reported to cause iron chlorosis in sugarcane in Tamil Nadu (Ramanathan *et al.*, 1987). Bicarbonate waters are mildly alkaline and therefore, less injurious than carbonates. Whereas, the saturated solution of sodium bicarbonate has pH of 8.4, that of sodium carbonate may have pH as high as 12. The presence of carbonate ions is least desirable in waters because they tend to virtually eliminate calcium ions from the water and cause strong alkalization in the irrigated soils, whereas bicarbonates precipitate calcium ions only partly. Absolute concentration of carbonate ions in amounts ranging from 2 to 5 meqL^{-1} should be considered seriously.

Silica: Silicon is the more abundant element in igneous rocks than some other types of deposits. Most silica in water is probably derived from the decomposition or metamorphism of silicate minerals rather than from solution of quartz, as quartz is one of the most resistant rock minerals to be attacked by water. Many waters contain less than 10 mgL^{-1} of silica; those that contain deposits high in silicate minerals, particularly feldspars, often contain up to 60 mgL^{-1} and are not commonly found. It is believed that most silica is present in a non-ionised form, but ionised silicate(s) is undoubtedly present in some waters.

The ground waters in western Rajasthan have been observed to contain high amounts of silica. The average concentration of silica in waters derived from alluvium, phyllite and granite formations have been reported to contain 31, 20 and 32 mgL^{-1} silica, respectively (Table 2.28; Goyal, 1985). The use of irrigation water containing silica from 5 to 75 mgL^{-1} has been observed to cause increase in pH_2 of the soil from 8.9 to 9.4 and ESP from 5.4 to 21.8. It ha also resulted in increase in stickiness, decrease in infiltration rate, and formation of wide and deep cracks during summer months (Table 2.29; Goyal, 1984). The importance of silica has usually not been recognized in evaluating quality of waters for irrigation but it seems to be necessary in view of the fact that it tends to cause alkalinity hazard.

Table 2.28. Silica in ground waters in relation to geological formation

Geological formation	No. of samples	EC (dSm^{-1})		pH	SiO_2 (mgL^{-1})	
		Range	Average		Range	Average
Alluvium	7	4-16	9	7.2-7.7	13-41	31
Phyllite	9	7-20	12	7.3-7.7	10-35	30
Granite	28	2-23	11	7.0-8.7	13-71	32

Table 2.29. Effect of siliceous waters on soil properties

SiO_2 (mgL^{-1})	EC (dSm^{-1})	pH_2	ESP	Crust force (kg/cm^2)
5	0.16	8.9	5	3.2
15	0.21	9.0	6	6.3
25	0.30	9.1	11	8.9
75	0.48	9.4	22	21.2

Nitrates: Nitrates in water may result from leaching of soils and rocks, from fertilizers, normal decomposition of plants and animals, sewage, industrial effluents, and living organisms. Although nitrates are highly soluble, natural waters contain only small quantities, usually less than 10 mgL^{-1}. In Rajasthan and Uttar Pradesh, ground waters containing nitrates from 400 to 700 mgL^{-1} (13-23 meqL^{-1}) have been reported. The CGWB report also states that in many parts of the country ground water from shallow aquifers has nitrate content which far exceeds the prescribed standards for drinking. A study conducted in Ludhiana district of Punjab revealed that during the post monsoon period about 35% of the water samples had nitrate concentration of more than 15 mg L^{-1}. Alarmingly increasing trend has been observed in some wells in Krishna and Godavari basins (Rajan *et al.*, 1997). The nitrates occurring in saline waters from 2 to 5 meq L^{-1} have been observed to cause beneficial effect on the plant growth and help partly in the substitution of nitrogenous fertilizer. About 10 kg N ha^{-1} is added to the soil per irrigation (7.5 cm depth) if irrigation water contains nitrates equal to 1.0 meqL^{-1}. However, in excessive amounts greater than 10.0 meqL^{-1}, nitrates may cause specific ion toxicity on plant growth. Nitrates are more harmful than chlorides. Fertilizers help plants by giving them extra supplies of minerals they need to grow. But the plants can not use all these and the left over nitrates wash away with the rain into rivers and trickle down through the earth into the ground water reserves which supply most of our drinking water.

Minor Constituents

Boron: This is the characteristic element of such well-known compounds as borax and boric acid. It is not found in its elemental form in nature but as a sodium or calcium borate salt. The concentration of boron ranges usually from traces to 5 mgL^{-1} in natural ground waters. However, in very high salinity waters it can occur up to 10 mgL^{-1} (Gajbhiye *et al.*, 1973). It increases with an increase in EC and SAR of the waters (Table 2.30). In sea water, it occurs up to the extent of 4.8 mgL^{-1}. In salt affected soils, it may go up to 10 mgL^{-1} whereas in normal soils it is generally below 2 mgL^{-1} (Bhargava *et al.*, 1974).

It differs from the other constituents of waters in that it occurs as un-ionized boric acid. It is essential for plant growth but may prove exceedingly toxic at concentrations slightly above optimum. Plants usually do not show symptoms of boron deficiency if irrigated with water containing 1 mgL^{-1} of B, but injury may develop on the more sensitive plants when irrigated with water containing B in excess of 3 mgL^{-1}. The injury appears as a burning and browning of the leaf top followed by yellowing of the margins. Retarded growth and premature shedding of leaves are the other symptoms. When a water containing boron is used for irrigation, part of it is adsorbed by the soil with the balance remaining in the soil solution depending upon texture, clay mineral type, organic matter, $CaCO_3$, salinity and sodicity of the soil. Since the soils adsorb boron and the plant obtains its boron from soil solution, the greater is the ability of the soil to adsorb boron the lower the boron content in the plant. If a crop can tolerate 5 mgL^{-1} of boron in soil solution, irrigation water containing 5 and 10 mgL^{-1} can be practiced in sands and clay (medium black), respectively (Table 2.31). Boron does not however, affect soil properties.

Table 2.30. Boron in ground waters of Bikaner district

EC (dSm^{-1})	No. of samples	B range (mgL^{-1})
0-0.25	–	–
0.25-0.75	2	0.22-0.28
0.75-2.25	37	0.22-1.57
2.25-5.0	50	0.25-2.47
5.0-10.0	29	0.44-3.89
10.0-15.0	14	0.70-4.75
>15.0	5	1.30-10.62

Eaton (1935) was probably the first Scientist who carried out a sand-culture experiment to determine the relative boron tolerance of different plant species. The numerical values for the boron tolerance of the plants were derived by dividing the average of the dry weights of the plants growing in nutrient solutions containing a trace or one mgL^{-1} (whichever was greater) and multiplying the quotient by 100. These values ranged up to 50 for the boron sensitive group of the plants and from 100 to 217 for the boron tolerant group of the plants. The plants were tentatively listed in each group in the order of increasing tolerance to boron.

The suggested critical limits of boron in irrigation water were 1, 2 and 4 mgL^{-1} for sensitive, semi-tolerant and tolerant plants respect-ively. The sensitive crops comprised most of the horticultural plants, with lemon, grape fruit and orange being the most sensitive. Semi-tolerant crops included sweet potato, tomato, oats, sorghum, maize, wheat, barley, radish, sweet pea, cotton, potato and sunflower. Carrot, cabbage, onion, sugar beet, palm and date palm were classified as tolerant (Wilcox, 1960).

Table 2.31. Adsorption of boron as affected by texture of the soils

B added (mgL^{-1})	C*	E**	C/E
Dune sand			
5	3.8	1.2	3.17
10	7.8	2.2	3.54
20	16.5	3.5	4.71
30	25.0	5.0	5.00
40	34.0	6.0	5.67
50	43.0	7.0	6.14
Sandy loam			
5	3.0	2.0	1.50
10	6.4	3.60	1.78
20	14.8	5.2	2.85
30	24.0	6.0	4.00
40	33.0	7.0	4.71
50	42.0	8.0	5.25
Clay (Medium black soil)			
5	1.2	3.8	0.32
10	3.8	6.2	0.61
20	8.5	11.5	0.74
30	14.0	16.0	0.87
40	19.0	21.0	0.90
50	25.0	25.0	1.00

*C=Boron in solution; **E= Boron adsorbed Source: Gupta (1980)

It is somewhat curious that the relative boron tolerance was based on leaves showing boron injury symptoms. Since the sand culture experiments were terminated after 45 days and conclusions were based on dry matter weight, the critical limits did not necessarily indicate significant reduction in yield. Since the observations were made in the climatic conditions prevailing in San Joaquin valley, California, where rainfall is virtually nil (<100 mm), the relevance of the critical limits as suggested above to be used as criteria in evaluating the quality of irrigation water have become of late seriously questionable. Interestingly, Eaton (1935) also pointed out that some soils have remarkable boron fixing power. He cited an example of a soil near Hollister, California, where crops did not show conspicuous injury even when irrigated with waters containing 10 mgL^{-1} of boron for a number of years.

Fig. 2.3. Effect of Li, F and B on wheat crop

Gupta (1983a) reviewed the work pertaining to tolerance of plants to boron in irrigation water and concluded that whereas citrus and some horticultural plants, oilseeds and leguminous plants may be affected adversely by 1 to 3 mgL^{-1} of boron, the growth of sugar beet, oats, cotton, milo and carrots is not affected adversely even if the irrigation waters contained boron in the range of 5 to 10 mgL^{-1}, depending upon the soil texture. Gupta (1983b; 1984b) examined the effect of irrigation with saline water (EC 5 dSm^{-1}) prepared by dissolving pure NaCl in tap water (EC 0.4 $dSm^{-1}$and Ca 2 $meqL^{-1}$ and B in different concentrations). Although definite yield decline occurred in grain and straw yields of wheat with increasing concentrations of boron, the differences in grain yields were small (Fig. 2.3; Table 2.32). However, the adverse effect on the straw yields was more than on the grain and more significant. The results suggest that where waters contain boron in the range of 5 to 10 mgL^{-1}, regular monitoring should be done. The suppressing effect of boron on plant growth is more severe in soil irrigated with high salinity water than normal water (low salinity) (Table 2.33). The highest yield of tomato was obtained by 4 mgL^{-1} B and total fruit weight was not affected significantly even by 10.0 mg^{-1} of boron contained in soil water and the yields of broccoli, cauliflower and radish were not affected by more than 10% by 6, 9 and 8 mgL^{-1} of soil water boron (Francois, 1984; 1986). Interestingly, boron tolerance limits established in the past for wheat and tomato ranged between 1 and 2 mgL^{-1} and for cabbage the limits were from 2 to 4 mgL^{-1}. It is obvious that the criteria for evaluating quality of irrigation water with respect to boron as suggested by Wilcox (1960) needs to be treated with care.

Table 2.32. Grain and straw yields of wheat (g/pot)

Treatment	Grain				Straw			
(mgL^{-1})	1982	1983	1984	Mean	1982	1983	1984	Mean
Control								
0	36	37	21	31	40	60	30	43
Boron								
5	33	31	19	27	37	48	25	37
10	31	32	18	27	38	51	27	38
15	32	34	13	26	37	54	21	37
20	33	30	16	26	37	47	28	37
25	33	31	15	26	38	52	25	38
CD (0.05)		7	6			7	5	
Lithium								
5	39	33	24	32	44	51	31	42
10	35	34	24	31	40	55	30	41
15	38	28	21	29	42	51	31	41
20	37	26	19	27	39	53	30	40
25	33	34	19	28	37	54	27	39
CD (0.05)		9	6			6	6	
Fluoride								
5	35	29	22	32	42	55	31	42
10	37	31	21	29	3	54	28	41
15	35	29	18	27	38	55	27	40
20	37	31	24	30	42	56	32	43
25	33	34	21	29	37	57	28	40
CD (0.05)	4*	9	7		3*	5	5	

*All treatments of B, Li and F considered together

As pointed out earlier, even the high salinity waters seldom contain boron greater than 5.0 $meqL^{-1}$. The irrigation waters containing B higher than 5.0 $meqL^{-1}$ do not constitute a significant area in India. Neither are there sufficient reports that reduction in yield in respect of field crops irrigated with saline waters is occurring anywhere due to boron, in concentrations normally observed. This is presumably due to appreciable and effective rainfall (>400 mm) in the areas where irrigation with saline water containing B is practised.

Table 2.33. Effect of boron in irrigation water on the dry matter (Five week old seedlings of barley (g/pot))

Salinity (dSm^{-1})	B (mgL^{-1})					Mean
	0	10	20	30	40	
Low (0.4)	1.69	1.62	1.51	1.56	1.55	1.59
High (8.5)	1.67	1.59	1.46	1.28	1.26	1.45
Mean	1.68	1.61	1.49	1.42	1.41	
CD at 5%	Salinity	0.07				
	Boron	0.11				
Salinity x Boron		0.5				

It was believed in the past that boron can be leached from the soil but if concentrations are high initially, boron in sufficient quantities to cause trouble may remain in the soil even after the concentration of other salts is reduced to a safe level. Reeve *et al.* (1955) conducted a leaching experiment on a clayey loam virgin saline soil containing 51% exchangeable sodium, 50 $meqL^{-1}$ boron and high amounts of gypsum (58 meq/100 g $CaSO_4$) in the surface layer and higher salt concentrations in the profile. The leaching was effective in reducing soluble salts, exchangeable sodium and boron but leaching of boron was slow as compared to soluble salts. Thus, it was concluded that though leaching with 120 cm of water reduced the salinity to a safe level, the boron content was still too high for good growth of plants sensitive to boron. Wilcox (1960) stated that there is no economically feasible method for removing boron from irrigation water. Similarly, there is no chemical or soil amendment that can be added to the soil to render boron non-toxic.

The gypsum has been found effective in reducing boron content of saline waters and soils (Gupta and Chandra, 1972). Reduction of boron was explained on the basis of solubility of salts. The boron occurs in the saline-sodic soils in the form of sodium metaborate, particularly in soils which have pH higher than 9.2. Sodium metaborate reacts with gypsum to form calcium metaborate and sodium sulphate. The solubility of calcium metaborate is very low (0.4%) as compared to sodium metaborate (26 to 36%). The large difference in the solubility renders boron to precipitate. Based on a laboratory experiment, Gupta (1979 b) reported that boron in saturation extract of a saline-sodic-soil (pH_2 10.5) decreased from 7 mgL^{-1} in untreated soil to 0 mgL^{-1} when gypsum of increasing fineness was mixed with the soil.

Boron content of a saline-sodic soil is markedly reduced by rainfall (740 mm) (Bandyopadhyay, 1974). There was remarkable leaching of boron, even more efficacious than soluble sodium (Table 2.34). Gupta (1979b) reported that after leaching with water, initial high boron concentration (6.5 to 10 mgL^{-1}) was greatly reduced (3 to 4.4 mgL^{-1}). The boron content decreased further with the application of gypsum. In the treatment where gypsum was applied at the rate of 100% GR, the boron content was reduced to 0.8 and 1.3 mgL^{-1} in 0 to 15 and 15 to 30 cm depth, respectively. The decrease of boron was followed simultaneously with the reduction in exchangeable sodium (Table 2.35).

Table 2.34. Effect of leaching through rainfall on sodium and boron content of the soil

Depth of soil (cm)	Sodium ($meqL^{-1}$)		B (mgL^{-1})	
	A*	B*	A	B
0-30	443	150	13.0	3.0
30-60	135	121	10.0	2.7
60-90	65	61	6.0	4.0
90-120	57	29	4.0	2.0
120-150	21	9	9.0	2.0
150-180	14	8	1.0	1.8
180-210	14	7	1.0	1.0

*A = Before rainfall (April, 1970); B = After rainfall (November, 1970)

Table 2.35. Boron and ESP as affected by gypsum treatment

Gypsum treatment	Boron (mgL^{-1})		ESP	
	0-15	15-30	0-15	15-30 cm
0	3.0	4.4	84	95
25% GR	2.1	3.9	52	87
50% GR	1.7	2.5	37	82
75% GR	1.4	1.8	17	83
100% GR	0.8	1.3	17	43

Source: Gupta (1979b)

Lithium: Lithium is present in some minerals but is not abundant in nature. The more common minerals which contain from 2 to 10% lithium oxide are petalite, spodomene, lepidolite and amblygenite. It is generally found in soils developed from mica minerals and is more frequent in arid regions. Although, surface waters generally do not contain lithium more than 0.3 mgL^{-1}, lake brines are reported to contain Li up to 16 mgL^{-1}. The sea water contains only 0.2 mgL^{-1} whereas saline ground waters have been reported to contain Li from 0.02 to 1.9 mgL^{-1} (Table 2.36). The ground waters rarely contain Li in excess of 5 mg L^{-1}. It increases with the increase in EC of waters. The normal and contiguous salt-affected soils have been observed to contain Li up to 0.5 and 2.5 mgL^{-1}, respectively (Gupta *et al.*, 1974).

Table 2.36. Distribution of lithium in saline ground waters

EC_{iw} (dS m^{-1})	No. of samples	Lithium (mg L^{-1})	
		Range	Average
1.7-2.0	6	0.02-0.05	0.04
1.9-3.7	3	0.04-0.05	0.05
4.0-5.6	9	0.02-0.14	0.06
6.4-24.8	17	0.05-1.90	0.35

Source: Gupta (1972)

Bingham *et al.* (1964) reported that though isolated citrus orchards exhibited lithium injury associated with irrigation waters containing 0.05-0.1 mgL^{-1}, lithium concentrations of 5 mgL^{-1} were required to produce Li injury in sand cultures. Sand culture experiments showed that gram seedlings were seriously affected with increasing concentrations of Li (Fig. 2.4; Gupta, 1974). Gupta (1983 b; 1984 b) examined the effect of irrigation with saline water (EC 5 dSm^{-1}) prepared by dissolving pure NaCl in tap water (EC 0.4 dSm^{-1} and Ca 2 $meqL^{-1}$) and Li in different concentrations. Although, a definite yield decline occurred in grain and straw of wheat during the first year, differences were very small and non-significant during the second and third year. However, lithium seems to be less toxic as compared to boron. There is no sufficient evidence of the excessive occurrence of lithium in irrigation water and adverse effect if any, on plant growth in India or any other region of the world. Therefore, any monitoring related to lithium in waters and soils is not required at present as a matter of routine unless sufficient data warrant its justification.

Fluoride: Unlike chlorides, fluorides are only sparingly soluble and are present in most natural waters in only small amounts. Calcium fluoride (fluorite) is the principal source of fluoride, but there are some other complex fluoride bearing minerals. In Andhra Pradesh, all the fluoride rich areas are underlain by crystalline rocks at places known to contain fluorspar and appatite, the main sources of fluoride in ground waters. The element is often characteristic of waters from deep strata and is frequently found in salt water from oil wells and in water from areas that have been subjected to the recent volcanism. The concentration of fluoride ranges from traces to more than 10 mg L^{-1} in natural waters, although in surface water, its content generally does not exceed 0.3 mgL^{-1} unless they are polluted from other sources. Analyses of ground water samples collected from districts falling in the arid zone of Rajasthan has shown maximum concentration of F as 90.0 mgL^{-1} in waters of Nagaur district in Rajasthan.

Ground waters with high fluoride content have been observed to occur in several villages in the Parkasam, Nalgonda and Guntur districts of Andhra Pradesh. The problem of high fluoride concentration has also been observed in several districts of Haryana and Tamil Nadu as well (CGWB, 1991). Thirteen states in India have been identified as endemic to fluorosis due to abundance of naturally occurring fluoride bearing minerals.

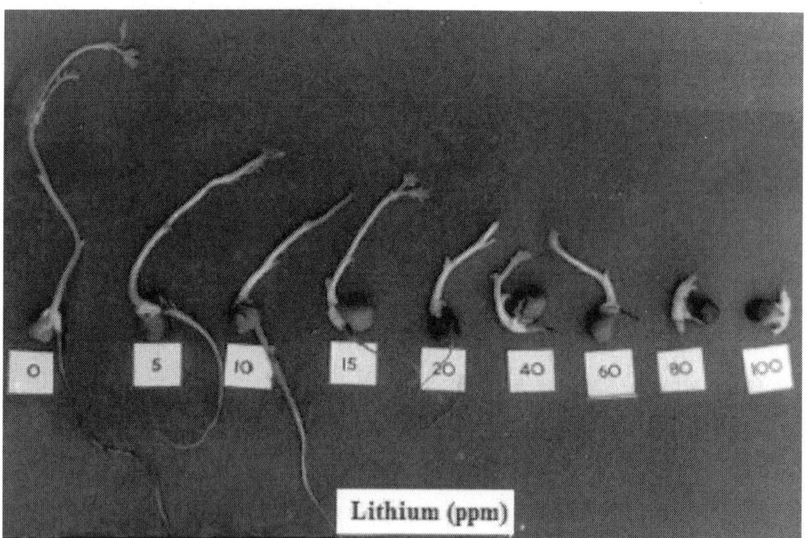

Fig. 2.4. Effect of lithium waters on the seedling growth of gram

Robinsons (1978) suggested tentatively that where water containing more than 5 to 10 mgL^{-1} F is regularly applied to crops, it will be desirable to monitor them by checking for damage from time to time through plant and soil analysis. Fluoride, if present in excessive concentration may cause specific ion toxicity to plants but in any case it does not affect soil properties. However, it is adsorbed by the soil.

Irrigation with saline waters containing fluoride up to 25 mgL^{-1} has not been found to affect significantly either the grain or straw yields of wheat crop. Fluoride seems to be less toxic as compared to lithium and boron. Three elements could be placed as B > Li > F in that order of decreasing toxicity. Therefore, it is doubtful if fluoride requires any monitoring in India at present because the average concentration of fluoride has not been observed to be very high (10 mgL^{-1}).

CRITERIA

1. United States Salinity Laboratory (USSL)

USSL proposed a comprehensive criteria for the first time to evaluate quality of irrigation waters as early as in 1954 (Richards, 1954). The criteria was primarily based on EC and SAR. Salinity standards were based primarily on two observations viz., (1) The percentage of waters having EC lesser than 0.25 dSm^{-1} was 10, between 0.25 and 0.75 dSm^{-1} was 47.5 and between 0.75 and 2.25 dSm^{-1} was 33.0. Only 9% waters had EC between 2.25 and 5.0 dSm^{-1} and percentage of waters having EC higher than 2.25 dSm^{-1} was nil, and (2) Nearly all irrigation waters as had been used successfully for a considerable time had EC lesser than 2.25 dSm^{-1}. Accordingly, four classes of sodicity based on SAR were: Low (<10), medium (10-18), high (18-26) and very high (>26).

A diagram was also devised for the classification of irrigation waters inter-linking salinity (EC) and sodicity (SAR) hazards. It was based upon the system by which waters at a constant SAR value were given a higher sodium hazard rating with an increase in total salt concentration. Although it was stated to be arbitrary and tentative but it was used very widely in different countries of the world for about three decades. In fact, this assumption proved to be the biggest drawback because later work proved it to be otherwise. The permissible limits of SAR do not decrease but rather increase with increase in EC. This diagram has therefore, become obsolete (Gupta, 1979).

2. Central Pollution Control Board (CPCB)

According to primary water quality criteria as laid down by CPCB (1978-79), the water for irrigation or controlled waste disposal should not have EC more than 2.25 dSm^{-1} and SAR greater than 26. As parameter of SAR and standards are based upon USSL criteria which is obsolete, these guidelines are not tenable.

3. Bureau of Indian Standards (BIS)

According to Indian Standards IS : 2490 Part 1-1981 (BIS 1988) and Part II (BIS 1991), treated effluent quality standards for irrigation are: Dissolved solids (inorganic) 2100 mg L^{-1}, RSC max. 5 meq L^{-1} and Percent Sodium 60. Again as all these parameters are obsolete, these standards are untenable. According to Indian Standard (IS: 11624, 1986), guidelines for quality of irrigation water are based upon the same parameters as used by USSL. Therefore, these guidelines are also not tenable.

4. FAO Guidelines

Based upon recommendations of University of California Committee of Consultants (1974), Ayers and Westcot (1985) devised new guidelines only for sensitive crops, low rainfall (LF 0.15) medium textured soils and normal management conditions. The drawback of U.S. Salinity Laboratory Diagram has been eliminated in these guidelines as permissible limits of EC increases with increase in SAR. However, the manner in which this has been done by linking SAR with EC is cumbersome and may not be necessary because in nature particularly in arid and semi-arid zones, SAR generally increases with increase in EC of waters (Dhir, 1977). Gupta (2000) examined relationship of SAR with EC in 165 ground water samples of Jalore district in western Rajasthan and observed that water of a particular SAR group generally did not have lesser EC than recommended (Table 2.37).

Table 2.37. SAR and EC relationship of ground waters of Jalore district

SAR	Obs. EC range (dSm-1)	Minimum EC range required
<3	0.43-2.58	0.2-0.7
4-6	0.75-6.0	0.3-1.2
7-12	1.14-6.4	0.5-1.9
13-20	2.8-10.4	1.3-1.2.9
21-30	2.6-11.7	2.9-5.0
31-40	5.5-27.0	

Therefore, it does not seem to be necessary in all situations to subject data to test this relationship. These guidelines have also been worked out for such waters which have Mg/Ca ratio as 0.25 to 1.0 and obviously this will not work where this ratio is 2 to 5. The upper limit of EC has been suggested as 3.0 dSm^{-1}. Infiltration rate of water into the soil will not be affected adversely even with a water having SAR as high 20-40 provided that EC is greater than 2.9-5.0 dSm^{-1}. In fact, waters having SAR in the range of 30-40 generally have EC in the range of 3.0-10.0 dSm^{-1}.

5. New Guidelines (Gupta, 2001)

The new guidelines (Gupta, 2001) are based upon the type of ground waters occurring in western Rajasthan (Gupta, 1991) (Table 2.38) and the observation that the waters having EC upto 10 dSm^{-1}, SAR upto 10.0 meq L^{-1} have been used successfully for several decades on light and medium textured soils having leaching fraction greater than 0.3 (Gupta, 1979, 1980, 1990; Sharma and Minhas, 1998). Gupta and Yadav (1984) defined sensitive crop which would not suffer any reduction in yield and quality at EC 1.5 dSm^{-1}. Likewise, permissible and relaxable limits for semi-tolerant crops stand at 3.0 and 5.0 dSm-1 and 5.0 and 10.0 dSm^{-1} for tolerant crops, respectively. The crops, which preform well at EC 10.0 dSm-1, are called as highly tolerant. As these critical limits correspond with LF 0.3, the yields may reduce by more than half at LF 0.1 and increase by more than one third at LF 0.5.

Table 2.38. Per cent distribution of ground waters of western Rajasthan

EC (dSm^{-1})	%	SAR	%	RSC	%
<1.5	21.8	<10	46.1	<2.5	67.2
1.5-3.0	27.9	10-18	26.2	2.5-5.0	10.4
3.0-5.0	21.8	18-26	15.2	5.0-10.0	12.4
5.0-10.0	20.9	26-34	6.4	>10.0	10.0
>10.0	7.6	>34	6.1		

The following guidelines are also based on the observations that at a given salinity, increase in SAR of irrigation waters tends to cause increase in ECe, SAR and pH and decrease in hydraulic conductivity of the irrigated sols and reduction in crop yields. With increase in salinity, same levels of SAR of waters accentuated adverse effect on the crop yield (Anoymous, 1986). This leads to conclude that for a water of given salinity, the lower the SAR, better it is. Also, at a given salinity and SAR, increase in Mg/Ca ratio in irrigation water, causes increase in ESP and reduction in hydraulic conductivity and crop yield (Yadav and Girdhar, 1981; Girdhar and Yadav, 1982).

In view of above, water quality rating based on salinity and sodicity is suggested as below (Table 2.39).

Table 2.39. Water quality rating based on salinity and sodicity

Salinity	Class	EC (dSm^{-1})	Sodicity	Class	SAR*
Normal	A	<1.5	Normal	I	5-10
Low	B	1.5-3.0	Low	II	10-20
Medium	C	3.0-5.0	Medium	III	20-30
High	D	5.0-10.0	High	IV	30-40

* Calculate SAR if Ca/Mg >1 and Ca/HCO_3 >1
SCAR if Ca/Mg <1 and Ca/HCO_3 >1
adj.SAR if Ca/Mg >1 and Ca/HCO_3 <1
adj.SCAR if Ca/Mg <1 and Ca/HCO_3 <1

For the sake of brevity, water quality combined rating on the basis of above two parameters may be expressed as AI, AII, BII, BIII, etc. As lower the salinity, the better it is and at a given level of salinity, the lower the SAR, all the more better it is. A1 will be called as the best followed by A II, AIII and A IV. Likewise, salinity classes B, C and D may be graded as B I, B II, B III, etc.

As quality of water depends upon leaching fraction and salt tolerance of the crops, guidelines for evaluation of quality of irrigation waters are suggested as below (Table 2.40). The fraction of applied water that passes through the entire ranging depth and percolate is called leaching fraction.

Table 2.40. Guidelines for evaluation of quality of irrigation waters.

Nature of crop	EC (dSm^{-1})		SAR	
	LF 0.15	3.0	0.15	0.3
Sensitive	1.5	3.0	<5	10
Semi-tolerant	3.0	5.0	10	20
Tolerant	5.0	10.0	20	30
Highly tolerant	10.0	>10.0	30	40

Although normally high sodicity waters do not have low salinity but if any, these may be treated with calcium amendment like gypsum. Likewise, water having high sodicity may be treated with such amendments in case values exceed permissible limits.

REFERENCES

Abrol, I.P. 1977. Exchangeable sodium and soil water behaviour, Proc. Indo-Hungarian Sem. Mgmt. Salt Affected Soils, Karnal, 103-19.

Agarwal, R.R. and Yadav, J.S.P. 1956. Diagnostic techniques for the saline and alkali soils of the Gangetic alluvium in Uttar Pradesh. J. Soil Sci. 7 : 109-21.

Anonymous, 1986. Coordinated Project on Management of Salt Affected Soils and Use of Saline Water in Agriculture. Ann. Rep. RBS College, Agra.

Ayers, R.S. and Westcot, D.W. 1985. Water Quality for Agriculture. Irrigation and Drainage Paper 29 Rev. 1, FAO, Rome, 174 p.

Bahri, A. 1988. Present and future state of treated wastewaters and sewage sludge in Tunisia. Proc. Regional Sem. Wastewater Reclamation and Reuse, Cairo.

Bajpai, P.D. and Gupta, B.R. Effect of salinity and alkalinity of soil and some important microbial activities. J. Indian Soc. Soil Sci. 27 : 197-8.

Bajpai, P.D., Arya, R. and Gupta, B.R. 1980. Comparative studies on decomposition pattern of some plant materials in two different soil conditions during winter season. Indian J. Agric. Res. 14 : 91-102.

Bajwa, M.S., Choudhary, O.P. and Josan, A.S. 1998. Sodic waters and their management. *In*: Agriculture Salinity Management in India (Ed. N.K. Tyagi and P.S. Minhas), CSSRI, Karnal, pp. 431-450.

Bajwa, M.S., Hira, G.S. and Singh, N.T. 1983. Effect of sodium and bicarbonate irrigation waters on sodium accumulation and on maize and wheat yields in northern India. Irrig. Sci. 4: 191-199.

Bandyopadhyay, A.K. 1974. Seasonal variation of boron in saline sodic soils. Ann. Arid Zone 13: 125-128.

Bhandari, L.M., Purohit, A.D., Bhargava, T.N. and Gupta, I.C. 1970. Potability classification of ground waters in the arid zone of western Rajasthan. Ann. Arid Zone 9: 221-228.

Bower, C.A. 1961. Prediction of the effect of irrigation water on soil. Proc. Tehran Symp. on Salinity Problems in Arid Zones 1-8.

Bresler, E., McNeal, B.L. and Carter, D.L. 1982. Saline and Sodic Soils. Springer Verlag, New York. 238p.

Bureau of Indian Standards, 1981 and 1991. Tolerance limits for industrial effluents IS: 2490 (Part I) Part II.

Bureau of Indian Standards, 1986. Guidelines for quality of irrigation waters. IS: 11624.

Central Pollution Control Board, 1979. Primary water quality criteria for various uses. ADSORBS/3/1978-79.

Chaturvedi, A.N. 1984. Fire wood crops in area of brackish water. Indian Forester 110: 364-365.

Dhir, R.P. 1977. Saline waters - their potentiality as a source of irrigation. *In:* Desertification and its Control. ICAR, New Delhi. pp. 130-148.

Doneen, L.D. 1963. Water Quality for Agriculture. Department of Irrigation, University of California, Davis. 48p.

Doneen, L.D. 1975. Water quality for irrigated agriculture. In: Plants in Saline Environments (Mayber and Gale, Eds.), Springer-Verlag, New York. 56-76.

Eaton, F.M. 1935. Boron in Soils and Irrigation Waters and Its Effect on Plants with particular Reference to the San JoaquinValleyof California. USDA Tech. Bull. 448.

Eaton, F.M. 1950. Signficance of carbonates in irrigation waters. Soil Sci. 69:122-123.

Gajbhiye, K.S., Kolarkar, A.S. and Bhatia, O.P. 1973. Quality of ground water in Bikaner district of western Rajasthan. Ann. Arid Zone 12:65-70.

Garrels, R.M. and Christ, C.L. 1965. Solutions, Minerals and Equilibria. Harper and Row, New York.

Girdhar, I.K. and Yadav, J.S.P. 1981. Role of magnesium in varying quality irrigation waters in influencing soil properties and wheat crop. Agrokem. Talajt. 30 (Suppl.) : 148-57.

Girdhar, I.K. and Yadav, J.S.P. 1982. Effect of different Mg/Ca ratios and electrolyte concentrations in irrigation water on the nutrient content of wheat crop. Plant Soil. 65: 63-71.

Goyal, R.S. 1985. Effect of irrigation with saline ground water containing silica on crust formation in soils. J. Indian Soc. Soil Sci. 33:235-242.

Goyal, R.S. and Jain, B.L. 1982. Use of gypsum in modifying crust conductive condition in saline water irrigated soils. J. Indian Soc. Soil Sci. 30 : 447-54.

Gupta, I.C. 1968. Influence of irrigation with Saline Water on Physical and Chemical Properties of Desert Soils. Ph.D. Thesis, University of Jodhpur, Jodhpur.

Gupta, I.C. 1972. Note on the relationship between pH of saturated soil paste and 1 : 2 soil water suspension. Indian J. Agric. Sci. 42 : 962-3.

Gupta, I.C. 1974. Lithium tolerance of wheat, barley, rice and gram at germination and seedling stage. Indian J. agric. Res. 8:103-108.

Gupta, I.C. 1979. Note on the effect of leaching and gypsum on the detoxification of boron in a saline - sodic soil. Curr. Agric. 3 : 231-4.

Gupta, I.C. 1979. Use of Saline Water in Agriculture in Arid and Semi-arid Zones of India. Oxford and IBH Publishing Co., New Delhi, 210 p.

Gupta, I.C. 1980. Effect of irrigation with high sodium waters on soil properties and growth of wheat. *In*: Intern. Symp. Salt Affected Soils, Karnal. 382-384.

Gupta, I.C. 1983. Quality of irrigation water and the concept of residual sodium carbonate. Curr. Agric. 7: 150-155.

Gupta, I.C. 1983a. Irrigation water quality and boron toxicity. Curr. Agric. 7:1-12.

Gupta, I.C. 1983b. Effect of boron, lithium and fluorine in irrigation water on plant growth. Annual Report. CSSRI, Karnal.

Gupta, I.C. 1984. Effect of boron, lithium and fluorine in irrigation water on plant growth. Annual Report. CSSRI, Karnal.

Gupta, I.C. 1984. Reassessment of irrigation water quality: criteria and standards. Curr. Agric. 8: 109-121.

Gupta, I.C. 1987. Prediction of sodification in saline water irrigated soils. J. Indian Soc. Soil Sci. 35:169-170.

Gupta, I.C. 1990. Use of Saline Water in Agriculture. Revised Edition. Oxford and IBH Publishing Co., New Delhi, 308 p.

Gupta, I.C. 2000. A new criteria for the evaluation of quality of irrigation waters. Proc. 8th ICD International Drainage Workshop, New Delhi, III: 363-376.

Gupta, I.C. and Abichandani, C.T. 1967. Seasonal variations in the composition of some saline irrigation waters of western Rajasthan. India Ann. Arid Zne 6: 109-116.

Gupta, I.C. and Abichandani, C.T. 1970. Use of salt waters predominant in divalent cations for irrigation in the semi-arid zone of western Rajasthan. Indian J. Agric. Sci. 40: 711-715.

Gupta, I.C. and Chandra, H. 1972. Effect of gypsum in reducing boron hazard of saline irrigation waters and irrigated soil. Ann. Arid Zone 11:228-230.

Gupta, I.C. and Jain, B.L. 1992. Salinisation and alkalinization of ground waters polluted due to textile hand processing industries in Pali. Curr. Agric. 16: 59-62.

Gupta, I.C. and Yadav, J.S.P. 1986. Crop tolerance to saline irrigation waters. J. Indian Soc. Soil Sci. 34: 279-286.

Gupta, R.K. and Verma, S.K. 1983. Water behaviour of black clay soil as influenced by degree of sodicity. Curr. Agric. 7 : 114-21.

Gupta, S.C. 1991. Quality classification of ground waters of western Rajasthan. Ann. Arid Zone 30: 315-321.

Gupta, S.K., Bansal, K.N. and Verma, G.P. 1970. Distribution of nutrient cations in saline and sodic soils of Chambal commanded area of Madhya Pradesh. J. Indian Soc. Soil Sci. 18 : 77-81.

Jain, B.L., Goyal, R.S. and Muthana, K.D. 1983. Performance of some tree species in relation to irrigation with saline waters. Ann. Arid Zone. 22: 233-238.

Kanwar, J.S. and Deo, R. 1969. Potassium and magnesium in irrigation water and their effects on the physico-chemical properties of soil. J. Indian Soc. Soil Sci. 17 : 217-26.

Karen, R. 2000. Salinity. *In*: Handbook of Soil Science (Ed. M.E. Sumner). CRC Press, New York, G1-G25.

Kelley, W.P. 1962. Sodium carbonate and adsorbed sodium. Soil Sci. 90:1-5.

Kovda, V.A., Vanden Berg, C. and Hagan, R.M. (Ed.) 1973. Irrigation, Drainage/Salinity, FAO Rome.

Lal, Sohan, Deshpande, S.R. and Sehgal, J.L. (Ed.) 1994. Soil Series of India. Publication No. 40 NBSS & LUP, Nagpur, 684 p.

Laura, R.D. 1974. Effects of neutral salts on carbon and nitrogen mineralisation of organic matter in soil. Plant Soil 41 : 113-27.

Manchanda, H.R., Garg, R.N., Sharma, S.K. and Singh, J.P. 1985. Effect of continuous use of sodium and bicarbonate-rich irrigation water with gypsum and farm-yard manure on soil properties and the yield of wheat in a fine loamy soil. J. Indian Soc. Soil Sci. 33: 876-883.

McNeal, 1976. Managing salt affected soils - recent dissolution of some myths. Crops Soils 28 (4). 122-125.

McNeal, B. L. 1974. Soil salts and their effects on water movement. In: (J. van Schilfgaarde Ed.) Drainage for Agriculture. Am. Soc. Agronomy, Madison, Wisconsin.

Meiri, A. and Shalhevet, J. 1973. Crop growth under saline conditions. In: Ecological Studies-Analysis and Synthesis (Yaron *et al.,* Eds.). Vol. 5. Springer-Verlag, New York.

Muhr, R.G. *et al.,* 1963. Soil Testing in India. USAID, New Delhi.

Neelakantan, S. and Pahwa, M.R. 1977 Note on the distribution of microorganisms in normal, saline and alkali soils. Curr. Agric. 1 : 77-8.

Paliwal, K.V., Srivastava, K. and Prasad, M. 1975. Magnesium hazard in well waters of Rajasthan J. Indian Soc. Soil Sci. 23: 113-123.

Patel, J.C. and Shah, R.K. 1979. Effect of exchangeable cations on hydraulic conductivity of soils. Curr. Agric. 3 : 81-5.

Pratt, P.F., Branson, R.L. and Chapman, H.D. 1960. Effect of crop, fertilizer and leaching on carbonate precipitation and sodium accumulation in soil irrigated with water containing bicarbonate. Trans. 7th Intern. Cong. Soil Sci. 2:185-192.

Rahman, W.A. and Rowell, D.L. 1972. The influence of magnesium in saline and sodic soils: A specific effect or a problem of cation exchange? J. Soil Sci. 30: 535-546.

Rajan, V., Venugopal, T. and Biswas, D.K. 1997. Water Quality. In: National Water Policy-Agricultural Scientists' Proceedings. National Acad. Agricul. Sci., New Delhi. 89-123.

Ramanathan, G., Nagasundaram, N., Bhaskar, A. and Selvaseelan, D.A. 1987. A note on iron chlorosis in sugarcane as related to soil factors and water quality. Madras Agric. J. 74: 57-60.

Rhoades, J.D. 1972. Quality of water for irrigation. Soil Sci. 113:277-284.

Richards, L.A. (Ed.) 1954. Diagnosis and Improvement of Saline and Alkali Soils. USDA. Hand book No. 60. Washington, DC. 160 p.

Robinsons, J.B.D. 1978. Fluorine: its occurrence, analysis, effect on plants, diagnosis and control. Special Pub. 6, CAB, England.

Sharma, D.R. and Minhas, P.S. 1998. Effect of irrigation with sodic waters of varying EC, RSC and SAR/adj.SAR on soil properties and yield of cotton and wheat. J. Indian Soc. Soil Sci. 46: 116-119.

Shivakumaraswamy, K.S. and Satyanarayana, T. 1982. Effect of soil compaction and solution composition on hydraulic conductivity. J. Indian Soc. Soil Sci. 30 : 200-1.

Singhla, S.K. Bhargava, G.P. and Gupta, I.C. 1974. Relationships between certain diagnostic indices of salt affected soils of Indo-Gangetic Alluvial region. Ann. Arid Zone 13 : 218-23.

Suarez, D.L. 1981. Relationship between pH and SAR and an alternative method of estimating SAR of soil or drainage water. Soil Sci. Soc. Am. J. 464-475.

Thorne, D.W. and Peterson, H.B. 1954. Irrigated Soils. Balkiston Co. Inc.

Thorne, D.W. and Thorne, J.P. 1954. Changes in composition of irrigated soils as related to the quality of irrigation waters. Soil Sci. Soc. Am. Proc. 18:92-97.

Varallayay, G. 1977. Moisture status and flow phenomenon in salt-affected soil. Proc. Indo-Hungarian Seminar, CSSRI, Karnal, pp, 85-102.

Wilcox, L.V. 1960. Boron Injury to Plants. USDA Information Bull. 211: 7p.

Wilcox, L.V., Blair, G.Y. and Bower, C.A. 1954. Effect of bicarbonates on suitability of waters for irrigation. Soil Sci. 77:259-266.

Yadav, J.S.P. and Girdhar, I.K. 1981. Effect of different magnesium and calcium ratios and sodium adsorption values of leaching water on the soil properties of calcareous and non-calcareous soils. Soil Sci. 131: 144-148.

■■■■

Chapter - 3

RECLAMATION AND MANAGEMENT PRINCIPLES

INTRODUCTION

The methods to be adopted for the reclamation and the practices for the subsequent management have necessarily to be based upon a proper understanding of the causative factors responsible for the development of the saline or saline alkali soil. The careful diagnosis of the problem is therefore, a very important and valuable pre-requisite for the reclamation and management. For any reclamation technique to be successful, salt must be completely removed from the root zone. This is achieved through the process of leaching defined as the application of excess amount of water to the soil surface for the purpose of pushing the salts down through the soil profile along with the water. To enforce leaching of salts effectively, it is essential to develop the land.

LAND DEVELOPMENT

It is essential to have reconnaissance survey of the area followed by land clearing, topographic and soil survey. The reconnaissance survey will help in identifying the location of field boundaries, necessity and extent of land clearing operation, extent and type of salinity problem, depth of the soil and position of the water table. Prior to making a detailed survey of the area, it is useful to remove undesirable vegetation, if any, from the land. After the clearing operation, a topographic survey of the area is completed to prepare a contour map. The map indicating the permanent structures in the area will be helpful in the design of the irrigation and drainage system, levelling design and field layout etc. Soil Survey on the other hand will help in identifying the salt affected areas from the normal lands, severity and kind of the salt problem, soil texture, depth of the soil and depth to impermeable layer.

Ground level in its natural state can seldom be suited for efficient and uniform application of water. Land levelling therefore, must precede any reclamation programme. Land levelling usually means gently grading the land surface along the length with nil or too little slope along the width. This will help in the movement of water along the length as a uniform sheet. With proper land levelling, depth of application of water can be controlled.

LEACHING

1. Salts

After levelling and ploughing, the field is bunded in small plots, to effect the uniform and efficient leaching. In situations where water is not constraint, leaching is carried out in the periods of low evaporative demands. However, in water constraint situations, the time of leaching just precedes the onset of monsoon so that ensuing monsoon helps in the leaching of salts with increased efficiency. Also, water table level is the lowest at this time of the year and hence the salts are leached to a greater depth.

Soil leaching is less efficient in continuous ponding compared to intermittent ponding due to saturation and the slow rate of water movement. In a continuous leaching case, about 0.5, 0.75 and 1.0 cm of water is required to remove 90 per cent of the soluble salts from each unit (cm) of the soil depth in coarse, medium and heavy textured soil, respectively. Water requirement is although less by 30-35 per cent in case of intermittent ponding but it takes more time. Nielsen *et al.* (1966) reported that intermittent sprinkling with 25 mm water reduced the salinity of the upper 60 cm of soil to the same degree as 75 cm of ponded water, thereby saving 66 per cent of water.

2. Boron

Sometimes saline soils may contain excessive concentration of boron, a toxic element. Reeve *et al.* (1955) postulated that leaching of boron to safe level is very difficult and it may take three times more water to remove boron as compared to soluble salts. Wilcox (1960) stated that there is no economically feasible method for removing boron from irrigation water and further that there is no chemical or soil amendment that can be added to the soil to render boron non-toxic. Gupta and Chandra (1972) reported that gypsum was effective in reducing boron content of waters and soils. Gupta (1979) reported that application of gypsum to alkali soil reduced ESP from 84 to 17 simultaneously with the decrease of boron from 3.0 to 0.8 mgl^{-1}. Bandyopadhyay (1974) observed that natural leaching due to rainfall (740 mm) of a saline alkali soil reduced soluble sodium from 443 to

150 mmol/l alongwith decrease of boron from 13 to 3 mgl^{-1}. Thus, although boron may occur in saline or saline alkali soils but any method used for the reclamation of such soils takes care of boron and it does not present any additional problem.

Leaching Requirement

Leaching requirement (LR) is an estimate of leaching required to maintain soil-water salinity within tolerable limits while leaching fraction (LF) is the fraction of applied water that actually passes the lower boundary of the root zone. Since the leaching requirements for waters with EC less than 3 dSm^{-1} and for yield reductions less than 10% seldom exceed 20%, inefficiency in water application with surface irrigation which exceeds 30%, eliminates the need to care for leaching requirement. However, with increasing use of saline waters and of more efficient irrigation methods, the concept of LR will be meaningful.

Rain Water Management for Leaching

Rainfall is an important source of water for the leaching of saline soils. Rainfall occurring during monsoon season can be beneficially exploited for this purpose. The effects on salt composition due to leaching by rainfall were studied by Gupta and Abichandani (1970) in western Rajasthan. After the rains (350-450 mm) high salt concentrations were reduced so that the top 40 cm of the soil was completely non-saline. The hydrolysis of Na-clay was of small significance and leaching did not result in high alkalinity. In calcareous soils exchangeable sodium was also reduced. Surface soils of the type of Na-Mg-Cl-SO_4 before the rains, became Na-Ca-Cl-HCO_3 after rainfall. These observations show that desalinisation and dealkalisation processes proceed simultaneously indicating thereby that the application of amendment is not necessary in such type of saline soils. On the basis of these data, Vander Molen (1975) reported that calculated leaching from the upper 40 cm of the uncultivated (fallow) soil during monsoon was about 30 per cent of the rainfall. However, Pandey (1978) pointed out that leaching efficiency worked out as 50 to 60 per cent, and that the leaching in such regions could be enhanced by proper rain water management.

Analysis of data for the last 30 years indicated that rainfall at Sampla in Rohtak district which is expected to occur in excess once in two years is sufficient to leach 90 and 80 per cent of the salts from the upper 40 and 55 cm of the soil profile, respectively. Bunds of 25 cm height conserved all the rain water in fields. In the absence of bunds, 30-35 per cent rain water was lost as run-off (Gupta and Gupta, 1987).

DRAINAGE

Waterlogging defined as congestion of drainage is generally linked with the saline soils. A soil is considered as waterlogged, if the water table rises to an extent that root zone becomes saturated, diffusion of air is curtailed and amount of oxygen is reduced with increase in carbondioxide. Generally, critical depth of water table ranges between 1.5 and 3.0 m, depending on soil characteristics, root zone depth, salinity of ground water etc. Artificial drainage, surface, subsurface or both are essential to reclaim the waterlogged saline soils. Though the problem of waterlogging many a times occurs without the development of soil salinity, the main objective of a drainage system is to remove excess water and to control soil salinity.

The factors responsible for the development of waterlogging are the following :

(i) Seepage from unlined canals, distributaries and field water courses, amounting to 25 per cent loss of the discharge at the head, causing recharge to the ground water.

(ii) Borrow pits dug to take earth for development schemes accumulate rain/flood water which in the long run is added to the water table seepage.

(iii) Mismanagement of water at field level resulting in about 33 per cent loss of the discharge at the head, due to deep percolation. Thus, nearly 50 per cent of water from the canal irrigation system is added to the ground water naturally causing waterlogging.

(iv) Transport of water from one basin to another to introduce irrigation, without planning conjuctive use of surface and ground waters, also increases the ground water recharge.

(v) Flodding of an area makes a sizeable contribution to ground water, and

(vi) Cultivation of crops requiring large amounts of irrigation water.

Surface Drainage

The fields in which water remains standing after heavy rainfall, require surface drainage. To check the entry of excess water in the soil, ditches are provided to runoff excess water, after maximum possible storage in the profile. Field ditches carry water into outlet ditches made to follow a natural water course. Field and outlet ditches should be designed in such a way as could remove at least 5 cm of water within 24 hrs. Field ditches are usually 30 to 45 cm deep laid parallel to each other at a space from 20 to 60 m.

Sub-surface Drainage

The rising water table can be controlled by a suitably designed, subsurface horizontal drainage system or a system of a shallow skimming tubewells. In horizontal sub-surface drainage system either perforated corrugated PVC pipes (about 10 cm diam) or open jointed pipes with an envelope of gravel are laid at about 2-3 m depth below the land surface. These pipes are either led to open drains or to covered collector drains with the diameter varying from 15 cm to 20 cm. The collector pipes or small open drains are then connected to surface drains which can ultimately be led to points of disposal either in rivers or canals. This system enables the depth of water table below 1.5 m and with crop stand in the field. Tendency of evaporation and accumulation of salts is practically eliminated.

Vertical drainage system consists of constructing small tubewells on a large scale with the care to design them such that deeper saline layers of waters do not come up and get pumped out. The waters pumped from the shallow tubewells require a proper system of gravity flow to points of disposal. The main requirement of this system is the availability of aquifer at shallow depth. Also, the quality of water in upper layers should be considerably better than in deeper layers. The existing cavity tubewells may also work as skimming wells.

Selection of drainage system, whether horizontal or vertical, depends primarily upon the availability of aquifer, cost and availability of disposal point for extracted sub-surface waters. Generally, there is only limited disposal possibility through drains into rivers and existing canal system. In later case, however, the quality of water may deteriorate. Because it is difficult to dispose off saline waters, only top layers of sub-surface waters having comparatively better quality upto limited depth are required to be extracted. Horizontal system may, therefore, be preferable in such condition.

IRRIGATION PRACTICES

The salinity regime of a plant to which it is exposed depends upon a number of factors, such as stage of the growth when salinisation occurs, the rate of salinisation, the length and time of exposure to salinity, periods of low salinity levels and salinity distribution with depth in the profile. Under field conditions it is difficult to control these factors and the variability among these factors could be responsible for wide variations in the salinity tolerance data reported from various locations. The problem of integrating the entire salinity regime of the plant during its growth period is a complex issue and weighted mean salinity values are commonly recommended as an index of salinity. Complete reclamation of a saline soil is often difficult because of limited fresh water supplies and erratic and undependable rainfall. Thus, one may be obliged to find out management alternatives which might reduce the damage to plant growth. The management practices which modify one or more factors favourably will also help in the use of saline water for irrigation and management of saline soils. Adapting to the problem rather than to solve it in its entity is the principle behind this approach.

Surface Irrigation Methods

Surface irrigation methods have an overall low water application efficiency resulting in increased waterlogging and thereby leading to more problems of salinity and alkalinity. Precise land levelling and smoothening help in increasing the uniformity of water application and in decreasing deep percolation losses. An efficient hydraulic design of the system resulting in optimum lengths of furrows or borders in relation to hydraulic gradient, infiltration rate, unit discharge and water requirement of the crop can further enhance the application efficiency. Design modification for saline soils in the case of borders would require open ended borders rather than the closed ended one. The increase in the cut-off ratio would also be required in comparison to a normal soil-fresh water combination as it will reduce the deleterious effects of salinity at the tail end. The size and shape of furrows could help seed germination and seedling growth (Fig. 3.1 & 3.2). As such it important to those crops which are tolerant to salinity at the later growth stages but are sensitive to salinity at germination. The sloping bed method of furrow irrigation provides low salinity levels in the vicinity of the seed. Seeding in the furrow itself may also be helpful as it will place the seed in a zone which has lowest salt content at seed germination stage. The old furrows later on can be filled and new furrows developed in between the rows of plants to follow the usual method of furrow irrigation. Initial beds will now become furrows in this system of management.

High-Frequency Irrigation

The soil water salinity continually changes following an irrigation. Frequent irrigations with lesser depth of water per irrigation are supposed to maintain better water availability in the upper part of the root zone, even if the total amount of water applied is more or less the same. With each irrigation, this upper area is more thoroughly leached than the lower root zone, thus reducing the osmotic effects and increasing the soil wetness. By using controlled amount of irrigation water, the soil remains unsaturated so that water moves down through the entire soil mass, rather than flowing ineffectively through large soil pores. The frequency of irrigation can be adjusted to meet seasonal crop demands.

Patel and Dastane (1969) reported that frequent irrigations increased plant height and length of ears in case of barley grown with saline water on the Yamuna alluvial soil of sandy clay loam txture. It has been observed that thin and frequent irrigations with saline water gave higher yields in case of wheat (K-65) grown in Jodhpur district (Rajasthan). Irrigation frequency of 10 days gave a yield of 31 q/ha as compared to 27,23.6 and 21 q/ha, with 15, 20 and 25 days frequency, respectively. Beneficial effects were observed not only in saline water irrigated lands but also where the irrigation water is saline sodic. Thin and frequent irrigations are supposed to perform better than the conventional practice.

SELECTION OF CROPS AND CROP ROTATIONS

Depending on the quality and quantity of water available for irrigation, selection of a crop and its variety, which can be best suited to the soil environment is very important. On the basis of crop tolerance to quality of irrigation water or soil salinity, the crops can be classified in four groups, viz., (a) highly tolerant, (b) tolerant, (c) semi-tolerant and (d) sensitive. When leaching fraction is nearly 0.30, EC_e of the irrigated soil after three to four irrigations, each of 5 to 6 cm depth is approximately equal to EC_{iw} of water (Table 3.1).

Table 3.1. Effect of leaching fraction on the value K

Leaching fraction	Fraction of irrigation water leached	K
Low	0.1	2.40
Medium	0.3	1.00
High	0.5	0.67

EC = K ECi

The highly tolerant crop can be defined as one which does not suffer any reduction in yield (and quality) with EC_{iw} equal to 10.0 dSm^{-1}. The tolerant crop does not suffer any reduction up to EC_{iw}

equal to 5 dSm^{-1} but yields may start to decline beyond this level of salinity. However, crops in this group can be irrigated with waters having EC_{iw} up to 10 dSm^{-1} with yields reduction of 10 to 25 per cent. Semi-tolerant crops do not suffer reduction in yields with water having EC_{iw} up to 3.0 dSm^{-1} but can be grown with water having EC_{iw} up to 5.0 dSm^{-1}, with some reduction in the yield. The crop which is adversely affected in yield with water having EC_{iw} equal to 1.5 dSm^{-1} is defined as sensitive. The sensitive crops can, however, be grown with waters having EC_{iw} up to 3.0 dSm^{-1} with some reduction in yield. It should be kept in mind that these limits of EC_{iw} hold good for such soils which have ECe more or less equal to EC_{iw} at LF equal to 0.30. It is also presumed here that water table is below the permissible limits and does not contribute to soil salinisation in the root zone of the soil during crop growth period. A general listing of crops on the basis of the degree of tolerance to salinity is, presented in Table 3.2.

Table 3.2. Crop tolerance table

Highly tolerant	Tolerant	Semi- tolerant	Sensitive
		Cereals	
Barley	Wheat Oats	Rice	
		Millets	
	Pearl millet	Finger millet	
	Sorghum	Maize	
		Minor millets	
		Pulses	
		Cluster bean	Black gram
		Pigeon pea	Bengal gram
		Cowpea	Green gram
			Lentil
		Oilseeds	
	Indian mustard	Niger	
	Castor	Groundnut	
	Indian rape	Sunflower	
	Linseed	Sesame	
	Safflower		
	Soybean		
	Coconut		

Fibre crops			
	Cotton		
	Sunhemp		
Sugar and starch crops			
Sugarbeat	Tapiaco	Sugarcane	
Vegetables			
	Spinach	Ash gourd	Radish
	Amaranthus	Bitter gourd	Carot
		Bottle gourd	
		Brinjal	
		Cabbage	
		Cluster bean	
		Pea	
		Lady's finger	
		Musk melon	
		Onion	
		Potato	
		Ribbed gourd	
		Sponge gourd	
		Dolichos	
		Sweet potato	
		Tomato	
		Turnip	
		Water melon	
Condiments and spices			
		Chillies	Coriander
		Fenugreek	Cumin
		Garlic	Mint
Narcotics			
		Tobacco	Opium
Fruits			
Datepalm	Pomegranate		Grape
	Guava	Custard apple	
	Ber	Kagzi lime	Sweet orange

Mehta *et al.* (1973) suggested that one of the two rotations, viz. (a) fallow-wheat, green manuring-wheat *bajra-* barely and (b) fallow-wheat, *bajra-* barley, green manuring-wheat could be practised in western Rajasthan, where irrigation with saline water is commonly practised. Jain *et al.* (1976) conducted a field experiment for three years at Jobner in Rajasthan on wheat irrigated with saline water (EC 4.8 dSm^{-1}, RSC 7.4 mel^{-1}, SAR = 32). The yields of wheat after fallow, a *Kharif* legume or sesame or pearl millet were not significantly different. *Guar*-wheat rotation proved to be the most economical for the area.

MANURES AND FERTILISERS

The fact that manures and fertilisers could compensate to some extent the adverse effects of saline waters has been recognised widely. Maliwal and Paliwal (1972) reported that dressing of FYM and fertilisers was beneficial in case of wheat and pearlmillet grown with saline water of moderate salinity and SAR, Application of manures increased nutrient uptake. Similar results were obtained in case of maize and barley (Paliwal and Maliwal, 1971). The yield of wheat, grown on poorly drained sandy clay loam and irrigated with HCO_3 dominant water has been reported to increase by the addition of 5-20 tonnes/ha of FYM (Puntamkar *et al.,* 1972).

Application of CAN gave average response of 13 kg of grain for every one kg of nitrogen on net harvested area basis in case of wheat (Dhir *et al.,* 1975). Increased application of nitrogen (180 kg ha^{-1}) did not increase the yield over 150 kg/ha in wheat irrigated with water of EC 6.3 dSm^{-1} (Sharma and Lal, 1975). The response of fertilisation decreased at higher levels of EC and SAR of water (Lal and Singh, 1973). In case of barley, response to higher doses of fertilisers was obtained only when EC and ESP of the soils did not exceed 6.5 dSm^{-1} and 32, respectively. The application of nitrogen beyond 90 kg/ha did not increase grain and stover yields of maize to any appreciable extent when irrigated with saline water (Verma *et al.,* 1977). Green manuring is also a useful practice in the managment of saline waters and therefore, should find an important place in crop rotation. Amongst several green manuring crops the efficiency of *Sesbania aculeata (dhaincha)* has been observed to be higher than sunnhemp and guar (Mehta *et al.,* 1975). Growing of a green manuring crop in reclaimed saline soils during summers is important in the following two ways :

(i) It provides a crop cover to the soil and therefore, resalinisation of reclaimed soil is minimised. The irrigation water applied to this crop also manages to keep the salt down.

(ii) The crop when turned over into the soils adds nutrients and organic matter which invariably helps in increasing crop yields.

Gupta (1992) reported that *dhaincha* grows well and provides complete cover to the soil compared to sunnhemp which could not grow well in saline soils at Sampla with a highly saline ground water underneath. The application of phosphatic fertilisers so that the phosphorus availability is maintained at a level higher than normally recommended, is also reported to enhance the plant growth under saline and/or sodic water irrigations. It is advisable to use high grade fertilisers in saline soils or soils irrigated with saline waters.

USE OF AMENDMENTS

In areas where saline-sodic (high EC-SAR) or alkaline (high RSC) waters are being used for irrigation, the physical properties of the soil get deteriorated resulting in permeability problem. The irrigation water does not enter the soil rapidly enough during an irrigation to replenish the soil water. This resulted in decreased water supply to the crop just as a salinity problem does, but for a different reason. Permeability reduces the amount of water in storage while salinity reduces the availability of the water although water may be stored in the soil. Permeability refers to the ease with which water enters and percolates down through the soil and is usually measured and reported as infiltration rate. An infiltration rate of 2.5 mm/hr is considered low while 12 mm/hr is relatively high (Ayers and Westcot, 1976). The adverse effect of sodic water is more to the shrinkage-swelling type soils (montmorillonite) than to the non-swelling types (illite-vermiculite and kaolinite). Also, the adverse effect is less due to high SAR of the waters (which have usually high EC) than the water with high RSC (which have low EC) because of the differences in electrolyte concentration. It is well established that physical properties of the soil are better in soils with high electrolyte concentration compared to the ones with low electrolyte concentrations at the same level of SAR.

Fallowing

In some situations or in some years, rainfall may be unduly low and far below the average. Fallowing of the lands in such events for one season or more in arid and semi-arid regions may be beneficial practice, in regenerating saline soils.

Planting Practices

Salinity reduces or retards germination thereby hindering a satisfactory stand of furrow irrigated crops. In some cases, farmers use higher quantity of seed to offset the adverse effect. Although this practice may enable higher plant density but apart from increased labour costs involved in thinning, uniform plant population can not be ensured. This necessitates modification in planting practices, bed shapes and irrigation management (Ayers and Westcot, 1985).

With furrow irrigated crops planted on raised beds, water moves from furrow into the bed. Since water moves from the two furrows towards the centre of the bed, salts move alongwith and concentrate in the upper centre of the bed. Placing seeds in a single row in the centre of a raised bed locates the seed in salt zone (Fig. 3.1 A). Double-row raised planting bed system will locate the seed in salt free zone (Fig. 3.1 D). However, in this system, the salts accumulate near the top on centre of the bed (Fig. 3.1 A & D). But if alternate rows are irrigated, the salt can be moved beyond the single seed row (Fig. 3.1 B). The salts may still accumulate but in smaller concentration. Off centre, single row planting on the shoulder of the bed closest to the watered furrow (Fig. 3.1 E) has also been used and aids germination under saline soil conditions. Double-row planting with alternate row irrigation is not recommended as salts would accumulate in the second seed row from the wet furrow (Bernstein *et al.,* 1975). If salinity still poses problem with either single or double row planting, depth of water in the furrow could be increased (Fig. 3.1 C and F). Still better salinity control could be achieved by using sloping beds with seeds planted on the sloping side and the seed row placed just above the water line (Fig. 3.2) (Bernstein and Fireman; 1957). Irrigation is continued until the wetting front has moved well past the seed row. A correct configuration for a single row sloping bed that is easy to cultivate and convert back to a conventional raised bed is shown in Fig. 3.3. This reshaping is usually done after germination and often early growth period.

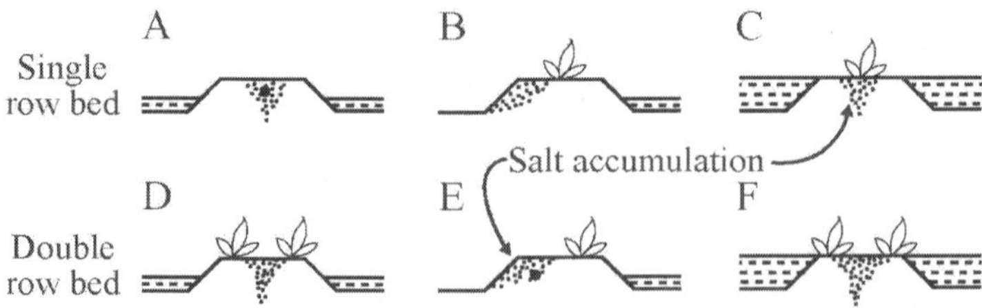

Fig. 3.1. Flat top beds and irrigation practice

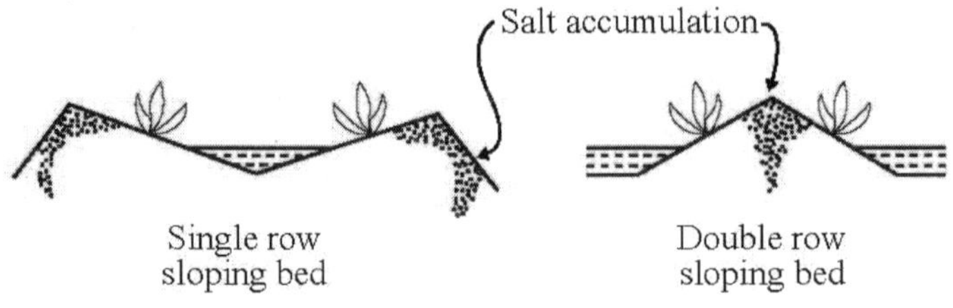

Fig . 3.2. Salinity control with sloping beds

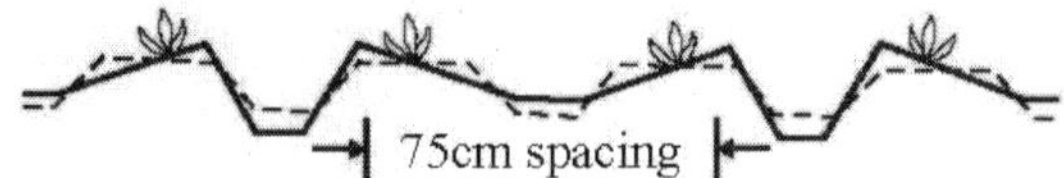

Solid line indicates profile of sloping bed shape. Dotted line indicates standard flat-topped bed which may be formed by cultivation, without moving soil across the plant rows.

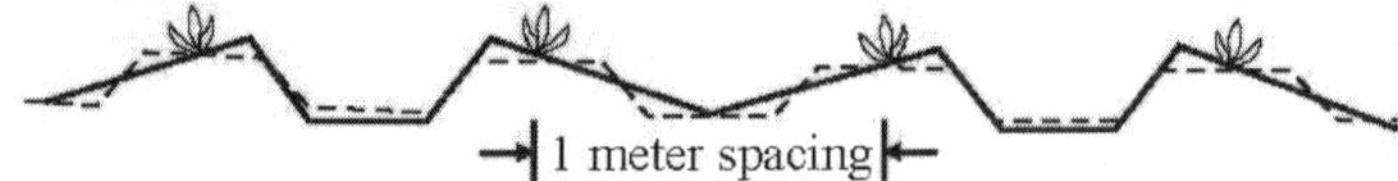

Fig. 3.3. Sloping seeds beds

Another widely used modification is followed for both salinity and temperature control (Fig. 3.4). The seeds are placed just above the water line in the furrow. For a crop planted in winter or early spring, a soil temperature a few degrees warmer is important, the sloping bed is oriented towards the south in northern hemisphere. In hot climates, where cooler soil temperature is desired, reversing this slope (facing away from the sun) has been beneficial.

Fig. 3.4. Sloping seeds beds used for salinity and temperature control

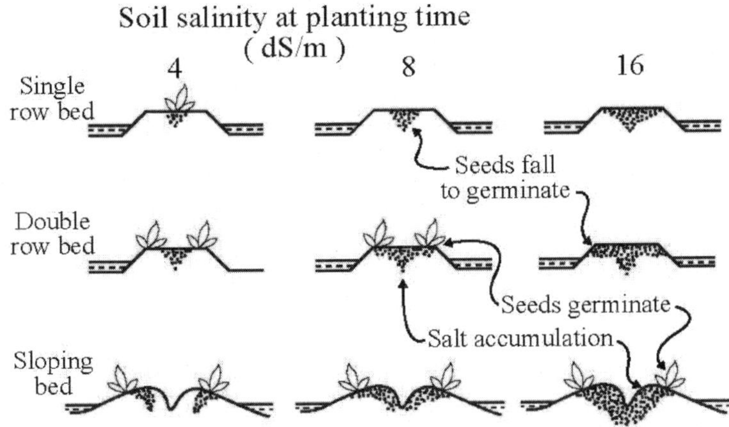

Fig. 3.5. Bed shapes and salinity effects

At given salinity levels, the effectiveness of modifying the shape of the planting beds is shown in Fig. 3.5 (Bernstein *et al.,* 1955). Actual response will depend on the initial soil salinity, the irrigation method, the irrigation water salinity and the crop tolerance during germination. Since salinity slows germination of many crops, retaining water in the furrows for a longer period sometimes improves emergence by maintaining moist soils, reducing crusting and also the salinity.

IRRIGATION METHODS AND PRACTICES FOR SALINE WATER USE

Surface Irrigation Methods

Surface irrigation methods have an overall low water application efficiency and thus aggravate the problems of salinity and alkalinity. Precise levelling and smoothening help in increasing the uniformity of water application and allow application of shallow depth of water per irrigation. Proper hydraulic design of the system resulting in optimum lengths of furrows or borders in relation with hydraulic gradient, infiltration rate, unit discharge and water requirement of the crop can further enhance the application efficiency of these methods of irrigation. Design modification for saline soils in the case of border irrigation may include the use of open ended border rather than the close ended one. The increased cut-off ratio in comparison with the normal soil will also reduce the deleterious effects of salinity at the tail end. The distribution of salts resulting from wetting of the ridge between the furrows is of particular importance in seed germination and seedling growth. It is of particular importance to those crops which have a high tolerance to salinity in the later growth stages but are sensitive to salinity at germination. The sloping bed method of furrow irrigation provides low salinity levels in the vicinity of the seed. Seeding in the furrow itself may also be helpful as it will place the seed in the lowest salt content zone during germination. The old furrows later on can be filled and new furrows developed in between the rows of plants to follow the usual method of furrow irrigation. Initial beds will now become furrow in this system of management.

Sprinkler Irrigation

A total of 6,58,500 ha area has been covered so far in India under sprinkler method of irrigation mostly using good quality water. The increase in crop yield compared to traditional methods of irrigation was 15% for wheat, 21% for groundnut and oilseed crops and 25% for low water requiring crops such as pearl millet, sorghum, barley and gram (Kumar and Sharma, 1999). In a field experiment conducted during a four year period (1975 to 1979), application of saline water (EC 11 dSm^{-1}) through sprinkler reduced yield of pearl millet by 33 per cent, but that of wheat only by 3 per cent (Agarwal and Khanna, 1983). Saline water upto 6 dSm^{-1} was safe for pearl millet if applied through a sprinkler in the late evening hours or during the night. The marked reduction in pearl millet yield was due to burning of leaves caused by excessive sodium content in the water, unlike wheat because of lower temperature and higher relative humidity prevailing during the winter season. Accumulation of salts on the leaves to depress photosynthetic activities when highly saline waters are used with sprinkler irrigation could also cause reduction in yield. The application of saline water (EC = 12 dSm^{-1}) to cabbage and cotton resulted in severe leaf burning due to high day time temperature. The salt accumulation in sprinkler irrigated plots was 30 to 40 per cent less as compared to surface irrigation (Agarwal *et al.*, 1982). Although the sprinkler irrigation may have operational cost 1.5 to 2.5 times more than the surface irrigation, there is a saving of water by 10 to 55 per cent (Kumar and Sharma, 1999).

This method is more effective in leaching of salts as compared to flood irrigation due to controlled rate of water application.

Drip Irrigation/High Frequency Irrigation

About 0.2 mha area has been covered so far in India under drip irrigation. Drip irrigation has given good results even when using relatively high salinity waters. Owing to continuous supply of water, the high salt concentration which would have built up with conventional irrigation is avoided. Low moisture tension levels can be maintained throughout the growing season. After the harvest, leaching is required to lower the salt content in the soil before sowing the new crop. Drip is more suitable for widely spaced fruit and vegetable crops. Crop water requirement is almost the same as in flood irrigation but water economy is mainly at the early stages of growth and as a result of lesser coverage of the actual irrigated area.

Table 3.3. Effect of furrow and drip methods of irrigation at different ET levels and water salinity on the yield of potato (Singh *et al.*, 1978)

Treatment	Water use (cm)	Yield (t ha^{-1})	Tubers No./m^2	Wt/tuber (g)
		1972-73		
Drip ET_{100}	36.6	33.4	48	98.0
Drip ET_{75}	27.4	27.8	33	100.0
Drip ET_{50}	18.3	20.5	31	86.8
Drip ET_{200} (S_1)	36.6	26.4	36	89.0
Furr. ET_{100}	36.6	20.2	34	62.5
SEM ±	–	1.0	3.5	5.6
CD (0.05)	–	3.2	10.8	17.3
		1973-94		
Drip ET_{100}	28.8	27.5	55	59.0
Drip ET_{75}	21.4	21.1	40	53.0
Drip ET_{50}	14.3	14.7	42	36.3
Drip ET_{200} (S_1)	28.6	14.4	42	39.0
Furr. ET_{100}	34.0	18.1	32	65.8
SEM ±	–	0.6	8.2	3.3
CD (0.05)	–	1.8	6.7	10.1

S_1 & S_2 = Salinity of water, 3 and 10 dSm^{-1}, respectively

$ET_{50,\ 75,\ 100}$ = Percentage of evapotranspiration of the crop;
Furr. = Furrow irrigation.

Drip irrigation at a rate less than the ET rate decreased the yield of potatoes compared with a rate that was equal to ET (Singh *et al.*, 1978). To obtain identical yields drip irrigation required 50 per cent less water than furrow irrigation. Water was applied daily by drip irrigation and at intervals of about seven days by furrow irrigation in amounts equal to 68 per cent of class A pan evaporation. Saline water of EC 3 dSm^{-1} applied by drip irrigation did not limit yields but with that of EC 10 dSm^{-1}, potato yields were reduced by 91 per cent (Table 3.3) and tomato yield by 35 per cent. The soil water content was about 15 per cent beneath the emitter, 7 per cent at a point 20 cm from the lateral, and 3 to 4 per cent near the wetting front located 40 cm from the lateral. The wetted zone, extending 20 cm on either side of the lateral, could be used for twin-row configuration. Salts were concentrated in the surface - 15 to 20 cm of soil at the mid-point between the emitters and towards the wetting front. Salts were not leached to lower soil horizons with the above treatment.

Gupta and Tyagi (1984) conducted a laboratory experiment to study the effect of emitter application rate (q) and irrigation frequency (IF) on the soil moisture and salt distribution pattern in a saline soil. A constant amount of saline water (10 L, EC 4.5 dSm^{-1}) was applied with two application rates (0.75 and 1.50 L hr^{-1}) and three irrigation frequencies (daily, one day and two days irrigation intervals). The stored amount of irrigation water in the soil profile showed that with daily irrigation, about 68 per cent of the applied water was stored in the top 30 cm depth as compared to 59 per cent in the two days irrigation frequency with q = 0.75 L h^{-1}. There was no loss of water below 60 cm in the daily applications compared to 7.1 per cent in two days irrigation frequency. In daily irrigation, 62.3 per cent of the applied water was stored in top 30 cm depth compared to only 48.5 per cent in two days irrigation frequency with q = 1.50 L hr^{-1}. Therefore, daily application at a lower emitter application rate seems to be beneficial in retaining a large percentage of applied water within upper 30 cm soil depth and in reducing the loss of irrigation water due to deep percolation (Table 3.4).

Table 3.4. Measured amount of water in the soil profile, expressed as a percentage of the total amount of applied water (Gupta and Tyagi, 1984)

Soil depth (cm)	Irrigation frequency					
	Daily		1-day interval		2-day interval	
	*0.75	1.5	0.75	1.5	0.75	1.5
0-15	36.2	34.2	35.2	32.5	30.9	25.4
15-30	31.6	28.1	30.1	26.5	28.1	23.1
30-45	19.1	16.5	20.5	14.5	19.1	18.0
45-60	13.6	16.8	10.8	16.4	14.8	16.2
>60	-	4.4	3.3	10.5	7.1	17.3

*Irrigation frequency (L h^{-1})

Agarwal and Khanna (1983) reported that the water use efficiency was higher for drip irrigation as compared to surface irrigation both for good and saline water. Subsurface drip performed better than surface drip in case of radish crop (Table 3.5).

Table 3.5. Water use efficiency under different methods of irrigations and quality of water (Agarwal and Khanna, 1983)

Method of irrigation	Good quality water (EC 0.25 dSm^{-1})		Saline water (EC 6.5 dSm^{-1})	
	Yield (q/ha)	WUE (q/ha-cm)	Yield (q/ha)	WUE (q/ha-cm)
Subsurface drip	268	30	236	26
Surface drip	175	19	257	18
Surface (35 mm CPE)	164	14	99	9
Surface (60 mm CPE)	139	12	67	6

WUE = Water use efficiency; CPE = Cumulative pan evaporation

Singh (2001) has comprehensively reviewed the work done on this aspect in India and reported that yield increased from 41% (grape) to 141% (pomegranate) compared to average yields in Maharashtra. A water saving of 36-79% has been observed in drip as against the conventional systems.

Suction Irrigation/Double Walled Earthen Pot

Gupta *et al.* (1988) have innovated a novel device for the early establishment of saplings with very limited water. The device consists of a double walled earthen pot. The diameter of the outer pot is kept approximately 25 cm at the top and 18 cm at the base. The diameter of the inner pot on top and at the base is kept as 15 and 12 cm respectively. Both the pots are joined together at the base allowing the basal part of the inner pot to remain almost open so that during growth the roots may easily go deep in the soil beneath (Fig. 3.6). The outer wall of the external pot is made impervious with the help of cement or some other impervious material, to check the loss of water outside. A tree sapling along with soil received from the nursery is placed in the inner pot. The device is fixed in a small pit so that its brim lies in line with the surface of the earth. The annular space between the two walls of the pot is filled up with water and it is covered with a polythene sheet or earthen lid with a circular hole of 15 cm diameter so as to check the loss of water due to evaporation. Whereas in drip irrigation, the water is added to the soil in drops by gravity, in this method soil energy is used to extract the water by suction. A suction force is created by the dry soil and plant roots requiring water draw moisture towards it from the neighbouring high moisture zones.

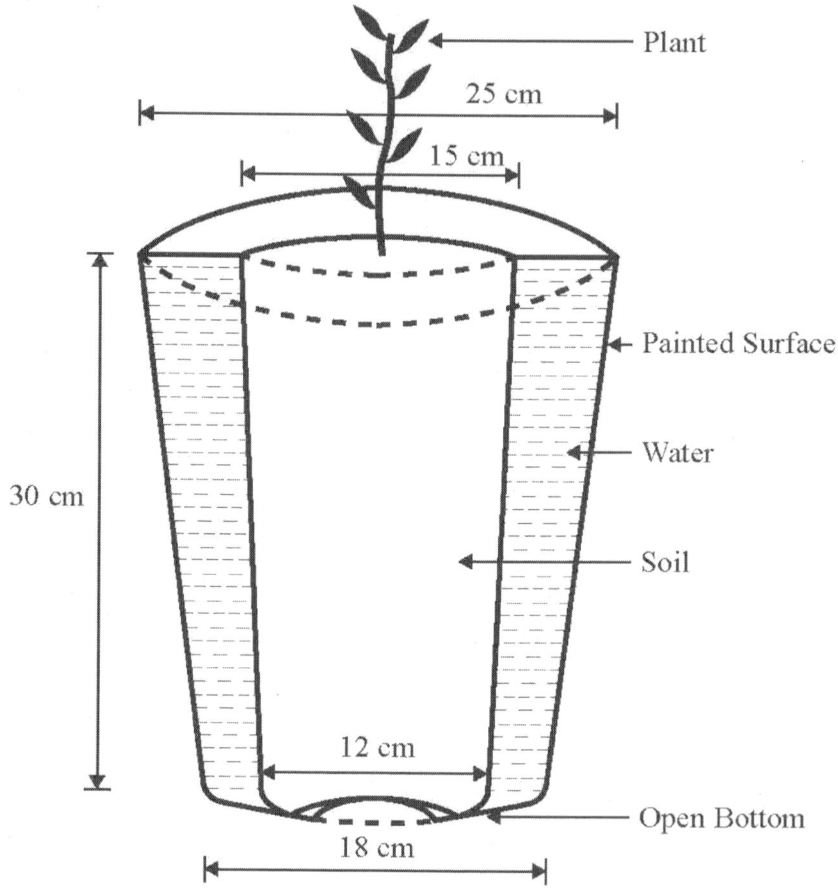

Fig. 3.6. Jaltripti (double-walled earthen pot) for the early establishment of saplings with limited water.

Several experiments have been conducted on the establishment of saplings of ber and neem and the relevant data are shown in Tables (3.7 to 3.9) showing a remarkable performance of the device. This device is a boon for the sandy soils in the desert areas where water is scarce and more than 90 per cent is wasted due to deep percolation. Since the water is to be filled at weekly or fortnightly intervals it saves a lot of labour. Fertilizers, insecticides and pesticides may also be mixed in the water and thus their amount can also be saved. The device can be used beneficially for forestry and horticultural plantation.

Table 3.7. Height and collar diameter of ber plants (6 months old)

Method of planting	Plant height			Collar diameter		
	Initial	Final	Increase (%)	Initial	Final	Increase (%)
Improved method (Jaltripti)	28	60	144	00.32	0.65	103.1
Conventional method	28	40	42.8	0.20	0.37	85.0
RGR		2.7:1			1.9:1	

Note: Rate of water application : 2 L/week, Total water applied 34 L

Table 3.8. Root length, fresh weight and dry weight of ber plants (6 months old)

Method of planting	Root length(cm)			Weight of plants (g)	
	Initial	Final	Increase (%)	Fresh weight	Dry weight
Improved method (Jaltripti)	20	86	330	17.9	10.1
Conventional method	20	25	25	1.5	0.8
RGR		13.2:1		-	-

Table 3.9. Height and collar diameter of neem plants (about 6 months)

Method of planting	Height (cm)			Weight of plants (g)		
	Initial	Final	Increase	Initial	Final	Increase
Improved method (Jaltripti)	9.3	17.0	82.8	0.38	0.45	60.7
Conventional method	9.5	11.4	20.0	0.28	0.32	14.3
RGR		4.1:1			4.3:1	

Note: Rate of water application: 3L/week, Total water applied = 48 L

Solar Desalinating Auto Irrigation System

A new device has been innovated for the development of saplings in a nursery for such regions as have only saline waters (Gupta *et al.*, 1989). The saline waters are desalinised using solar energy and simultaneously the waters are used for irrigation *in situ*. The device consists of a double-walled earthen pot, a coupling of two earthen pots of different diameters and heights joined at the base so that the bottom of the inner pot is open. The diameter of the outer pot is approximately 25 cm at the top and 18 cm at the base whereas the corresponding diameter of the inner pot is 15 and 12 cm, respectively.

The saline water (3 to 4 L) is filled up in the annular space between two pots. The device can be made of hard polythene sheet also. The circular surface of water is covered with funnel shaped transparent polythene cap so that it is sloped at 45°C and embedded upto 2.5 cm depth in the nursery soil mixture contained in the polythene bag in the inner pot. The device works on a simple principle of condensation. The water evaporates due to solar energy from the annular space and condenses on the polythene cap. The condensed droplets slide down around the seed in the nursery bag. Thus, moisture to the soil is supplied through the condensate, which is completely free of salts.

Pitcher Irrigation

The pitcher irrigation has been appropriately named to highlight the basic component of the system, the pitchers (Fig. 3.7). It is an ordinary earthen pitcher commonly used in India to cool water during the summer seasons, also referred to as poor man's refrigerator. A unglazed pitcher of 5-8 L capacity would be appropriate for use in this system. This method of irrigation would be most suited to the following areas: water is either scarce or expensive; soils are difficult to level; soils are light and limited amounts of available water cannot be spread over the area easily; water is highly saline and cannot otherwise be used to grow crops; the remote areas where transport of vegetables is expensive and uneconomical; and for initial establishment of horticultural and forest plantations. Such situations are of common occurrence in the arid and semi-arid regions of India, indicating that this system can be employed over large areas in this country.

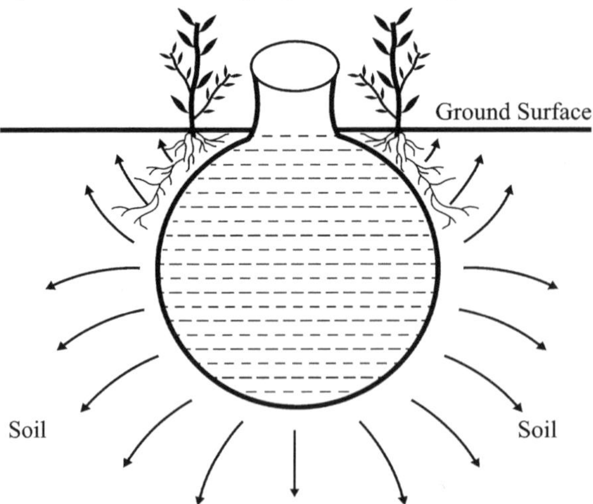

Fig. 3.7. A view of the pitcher irrigation technique for conserving water and use of saline water.

The water requirement in the pitcher irrigation technique depends upon number of pitchers per ha, crop grown, quality of water and the filling schedule. A simple calculation would reveal that water requirement could be as low as 21.8 cm ha^{-1} to 12.5 cm ha^{-1} as the number of pitchers ha increase from 800 to 5000. It means, in pitcher irrigation water requirement may not exceed a maximum of 2 irrigation equivalent in the surface irrigation method.

A large number of crops could be grown with the help of pitcher irrigation technique. The most useful application as already discussed is to grow vegetables or horticultural crops. Creeping types of crops can be grown with lesser number of pitchers as distance from one pitcher to another pitcher could be kept more. The yield of several crops when fresh water was used to irrigate the crop is reported in Table 3.10. It may be seen that even the grape vine could be established well with this technique.

Table 3.10. Yield of various crops with fresh water (EC-0.5 dS/m)

Crop	Yield (kg/pitcher)	Crop	Yield (kg/pitcher)
Watermelon	11.3	Tomato	5.8
Muskmelon	7.4	Cauliflower	5.2
Bottlegourd	21.5	Brinjal	5.1
Bitter gourd	7.5	Cabbage	4.8
Ridge gourd	4.5	Radish	8.0
Cucumber (Kakri)	14.0	Grapes	3.5

The vegetable crops are quite sensitive to salts. Data in Table 3.11 reveal that most of the crops can be grown with waters having EC more than 5 dSm^{-1} except ridge gourd and grapes (Gupta and Dubey, 2001). At these levels of water salinity the yield of crops would be about the same as that with fresh water (Table 3.10). Cauliflower could be grown even with a water as saline as 15 dSm^{-1}. Although, the experimental conditions are not the same, yet it can be seen that with surface irrigation, in general, water of 2-4 dSm^{-1} could only be used to grow vegetable crops (Minhas and Gupta, 1992).

Table 3.11. EC of the irrigation water for equivalent crop yield as with fresh water

Crop	Irrigation water salinity (EC dS/m)
Tomato	5.7
Brinjal	9.8
Cauliflower	15.0

Crop	Irrigation water salinity (EC dS/m)
Ridge gourd	3.2
Cabbage	9.7
Water melon	9.0
Musk melon	9.0
Grapes	4.0

AGRONOMIC AND CULTURAL PRACTICES

Pre-sowing Irrigation

The soluble salts tend to accumulate in the top few centimeters of the soil during fallow periods, more conspicuously in situations where water table is high. Under such conditions, both germination and yield are adversely affected. A heavy pre-sowing irrigation with the best quality water to free the root zone from excess salt concentration and to improve germination and emergence of seeds of winter season crops is essential. Early sowing just after the cessation of the monsoon rains may be advantageous because salts will not get time to accumulate in the root zone (Gupta *et al.*, 2002). Obviously, a heavy pre-sowing irrigation will be very beneficial where sufficient time has elapsed between the last rainfall and sowing date of the *rabi* crop. This practice also assures the availability of sufficient moisture in the root zone. It is also important to add here that crop tolerance limits with respect to salinity of irrigation water hold good only when germination and emergence are successfully established.

Sharma (2000) has reported the beneficial effects of pre-sowing irrigation with good quality water followed by all the irrigation with mixed water throughout the crop growth. The yields at the same average salinity were higher in each case when pre-sowing irrigation was given with good quality water (Table 3.12).

Table 3.12. Comparative effect of pre-sowing irrigation with canal water and blended saline water on wheat yield (tha^{-1})

EC_{iw} of blended water (dS m^{-1})	All irrigations with blended water	Only post-plant irrigation with blended water
0.5 (Canal water)	100.0	100.0
3.0	90.0	- - -
6.0	80.4	95.8
9.0	72.5	90.3
12.0	72.5	83.7
18.8	- - -	78.0

Fallowing

If rainfall is unduly low or below average then fallowing for one season or even more than one season in arid or semi-arid regions may be a beneficial practice. It would help in regenerating the saline-water irrigated soils. Sometimes, fallowing has been practised due to some constraint for example scarcity of irrigation water. As a rule, therefore, soils and irrigation water must be periodically analyzed particularly after rainfall, so as to examine their suitability for re-use. In the event of salinity of the soils remaining very high even after rainfall and good quality water not being available for pre-sowing irrigation, it is advisable to keep the lands fallow for one or two seasons as may be necessary.

Dhir and Bhatia (1975) observed that lands irrigated with waters below EC 5 dSm^{-1} are generally double-cropped in western Rajasthan, whereas those irrigated with waters of higher salinity are used in rotation with a fallow period of different durations but extended mostly to 18 months in between for natural amelioration. Dhir (1977) examined salinity changes during the fallow phase at several locations in western Rajasthan having soils of different textures in relation to natural rainfall for a period of five years. It was observed that the huge salinity build-up during the irrigation cycle is taken care of by one above average yearly rainfall or two subnormal rainy seasons. Some salinity is retained at the 40 to 60 cm soil depth, particularly in clayey loam and heavier soils but this condition has not been found to interfere with re-use of the lands. It is also reported that it is economical to follow fallowing than to double-crop the lands in such cases.

An alternate practice for fallowing would be to divide the land into 2-3 parcels and to crop in different years depending upon the rainfall and leaching. For example in a wet year all the parcels could be cultivated, while in the other cases only one/two parcel which were fallow during the previous years might be cultivated.

In shallow water table area, where use of saline water is made to augment the canal water and/or to minimize the disposal related problems, it is not recommended to keep the land fallow as it may lead to secondary salinization of the reclaimed lands.

Cropping System

The saline waters could be used very effectively by growing two different crops simultaneously in strips. The irrigation of the main crop helps the unirrigated intercrop and thus, both the crops play a complementary role. Intercrop of unirrigated gram and raya could be raised successfully with wheat irrigated with low and high salinity water, respectively (Goyal, 1984). Cultivation of 2 m wide strips of wheat (var. Kalyan sona) alternated with 80 to 160 cm wide strips of raya has given 5 to 7 qha^{-1} extra production of raya along with 22 to 24 q of wheat from just 43.7 $cmha^{-1}$ water as compared to 20 to 21 q of wheat alone under conventional method with the same amount of water (Table 3.13).

Table 3.13. Crops yield as affected by cropping system

Cropping system	Yields (qha^{-1})			Total production (q)	Water use efficiency (kg $cm^{-1}ha^{-1}$)
	Irrigated wheat main crop	Unirrigated (control)	Unirrigated intercrop of raya		
Conventional system	21.3	1.40		22.7	25.9
New system					
Wheat + raya*(40%)	23.4	0.83	6.0	30.23	24.6
+ raya (60%)	23.3	0.55	5.6	28.45	32.5
+ raya (80%)	24.5	0.27	7.0	31.77	36.7

* Cropped area extended under irrigated intercrop of raya

Seed Rate

Nalamwar and Dastane (1966) reported that sowing of barley by ridge and furrow method at high seed-rate (90 $kgha^{-1}$) in saline soil (EC_e 4.7-5.2 dSm^{-1}), with a high water table (69 to 110 cm) gave a yield of 22.2 qha^{-1} as compared to 17.1 q ha^{-1} by flat method of sowing and low seed-rate (60 $kgha^{-1}$). In this method, the crop was sown near the bottom of the furrow on both sides of the ridge. The salts moved on the top of the ridge resulting in lower salt concentration near the bottom of the ridge. Consensus is now emerging to increase the seed rate by 25% over and above the seed rate recommended for normal lands and normal irrigation waters. For transplanted crops, this effect can be achieved by decreasing the plant to plant or row to row distance.

Manures and Fertilisers

Manures and fertilisers enhance the suitability of saline waters for irrigation has been recognised widely. Maliwal and Paliwal (1971a, 1971b) reported that dressing of FYM and fertilisers was beneficial in case of wheat and pearl millet grown under saline water irrigation, up to a moderate level of salinity and SAR. Application of manures enhanced nutrient uptake. Similar results were obtained in the case of maize and barley (Paliwal and Maliwal, 1971b; Maliwal and Paliwal, 1971c). The yields of wheat grown on poorly drained sandy clay loam and irrigated with HCO_3 dominant water have been reported to increase by the addition of 5 to 20 tha^{-1} of dung (Puntamkar *et al.*, 1979).

Application of CAN gave an average response of 13 kg of grain for every 1 kg of nitrogen on net harvested area basis in the case of wheat (Dhir *et al.*, 1975). Increasing application of nitrogen to 180 $kgha^{-1}$ did not give response over 150 $kgha^{-1}$ in wheat irrigated with water of EC 6.3 dSm^{-1} (Sharma and Lal, 1975). The response of fertilisation decreased as the level of EC and SAR of water increased (Lal and Singh, 1972). In barley, good response to fertilizers was obtained when EC and ESP of the soils did not exceed 6.5 dSm^{-1} and 32 respectively. The application of nitrogen beyond 90 $kgha^{-1}$ did not increase grain and stover yields of maize to any appreciable extent under saline water irrigation (Verma *et al.*, 1971). Nitrogen responds well also under conditions of high water table-cum-salinity. Panwar (1978) reported 62.6 kg N ha^{-1} as the economic optimum dose for *Brassica campestris* var. Sarson. Interestingly, higher dose (80 kg N ha^{-1}) had a depressing effect on the oil content of seed.

Singh and Bajwa (1990) reported that NH_3 volatilisation losses increased with increase in EC, RSC and SAR of irrigation waters. Losses as high 37% and 40% were observed at RSC 15 $meqL^{-1}$ and EC 4 dSm^{-1}, respectively. Application of higher doses of N may not alleviate the adverse effect. However, efficiency could be increased through splitting of doses so as to match crop demands, deep incorporation, use of slow release fertilizers and urease inhibitors (Minhas and Gupta, 1992).

For improving the N-use efficiency, a better strategy seems to substitute a part of inorganic fertilizer requirement through organic sources. Addition of FYM in this respect helps in improving the soil physical properties, increases moisture retention, decreases nutrient losses and increases nutrient availability either directly or indirectly. Addition of FYM + NPK compared to NPK alone increased the potato yield (Table 3.14).

Phosphorus has been reported to mitigate the adverse effects due to higher salinity. The grain yield of wheat increased by 6, 15 and 32% under irrigation with waters of EC 3, 6 and 12 dSm^{-1}, respectively, where P dose was increased from 30 to 90 kg ha^{-1} (Minhas and Gupta, 1992).

Table 3.14. Effect of FYM application on potato yield (qha^{-1})

EC_{iw}	FYM (tha^{-1})		
	0	32	64
BAW*	174	215	232
6	144	202	218
9	122	158	178

*Best available water. Source: Chauhan (1998)

Green manuring is also one of the useful practices in the management of saline water irrigated agriculture and therefore it should find an important place in crop production strategies. The efficiency of *Sesbania aculeata* was observed to be higher than sunn hemp and cluster bean (Mehta *et al.*, 1977). Growing of a green manure crop in reclaimed saline soils during summer is important for the following two reasons:

- It provides a crop cover to the soil and therefore, resalinisation of reclaimed soil is avoided. The irrigation water applied to crop also manages to keep the salts down.
- The crop when added to the soils adds nutrients and organic matter which invariably help in better crop yields.

Gupta (1985) reported that *dhaincha* grows well and provides complete cover to the soil compared to sunn hemp which could not grow well in saline soils at Sampla with highly saline ground water underneath. The application of phosphate fertilizers to keep the phosphorus availability at a level higher than normally recommended has also been reported to enhance the plant growth under saline and/or sodic water irrigation. Phosphorus and probably sulphur and potassium requirement is more for waters which are dominated by chloride than sulphate (Manchanda, 1995). It is advisable to use high grade fertilizers in saline soils or soils irrigated with saline waters.

Application of gypsum and green manuring crops alone or in combination increased the yield of wheat (K-65) from 18 to 37.4 qha^{-1} on saline-alkali soil (EC_e 9 dSm^{-1}; ESP 33.5) irrigated with highly saline water (EC 10.5 dSm^{-1}) at Bilara (Jodhpur). The highest yield was obtained where gypsum was applied with *Sesbania aculeata* (Table 3.15).

Micronutrients

The application of zinc at the rate of 20 $kgha^{-1}$ $ZnSO_4$ has been reported to increase grain yield of wheat from 28.6 to 33.1 qha^{-1} on a normal loamy sand soil irrigated with saline-sodic water (EC dSm^{-1};

SAR 26) at Jobner (Rajasthan) (Lal *et al.*, 1980). Singh (1999) found that with use of sodic waters (RSC-10 and 20 meqL^{-1}), zinc application @ 20 kgha^{-1} improved the grain yield of rice under no gypsum treatment. In presence of gypsum, increase in yield was non-significant.

Table 3.15. Grain yield of wheat and soil properties as affected by treatments

Treatments	Average grain yield (1970-1972) (qha^{-1})	pH	EC (dSm^{-1})	SAR	ESP
Control	15.76	8.6	10.5	33	35
Gypsum	22.60	8.2	12.5	25	25
Cluster bean	19.50	8.4	10.5	31	30
Sunn hemp	19.96	8.4	10.0	29	30
Sesbania	20.83	8.4	10.0	29	28
Gypsum + cluster bean	23.00	8.4	13.0	25	24
Gypsum + sunn hemp	24.23	8.3	12.5	24	24
Gypsum + sesbania	26.66	8.2	11.0	22	22

C.D. 5% 1.87 ; Source : Puntamkar *et al.* (1979)

Blue Green Algae

Effect of blue green algae soil culture at the rate of 125 kgha^{-1} along with 40 kg N ha^{-1} was compared with 80 kg N ha^{-1} to sorghum and wheat crops irrigated with waters of EC from 2.7 to 9 dSm^{-1} at Pali. Application of blue green algae culture + 40 kg N ha^{-1} was observed to be effective in giving higher grain yields of both sorghum and wheat when compared with the yields under 80 kg N ha^{-1} (Table 3.16; Anonymous, 1986b).

Table 3.16. Effect of blue green algae soil culture on yields of sorghum and wheat (qha^{-1})

EC_{iw} (dSm^{-1})	Sorghum		Wheat	
	N1	N2+ BGA	N1	N2+ BGA
2.7	9.8	11.4	16.4	17.9
6.0	7.9	7.4	15.9	20.5
9.0	5.9	6.9	13.8	14.1

N1 = 80 kg N ha^{-1}; N2 = 40 kg N ha^{-1}; BGA = Blue green algae 125 kgha^{-1}

Azospirillum Culture - Bio-fertiliser

Besides leguminous crops, *graminaceous* plants like pearl millet and a number of desertic grasses have been observed to have nitrogen fixing ability associated with their roots, predominantly through the bacterium, Azospirillum. Efficient strains of this micro-organism have been identified which upon inoculation have increased the growth, nitrogen uptake and yields of pearl millet. Different strains of Azospirillum, significantly differed in their N_2-fixing ability. Under saline water (EC 7 dSm^{-1}) irrigation, Azospirillum culture S-14 was effective in increasing the grain and fodder yield of pearl millet where culture S-54 was observed effective in increasing the yield of sorghum at Pali (Rao and Bala, 1986).

Treatment of Seeds/Seedlings

The plants which are able to maintain chemo-osmotic potential in their cells under salt stress have the capacity to withstand salinity and alkali. The chemical, physical and agro-physical treatments of seeds and seedlings have been reported to induce salt tolerance (Henckel, 1954; Strogonov, 1964). One of the main causes for the poor germination or stand of the crop in sodic soils is the disruption in the permeability of roots due to which it fails to control balanced ion absorption. The sodium ions which are predominant in such soils get free access inside the cell due to the deranged membrane permeability and become toxic to plants. The calcium ions in alkali soils have been reported to maintain the cell membrane integrity vis-a-vis permeability and antogonise the toxic effect of sodium in plant tissues and the growing media (Mangel and Kirkby, 1979; Clarkson and Hanson, 1980).

Chaudhuri and Wiebe (1968) reported that the pre-treatment with calcium salts increased the salt tolerance of wheat in saline conditions. Shannon and Francois (1977) did not find any evidence in increasing the salt tolerance at germination in the case of cotton. The germination was hastened by one day only by the pre-treatment of cotton seeds with salt solution as well as with distilled water.

In a field experiment on saline soil (EC_e 12.6 dSm^{-1}), irrigated with saline water (EC 10.5 dSm^{-1}), the highest yield of wheat was obtained when seeds were pre-soaked in 3% Na_2SO_4 solution (Puntamkar *et al.*, 1971). Manchanda and Bhandari (1976) observed that grain yield of wheat and barley was not significantly affected when their seeds were pre-soaked in salt solution. However, pre-soaking in the low salinity tube well water usually resulted in a higher grain yield than seeds pre-soaked in NaCl or Na_2SO_4 solution. It seems, therefore, that treatment of seeds and or seedlings may not cause significant response on plant growth. However, early germination and seedling development may be possible.

USE OF AMENDMENTS

In areas where saline-sodic (high EC-SAR) or alkali (high RSC) waters are used for irrigation, the physical properties of the soil are likely to get deteriorated resulting in a permeability problem. The irrigation water does not enter the soil rapidly enough during an irrigation to replenish the soil with water needed by the crop before the next irrigation. It results in reduced yields due to decreased water supply to the crop just as a salinity problem does, but for a different reason (Table 3.17). Permeability reduces the quantity of water placed into storage while salinity reduces the availability of the water into storage. An infiltration rate of 2.5 mmh^{-1} is considered low while 12 mmh^{-1} is relatively high (Ayers and Westcot, 1976). The average effect of high sodicity of waters is more to the shrinking-swelling type soils (montmorillonite) than to the non-swelling types (illite-vermiculite and kaolinite). Besides, average effect is less due to high SAR which has usually high EC values than high RSC which have low EC. It is due to differences in electrolyte concentration, as the physical properties of the soil containing higher electrolyte concentration are better than the lower ones at the same level of SAR.

Table 3.17. Crop yields as affected by irrigation with alkali waters in a sandy loam soil at Agra

RSC (meq/l)	Crop yield (tha^{-1})			
	Wheat	Pea	Gram	Lentil
Control	4.04	1.33	0.76	1.38
05	3.98	1.22	0.69	1.25
10	3.87	0.93	0.50	1.11
15	3.43	0.86	0.34	1.05
20	3.04	-	-	-
CD 5%	0.2	0.34	0.22	0.15

For western Rajasthan, where high EC-SAR waters are used for irrigation, Gupta and Abichandani (1977) suggested that application of gypsum on the surface of the soils in May and June, before the onset of the monsoon would play an important role in the effective amelioration of the saline water irrigated soils. Goyal and Jain (1982) reported that application of gypsum to soils irrigated with saline water (EC = 6 to 12 dSm^{-1}; SAR = 20 to 27) reduced soil dispersion, water stagnation, translocation of silt and clay and also decreased crust strength besides lowering the suspense load in ponded waters. Row application of gypsum was found to be equally effective even at its low application rate (1.0 tha^{-1}) in comparison with broadcast method. The flat-bed system performed better than ridge-furrow system. Gypsum application ameliorated the crusting properties of the soils.

Application of Amendments

The sodicity hazard of the irrigation waters on the soil can be mitigated by neutralization of the RSC of irrigation water with gypsum. If the RSC of the irrigation water is 2.5 or less, the water is considered to be of good quality and it would not be necessary to add gypsum or any other amendment. However, for every additional 1 $meqL^{-1}$ RSC to be neutralized, agriculture grade gypsum (70% purity) @ 12 $kgha^{-1}$ should be added for each cm of alkali water applied. The quantity of gypsum to be added is, thus determined by the RSC of irrigation water and quantity of water required for irrigation during the growing season or on yearly basis. Besides several other amendments such as acid forming substances (notably pyrites), distillery spent wash (DSW), phospho-gypsum (PG) and sulphitation press mud (SPM) have been used as soil application or through gypsum beds or mixing in water as in the case of DSW (Anonymous, 2007, Anonymous, 2009, Gupta, 2010). The benefits of acid forming amendments would depend upon the calcareous nature of the soils.

Example: Calculation of quantity of gypsum: The water of a tube well has a RSC of 10.5 $meqL^{-1}$. The water is to be used to irrigate wheat crop, which needs about 5 irrigations of 6 cm each. Calculate the gypsum requirement to treat the soil to avid adverse effects.

RSC to be neutralized = 10.5-2.5 = 8.0

Number of irrigation = 5 (6 cm each)

Gypsum to be added (t/ha)= (12 x 8 x 5x6)/1000 = 2.88 tha^{-1}

The beneficial effects of soil application of gypsum on rice and wheat yield are depicted in Table 3.18.

Gupta (1980c) pointed out that application of gypsum prevented the rise in pH and SAR of the soils irrigated with high RSC waters. Gypsum added at the rate of 4 qha^{-1} to the soil takes care of the RSC equal to 10 $meqL^{-1}$. It was top dressed after pre-sowing irrigation.

However, there was no response to the crop growth of wheat. He concluded that crop growth of wheat was not affected significantly under irrigation with water having RSC up to 10 $meqL^{-1}$ provided that EC (<2 dSm^{-1}) and SAR (<10) were low for areas such as Karnal where the average total (700 mm) and the winter season rainfall (100 mm) were appreciable and effective and evaporation was less than 2,000 mm. Indiscriminate use of gypsum is therefore, to be avoided. It should be applied only where crop growth is adversely affected, in areas where rainfall is very low and evaporation is very high.

Table 3.18. Effect of gypsum application on grain yield of rice and wheat in farmer's fields

Wheat 1973-74			Rice 1974		
RSC in water	Gypsum (tha^{-1})	Wheat yield (qha^{-1})	RSC in water	Gypsum (tha^{-1})	Rice yield ($qtha^{-1}$)
12.5	Nil	9.45	10.0	Nil	6.13
7.5	2.0	21.0	7.5	2	9.93
5.0	3.0	26.25	3.0	4	16.24
2.5	4.0	26.25	2.5	6	20.22

Assuming application of 50 cm of water. Gypsum applied would neutralize the RSC to the level indicated. Source: Sharma *et al.* (1975)

Maize yields at Ludhiana were reduced to 50% while wheat yields were unaffected under irrigation with water containing excessive $NaHCO_3$ (RSC = 8.0 $meqL^{-1}$). Application of gypsum decreased the rise in pH and ESP. Gypsum applied with each irrigation was more effective than when applied on a cumulative basis (Bajwa *et al.*, 1983). Application of pyrites equivalent to reduce RSC to 15 $meqL^{-1}$, dressed on the soil before sowing was sufficient for normal wheat production at Agra, but a higher amount (to reduce RSC to 10 $meqL^{-1}$), was necessary to check the rise in pH and ESP (Chauhan *et al.*, 1986). The crop growth of wheat was not affected adversely even with the use of highly saline-sodic water having high RSC (EC 4 dSm^{-1}, SAR 26, RSC 15 $meqL^{-1}$) at Hisar under wheat-fallow crop sequence (Manchanda *et al.*, 1985a). The use of gypsum checked the rise in ESP. Irrigation with water having TCC 25 $meqL^{-1}$, SAR 16 and RSC upto 12.5 $meqL^{-1}$ depressed pearl millet fodder yield but crop growth of wheat was unaffected (Manchanda *et al.*, 1985b).

The most severe effect of RSC takes place where waters have high values of EC and SAR. It is doubtful if any chemical amendment can counteract economically the adverse effect due to the use of such waters for irrigation for long.

Application of gypsum at the rate of 30 tha^{-1} or double dose of G.R. was required to have satisfactory yields of wheat (4.2 tha^{-1}) on a barren sandy clay loam sodic soil (pH_2 9.5, EC 13 dSm^{-1}, ESP 42.6) irrigated with high SAR and RSC water (SAR 25, RSC 10.6 $meqL^{-1}$, EC 2.4 dSm^{-1}). Unfortunately, crop yields were observed to decrease in subsequent years and effectiveness lasted for three years (Mehta *et al.*, 1977). Chhiba and Goswami (1985) reported that a dose of gypsum higher than generally recommended would be required if high RSC waters have to be used for reclamation of sodic soil having high ESP (50-60).

Field trials have shown that gypsum application is required to maintain yields of crops irrigated with alkaline waters (RSC 7.5 meq L^{-1}) especially when rice is grown and rainfall is <500 mm. Keshwa and Singh, (1988) observed that wheat and pearl millet yields were significantly superior with gypsum applied @ 50% GR whereas crops did not respond to application of FYM @ 25 tha^{-1} (RSC 7.3 $meqL^{-1}$, SAR 15). However, a significant increase in wheat yield was obtained with gypsum alone or in combination with FYM on a soil irrigated with alkaline water (EC 1.95 dSm^{-1}, RSC 13 $meqL^{-1}$ and SAR 19.0) in Rajasthan.

SALT BALANCE AND LEACHING REQUIREMENT

The salt balance in the soil, as affected by the quality and quantity of irrigation water and the effectiveness of leaching and drainage, is of paramount importance. If irrigated agriculture is to remain sustainable, soil salinity must be controlled. The importance of the related concepts lies in the maintenance of salt balance in the root zone depth of the irrigated soil, so that the amount of salt added during irrigation is equal to the amount of salt drained. Salt balance is determined by accounting for all the processes which contribute to the inflow, outflow and changes in salt in the profile (ΔSg). This can be expressed mathematically as a mass conservation equation.

$$\Delta Sg = D_r\, C_r + D_g\, C_g + D_i\, C_i + S_m - D_d\, C_a - S_p - S_c \qquad (3.1)$$

Where, D is the amount of water (in terms of depth of water spread uniformly over the soil surface), C is the salt concentration, R is rain water, i is the irrigation water, d is the drainage water, g is the upward movement of water from the ground water table, S_m is the amount of salt dissolved from soil minerals, S_p is the amount of salt that precipitates in the root zone, S_c is the amount of salt removed in harvested crop.

Assuming no appreciable contribution of salts from the dissolution of soil minerals, no loss of soluble salts by precipitation and crop removal, no rain and a water table depth sufficient to prevent upward movement of salt from ground water, equation (1) above is reduced under a steady state to

$$D_i C_i - D_d C_d = 0 \tag{3.2}$$

The equation states that, under the assumed conditions, in order to maintain salt balance the amount of salt added during irrigation must be equal to the amount drained. It is implied that drainage must be provided in irrigated agriculture failing which the salt concentr-ation in the soil solution will increase to a point where no crop growth will be possible. Based on this concept, simple equations have been developed that relate the water needs of crops and the salinity of irrigation water to the extent of leaching required. Leaching fraction is the ratio between the amount of water drained below the root zone and the amount applied through irrigation. While leaching fraction (LF) is the fraction of applied water that actually passes the lower boundary of the root zone, leaching requirement (LR) is an estimate of leaching required to maintain soil-water salinity within tolerable limits. Under the conditions of insignificant rainfall and no salt precipitation, the EC of drainage water is controlled by the leaching fractions. It follows from eqn. (3.2) that this ratio is related to salt concentration in irrigation and drainage water.

$$LF = \frac{D_d}{D_i} = \frac{C_i}{C_d} = \frac{EC_i}{EC_d} \tag{3.3}$$

To estimate the minimum leaching requirement of an irrigated soil, extra depth of irrigation water that has to be applied to maintain the average salt concentration of soil water below a level that would result in a significant yield reduction has to be estimated, by the following equation.

$$LR = \frac{EC_i}{EC_d{}^*} = \frac{D_d{}^*}{D_i} \tag{3.4}$$

Where,

EC_d^* = maximum permissible salinity,

D_d^* = minimum permissible amount of drainage water to maintain high yields.

The evaluation of LR requires selection of proper EC_d^* values which can be estimated from crop tolerance tables.

Assuming uniform distribution of irrigation water, with no losses due to runoff, the amount of water needed for irrigation is equal to the amount required for evapotranspiration (E_{et}) and leaching or

$$D_i = D_{et} + D_d \tag{3.5}$$

By using equation (3.4) to eliminate D_d from equation (3.5), the following equation is obtained.

$$D_i = \frac{D_{et}}{1 - LR} \tag{3.6}$$

The irrigation requirement calculated from equation (3.6) is an estimate based on salt tolerance and water requirement of the crop and the salinity of irrigation water. It does not take into account the variability in water infiltration into the soil. With current practices, irrigation efficiencies remain far below 70 per cent with surface irrigation; it may be as low as 50 per cent. Inefficiency in water application therefore naturally takes care of the leaching requirement of the crops, which varies from 3 to 17 per cent (Shainberg and Oster, 1978). It is doubtful therefore, if there is some significance of the leaching requirement concept under the current irrigation practices. Jain *et al.* (1976) reported that the yield of wheat under irrigation with saline-sodic water (EC 4.8 dSm^{-1}, RSC 7.4 $meqL^{-1}$, SAR = 32) did not increase by application of 25 per cent excess water over field capacity. Similarly, Poonia *et al.* (1974) observed that level of leaching (0, 25 and 50 per cent) did not have much effect on wheat and pearl millet grown with saline waters. The problem of leaching for salinity control under practical field conditions is complicated by the variability of soil characteristics and water distribution and the inevitable loss of water below the root zone, except under high frequency or deficit irrigation. In many cases, for annual crops, leaching may be harmful. Research findings on leaching requirement for specific crops, obtained under steady state and or high frequency irrigation, may not be directly applicable to current irrigation practices (Shalhevet, 1984). However, with increasing use of saline waters and increased efficiency in irrigation methods, the use of equation (3.5) may acquire practical importance.

It follows from equation (3.3) that the salinity of drainage water can be estimated as:

$$EC_d = \frac{EC_i}{LF} \tag{3.7}$$

Where, EC_d is the salinity of the drainage water percolating below the root zone (equal to salinity of soil water, EC_{sw}) and EC_i is the salinity of applied irrigation water.

Thus, equation (3.7) can also be used to predict average soil water salinity (EC_{sw}) in the rooting depth if certain assumptions are made regarding water use within the root zone. Under normal irrigation practices, 40, 30, 20 and 10 per cent of water used by the crop comes, respectively, from the upper to lower quarter of the rooting depth. The same water use pattern has been used to estimate average soil water salinity by Ayers and Westcot (1985).

The soil water salinity (EC_{sw}) is the average root zone salinity to which a plant is exposed. It is difficult to measure. Salinity measurement is usually done with respect to saturation extract of the soil and referred to as soil salinity (EC_e). EC_e is approximately equal to one-half of EC_{sw}. With a 15 per cent leaching fraction and a 40, 30, 20 and 10 water use pattern the average soil water salinity (EC_{SW}) is approximately 3.2 times more concentrated than the applied irrigation water. At a leaching fraction of 20 per cent, the average EC_{SW} is 2.7 times the salinity of the applied irrigation water (EC_i). If irrigation practices result in greater or lesser leaching than 15 to 20 per cent LF, a more correct concentration factor can be calculated using a new estimated average LF, according to the following equation (Rhoades and Merill, 1976):

$$EC_e = 0.2\ EC_i \left[1 + \frac{1}{LF}\right] \tag{3.8}$$

The concentration factors (EC_e/EC_i) vary from 2.2 to 0.4 as LF changes from 0.1 to 1.0 assuming crop water use pattern of 40-30-20-10% in the root zone depth. The predicted EC_e is estimated by multiplying EC_i with the appropriate concentration factor for the estimated LF. These predicted EC_e values reflect changes due to long-term water use and not short-term changes that may occur within a season or between irrigations.

The quality of irrigation water is affected by the amount of rain and its seasonal distribution. When rainfall is distributed throughout the growing season, a weighted average of rainfall and irrigation water quality is generally computed as follows (Richards *et al.*, 1954).

$$EC_{i+r} = \frac{EC_i\ D_i + EC_r\ D_r}{D_i + D_r} \tag{3.9}$$

The value of EC_{i+r} is then substituted for EC_i in equation (3.4) to calculate the LR.

RAIN WATER MANAGEMENT

When the rainy season is short and concentrated as in most arid zones of India, rainfall may be sufficient to leach salts out of the root zone. When the rainfall is about 500 mm, the effective rainfall may be considered as 300 mm. To leach soluble salts in medium textured soils, 0.5 cm depth of water is required per unit cm depth of the soil. Therefore, to displace 80 per cent salts below 60 cm, the required depth of rain-water is 30 cm. Likewise, 20 cm depth of effective rainfall against 30 cm of total rainfall is required to leach 80 per cent salts from 40 cm depth of the soil. These calculations identify well with the field observations by Gupta and Abichandani (1970) who reported that after the rains amounting to 35 to 45 cm, the initial high salt concentrations in the soils were largely reduced making 40 cm of the surface completely non-saline. It can be generalised that the depth of the soil from which the soluble salts (>80 per cent) may be leached during the monsoon season, corresponds roughly with the total rainfall during the monsoon season. It is obvious, therefore, that if the root zone depth of the soil is 40 cm from which accumulated salts are intended to be leached, a minimum amount of rainfall equal to 40 cm is necessary. For light textured soils the rainfall requirement is low and for heavy textured soils the requirement will be high. Maximum utilisation of rain-water is the most important practice in the saline-water irrigated agriculture, because rainfall influences the development of soil salinity in two ways: (i) it has a modifying effect on the number of irrigations required to meet the evaporative demand as part of it is met by rainfall, and (ii) it helps in leaching of accumulated salts, because it is the best quality water available for leaching. Effect of rainfall and number of irrigations on salinity build-up is shown through regression equations in Table 3.19.

Table 3.19. Regression equations for maximum and minimum salinities at Agra as influenced by rainfall in loamy sand soil

Crop	Rainfall (mm)	No. of irrigations	Regression equation
Wheat	17.1	4	$EC_e = 2.437 + 1.2912\ EC_i$
Wheat	113.9	3	$EC_e = 1.759 + 0.6926\ EC_i$
Mustard	18.2	3	$EC_e = 2.758 + 1.1070\ EC_i$
Mustard	113.9	3	$EC_e = 1.795 + 0.4990\ EC_i$

Source: Gupta (1985a)

The annual variations in salinity build-up as influenced by winter rainfall leads to similar conclusions. The effect of rainfall on salt leaching/build up during the monsoon (kharif) is similar (Table 3.20).

Table 3.20. Average soil salinities after *kharif* as influenced by rainfall, number of irrigation and quality of irrigation water

Cropped conditions	No. of irrigations	Regression equations
Rainfed (wet year)	0	$EC_e = 1.742 + 0.093\ EC_i$
Irrigated (dry year)	3	$EC_e = 2.520 + 0.306\ EC_i$

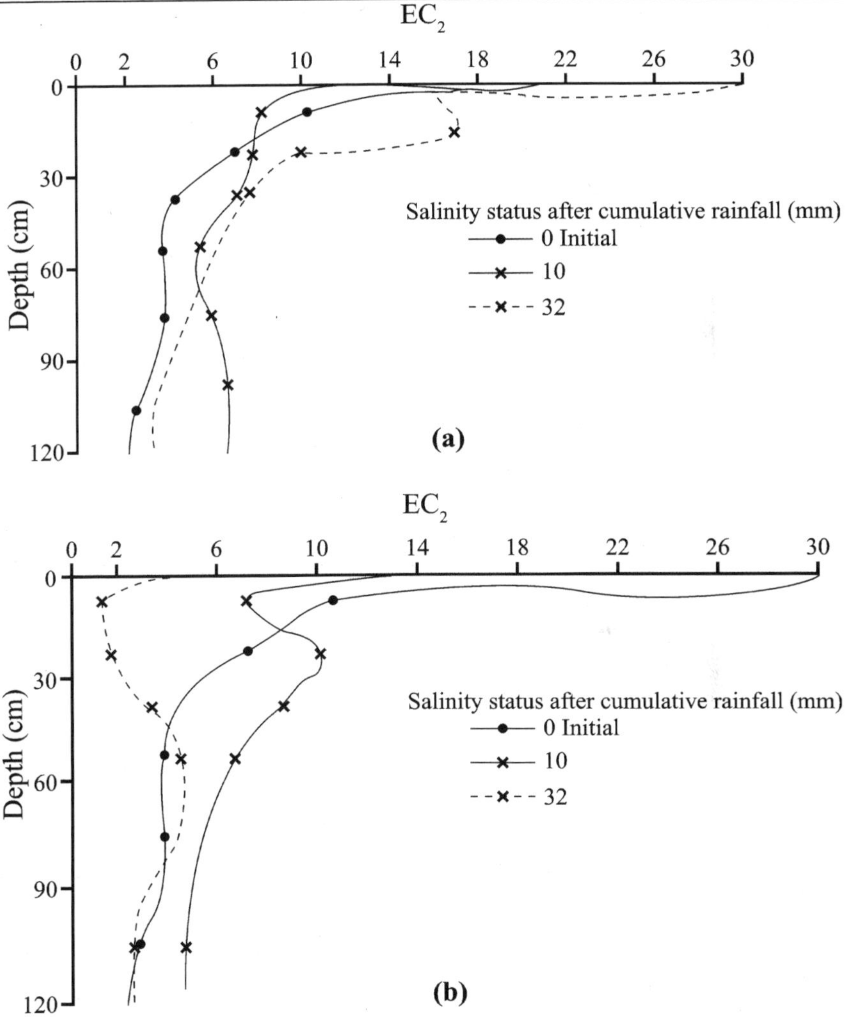

Fig. 3.8. Leaching of fields under natural rainfall (a) Unbunded unlevelled (b) bunded levelled. Source: Gupta, 1992.

Rainfall pattern during a year can completely change the saline water irrigation tolerance limits due to the reasons discussed above (Table 3.21). It is obvious that rainfall plays a very significant role in the process of saline water irrigation by controlling salinisation-desalinisation cycle. It is therefore, essential to manage rainfall by proper management of fields which include uniform levelling, bunding and cultivation before the monsoon season. Effect of bunding and levelling on rain water utilization shown in Table 3.22 indicates that by bunding and levelling the fields, the rain-water utilisation can be increased by 2 to 2.5 times as compared to the unbunded fields. In unlevelled fields considerable non-uniformity of salt leaching has been observed at Sampla. The relatively higher leaching of salts in bunded and levelled fields compared to unbunded and unlevelled fields at Sampla is clearly visible in Fig. 3.8 (a,b). Apparently higher runoff in unbunded/unlevelled fields caused these differences.

Table 3.21. Effect of rainfall on relative yield of wheat with saline water irrigation

EC_{iw} (dSm^{-1})	Relative crop yield		Remarks
	Dry year 1974-75	Wet year 1978-79	
0.6	100	100	The rainfall in the year 1974-75 was 188 mm and 18 mm compared to 701 mm and 114 mm in 1978-79 during *kharif* and *rabi* crops, respectively
2	100	100	
4	89	100	
6	78	100	
8	85	100	
12	26	100	
16	11	95	

*Steep decline due to residual salts of previous season

Table 3.22. Effective rainfall at different locations as influenced by land management parameters

Place	Soil type	Management practice	Effective rainfall as fraction of the total rainfall during the monsoon
Haryana			
Hansi	Coarse loamy to sandy loam	Unbunded Levelled	0.28
Hansi	Coarse loamy to sandy loam	Bunded Levelled	0.70
Sampla	Sandy loam	Bunded Levelled	0.70
Rajasthan			
Dantiwada	Sandy loam	Bunded Levelled	0.70
Kaparda	Sandy loam to silty loam	Bunded Levelled	0.55

Effective conservation of rainfall is important not only in the desalinisation of saline water irrigated soils but also in the desodification. Gupta and Abichandani (1970) reported that desalinisation of the profile was accompanied by simultaneous reduction of ESP (and SAR) (Table 3.23). The surface soils of Na-Mg-Cl-SO_4 type before rainfall became Na-Ca-Cl-HCO_3 type after rainfall, revealing thereby that these soils hardly need any amendment for their management.

Table 3.23. Desalinisation and desodification of saline water irrigated soils affected due to rainfall

Location	Soil depth (cm)	EC_e (dSm^{-1})		SAR	
		A*	B	A	B
	Sandy loam, $CaCO_3$, 2-8 per cent				
	0-20	11.0	2.6	26.9	14.5
	20-40	8.4	3.8	26.1	21.1
Dantiwada	40-60	7.2	12.3	19.5	29.0
(Jodhpur)	60-100	5.9	11.3	23.7	29.9
	100-150	5.0	8.3	26.8	24.2
	150-200	5.0	9.8	25.4	19.5
	Sandy clay loam to silt loam, $CaCO_3$ 4.3-18.5 per cent				
	0-20	11.8	2.6	30.8	8.5
Kaparda	20-40	9.3	3.8	31.1	13.8
(Jodhpur)	40-60	9.3	10.8	31.1	31.5
	60-80	9.3	12.8	32.1	34.4

* A = Before rainfall, B = After rainfall

CONJUNCTIVE USE WITH SURFACE WATERS

Various alternate strategies in the conjunctive use mode are mentioned in the following flow chart.

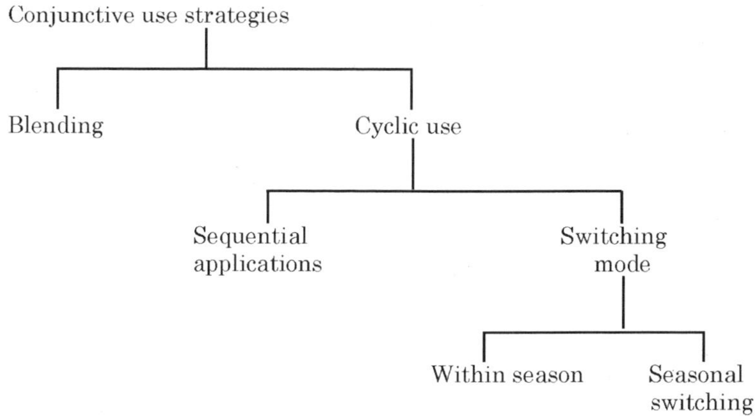

Blending with Good Quality Water

One of the practices for utilising water of high salinity consists in blending it with good quality water so that EC and/or SAR could be reduced to the desired extent. This practice may increase the total quantity of water available for irrigation but at the same time reduces the amount of good quality water. Shalhevet and Kamburov (1976) have outlined three points that should be borne in mind in this regard:

(i) The increase in yield through the increase in water supply should be at least as large as the potential decrease in yield caused by increased salinity.

(ii) When the decrease in water quality through mixing requires leaching, the additional water required for leaching should be taken into account. Furthermore, if removal of leaching water requires the construction of an artificial drainage system, this should also be taken into account in the evaluation.

(iii) The use of good quality irrigation water to dilute drainage water so that the latter may be re-used to increase the total supply, is a questionable practice. It should be used only when leaching is provided by rainfall.

Nath *et al.* (1982) reported that the practice of mixing saline tube-well water (EC = 6.6 dSm^{-1}) with good quality canal water (EC = 0.24 dSm^{-1}) resulted in increased yields of wheat and pearl millet (Table 3.24). This is obviously due to higher salt accumulation in treatment where saline water was used throughout and lowest in treatment which consisted of half saline water plus half canal water. Sharma (2000) intensively studied the mixing mode strategy at Sampla in Haryana. The results over a period of 6 years revealed that mixed waters of 9 dSm^{-1} can be used to grow winter wheat in tile drained lands on a long-term basis. It was also revealed that more saline water could be used in years with above average winter rainfall compared to years with average or below average rainfall.

Table 3.24. Effect of dilution with good quality water on the yields of wheat and pearl millet

Treatment	Av. yields (q ha^{-1})	
	Wheat (5 years)	Pearl millet (4 years)
Saline water (SW)	30.0	22.6
3/4 SW + 1/4 CW	36.2	23.3
1/2 SW + 1/2 CW	38.7	25.3
Canal water (CW)	40.0	29.8

Conjunctive use of good quality surface water with saline ground water, along with improvements in the existing poor irrigation system, has been suggested as the only feasible alternative for saline water table stabilisation in arid and semi arid parts of Haryana (Agarwal and Malik, 1984). However, the use of saline water, either in isolation or by mixing with canal water, threatens soil salinisation, unless steps are taken to protect the salt balance in the root zone (Kaushal *et al.*, 1985, Panda, 2000). A deterministic linear or dynamic programming model has to be developed, incorporating both leaching requirements and the salinity response function of crops, to find the optimal cropping pattern for optimum use of saline water.

Studies made in Haryana on mixing of saline ground waters with fresh water in canals and water courses have revealed that the distance for complete mixing of saline water in canal water varies from 300 to 500m. Tubewells installed along water courses will be very much under-utilised if only mixing with canal water courses is resorted to. It has also been concluded that much better utilisation can be achieved if salt tolerant crops are adopted on a large scale in a saline water belt of the state (Anonymous, 1984).

Cyclic Use with Good Quality Water

In some specific situations where good quality water is available but in limited quantity, high salinity waters can be used for irrigation in conjunction with the former. This practice consists in substituting the saline water irrigation with good quality water when required. The prerequisite of this practice is that good quality water should remain at the disposal of the user. The soil salinity in the root zone resulting from continuous use of saline water can be prevented to increase beyond the permissible limits when two to three cycles with saline water are followed by one cycle with good quality water. Alternately, saline water could be used at tolerant growth stages and good quality water at sensitive stages of plant growth. The knowledge of plant growth stages which are sensitive or tolerant, is therefore, of utmost importance for the judicious use of saline waters in cycle with good quality water. Germination stage, however, remains usually most sensitive and therefore water of best available quality must be used for pre-sowing irrigation. Likewise, selection of crop rotation may be made in such a manner that sensitive crops could be raised with good quality water and tolerant crops with saline water so that the annual salt balance remains stable (Seasonal switching mode). The timing and amount of substitution will vary with quality of the two waters, the cropping pattern, the climate and the availability of fresh water.

Several field experiments have been conducted in India, which bear testimony to the usefulness of this practice. Sarma (1980) reported that rice crop could be grown by giving two canal-water irrigations supplemented with one saline-water (EC 4 dSm^{-1}) irrigation with a reduction of only 17 per cent in grain yield as compared to 42.8 per cent when saline water alone was used for irrigation (Table 3.25). Similarly, the percentage reduction in the pod yield of groundnut was the lower (10.8 per cent) when two canal water irrigations were followed with one saline water irrigation (Sarma *et al.*, 1989) (Table 3.26). The build-up of pH and ESP in cyclic mode is much less than use of saline water throughout (Table 3.27, Singh and Brar, 1995). As a result yields are superior in conjunctive use than with saline water alone. Sharma (2000) tested cyclic use strategies for a number of crops on a tile drained field at Sampla. It was observed that two irrigations with drainage water could be substituted for fresh water if the salinity of the drainage water is in the range of 10-15 dSm^{-1} (Table 3.28). However, fresh-saline sequence was superior and is recommended for application in such a strategy.

Table 3.25. Grain and straw yield of rice as affected by different cycles of saline and canal water irrigation

Saline water irrigations	Canal water irrigations	Mean yield for two seasons (q ha^{-1})	
		Grain	Straw
-	14	37.1	39.4
4	10	30.8	37.4
7	7	29.2	34.0
9	5	29.2	33.6
14	-	21.9	29.5

Table 3.26. The pod yield of groundnut as affected by different cycles of saline and canal water irrigation

Saline water irrigations	Canal water irrigations	Mean pod yield (3 years) (q ha^{1})
-	10	16.28
3	7	14.33
5	5	13.72
7	3	13.12
10	-	8.23

Table 3.27. Average pH and ESP of 0-60 cm soil profile after 6 years of sodic water irrigation application

Treatments*	pH_2	ESP
Canal water (CW)	8.2	3.3
Saline water (SW)	9.6	39.9
2 CW-SW	8.9	12.4
SW-CW	9.1	15.3
CW-2 SW	9.3	20.6

*SW: RSC 10 meq L^{-1}, EC 1.4 dSm^{-1}, SAR 13.5

Table 3.28. Effect of various cyclic modes of irrigation with canal and drainage waters on wheat and succeeding *kharif* crops (Mg/ha)

Mode of application	Wheat	Pearl millet	Sorghum
4 CW*	6.1	3.3	43.3
CW:DW (alternate)	5.8	3.2	39.8
DW:CW (alternate)	5.6	3.2	39.5
2 cW + 2 DW	5.7	3.2	40.2
2 DW + 2 CW	5.4	-	39.5
1 CW + 3 DW	5.1	3.1	37.8
4 DW	4.5	2.8	34.1
Rainfall (mm) during the growth period	64	460	570

*CW = Canal water, DW = Drainage water

Wheat, cotton and sugar beet were grown with drainage saline water (EC 6 dSm^{-1}) in conjunction with Colorado river water (EC 1 dSm^{-1}). Highest yields were actually obtained in all cases with the treatments which received the largest amount of drainage water substituted for river water, that is 75 and 100 per cent, respectively (Rhoades, 1983). Little or no response of river water used in conjunction with saline water seems to be due to the fact that all the crops tried were tolerant and are known to grow well with water having EC upto 5 dSm^{-1}. Sarma *et al.* (1978) reported that there was no significant difference in yield of finger millet in treatments where saline water (EC 4 dSm^{-1}) was used in conjunction with waters on sandy soils. Since finger millet could be grown well on sandy soils with water having EC upto 4 dSm^{-1}, no response could be expected for the conjunctive use with canal waters. This leads to conclude that conjunctive use with canal waters should be practised only where tolerance limits of the particular crop to be grown are higher than EC of the irrigation water.

Blending versus Cyclic Use

Blending and cyclic use strategies have been adopted as a means to achieve conjunctive use of saline and fresh waters. Recently, it has been said that between the two strategies, cyclic use has an edge over the blending strategy. Contrary to this, earlier view has been that both the strategies are at par as long as the threshold salinity of the blended water is kept at or below the threshold salinity of the crop to be grown. The various blending and cyclic use combinations were tried to assess the impact on mustard yield over a 3 years period (Table 3.29). It seems that in mustard crop with a threshold salinity of the irrigation water of 7 dSm^{-1}, the yield differences between various cyclic and blending modes were non-significant (Anonymous, 1998). Numerically, best yield had been obtained in a treatment where one canal water was mixed with two saline water irrigations.

Table 3.29. Crop yields (q/ha) with different modes of irrigation of canal/saline waters

Treatments	1994-95		1995-96		1996-97		Mean yield
	EC_{iw}**	Yield	EC_{iw}**	Yield	EC_{iw}**	Yield	
All canal (C)	0.9	22.3	0.9	24.8	0.9	26.1	24.2
IC:RTS*	4.9	21.0	5.6	23.5	5.4	23.1	22.5
IS:RTC*	2.9	22.2	3.3	24.8	3.2	24.8	23.9
IC:IS	2.9	24.7	3.3	25.5	3.2	26.0	25.4
2C:IS	2.9	24.8	3.3	25.6	5.4	23.8	24.7
IS:IC	4.9	22.0	5.6	24.6	3.2	22.9	23.1
Mixing 1:2	4.8	26.2	5.5	27.7	5.4	24.2	26.0
Mixing 2:1	2.8	25.8	3.2	25.8	3.2	23.4	25.0
All saline (S)	6.9	24.0	7.9	25.3	7.7	25.5	25.2
LSD (p=0.05)	-	NS	-	NS		NS	

If we consider the second data set for cotton, wheat, pearl millet and mustard reported in Table 3.30, it is apparent that differences in the yield due to various treatments are non-significant (Anonymous 1998). Numerically, the yields are in the order of canal water > fresh - saline cyclic use > mixing mode > saline-fresh cyclic use > saline water. On this basis the following could be inferred:

Table 3.30. Crop yields (q ha^{-1}) under varying canal/saline irrigation modes

Treatments	Cotton	Wheat	Pearl millet	Mustard
All canal	16.3	48.8	31.5	20.7
1C : RTS	9.8	40.5	29.9	18.8
1T : IC : RTS	7.2	40.8	28.0	16.7
1C : 1T	12.3	47.2	29.6	19.6
2C : 2T	12.8	46.2	—	—
1T : 1C	7.6	40.2	—	—
2T : 1C	—	—	29.1	14.1
Mixing	10.4	43.7	28.0	18.1
All saline	4.6	35.9	29.1	11.8
LSD (p = 0.05)	3.2	3.5	ns	3.6

EC_{iw} 9 dSm^{-1} for cotton-wheat and 12 dSm^{-1} for pearl millet mustard

- If the salinity of the blended water is less than the threshold salinity, blending and mixing mode would give about the same performance. It may be pertinent to repeat here that the threshold salinity of most crops is less at the germination stage than the average or at later stages of growth. In mixing mode, if this point is considered, than the available differences in the results of the data sets can be explained easily.
- In case the salinity of the blended water is more than the threshold salinity, the fresh-saline sequence in cyclic mode is superior to the saline fresh sequence. This sequence, therefore, requires that fresh water in adequate quantities be available at the time of pre-sowing/first irrigation which may put an additional constraint in this strategy.

Rhoades (1983) has contested the practice of blending of saline waters with good quality waters on theoretical grounds. He points out that if the water being diluted is so saline that its use for crop production is already spent, then diluting it with good quality water and using the water mix for irrigation of crops, of lower salt tolerance does not add to or contribute to the usable water supply for crop production. His conclusion is based upon the logic that a plant must expend bio-energy (that would otherwise be used in biomass production) to extract water from a saline (low osmotic potential) soil solution. When for crop production, water of excessive salinity is mixed with low-salinity water and used for irrigation, the plant removes the 'good water; fraction from the mix until the fraction of the mix made up of the excessively saline portion is left. This fraction of water is just as unusable at this point as it was before mixing because it requires more energy to separate the pure water from the low osmotic potential solution than the plant can muster. Thus, blending, that is, diluting excessively saline water with low-salinity water does not increase the water supply for crops of the same or lower salt tolerance.

There is, in general, complete agreement with the above views as these are developed on the basis of steady state salt balance theory (Rhoades *et al.*, 1992). However, in the monsoon climatic conditions as are prevalent in India, steady state situation never reaches and salt status in the root zone every year is dynamic. The salt build-up occurs during winter and build-up increases with increasing number of irrigations (Gupta, 1985). Moreover, the leaching fractions assumed by Rhoades *et al.* (1992) are as low as 0.09. Again, such leaching fractions are uncommon in India. Application efficiencies as low as 40% are quite common and do not exceed 70% in all cases. Thus, leaching fraction exceeding 0.3 would in most cases be obtained. At this leaching fraction, even under steady state conditions such a high salinity of the drainage water is not expected. Thirdly and finally most of the salts would be leached down during the monsoon season whether it is applied in a cyclic or mixing mode. As long as the total salt applied is the same, equal amount of salt would be added to the system below the root zone. In view of this, cyclic use strategy has also been termed as mixing strategy where mixing of the water occurs in the soil profile.

TOXIC IONS MANAGEMENT

There are some elements, for example, boron, lithium and fluorine, the presence of which in saline water has been recognised to be harmful to plant growth. The toxicity due to these ions is not limited to the tree crops and other woody-type perennial crops unlike sodium, chloride and bicarbonate ions, but covers almost all the annual crops, the degree of toxicity changing from one to another.

Although boron, lithium and fluorine have been observed to be associated with high salinity waters, the instances where these elements could be held responsible for the adverse effect on the crop growth, if any, are not known widely in natural field conditions. The reason is that boron, lithium and fluorine occur in saline waters in amounts less than 5, 5 and 15 mgL^{-1}, respectively and these concentrations are not known to limit the crop growth.

An amendment such as gypsum is very beneficial to counteract the toxicity due to boron, if any. The application of gypsum not only improves soil permeability and allows increased leaching but also lowers the availability of soluble boron. Gupta and Chandra (1972) explained that boron occurs in toxic concentration in the form of sodium metaborate. Sodium metaborate reacts with gypsum to form sodium sulphate and calcium metaborate. The solubility of calcium metaborate is very low, 0.4 per cent as compared to sodium metaborate the solubility of which is 26 to 36 per cent in temperature range of 20°C to 35°C. The large difference in the solubility renders boron precipitated. The toxicity due to presence of lithium and fluorine, if any, may also be counteracted with the application of gypsum.

Gupta (1979) reported the changes in the boron content of a saline-sodic soil affected by leaching and gypsum treatment in the field condition. After leaching with water amounting to 95 cm, initial high boron concentration (6.5 to10 mgL^{-1}) was largely reduced (3 to 4.4 mgL^{-1}). The boron content decreased further with the application of gypsum in the soil increasing from 25 to 100 per cent G.R. In the treatment where gypsum was applied at the rate of 100 per cent GR, the boron content was reduced to 0.8 and 1.3 mgL^{-1} in 10 to 15 and 15 to 30 cm soil depths respectively.

Alfalfa, wheat, barley, Lucerne, oats, cotton, sugar beet, sorghum and maize are some of the crops which have been reported to be tolerant to boron (5 to 10 mgL^{-1}). The oilseed, legumes, citrus and horticultural plants in general are sensitive to boron. The boron tolerance increases in the presence of soluble calcium, nitrogenous and phosphatic fertilisers and decreases with increase in salinity of the soil (Gupta, 1980).

Specific ion toxicity due to sodium chloride and bicarbonate ions has often been a subject of discussion while dealing with irrigation with saline and/or sodic waters. Since most of the annual crops, which have been dealt with here, are not affected at all due to specific sodium, chloride/bicarbonate ions, toxicity due to these ions will not be discussed. It is believed that most of the annual crops which are affected adversely under saline and/or sodic water, it is due to the overall concentration of salts than to any individual ion. The saline/sodic waters which occur in natural conditions in arid and semi-arid zones in India have not been observed so far to cause specific ion toxicity to any annual crop.

REFERENCES

Agarwal, M.C. and R.K. Malik 1984. Stabilization of rising brackish water table in arid and semi-arid irrigated areas. Trans. Indian Society Desert Technology 9: 89.

Agarwal, M.C. and S.S. Khanna 1983. Efficient Soil and Water Management in Haryana. CCSHAU, Hisar Bulletin, 118p.

Agarwal, M.C., Singh, R. and Verma, S.K. 1978. Performance of *bajra* and wheat in relation to water quality under sprinkler and surface irrigation. Proc. Symp. Arid Zone Res. & Develop., Jodhpur.

Anonymous 1984. Appraisal Document (Summary) on studies for the Use of Saline Water in the Command Areas of Irrigation Projects, Haryana, India, HSMITC, Chandigarh.

Anonymous 1998. Biennial Report 1996-98. AICRP on Mgt. of Salt Affected Soils and Use of Saline Water in Agriculture. CSSRI, Karnal.

Ayers, R.S. and Westcot, D.W. 1976. Water Quality for Agriculture. Irrigation and Drainage Paper, 29. FAO, Rome.

Ayers, R.S. and Westcot, D.W. 1985. Water Quality for Agriculture. Irrigation and Drainage Paper 29, Rev. 1, FAO, Rome, 174 p.

Bandyopadhyay, A.K. 1974. Seasonal variation of boron in saline-sodic soils. Ann. Arid Zone 13 : 125-8.

Bernstein L. and Ayers, R.S. 1955. Sloping seed beds. Calif. Agric. Nov. issue, 8 p.

Bernstein, L. and Fireman, M. 1957. Laboratory studies on salt distribution in furrow irrigated soil with special reference to the pre-emergence period. Soil Sci. 83: 249-63.

Bernstein, L., Fireman, M. and Reeve, R.C. 1955. Control of salinity in the Imperial Valley. USDA, ARS, Bull. 41 (4): 15.

Chaudhuri, I.I. and Wiebe, H.H. 1968. Influence of calcium pretreatment on wheat germination on saline media . Plant Soil 28: 208-216.

Dhir, R.P. 1977. Saline waters-their potentiality as a source of irrigation. In: Desertification and its Control. ICAR, New Delhi, 130-148.

Dhir, R.P. and Bhatia, O.P. 1975. Use of saline water in agriculture. I-Description of the system. Ann. Arid Zone 14: 206-211.

Dhir, R.P., Kolarkar, A.S. and Bhola, S.N. 1975. Use of saline water in agriculture. 2-Crop growth and response to fertiliser application under saline water use in cultivators fields. Ann. Arid. Zone 14 : 277-84.

Gupta, I.C. 1979. Note on the effect of leaching and gypsum on the detoxification of boron in a saline-sodic soil. Curr. Agric. 3: 231-34.

Gupta, I.C. 1980. Effect of irrigation with high-sodium waters on soil properties and the growth of wheat. Intern. Symp. Salt. Affected Soils, Karnal. 287-288.

Gupta, I.C. 1980. Soil salinity and boron toxicity. Curr. Agric., 4: 1-16.

Gupta, I.C. and Abichandani, C.T. 1970. Seasonal variations in salt composition of some saline water irrigated soils of western Rajasthan, I - Effect of rainfall. J. Indian Soc. Soil Sci. 18 : 528-535.

Gupta, I.C. and Abichandani, C.T. 1977. Use of saline waters for irrigation and their influence on soil characteristics. In: Natural Resources of Rajasthan. (Roonwal, Ed.), Univ. of Jodhpur, Jodhpur.

Gupta, I.C. and Chandra, H. 1972. Effect of gypsum in reducing boron hazard of saline irrigation waters and irrigated soils. Ann. Arid zone 11 : 228-30.

Gupta, I.C. and Gupta, S.K. 1987. Management of Saline Soils and Waters. Oxford and IBH Publishing Co., New Delhi, 339 p.

Gupta, I.C., Singh, P.M., Yadav, N.D. and Sharma, B.D. 1988. Device for succe-ssful establishment of tree plants with limited water boon. Invention Intelligence 23: 283-83.

Gupta, I.C., Yadav, N.D. and Sharma, B.D. 1989. 'Jalshuddhi' - a new device for growing nursery plants in desert. Invention Intelligence 24: 555-57.

Gupta, R.K. and Tyagi, N.K. 1984. Studies of irrigation techniques for use of saline water. Ann. Rep. CSSRI, Karnal, p. 114.

Gupta, S. K., Sharma, D. P., Tyagi, N. K. and Dubey, S. K. 2002. *Lavaniya-Kshariya Mirdaon ka Sudhar aur Parbandh.* CSSRI, Karnal. 153 p.

Gupta, S.K. 1985a. Dynamics of salts in saline water irrigated soils. J. Instt. Engrs. 66(AGRI): 17-22.

Gupta, S.K. 1985b. Leaching saline soils through rainfall. J. Ind. Soc. Soil Sci. 33: 128-136.

Gupta, S.K. 1992. Dhaincha is a better green manure and cover crop than sunhemp in saline soils. Indian Fmg. 42 (6) : 29-31.

Jain, T.C., Saxena, G.C., Jain, K.C. and Totawat, K.L. 1976. Effect of crop rotation, farmyard manure and quantity of irrigation water on the yield of wheat and physico-chemical properties of soil Indian J. Agric. Res. 10 : 1-9.

Jain, T.C., Saxena, G.S., Jain, K.C. and Totawat, K.L. 1976. Effect of crop rotation, farm-yard manure and the quantity of saline irrigation water on the yield of wheat and physico-chemical properties of the soil. Indian J. Agric. Res. 10: 1-9.

Kaushal, M.P., Khepar, S.D. and Panda, S.N. 1985. Saline ground water management and optimal cropping pattern. Water International 10: 86-91.

Kumar, A. and Sharma, B.R. 1999. Micro-irrigation scenario in India : an overview. Proc. Workshop on Exchange of Information and Expertise with Aral Sea Basin States (Unpublished). CSSRI, Karnal 133-140.

Lal, P. and Singh, K.S. 1972. A comparative study of the interactive effect of qualities of irrigation water and fertilizer levels on the yield of wheat grown on different soils. Z. Acher. Pflan. 136: 302-308.

Lal, P. and Singh, K.S. 1973. Effect of qualities of irrigation water and fertilizer on soil properties, yield and nutrient uptake by wheat. Indian J. Agric. Sci. 43 : 392-400.

Lal, P., Singh, M. and Lal, F. 1980. Response of wheat to micro-nutrient fertilization under different qualities of irrigation water. Curr. Agric. 4: 67-71.

Maliwal, G.L. and Paliwal, K.V. 1971a. Effect of manure and fertiliser on the growth and chemical composition of pearl millet (*P. typhoides*) irrigated with different qualities of water. Indian J. agric. Sci. 41: 136-142.

Mangel, K. and Kirkby, E.A. 1979. Principles of Plant Nutrition. International Potash Institute. Berne, Switzerland.

Mehta, P.C., Puntamkar, S.S. and Seth, S.P. 1975. Quality of tubewell water in Rajasthan and its utilization for crop production. Rajasthan J. Agric. Sci. 4 : 113-9.

Minhas, P.S. and Gupta, R.K. 1992. Quality of Irrigation Water - Assessment and Management. Indian Council of Agricultural Research. 123p.

Nalamwar, R.V. and Dastane, N.G. 1966. Cultural practices for crop production under soil salinity-cum-high water table conditions. Proc. Symp. Water Manangement. Udaipur (1963). 251-255.

Nath, J., Raj, M., Singh, A., Dev, S. and Singh, R. 1982. Performance of sorghum, sunflower and wheat as affected by salinity of irrigation water. Trans. ISDT 7: 59-61.

Nielsen, D.R., Biggar, J.W. and Luthin, J.N. 1966. Desalinization of soils under controlled unsaturated conditions. ICID, 6th Congr. New Delhi, India.

Paliwal, K.V. and Maliwal, G.L. 1971. Effect of fertilizers and manure on the growth and chemical composition of maize irrigated with different quality waters. Indian J. Agron. 16 : 316-21.

Paliwal, K.V. and Maliwal, G.L. 1971a. Cation exchange equilibria II. Na-Ca-Mg system of soils and clay minerals. Proc. Indian Nat. Sci. Acad. 37A: 114-121.

Panda, S.N. 2000. Optimal water and crop management in regions underlain with sodic ground water. In Salinity Management in Agriculture (Gupta, S.K., S.K. Sharma and N.K. Tyagi (Eds.). pp. 253-271.

Pandey, R.N. 1978. A prediction equation for leaching of uniformly incorporated salts to a known depth. Curr. Agric. 2 (1-2) : 7-12.

Panwar, K.S. 1978. Response of Sarson (*Brassica campestris* var. *Sarson*) varieties to nitrogen on saline soil with higher groundwater table. Curr. Agric. 2: 49-54.

Patel, A.S. and Dastane, N.G. 1969. Effect of layout, irrigation frequency and levels of irrigation salinity on radish. Indian J. Agric. Sci. 39 : 706-12.

Poonia, S.R., Jhorar, L.R., Nath, J., Khanna, S.S. and Singh, A. 1974. Salt balance in soil profiles as affected by different combinations of canal and saline water irrigation. J. Res. Hissar. 4: 226-70.

Puntamkar, S.S., Mehta, P.C. and Seth, S.P. 1971. Note on the inducement of salt resistance in two wheat varieties by pre-soaking with different salts of varying concentrations. Indian J. agric. Sci. 41: 717-718.

Puntamkar, S.S., Mehta, P.C. and Seth, S.P. 1972. Effect of gypsum and manure on the growth of wheat irrigated with bicarbonate rich water. J. Indian Soc. Soil Sci. 20 : 281-5.

Reeve, R.C., Allison, L.F. and Peterson, D.F. 1984. Reclamation of saline-alkali soils by leaching. Bull 355, Utah Agril. Expt. Stn. Logan, Utah.

Rhoades, J.D. 1983. Using saline water for irrigation. Proc. Intern. Workshop Salt Affected Soils Latin America, Maracay, Venezuela, pp. 233-64.

Rhoades, J.D., Kandiah, A. and Mashali, A.M. 1992. The Use of Saline Waters for Crop Production. Irrig. Drain Paper 48. FAO, Rome.

Sarma, D.A. 1980. Effect of using saline water to supplement canal water irrigation on the crop growth of rice. Curr. Agric. 4: 79-82.

Sarma, D.A., Reddy, D.R., Raju, K.V. and Prasad Rao, D.M.V. 1980. Effect of using saline water to supplement canal water irrigation on crop growth of groundnut. Curr. Agric. 4: 73-78.

Sarma, D.A., Venkataraju, K. and Ramayya, B. 1978. Performance of finger millet and changes in soil properties under saline water irrigation to supplement canal water. Andhra Agric. J. 25: 57-60.

Shainberg, I. and Oster, J.D. 1978. Quality of Irrigation Water. IIIC Publication No.2, International Irrigation Information Center, Israel.

Shalhevet, J. 1984. Management of irrigation with brackish water. In: Ecological Studies. Analysis and Synthesis. Vol. 18 Arid Zone Irrigation. B. Yaron *et al.* (ed.), Springer-Verlag, New York.

Shalhevet, J. and Kamburov, J. 1976. Irrigation and Salinity - A Worldwide Survey. ICID, New Delhi, p. 106.

Shanon, M.C. and Francois, L.E. 1977. Influence of seed pre-treatment on salt tolerance of cotton during germination. Agron. J. 69: 619-622.

Sharma, G. and Lal, P. 1975. Effect of nitrogen levels and leaching regimes on the use of saline waters for wheat grown on sandy and clay loam soils. J. Indian Soc. Soil Sci. 23 : 302-9.

Sharma, S.K. 2000. Ground water management. In: Conf. Proc. 3rd Water Asia 2000. Interads Ltd. New Delhi.

Singh, B. and Brar, J.S. 1995. Status of ground water quality and manageme-nt. Proc. Symp. held on Water Resources Day. PAU, Ludhiana, pp. 107-120.

Singh, H.P. 2001. Emerging scenario of microirrigation in India. In Micro irrigation (Ed. Singh, H.P. *et al.*), CBIP. 18-30

Strogonov, B.P. 1964. Physiological Basis of Salt Tolerance of Plants. IPST, Jerusalem. 279p.

Vander Molen, W.H. 1975. The Zuur-varhoevan method for calculating water movement in saline soils and its application to some Indian data. J. Res. Hisar 5 : 195-9.

Verma, K.S., Bhandari, G.S. and Singh, B.R. 1977. Response of maize variety Ganga, 101 to nitrogen under brackish water irrigation. J. Res. Hissar, 1 : 29-32.

Verma, S.K., Gupta, R.K. and Sharma, R.A. 1987. Hydraulic properties of a sodic clay soil as modified by quality of irrigation water. J. Indian Soc. Soil Sci. 35: 1-4.

■■■■

Chapter - 4

PLANTS IN RELATIONS TO SALINITY

INTRODUCTION

Salinity is a major factor causing reduction in plant growth and productivity throughout the world. Approximately 10 per cent of the world's 7000 m ha arable land surface consists of saline or sodic soils. Of the 1500 m ha cultivated lands, 23% are considered saline and another 37% are sodic (Tanji, 1990). Despite the advanced management technologies available today, salinization of millions of hectares of land continues to severely reduce crop production worldwide. Natural salinity is a widespread phenomenon on earth and the evolution of living organisms has resulted in numerous species that show special adaptive mechanisms of growth in saline environments. The majority of plants are relatively salt sensitive. In particular, almost all crop plants are unable to tolerate permanently saline conditions.

World agriculture is facing a lot of challenges like producing 70% more food for the growing population and the productivity of crops is not increasing in parallel with the food demand. The lower productivity in most of the cases is attributed to various abiotic stresses. Curtailing crop losses due to various environmental stressors is a major area of concern to cope with the increasing food requirements (Shanker and Venkateswarlu, 2011). The major abiotic stresses like high salinity, drought, cold and heat negatively influence the survival, biomass production and yield of staple food crops up to 70% (Vorasoot *et al.,* 2003; Mantri *et al.,* 2012; Ahmad *et al.,* 2012). The adverse effect of excess minerals such as Na^+ and/or Cl^- on plant is called salt stress (Munns, 2005). It has been shown that soil salinity subsisted long before humans and agriculture; however, the problem has been arisen by agricultural practices such as over irrigation. Salt stress is one of the most serious limiting factors for crop growth and production. Based on the nature, characteristics, and plant growth relationships in salt affected sols, two main types of soils have been coined by Szaboles (1974). These are (1) saline soils - the soluble salts are chiefly NaCl and Na_2SO_4 and sometimes also contain appreciable quantities of Cl^- and SO_4^- of Ca^{2+} and Mg^{2+}; these soils contain sufficient neutral soluble salts to pose negative effect on growth of most crop plants, and (2) sodic soils - these soils contain Na^+ salts capable of alkaline hydrolysis, mainly Na_2CO_3.

Sustained and profitable production of crops on salt-affected soils is possible if appropriate on farm management decisions are made. To be successful one needs an understanding of as to how plants respond to salinity, the relative tolerance of different crops and their sensitivity at different stages of growth and the manner in which different soil and environmental conditions affect salt stressed plants.

TYPES AND CAUSES OF SALINITY

Natural or Primary Salinity

Primary salinity results from the accumulation of salts over long periods of time through natural processes in the soil or groundwater. It is caused by two natural processes. The first is the weathering of parent materials containing soluble salts. Weathering processes break down rocks and release soluble salts of various types mainly chlorides of sodium, calcium and magnesium, and to a lesser extent, sulfates and carbonates. Sodium chloride is the most soluble salt. The second is the deposition of oceanic salt carried in wind and rain. "Cyclic salts" are ocean salts carried inland by wind and deposited by rainfall and are mainly sodium chloride.

Secondary or Human-induced Salinity

Secondary salinization results from human activities that change the hydrologic balance of the soil between water applied (irrigation or rainfall) and water used by crops transpiration (Garg and Manchanda, 2008). The most common causes are (a) land clearing and the replacement of perennial vegetation with annual crops and (b) irrigation schemes using salt-rich irrigation water or having insufficient drainage.

According to the FAO Land and Plant Nutrition Management Service, over 6% of the world's land is affected by salinity. Of the current 230 million hectares of irrigated land, 45 million hectares is salt-affected (19.5%), and of the 1,500 million hectares under dry land agriculture, 32 million are salt-affected to varying degrees (2.1%). Table 4.1 shows that the proportion of salt-affected irrigated land in various countries ranges from a minimum of 9% to a maximum of

34%, with a world average of 20%. Irrigated land is only 15% of total cultivated land, but as irrigated land has at least twice the productivity of rainfed land, it may produce one third of the world's food.

Irrigation water adds appreciable amounts of salt even with good quality irrigation water containing only 200-500 mg/kg of soluble salt. Irrigation water with a salt content of 500 mg/kg (i.e., 500 mg/l) contains 0.5 tons of salt per 1,000 m^3. Since crops require 6,000-10,000 m^3 of water per hectare each year, 1 ha of land will receive 3-5 tons of salt. Because the amount of salt removed by crops is negligible salt will accumulate in the root zone and must be leached by supplying more water than is required by the crops. If drainage is not adequate, the excess water causes the water table to rise, mobilizing salts which accumulate in the root zone. When the crop is unable to use all the applied water, water logging occurs.

Table 4.1. Global estimate of secondary salinization in the world's irrigated soils

Country	Total land area cropped	Area irrigated		Area of irrigated land that is salt affected	
	Mha	Mha	%	Mha	%
China	97	45	46	6.7	15
India	169	42	25	7.0	17
Soviet Union	233	21	9	3.7	18
United States	190	18	10	4.2	23
Pakistan	21	16	78	4.2	26
Iran	15	6	39	1.7	30
Thailand	20	4	20	0.4	10
Egypt	3	3	100	0.9	33
Australia	47	2	4	0.2	9
Argentina	36	2	5	0.6	34
South Africa	13	1	9	0.1	9
Subtotal	843	159	19	29.6	20
World	1,474	227	15	45.4	20

DIVERSITY OF SALT-TOLERANT SPECIES

Salt tolerance has evolved in many orders of vascular plants. In fact, Aromon (1990) lists more than 100 plant families that contain salt-tolerant species. Most of the families contain less than 10 salt-tolerant species, but Chenopodiaceae contains over 350. Based on the diversity of evolutionary lineages of salt-tolerant species, salt tolerance traits have most likely evolved many times over the history of vascular plants. It is thought that salinity tolerance probably evolved in several families of higher plants growing in esturies and spread to inland areas (O'Leary and Glenn, 1994). The result of that evolutionary process is that more than one-third of extant plant families have halophytic members, and about 2500 species of halophytes are found in the world's salt marshes and desert flats (Glenn *et al.,* 1997).

HALOPHYTES AND GLYCOPHYTES

Plants may be categorized as halophytes or glycophytes depending upon their response to salinity.

Although a clear distinction is often made between them, but there is in reality a continuous spectrum of tolerance ranging from the most sensitive species which are severely affected by as little as 1/10 th of sea water (50 mol m^{-3} NaCl) to those species able to complete their life cycle in full strength sea water (500 mol m^{-3} NaCl).

Halophytes

In general the term halophytes refers to salt-tolerant plant species. Although there is no generally accepted definition of this term, we will use it to classify all those species that grow on saline soils and have some tolerance of elevated salinity. Ungar (1991) defined halophytes as those species that tolerate relatively high soil salinities and are capable of accumulating relatively high quantities of Na^+ and Cl^- in their tissues (Table 4.2). Glycophytes, on the other hand, are those species that show little tolerance to elevated root-zone salinity, and do not accumulate high concentrations of ions in their growing tissues. The most tolerant halophytes such as *Salicornia europea* and *Sueda maritima* can tolerate salinities well above that of seawater (0.5 M NaCl), while glycophytes commonly are intolerant of salinities greater than 10% that of seawater. Obligate halophytes reach their maximum growth under moderate salinity and have reduced growth in low salinity conditions (Ungar, 1991). However, most halophytes reach their optimal growth in low salinity conditions, but can tolerate higher salinity than glycophytes can. There are also a number of facultative halophytes such as the ice plant (*Mesembryanthemum crystallinum*) that gain the properties of halophytic metabolism when exposed to a pretreatment of moderate salinity.

Table 4.2. Best supported definitions of salt tolerance function types

Functional type	Definition
Glycophyte	A plant whose growth and fitness decreases when exposed to any level of salinity greater than approximately 10 mM.
Intolerant glycophyte	Plants that cannot tolerate a soil salinity greater than approximately 10 mM.
Halophyte	Plants that tolerate higher salinity than glycophytes (upto approximately 50 mM) before a reduction in growth or fitness occurs.
Obligate halophyte	Plants that do not attain their highest fitness at low soil salinity (less than 10 mM). A moderately high soil salinity (50 mM) is required to attain maximum fitness.
Facultative halophyte	Plants that attain a halophytic nature only after they experience moderate salinity in the soils or as an aerosol.

The term halophytes literally means salt plant, but is used specifically for plants that can grow in presence of high concentrations of sodium salts. Greenway and Munns (1980) adopted the ecological definition of Jennings (1976) i.e., ``the native flora of saline soils" and have assumed the salt concentrations with osmotic potential of at least -0.33 MPa. Plants that cannot grow in the presence of high concentration of sodium salts are called glycophytes or non-halophytes.

Because of the wide range of resistance found among halophytes, they have been subdivided into the extreme euhalophytes and the moderate oligohalophytes. Many halophytes are able to grow perfectly normally in low or non-saline environment and are therefore called facultative halophytes. The growth of these plants is favoured by NaCl but Na can be replaced by K (e.g. *Aster tripolium, Artemisia maritima, Plantago maritima* and *Suaeda maritima*). However, halophytes such as *Salicornia herbacea* and *Atriplex vesicaria* are called obligate halophytes because they can not grow without NaCl and are favourably affected by NaCl and NaCl alone.

Chapman (1966) grouped the halophytes according to the salinity levels in which they usually grow well. These are :

(1) Miohalophytes: Plants growing in habitats of low salinity (below 0.5% NaCl).

(2) Euhalophytes: Plants growing in highly saline habitats. These are further subdivided into three types according to salinity range as below :

(a) Mesohalophytes : 0.5% to 1%.

(b) Mesoeuhalophytes : 0.5% and higher.

(c) Euuhalophytes : 1% NaCl and higher.

Waisel (1972) classified halophytes putting more emphasis on salt resistance mechanisms and on internal rather than on external salt relationships. The classification and saline habitat has been considered where the minimal NaCl content of the saturated soil extract is 100 mmol (EC about 10 dSm^{-1}, osmotic potential of about -0.4 MPa) (Table 4.3).

Table 4.3. Classification of plants of saline habitats

Euhalophytes					Pseudohalophytes
Salt-requiring halophytes		Salt-resisting halophytes			
Obligatory halophytes	Preferential halophytes	Salt-enduring halophytes (salt-tolerant)	Salt-excluding halophytes	Salt-evading halophytes	
Plants dependent upon salts for their survival, e.g. *Salicornia* sp. and various bacteria and algae	Plant whose growth and development improve in the presence of salts e.g. *Arthrocnemum* sp., *Aster* sp., *Nitaria* sp., *Salicornia* sp. *Suaeda* sp.	Plants enduring a high protoplasmatic salt content e.g. *Suaeda monoica*	Plant accumulating salts in special hairs e.g. *Atriplex* sp. Plants secreting salts from their shoot e.g. *Aeluropus* sp., *Limonium* sp., *Tamarix* sp. Plant retransporting salts from the shoot into the root, e.g. *Salicornia* sp.	Plant evading salt uptake e.g. *Rhizophora* sp. Plant evading salt transport into the leaves e.g. *Prosopis foxia*	Ephemers Niche plants

Physiology of Halophytes

Halophytes are plants that can survive and reproduce in environments with high salt concentration (200 mM NaCl) and these plants constitute about 1% of the world's flora. Halophytes can be classified as "natural" and plants that tolerate salt but do not normally live in saline conditions (Patane *et al.,* 2012). Other classifications of halophytes are based on the characteristics of naturally saline habitats or the chemical composition of the shoots (Physiotypes) or the ability to secrete ions (recreto-halophytes).

The effect of salinity on growth varies amongst halophytes, which can be observed in figure 4.1 for different species (Patane *et al.,* 2012). According to this figure, *Sueada maritima* was the species that could grow in the highest NaCl concentration, however, the best condition was around 40 mM NaCl. The most affected species was *Thellungiella halophila,* which showed stronger growth reduction from 200 mM NaCl. In addition, it can be highlighted that dicotyledonous species showed higher tolerance to NaCl than monocotyledonous species, except for *T. halophila.*

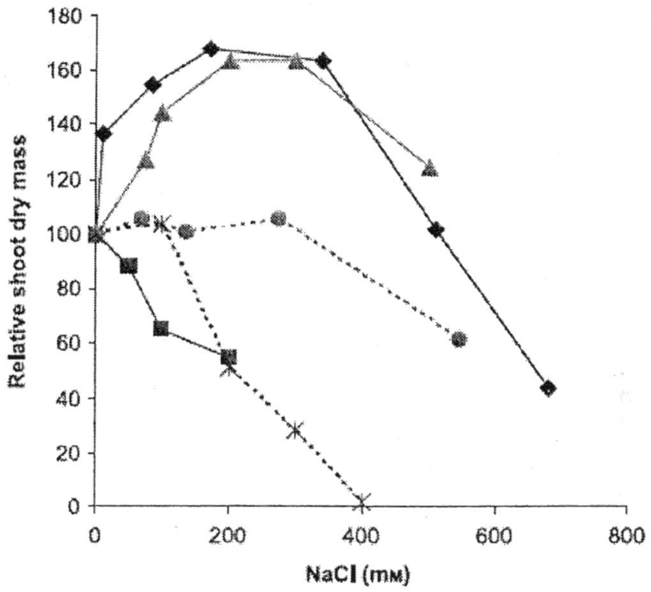

Fig. 4.1. Growth of halopytes of different species under salinity

Studies involving extreme dicotyledonous halophytes show the optimal growth in the presence of low or moderate salt concentrations; on the other hand, this stimulatory effect is not commonly found in monocotyledonous halophytes, nor in other salt-tolerant dicots, which grow best in the absence of salt (Flowers and Colmer, 2008). The salt tolerance degree usually varies among halophytes and, for a specific species, also at different developmental stages, but inhibition of plant growth is always observed at sufficienctly high salinity levels, in all investigated taxa (Breckle, 2002; Grigore *et al.,* 2012).

Salinity causes certain structural changes in halophytes. They have thicker leaves, smaller and fewer stomata, larger cells and presence of well developed water strong tissues. Many halophytes show highest degree of succulence (e.g. *Salicornia* and *Halocnemin*). The succulence has been reported to occur due to development of larger cells on the spongy mesophyll and presence of multilayer palisade tissue. On the other hand salt secreting halophytes which do not accumulate more salts are generally non-succulent.

The optimal NaCl concentration for the growth of halophytes in culture solutions ranges from 20 to 500 mM (Table 4.4).

Table 4.4. Optimal salinities (normally NaCl) for growth of a variety of halophytes (Flowers *et al.*, 1977).

Species	Optimal salinity (mM)	Relative growth (%)
Salicornia bigelowii	~ 170	–
S. europoea	136	560 (342)
	235-510	–
	340	1010 (616)
	255	938 (572)
Suaeda depressa	170	208 (Short days)
	170	112 (long days)
S. maritima	170-340	250
	400+	625
S. vulgaris	170	241
Atriplex halimus	120	1125
	100	161
	96	173
	24	158
	6	100
A. hastata	100	126
A. inflata	50	248 (246)
A. nummularia	200	216 (371)
	100	127
A. polycarpa	233	145
A. vesicaria	20	249
Avicennia marina	340	230
Aster tripolim	170	339
Mesembryanthemum crystallinum	(100) -200	256
Sarcobatus vermiculatus	50	115
Puccinella nuttaliana	128	157
Distichlis spicata	170	–
Diplanthera wightii	~680	–
Ruppia maritima	~60	332
Potamogeton pecifratus	~119	242
P. perfoliaus	~71	320

The values in parenthesis are for the relative weight of the organic material.

However, there are no rigorously determined optima since they vary with the age of the plant and with the environmental conditions (Flowers *et al.*, 1977). Research efforts during the last two decades have led to the identification of essential features of the cellular basis of salt tolerance in halophytes. For members of the Chenopodiaceae at least, the ability to withstand high external salt concentrations appears to depend on osmotic adjustment, most commonly with Na^+ and Cl^-. As these are toxic, their concentration in the cytoplasm is maintained substantially lower than in the cytoplasm is maintained substantially lower than in the vacuole. Osmotic adjustment in the cytoplasm is achieved with compatible organic solutes such as glycinebetaines and proline (Storey and Wyn Jones, 1977). It has been concluded by Flowers (1985) that transport and regulation of Na^+ and Cl^- ions is of primary importance in the physiology of salt tolerance of halophytes and in this regard limiting Na^+ and Cl^- concentrations in the xylem together with continued leaf expansion are particularly important.

Although halophytes are tolerant to salinity but it has been observed that at germination stage they are less tolerant as compared to later growth stages. Under natural conditions, seeds of halophytes may remain dormant till the onset of rains.

A number of studies have shown that germination even in the case of obligate halophytes takes place under conditions of reduced salinity (Chapman, 1966; Flowers *et al.*, 1977). However, the root growth of halophytes is not affected by the presence of salts. The roots of plants such as *Suaeda monica* and *Atriplex halimus* can reach upto 5 m depth. The roots of some of the halophytes (*Prosopis farcata* and *Alhagi maylocum*) have been reported to reach the depth upto 20 m.

Joshi *et al.* (1985) found that the root growth of alkali halophytes is not affected by the presence of high exchangeable sodium in the soil. The root growth of *Desmostachya bipinata* reaches to a depth of 1 m under sodic conditions.

Certain plant species can be used as indicators of degree of soil salinity. *Aster subulatus* in U.S.A. and as an introduced species in New Zealand is a recognized indicator of brackish conditions. Rana (1983) on the basis of plant population studies identified following plant species as plant indicators of saline-sodic soils of Karnal - Kurukshetra region.

(a) Saline soils : *Alhagi pseudoalhagi, Salsola baryosma, Suaeda fructicosa* and *Salvadora oleoides.*

(b) Sodic soils : *Sporobolus coromandelianus, S. marginatus, Diplache fusca, Chloris barbata, Kochia indica* and *Desmostachya bipinata.*

However, no detailed studies have yet been carried out to establish the existence of such other species and the range of salinity values which they can indicate. Recently, the value of halophytes towards their use as crop species has generated a great interest. Long term values based on organic dry weight for certain species such as *Salicornia virginica, Suaeda maritima* fall within the range normally reported for glycophytes. Therefore, halophytes seem to offer a viable alternative for future (Flowers, 1985).

Glycophytes

The glycophytes or non-halophytes to which most crop species belong, differ in response to salinity from very salt sensitive to moderately salt tolerant (Greenway and Munns, 1980). Almost all crop species are glycophytes. They have a selective advantage in non-saline soils over halophytes because, as a general rule, their rate of growth is faster. In view of the importance of crop plants, the influence of salt stress on them is discussed below in detail.

A few agricultural crops have moderate salt tolerance (Table 4.5) and there can be a wide variation in salt tolerance among varieties or genetic lines of one crop species. For example, tomato cultivars have a diverse salt tolerance and breeding programs have had some degree of success in promoting salt tolerance of tomato cultivars. Because of this and other successful breeding programs, there is extensive information on the salinity tolerance of crop varieties (Mass, 1987). Even though there is a broad variation in salt tolerance among crop species and cultivars, that tolerance is very limited compared with many native species from saline habitats. For example, wild varieties of tomato posses many traits that confer a degree of salt tolerance and that are not found in saline-sensitive cultivated varieties.

Table 4.5. The few crops with a relatively high tolerance of salinity

Common name	Species	Salinity at 50% reduction in yield		Tolerance rating
		dS m^{-1}	mM	
	Crops			
Barley	*Hordeum vulgare*	18	200	Tolerant
Cotton	*Gossypium hirsutum*	17		Tolerant
Sugar beet	*Beta vulgaris*	15		Moderately tolerant
Sorghum	*Sorghum bicolor*	14		Moderately tolerant
Safflower	*Carthamus tinctorius*	14		Moderately tolerant
Wheat	*Triticum aestivum*	13	140	Moderately tolerant
Beet, red	*Beta vulgaris*	9.9		Moderately sensitive
Cowpea	*Vinga unguiculata*	8.9		Moderately sensitive
Alfalfa	*Medicago sativa*	8.9		Moderately sensitive
Tomato	*Lycopersicon lycopersicum*	7.6		Moderately sensitive
Cabbage	*Brassica oleraceae capitata*	7.0		Moderately sensitive
Corn	*Zea mays*	5.9	60	Moderately sensitive
Lettuce	*Lactuca sativa*	5.2		Moderately sensitive

Common name	Species	Salinity at 50% reduction in yield		Tolerance rating
		dS m^{-1}	mM	
Onion	*Allium cepa*	4.3		Sensitive
Rice	*Oryza sativa*	3.6	40	Sensitive
Bean	*Phaseolus vulgaris*	3.6	40	Sensitive
	Native species			
Glasswort	*Salicornia biglovii*		>720	Tolerant
Salt bush	*Atriplex canascens*		700	Tolerant
Salt grass	*Distichlis palmeri*		600	Tolerant
Salt bush	*Atriplex canescaens* sp.*canascens*		350	Tolerant
Beetle grains	*Diplanche fusca*		313	Tolerant

CROP RESPONSES TO SALINITY

Salts in the soil water may inhibit plant growth for two reasons. Firstly, the presence of salt in the soil solution reduces the ability of the plant to take up water and this leads to reductions in the growth rate. This is referred to as the osmotic or water-deficit effect of salinity. Secondly, if excessive amounts of salt enter the plant in the transpiration stream, there will be injury to cells in the transpiring leaves and this may cause further reductions in growth. This is called the salt specific or ion-excess effect of salinity (Greenway and Munns, 1980). These salinity effects has threefold effects viz. it reduces water potential and causes ion imbalance or disturbances in ion homeostasis and toxicity; this altered water status leads to initial growth reduction and limitation of plant productivity. The detrimental effect is observed at the whole plant level as death of plants or decrease in productivity. Salt stress affects all the major processes such as germination, growth, photosynthetic pigments and photosynthesis, water relations, nutrient imbalance, oxidative stress, and yield. These are discussed under separate headings.

Although salinity affects plants in many ways physiologically, overt injurt symptoms seldom occur except under extreme salinization (Maas and Hoffman, 1977). Salt affected plants usually appear normal although they are stunted and many have dark green leaves which in some cases are thickner and more succulent.

The most common influence of salinization on crop plants is suppressed growth. Salinity usually varies over the field. Therefore, the main characteristic of crop on a salt affected land is the presence of a poor and patchy stand alongwith stunted plant growth. Under sodicity, sodium toxicity symptoms are found more clearly and yellowing of older leaves is followed by drying. Occasionally, nutritional imbalances caused by salinity, produce specific nutrient deficiency symptoms.

An increase in soil salinity commonly causes a reduction in growth and yield. The relationship between salinity and growth reduction is not uniform among ions; some are more toxic than others. It is also true that a combination of ions has a less detrimental effect on growth than does an equamolar concentration of one ionic species (Bernstein, 1975). Originally, it was thought that reduced growth was due to direct effects of ions on photosynthesis that led to reduced growth. However, the current theory is that the specific influence of salinity on photosynthesis alone is not larger enough to account for the magnitude of reduced growth. Munns (1993) explains that growth is reduced before photosynthesis decreases and growth decreases more than photosynthesis does. Allocation patterns and leaf turnover patterns in response to salinity are associated much more with reduced growth than with reduced photosynthesis.

The most common effect of salinity on plants is a general stunting of growth. The plants usually appear normal, although if compared with nonstressed plants, they may have darker green leaves that, in some cases, are thicker and more succulent. Visual symptoms, such as leaf burn, necrosis, and defoliation occur in some species, particularly woody crops, but these symptoms are rare in herbaceous crops unless plants are severely stressed. Consequently, it is difficult to diagnose a moderately salt-affected crop in the field without having a nonstressed crop nearby for comparison.

One immediate response of plants to elevated salinity is a decrease in the rate of leaf expansion. Consequently, whole-plant leaf area is reduced. In fact, the decrease in growth a during the initial buildup of salinity in soils can be attributed to a decrease in the leaf area ratio. Ball (1988) found that the diminution in growth of a mangrove species could be attributed entirely to a decrease in the leaf area ratio until very high salinities were reached. The common decrease in leaf expansion is not related to a loss in turgor pressure and is most likely a result of a change in hormonal signaling from roots to leaves. Even though salinity can induce a rapid reduction in root growth (Neumann, 1995), shoot growth decreases proportionally more than root growth, causing an increase in the root/shoot ratio. A general pattern in the plant response to salinity is that the larger the change in the root/shoot ratio, the greater the effect of salinity on productivity (Cheeseman, 1988).

Besides reducing plant growth and yield, salinity and sodicity affect germination, root growth, shoot growth and nodulation and N_2-

fixation in legumes also (Bernstein, 1975). Some of these aspects of salinity on plant growth processes are described below.

Germination

Salinity delays as well as decreases germination of most of the crops. Lower levels of salinity delay germination whereas higher levels, in addition, reduce the final percentage germination (Ayers and Hayward, 1948; Bernstein, 1975; Garg and Garg, 1981). Salinity effects are more severe at lower soil moisture content than at higher because soil suction and osmotic potential exert additive adverse effects. The imbibitional power of the seed is conditioned by the moisture content within the range of available soil moisture. Seed germination, therefore, would take place in shorter period of time at high soil moisture. It has been observed that at soil salinity greater than 12 dSm^{-1} (ECe), the germination of wheat was delayed by 20-25 days and decreased to about 35 per cent of the normal. Many other crops exhibited almost similar trends. Van Hoorn (1991) found that increasing salt concentration delayed germination but after 10 days a high germination percentage was attained for salt tolerant crops like safflower, sorghum, sunflower and wheat (Table 4.6). However, during early seedling growth these crops appeared to be less tolerant than germination and later growth.

Seed germination is one of the most fundamental and vital phases in the growth cycle of a plant that determines the growth and yield. However, it has been established that salinity adversely affects the process of germination in various plants like *Posidonia* (Fernandez-Torquemada and Sanchez-Lizaso, 2013), *Oryza sativa* (Xu *et al.*, 2011), *Triticum aestivum* (Akbarimoghaddam *et al.*, 2011), *Zea mays* (Khodarahmpour *et al.*, 2012), and *Brassica* spp. (Ulfat *et al.*, 2007). Salinity affects the germination process many-folds. It alters the imbibition of water by seeds due to lower osmotic potential of germination media (Khan and Weber, 2008), causes toxicity which changes the activities of enzymes of nucleic acid metabolism, alters protein metabolism (Dantas *et al.*, 2007), disturbs hormonal balance, and reduces the utilization of seed reserves (Othman *et al.*, 2006). The germination rates and percentage of germinated seeds at a particular time vary considerably among species and cultivars. Lauchli and Grattan (2007) proposed a generalized relationship between percent germination and time after adding water at different salt levels (Fig. 4.2). Kaveh *et al.* (2011) found a significant negative correlation between salinity and the rate and percentage of germination which resulted in delayed germination and reduced germination percentage in *Solanum lycopersicum.* Germination rate also decreased on increasing concentration of salinity levels. Compared with control, germination percentage, and germination speed were decreased by 38 and 33 per cent, respecitvely, at 200 mM NaCl. In a recent study, Khodarahmpour *et al.* (2012) observed drastic reduction in germination rate (32%), length of radicle (80%) and plumule (78%), seedling length (78%), and seed vigor (95%) in *Z. mays* seeds exposed to 240 mM NaCl.

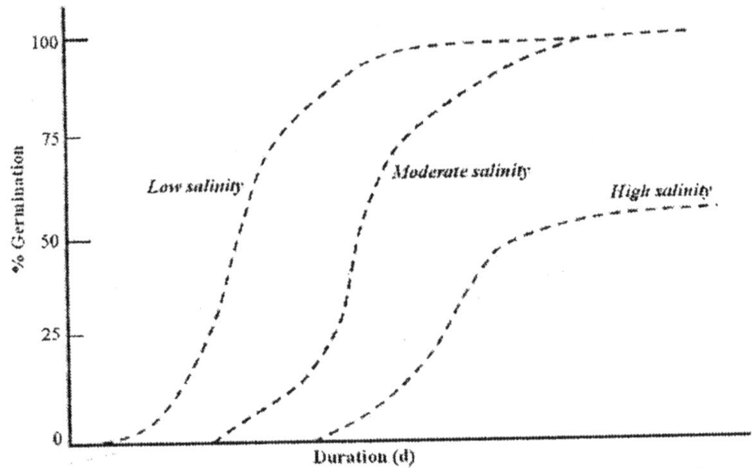

Fig. 4.2. General relationship between time and per cent germination in plants at low, medium and high salinities

Earlier investigators studied germination so as to determine the crop resistance to salinity. But subsequent studies have shown that it would be a futile attempt to correlate the performance at germination with the tolerance to produce grain yield, because the responses of crop species differed at germination and subsequent growth stages. It is now well known that tolerance at germination or emergence may bear no relationship to growth and yield of mature crops (Maas and Hoffman, 1977; Rogers and Noble, 1991; Francois and Maas, 1994). Sensitive crops like maize, pea and beans are able to germinate at higher salt concentration than sugarbeet, which is highly tolerant crop. Again, there is also differential response to percentage of germination with respect to the varieties of the same crop.

The reduction in germination under saline conditions could be attributed to increased osmotic potential of the soil solution which decreases the absorption rate leading to water stress in the seed. Poor and delayed germination could also be the influence of ions in quantities large enough to make them toxic to the seed embryo.

Table 4.6. Influence of increasing salt concentration on development of germination percentage after sowing in loamy sand soil (Van Hoorn, 1991).

Crop	Time (days)	Initial chloride concentration of soil water (mM)			
		0	30	60	90
Sorghum	4	59	38	32	20
	7	100	98	97	87
	10	100	99	99	93
Sunflower	4	84	89	70	58
	7	93	93	87	80
	10	76	93	87	81
Safflower	4	76	42	33	26
	7	96	95	94	95
	10	98	96	96	97
Wheat	4	92	68	54	28
	7	100	99	100	97
	10	100	100	100	98

Significant varietal differences in the salt tolerance of many crops at germination stage have been demonstrated by a number of investigators (Wahhab *et al.,* 1957; Paliwal and Maliwal, 1970). In certain cases tolerant varieties showed 75 to 90% germination at high salinity while sensitive varieties germinated only from 15 to 60%. However, as mentioned earlier, germination alone may not constitute a reliable criterion of salt tolerance of a crop. But, the knowledge of salt tolerance at germination and early seedling growth is of primary importance as the initial stand determines the final made to identify such varieties which give high yield rather than those which show only a high germination percentage.

Under sodic conditions, germination is not as adversely affected as in case of salinity. In sodic soils the emergence of plumule is restricted because of the poor physical properties of the soil. Certain crops like wheat, maize and some of the legumes are not affected by sodicity if soil is at proper moisture condition. In the absence of proper soil moisture, if these fields are irrigated prior to emergence, a hard crust restricts the emergence in case of germinating seeds. However, the excess of sodicity depresses germination and emergence of alfalfa more than the yield. Paliwal (1972) reported that the emergence time was delayed and percentage germination decreased as the degree of salinity and SAR or both were increased. The effect of SAR was less pronounced than that of salinity.

In alkali soils, emergence of plumule is rendered difficult because of poor physical properties particularly in the absence of proper moisture conditions. If these soils are irrigated prior to emergence, a hard crust is formed, which restricts the emergence of germinating seeds. Poor physical conditions also adversely affect the moisture availability and aeration, which influence the overall plant growth. Some of the attributes of vegetative growth that get reduced are plant height, leaf number, leaf area, tiller number and root and shoot dry weight.

Bicarbonate and carbonate are more harmful to germination than sulphate and chloride ions (Rao *et al.,* 1969; Somani, 1991). Garg and Garg (1981) showed that bicarbonate ion was more harmful than carbonate ion for the germination of maize, pea and mungbean at equal salt concentration (Table 4.7). Therefore, specific ion effects are also important besides the osmotic effects for the germination of seeds.

Table 4.7. Effect of Na_2CO_3 and $NaHCO_3$ on germination and dry weight of embryo axis in *P. sativum, P. aureus* and *Z. mays* at different hours (Garg and Garg, 1981).

Period (hrs.)	Percent germination															
	*C dS m^{-1}	Na_2CO_3 2	4	6	$NaHCO_3$ 2	4	6	CD 5%	C	Na_2CO_3 2	4	6	$NaHCO_3$ 2	4	6	CD 5%
P. sativum																
48	85	78	76	47	50	46	28	4.2	25	24	19	15	20	17	15	3.6
72	95	85	75	54	74	62	38	3.8	61	60	44	23	40	31	12	6.8
96	97	89	77	68	78	70	53	5.1	74	81	70	68	45	35	31	6.5
P. aureus																
24	100	100	86	37	100	74	27	5.4	28	27	20	15	26	19	13	3.0
48	100	100	95	56	100	85	43	5.0	78	76	55	16	70	46	+*	6.0
72	100	100	95	75	100	95	70	5.5	277	157	95	+*	151	47	+*	13.0
Z. mays																
48	77	75	66	61	80	72	42	78	54	60	40	39	54	37	33	5.0
72	95	91	84	80	95	87	70	6.5	280	288	256	211	281	252	140	8.0
96	98	95	91	90	98	90	70	7.0	–	–	–	–	–	–	–	–

(+)* - Seedlings did not survive. *C = Control

Salinity and Vegetative Growth

One of the initial effects of salt stress is the reduction of growth rate. Salts in soil water inhibits plant growth for two reasons. First, it reduces the plant's ability to take up water and this leads to slower growth. This is the osmotic or water deficit effect of salinity. Second, it may enter the transpiration stream and eventually injure cells in the transpiring leaves, further reducing growth. This is the salt-specific or ion-excess effect of salinity. The two effects give rise to a two-phase growth response to salinity given by Munns (1993, 2005).

Phase 1: The first phase of the growth response results from the effect of salt outside the plant. The salt in the soil solution reduces leaf growth and to a lesser extent root growth (Munns, 1993). The cellular and metabolic processes involved are in common to drought-affected plants. Neither Na^+ and Cl^- builds up in growing tissue at concentrations that inhibit growth: meristematic tissues are fed largely in the phlem from which salt is effectively excluded, and rapidly elongating cells can accommodate the salt that arrives in the xylem within their expanding vacuoles.

Phase 2: The second phase of the growth response results from the toxic effect of salt inside the plant. The salt taken up by the plant concentrates in old leaves: continued transport into transpiring leaves over a long period eventually results in very high Na^+ and Cl^- concentrations, and the leaves die. The cause of injury is probably the salt load exceeding the ability of cells to compartmentalize salts in the vacuole. Salts would then build up rapidly in the cytoplasm and inhibit enzyme activity. Alternatively, they might build up in the cell walls and dehydrate the cell. The excessive salt concentration correspondingly increases the osmotic potential of the soil that restricts the water uptake by plants. The Na^+ and Cl^- ions are the major ions that produce many physiological disorders and detrimental effects on plants. However, Na^+ is the primary ion as it interferes with the uptake of potassium (K^+) ion and disturbs stomatal regulation that ultimately causes water loss while the Cl^- ion disturbs the chlorophyll production and causes chlorotic toxicity. But, Cl^- is more dangerous than Na^+. Moreover, the need of Cl^- in plant is essential as well and is required for the regulation of turgor pressure and pH and enzyme activities in the cytoplasm. Dang *et al.* (2008), on the basis of number of field trials, concluded that Cl^- concentration in the soil was more important to growth and yield reduction than Na^+ and the critical level (defined as the concentration that reduces the growth or yield by 10%) of subsoil Cl^- concentration was estimated to be 490 mg Cl^- kg^{-1} soil. The Cl^- concentration in the youngest mature leaf of bread wheat, durum wheat, and chickpea showed greater variability with increasing levels of subsoil constraints than Na^+ concentration (Dang *et al.,* 2006). However, it is toxic to plants at high concentrations with critical levels for toxicity reported to be 4-7 mg g^{-1} for Cl^- sensitive species and 15-50 mg g^{-1} for Cl^- -tolerant species (White and Broadley, 2001).

The most common and conspicuous effect of salinity is growth retardation. As salt concentration increases beyond a threshold level, both the growth rate and ultimate size of most plant species progressively decrease. Not all plant parts are affected equally, however, any correlation between growth response and salinity must take this fact into account. Top growth is often suppressed more than the root growth (Meiri and Poljakoff-Mayber, 1970, Maas and Hoffman, 1977). Salinity also increases the leaf stem ratio of several crops and this affects quality in the case of forage crops.

In plants where the harvestable yield is composed of vegetable parts such as the forage crops, or in plants such as maize where yield is strongly linked with vegetative dry matter production, yields are generally reduced in production to the decrease in plant size (Bernstein, 1964). Stunted growth, restricted lateral shoot development, reduction ins size of leaves, fruit and grain, decrease in fresh and dry weight of different plant parts, restricted leaf number and area and finally diminished yield are some of the features of salinization. In case of wheat, barley and pearl millet tillering is drastically reduced. Under conditions of severe salinity these tillers may die before they are able to grow and bear ears and only main shoot produces grains. The reduction in leaf area by increasing salinity is a common feature which affects plant productivity by reducing the rates of total photosynthesis by the crop canopy (McCree, 1986). The salinity induced decrease in leaf area is particularly severe in salt sensitive crops like most of the legumes (Garg and Lahiri, 1986).

There is abundant literature indicating that plants are particularly susceptible to salinity during the seedling and early vegetative growth stage. One of the studies in rice showed a remarkable reduction in plant height, tiller number and leaf area index in plants grown in saline soil (Hasanuzzaman *et al.,* 2009). In *Suaeda salsa,* plant height, number of branches, length of branches, and diameter of shoot were significantly affected by salt stress which was due to the increased content of Na^+ and Cl^-. In *Glycine max,* Dolatabadian *et al.* (2011) observed that salinity stress significantly decreased shoot and root weight, total biomass, plant height, and leaf number. In one of the studies on *Foeniculum vulgare,* it has been shown that yields and plant growth parameters including plant height, fresh weight, yield and biomass were affected significantly by irrigation water salinities (Semiz *et al.,* 2012). However, there are many mechanisms that plants employ to combat the salt stress, retain homoeostasis, and overcome ion toxicity (Parida *et al.,* 2005). Some of these mechanisms include restriction of the mechanisms involved in salt uptake, control of long distance transport of salt, compartmentalization of salt, extrusion of salt from the plant, and prioritization of the maintenance of K/Na^+ ratio in the cytosol.

In several crops such as barley, wheat, cotton, cowpea and certain tolerant grasses, vegetative growth is much more affected by salinity than the seed or fibre production (Maas and Hoffman, 1977; West and Francois, 1982). On the other hand, grain yields of rice and corn are generally reduced without appreciably affecting the dry matter production (Pearson, 1959). Therefore, too often vegetative growth response to salinity is not a reliable guide for predicting fruiting or seed production. However, for root crops, storage root yields may be decreased much more than that of the fibrous roots (Hoffman and Rawlins, 1971).

Excess salinity affects crop growth in three ways. First and most important, as the amount of salts increases, the water in the soil becomes less available to the plants even though the soil may appear quite moist. This is because the osmotic pressure of the soil solution increases with the increase in salt concentration and the plants are unable to extract water as readily as they can from a relatively non-saline soil. In addition to the osmotic effect of salts in the soil solution, at high concentration, absorption of an individual ion may prove toxic to the plants. For example, some constituents of salts e.g. chloride, which at moderate concentration are harmless to most crops, may be toxic at higher concentrations to some crops, e.g. citrus and grapes. In addition, some elements like lithium, boron etc. are toxic to citrus and grapes even though they may be present only in very less quantities. Besides preferential absorption of one ion may also retard the absorption of other essential plant nutrients necessary for the normal growth of plants. It is believed that the adverse effects are usually due to cumulative effects of these factors although one may be dominating others in many cases.

Several investigators have reported varietal differences to salinity at the vegetative stage (Bernstein, 1975; Maas and Hoffman, 1977; Francois and Maas, 1994). In citrus significant differences among rootstocks for salt tolerance have been reported in a number of studies

(Bernstein, 1969; Shalhevet *et al.,* 1974). Increasing concentrations of NaCl caused dwarfing and reduced leaf number and fresh weight in radish, turnip, and carrot (Osawa, 1961). However, in leafy vegetables, like cabbage, spinach, chinese cabbage, celery, onion, lettuce and Japanese hornwart results indicated that NaCl at 1000 and 2000 mg l^{-1} improved plant growth and fresh weight of tips of cabbage and celery but other crops were dwarfed. Similarly, moderate levels of salinity and sodicity have been found to increase the vegetative growth of finger millet.

Root Growth and Salinity

There have been comparatively few investigations on the effect of salinity on root growth and function because root growth is usually less affected by salinity than shoot growth. However, salinity may induce morphological and anatomical changes to the roots in a range of species (Poljakoff-Mayber and Lerner, 1994). Solomon *et al.* (1986) reported thickening and curving of the roots and a reduction in the cross-sectional diameter of the roots in pea seedlings as well as of vascular cylinder because of the suppression of meristematic activity. Root elongation can be rapidly inhibited by salinity depending upon the type of salinity and concentration of salt (Cramer *et al.,* 1988). In many of the plants studied the effect of salinity on root growth is a function of salt concentration. Some concentrations can stimulate root growth while inhibiting shoot growth. For instance in pea plants, 4-5 days after imbibition the daily increment in root length, in presence of 120 mM NaCl, was higher than in roots of the control plants, but decreased after that period (Hasson and Poljakoff-Mayber, 1981). Higher NaCl concentration was inhibitory throughout. However, in shoots, no stimulatiion of growth by salinity was observed. Cramer and Bowman (1991) also found maize roots to be less sensitive to salinity than shoots. At levels that inhibited shoot growth, root growth was unaffected.

Roots are directly in contact with the salinity and are potentially the first line of defence. Several studies (Cramer *et al.,* 1990; Zidan *et al.,* 1982) have focussed on the mineral nutrition and root growth of salinized plants. In all cases, salinity inhibited long term root growth and disturbed the mineral nutrition of the plants. In general, macro-nutrient concentrations were increased. The growth zone of the roots completely adjusted osmotically to the saline conditions. Supplemental calcium to the root medium increased the root growth of salt-stressed maize plants (Evalgon *et al.,* 1990). Salinity can reduce water transport by inhibiting root hydraulic conductivity and the effect of calcium on root growth has been attributed to its ameliorative effects on root hydraulic conductivity.

It is often observed that response to salinity in the shoot is observed earlier than in the root. This occurs despite the fact that root is the organ exposed to the soil salinity. It is interesting that most sensitive processes such as, stomatal conductance and leaf growth show a high correlation with the salinity of the root substrate but not with the leaf salt content. Therefore, the root seems to be a mediator capable of monitoring the changing conditions in the soil and transmitting to the shoot (Gowing *et al.,* 1990). Studies indicate that root senses the soil environment by the effect of environmental factors on the level of phytohormones reaching the xylem and thus affects shoot growth accordingly. In recent years, the idea has developed that, independently of osmotic and ionic influences, chemical information advances from the root, which controls the physiology of the shoot (Chapin, 1990). Since roots produce hormones, such as auxin, gibberellin, cytokinin and abscisic acid; much of the evidence suggests that ABA may be more important than the other hormones in root shoot communications.

Flowering, Grain Setting and Yield under Salinity

Because salinity retards vegetative growth, it may be expected to delay flowering. However, very limited information is available on flowering in relation to salinity. A delay in flower bud formation of tomato and in anthesis of flax has been reported (Bernstein and Hayward, 1958). Osawa (1957) found that the rate of development of perfect flowers in egg-plant decreased with increasing salinity although the fruit setting was not affected. Flowering in peas and beans has been known to be delayed by NaCl at 0.2 per cent or higher concentrations. Pods did not develop in bean plants at 0.3% NaCl and did not mature at 0.2% NaCl. Pea pods did not develop at either of these salt concentrations and maturity was delayed in both the crops by NaCl at and above 0.15%. Salinization retarded flower bud formation in cotton while the opening of the flower was accelerated slightly which shortened the period between flowering and ripening.

Salinity leads to dwarfing of plants, delay in flowering and reduction in number and size of floral buds. Mild salinity stress at the sowing time or postponement of salinity treatments to the stage of florogenesis stimulated flower production. Salinity though did not affect microsporogenesis, reduced fecundity of pollen by affecting numerical production, viability, germination and tube growth. The extent of reduction varied with the genotype, type of salinity and stage of plant growth at which salinity stress is imposed. The decline in male fitness characters is a physiological phenomenon, resulting from metabolic shifts due to accumulation of larger quantities of non-nutrient ions like Na^+, Cl^- and SO_4^{-2}. These deleterious effects of salinity on male fitness characters can be partly mitigated by substances like $CaCl_2$, proline, putrescine and growth regulators. In contrast to water stress, salinity and growth regulators did not affect ovule number and their receptivity, but adversely affected number of fruits and seeds per plant. It also reduced the girth, number of secretory cell layers and development of vascular tissue in the style, which may lead to starvation of stylar tissue and thus growing pollen tubes. An increasing, proportion of stylar ovules fail to set seed under saline conditions, though these are positioned favourable with better chances of male gametes reception in comparison with their pedicular counterparts. Although under saline conditions many of the pollen tubes are lost in the upper half or two-third of style, the number reaching in the ovarian cavity are seemingly sufficient to fertilize a sizeable proportion of the stylar ovules. The embryo, however, in these ovules aborts during early stages of embryogenesis.

Foliar application of BAP and NAA induced the stylar ovules to set seeds, more so when higher salinities are applied at sowing time instead of prior to florogenesis. However, the resulting seeds are of smaller size. The plants receiving saline irrigation prior to florogenesis are visibly healthy and study, but unable to support the development of good number of pods with healthy seeds. Since resource limitation does not seem to be solely responsible for reduced yield, there is every likelihood that substantial number of fertilized ovules possibly fail to act as a strong sink due to their inability to incite adequate level of hormones. Poor vascular supply of the pistil and ovule may be the other intriguing factor.

In cereals salinity increased the spikelet sterility but decreased the number of bearing tillers per plant, 1000 grain weight and number of grains per panicle (Murthy and Rao, 1968). Similar results were obtained with pearl millet under increasing salinity (Garg and Lahiri,1986). Osawa (1960) found a progressive decrease in the number of flowers of beans, tomato and cucumber with increasing NaCl concentrations. Singh and Mengel (1970) also observed that both chloride and sulphate decreased number of flowers in okra and delayed the initiation of flowering. In general salinity retards flower bud initiation, anthesis of flowers and rate of setting.

The above mentioned effects of salt stress on plants ultimately lead to reduction of yield of crops which is the most important effect of salt stress in agriculture. Different yield components of *Vigna radiata* were significantly affected by salinity stress as reported by Nahar and Hasanuzzaman (2009). Number of pods per plant, seeds per pod, and seed weight were negatively correlated with salinity levels. The reproductive growth of *V. radiata* was also affected by salinity as the number of pods per plant substantially decreased with increasing salinity levels. An application of 250 mM NaCl reduced 77, 73, and 66% yield in *V. radiata* cv. BARI mung-2, BARI mung-5, and BARI mung-6, respectively over control (Nahar and Hasanuzzaman, 2009). The reduction of yield and its components under salt stress condition may be attributed to low production, expansion, senescence, and physiologically less active green foliage, and thus, reduced photosynthetic rate might be a supplementary effect. In *O. sativa* varieties, grain yield, which is the ultimate product of yield components, is greatly influenced by salinity levels. The loss of grain yield due to 150 mM salinity was 50, 38, 44 and 36% over control for the cultivars BR11, BRR1 dhan41, BRRI dhan44, and BRRI dhan 46, respectively. The severe inhibitory effects of salts on fertility may be due to differential competition in carbohydrate supply between vegetative growth and constrained supply of these to the developing panicles. Also reduced viability of pollen under stress condition could result in failure of seed set (Abdullah *et al.,* 2001). Grain yield reduction of rice varieties due to salt stress was also reported earlier by Gain *et al.* (2004). As reported by Greenway and Munns (1980), after some time in 200 mM NaCl, a salt-tolerant species such as sugar beet might have a reduction of only 20% in dry weight, a moderately tolerant species such as cotton might have a 60% reduction, and a sensitive species such as soybean might be dead. On the other hand, a halophyte such as *Suaeda maritima* might be growing at its optimum rate (Flowers *et al.,* 1986). In one of the recent studies of *F. vulgare,* it has been shown that yields and plant growth parameters including plant height, fresh weight yield, and biomass were affected significantly by increasing irrigation water salinities (Semiz *et al.,* 2012).

Literature abounds with reports on salinity mediated yield reductions in crop plants. Such a reduction in yield may be cumulated effect of various factors like decline in number of flowers, pod setting (Bishnoi *et al.,* 1990; Sharma, 1992), the number of ovules fertilized and nurtured into healthy seeds and thus number of seeds per pod and seed weight (Kumar *et al.,* 1980). Increased production of flowers alone does not help in achieving higher yield, both in terms of number of fruits and number of seeds per fruit. This is borne out from the fact that saline irrigation prior to florogenesis in chickpea though incited production of higher number of flowers, but is unaccompanied by yield improvement.

The checking of vegetative growth by salinity may, however, hasten maturation and final harvest dates especially of crops having indeterminate growth habits (Bernstein and Hayward, 1958; Bernstein, 1975). A general and progressive decrease in grain or seed yield of most of the crops with increasing salinity is well documented (Maas and Hoffman, 1977; Maas, 1992; Francois and Maas, 1994). However, the magnitude of yield reduction changes on the basis of crop species and their varieties. Table 4.8 lists the 0, 10 and 50% yield reduction in important agricultural crops at the salinity levels which shows a wide variation in crop response to salinity.

Table 4.8. The ECe (dSm^{-1}) at which 0, 10 and 50% yield reduction can be expected for various agricultural crops

Crops	Per cent yield reduction		
	0	10	20
Field crops			
Barley	8.0	10.0	18.0
Cotton	7.7	9.6	17.0
Sugarbeet	7.0	8.7	15.0
Wheat	6.0	7.4	13.0
Soybean	5.0	5.5	7.5
Sorghum	4.0	5.1	11.0
Groundnut	3.2	3.5	4.9
Rice (Paddy)	3.0	3.8	7.2
Corn	1.7	2.5	5.9
Flax	1.7	2.5	5.9
Cowpea	1.3	2.0	4.9
Beans	1.0	1.5	3.6
Vegetable crops			
Beets	4.0	5.1	9.6
Tomato	2.5	3.5	7.6
Spinach	2.0	3.3	8.6
Cabbage	1.8	2.8	7.0
Pepper	1.5	2.2	5.1
Lettuce	1.3	2.1	5.2
Radish	1.2	2.0	5.0
Onion	1.2	1.8	4.3
Carrot	1.0	1.7	4.6

Crops	Per cent yield reduction		
	0	10	20
Fruit crops			
Datepalm	4.0	6.8	17.9
Fig	2.7	3.8	2.4
Orange	1.7	2.3	4.8
Lemon	1.7	2.3	4.8
Grape	1.5	2.5	6.7
Plum	1.5	2.1	4.3
Avocado	1.3	1.8	3.7
Strawberry	1.0	1.3	2.5

Salinity may thus exert a variety of effects on plant development and yield depending upon such factors as the nature of the crop (vegetative organs or fruit and seed), the determinancy of growth, the differential tolerance to salinity at various growth stages, varietal behaviour and other interacting factors.

Salinity and Quality of Produce

Although yield is primary consideration but it is equally important that quality of the produce obtained under saline conditions is not adversely affected. However, very limited information is available on quality aspects of crops grown under such conditions.

Cereal crops are generally less adversely affected as compared to sugar crops, vegetable crops and oil seed crops. Although some citrus and other horticultural crops may apparently grow well under saline conditions but either fruit size and /or quality is impaired.

Salinity decreased the percentage of oil but increased the iodine value of rape and linseed (Deo and Ruhal, 1971) and sunflower (Raju and Ranganayakulu, 1978). In case of sunflower salinity decreased the oil content but the quality of the oil remained unaffected. The oil extractability became poor although fatty acid composition was not changed.

NaCl increased reducing sugars in okra pods while Na_2SO_4 had no effect (Singh and Mengel, 1970). However, Na_2SO_4 affected pods contained more ascorbic acid than NaCl treatment. Kim (1958) found that NaCl decreased ascorbic acid in radish and cabbage.

Mizrahi (1982) found that tomato fruits produced under saline conditions were smaller but developed a better colour due to higher pigment concentration in the pericarp had a much better taste.

However, their shelf life tended to be shorter, due to higher polygalacturonase activity. A chemical analysis of the fruit demonstrated higher values for TSS, reducing sugars and total acidity. Because of their better taste, the fruits from the salt treated plants could compensate for the lower yields.

In another study Mizrahi and Pasternak (1985) reported that fruits from a processing tomato cultivar exposed to various degrees of salinity had higher values for TSS and acidity (Table 4.9). The yield of fruit after saline water irrigation was lower, but this was offset by the higher fruit quality. Melon fruits subjected to saline water irrigation scored higher in taste when the fruits were analysed fresh. Iceberg lettuce grown with saline water did not significantly differ in taste from its control. The same was true for peanuts.

Table 4.9. Effect of salinity on quality parameters of juice of processing tomatoes cv. M-82-18 (Mizrahi and Pasternak, 1985).

EC of irrigation water (dSm^{-1})	Quality parameters			ECe (dSm^{-1})
	TSS (mg^{-1})	pH	Acidity (% Citric acid)	
1.2	4.76±0.15	4.26±0.02	0.36±0.01	6.3±0.2
4.5	5.11±0.12	4.31±0.03	0.37±0.01	7.4±0.3
7.5	5.76±0.19	4.30±0.02	0.40±0.02	7.6±0.3
10.0	6.10±0.14	4.16±0.02	0.53±0.02	8.9±0.6

Higher chloride levels are known to impair the quality of flue-cured virginia tobacco causing leaf burn, reduction in reducing sugars and nicotine content (Gopalchari *et al.,* 1968).

Garg and Gupta (1987) reported salinity mediated deterioration in the quality of fruits in terms of their protein and carbohydrate contents. Salinity mediated changes to protein and carbohydrate metabolism during seed development have also been reported in mungbean, chickpea and mustard (Dhingra and Varghese, 1997). Reduction in starch content with salt stress was probably due to non-availability of soluble sugars and reduced protein content due to decreased protein synthesis and increased hydrolysis of protein. Salinity induced decline in calorific food value of seeds in legumes was overcome by the application of growth regulators like BAP and NAA.

Effect of Salinity at Different Stages of Development

Information about the salt tolerance of crops at different stages of growth is rather limited. Most of the salt tolerance data have been obtained from studies in which salinity was relatively constant from seeding to harvest or from late seedling stage to maturity. These studies provided no information about the salt sensitivity or tolerance at individual stages of growth.

The information on the salt sensitivity of crops at different development stages is important for the management of saline waters. Many recent studies have shown that saline agricultural drainage waters can be used successfully to grow crops without detrimental long term consequences to either crops or soils. Different crops could be irrigated with relatively saline water during the later stages of their growth provided the salinity water below threshold salinity level be used for irrigation during the sensitive or initial stages of growth. Genetic variation for salt tolerance at specific growth stages needs to be further exploited for gainful crop production under saline water irrigation. Lines selected for tolerance at particular stages of growth could be recombined and their segregates screened for overall tolerance throughout plant growth (Garg and Gupta, 2001).

Sometimes, salt tolerance at specific growth stages may be used to advantage. Moderate salinity applied during fruit development can change the partitioning of photosynthetes and improve soluble solids in melons and tomato (Mizrahi and Pasternak, 1985). Any small yield decrease due to salinity is thus offset by the higher marketable quality of the fruit. Some grain crops such as sorghum, wheat and barley are extremely insensitive to high concentrations of saline water applied during or just prior to anthesis. Such tolerance could be exploited by substituting brackish water for irrigation water during later plant growth stages. This strategy has been used successfully for both field and vegetable crops (Grattan *et al.,* 1987, Shannon, 1997).

The available data generally indicate that the early seedling stage is the most sensitive for most crops (Ayers *et al.,* 1951; Bernstein, 1975; Maas and Poss, 1989). It is during this stage of growth with cereal crops that leaf and spikelet primordia are initiated and tiller buds are formed (Maas and Grieve, 1992). Consequently high salinity during this stage can severely affect final seed yield.

Some of the crop species at germination stage are not as salt tolerant as at later stages of growth. Sugarbeet, barley and cotton for instance are among the most salt tolerant crops but each is relatively sensitive during either germination or early seedling growth (Bernstein and Hayward, 1958). Rice may be sensitive during seedling and flowering stages (Pearson, 1966). Similarly, salt tolerance of alfalfa is poor during germination but is good subsequently. However, pea, gram and beans are not as sensitive during germination as during later stages of growth. Although salt stress delays germination and emergence but most crops are capable of germination at higher salinity levels than they would tolerate normally at the vegetative or reproductive stages of growth (Maas and Grieve, 1992). However, this high tolerance may be of limited value when the plants are highly sensitive during the early seedling growth as is normally found in many crop species.

Most of the major cereal and millets crops exhibit high tolerance to soil salinity. Wheat, triticale, rye, oats, barley, pearl millet and sorghum come in this group (Francois *et al.,* 1986, 1988, 1989). The only exceptions are corn and rice (Maas, 1990). The results suggest that cereals become increasingly tolerant after the booting stage and are not particularly sensitive during flowering. However, conflicting data have been obtained for rice. Although sensitivity of rice during the seedling stage is widely accepted, its sensitivity during the late boot and the subsequent flowering stages has been debated. According to Pearson and Bernstein (1959) and Pearson *et al.* (1966) rice is very tolerant to salt during germination, but very sensitive to it during the early seedling stages, gains tolerance progressively during the tillering stage, again becomes sensitive when flowering and is tolerant during the period of grain maturation. In contrast Kaddah (1963), and Kaddah *et al.* (1965) found no evidence that rice increased in sensitivity to salt at the flowering stage. Kaddah *et al.* (1975) further concluded that rice is not sensitive to salinity after the boot stage. According to Maas (1993) explanation of this discrepancy may be found in the work of Akbar and Yabuno (1977). They found that salinity caused panicle sterility in some rice varieties and not in others and suggested that resistance to panicle sterility was genetically controlled. Ramayya *et al.* (1976) reported the most sensitive growth stage to be two days after transplanting with 24% yield reduction followed by maximum tillering stage with 17% yield reduction.

Results of green house experiments have shown that wheat was sensitive during flowering and least sensitive during grain filling (Maas and Poss, 1989 a). Grain yields from plants of two wheat species (*T. aestivum* cv. Probred and *T. trugidum* cv. *Aldura*) stressed during either the vegetative, reproductive and maturation stages indicated that both species became less sensitive to salinity, the later plants were stressed. Salinity reduced grain yield by reducing seed number more than seed weight indicating that salt stress during vegetative stage affected spikelet differentiation.

Barley is one of the most salt tolerant crops having high degree of genetic variability in salt tolerance. However, it is less tolerant at germination and early seedling growth as significant decrease in germination occurs beyond 4 dSm^{-1} and 50% reduction at 12.3 dSm^{-1}. Husked varieties of barley have higher salt tolerance than huskless varieties. However, both suffered more when exposed to saline waters at germination followed by flowering and tillering stages.

Sorghum has been rated as moderately salt tolerant with a threshold level of about 6.8 dSm^{-1} and a slope of 16%. The crop is relatively more tolerant to salinity at the germination than other stages, being most sensitive at the vegetative stage (Francois *et al.,* 1984). Maas *et al.* (1986) further observed that sorghum was not sensitive to salinity during the vegetative stage and least sensitive during the maturation stage. It was concluded that sorghum can be irrigated with poor quality waters after flowering without reducing yields. However, Boursier and Lauchli (1990) found that sorghum dry matter production decreased substantially in response to moderate salinity.

Cluster bean is a moderately salt tolerant crop which can be successfully grown in salt-affected areas. However, there is paucity of information on the salt sensitivity of this crop at various growth stages Garg *et al.* (1997) found that salt sensitivity changed with the stage of growth and flowering stage was the most sensitive to NaCl for seed and pod yields of clusterbean (Table 4.10). However, dry matter production was affected more adversely by salinity at the seedling stage followed by vegetative and flowering stages. Salinity induced changes in the levels of certain leaf metabolites (starch aminoacids, reducing sugars, total chlorophyll, soluble protein, free aa and free proline) and activities of enzymes (NR, GS and GDH) also showed stage dependent variations. It thus appears that when salt stress is imposed the plant organ growing most rapidly at that time is affected the most. Francois *et al.* (1990) found that vegetative growth of guar was decreased 9.6% for each unit increase in salinity above 4.9 dSm^{-1}. However, emergence was unaffected by salt levels upto 8.5 dSm^{-1}. Work conducted at CSSRI, Karnal indicated pod development to be most important stage to salt stress in cluster bean, black gram and green gram crops.

It is generally agreed that after the seedling stage most plants become increasingly more salt tolerant as growth proceeds through the vegetative, reproductive and grain filling stages. Rice may be an exception. Pearson and Bernstein (1959) reported that rice yields are significantly reduced if salt stress is imposed at either the seedling stage or during the pollination and fertilization. Indian mustard is most sensitive during the emergence stage but becomes tolerant at later stages of growth (Kumar, 1984). Therefore, Maas *et al.* (1983)

suggested that poor quality water can be used during and after tillering without reducing yields provided water of acceptable quality is used during early vegetative growth. Thus, the level of tolerance at one growth stage can not be assumed to be the same at all growth stages. Relative salt tolerance of some crops at emergence and maturity illustrates this fact very clearly (Table 4.11).

Table 4.10. Seed, pod and shoot dry weight yield (g/plant) of cluster-bean as influenced by NaCl salinity at the seedling (S), vegetative (V) and flowering (F) stages of growth (Garg *et al.*, 1997).

NaCl	Seed yield			Mean	Pod weight			Mean	Shoot dry weight			Mean
(dSm^{-1})	S	V	F		S	V	F		S	V	F	
0	2.07	2.07	2.07	2.07	3.84	3.84	3.84	3.84	9.93	9.93	9.993	9.93
4	1.89	1.87	1.56	1.83	3.40	3.33	2.82	3.18	6.72	7.26	9.78	7.92
8	1.68	1.71	1.26	1.65	2.93	2.97	2.46	2.79	4.47	6.63	8.49	6.53
12	1.35	1.47	0.75	1.34	2.41	2.41	1.62	2.17	4.08	6.30	6.93	5.77
Mean	1.75	1.77	1.41		3.15	3.15	2.93		6.30	7.53	8.78	
LSD		0.05	0.01		0.05		0.01		0.05		0.01	
Stage		0.18	0.24		0.33		0.45		1.08		1.47	
Salinity		0.21	0.29		0.39		0.53		1.26		1.71	
St X SI		0.36	NS		0.66		NS		2.16		NS	

Table 4.11. Relative salt tolerance of various crops at emergence and during growth to maturity (Shannon, 1984).

Crop	ECe (dSm^{-1})	
	50% Emergence	50% Yield
Barley	16-24	18
Cotton	15	17
Sugarbeet	6-12	15
Safflower	12	14
Sorghum	13	15
Wheat	14-16	13
Beet root	13.8	9.6
Cowpea	16	9.1
Tomato	7.6	7.6
Cabbage	13.0	7.0
Corn	21-24	5.9
Lettuce	11	5.2
Onions	5.6-7.5	4.3
Rice	18	3.6
Beans	8.0	3.6

Strogonov (1964) stated that property of salt tolerance changes with the development of plant and may also change with the type of salinity. Garg and Garg (1980) reported a shift in sensitivity to the same anion with a change in the stage of plant growth in pea. During germination, bicarbonate was more depressive than carbonate but during vegetative growth carbonate was more detrimental than bicarbonate. In general, sodicity has more detrimental effects on tillering and later stages of growth as compared to germination stage in cereals.

Information regarding the stage of growth and development of crops and salt stress is important in irrigation management in problem soils. For example, if crop is more sensitive during one stage than another, the salinity or irrigation water could be regulated during the season to minimise salt injury during the sensitive stage. Lunin *et al.* (1963) showed with the help of the glass house experiments that five of six vegetable crops tomatoes, broccoli, spinach, beets, peppers and onions tested were especially sensitive to salinity during their early development stage. The yields of onion bulbs and beet roots were actually increased by mixed salinity stress imposed at later stages of development. Young rice seedlings of 25-35 days are less tolerant to salinity than 40-45 days seedlings. Therefore, older seedlings are recommended for transplanting in sodic soils.

Differences among Cultivars and Rootstocks

Most commercially grown cultivars are developed under nonsaline conditions and are not bred to endure salt stress. Therefore, their relative tolerance to salinity is often similar and difficult to measure. In addition, many cultivars developed in the past were derived from a narrow genetic base and thus possessed similar traits. However, currently developed cultivars are from a much more diverse genetic base and therefore possess a wide range of salt tolerance.

Among the crop species that already show diversity in salt tolerance are barley, wheat, rice, pearl millet, soybean, pigeonpea, chickpea, clusterbean, berseem clover, squash, bermuda grass, brome grass, creeping bent grass, muskmelon, strawberry, Indian mustard, cotton, sugarcane and several others (Greenway and Munns, 1980; Francois and Maas, 1994; Shalhevet, 1994). Variation in salt tolerance both between and within variety may be used for the improvement of salt tolerance of cultivated species. The mechanisms causing varietal differences in salt tolerance are not clearly understood. However, it has been found that salt concentrations (mostly sodium and chloride) in leaves reach toxic levels quickly in sensitive than in salt tolerant genotypes. This has been attributed primarily to the ability of roots to exclude the salt from the xylem sap flowing to the shoot. Rates of accumulation of sodium and/or chloride in the shoot are the critical processes determining varietal differences in salt tolerance. Evidences in this regard have been presented and discussed for a large number of crop (Garg and Gupta, 1996). In certain species (eg. tomato and chickpea) both halophytic mechanisms of Na^+ and Cl^- compartmentation in vacuoles, as well as glycophytic mechanism of salt exclusive have been simultaneously observed in different cultivars. However, maintenance of high K : Na ratio and potassium versus sodium selectivity seem important factors contributing to salt tolerance of crops. Metabolic processes responsible for varietal differences in salt tolerance, both at the whole plant level and at the cellular level have been implicated for such differences. In studies conducted over the past 40 years at United States Salinity laboratory, California and other places, significant varietal differences are now well established in a number of above mentioned crops. The causes of varietal differences have also been studied but mechanism is not yet clearly understood. It is generally found that salt concentrations in leaves reach toxic concentrations more quickly in salt sensitive than in salt tolerant varieties. Although some of the varietal differences have been utilized in breeding programmes but these efforts are still in infancy.

Likewise rootstock difference is an important factor in the salt tolerance of fruit and vine crops. Fruit crops are not only sensitive to salinity per se but are particularly susceptible to toxic effects of Na^+ and Cl^- ions. Varieties and rootstocks that differ in the absorption and translocation of Na^+ and/or Cl^- ions have different salinity tolerances. The salt tolerance of avocado, grape fruit and orange is closely related to the Cl^- accumulation properties of the rootstocks. The differential ability of the rootstocks to transmit chloride to the top of the plant determines the salt tolerance capacity in the fruit trees (Cooper *et al.,* 1951; Walker and Douglas, 1983). Similar effects of rootstocks on salt accumulation and tolerance have been reported for stone fruit trees. Large differences in the salt tolerance of grape varieties have been linked with rootstock effects on Cl^- accumulation (Bernstein *et al.,* 1969). Although citrus is not considered very salt tolerant, there are differences in salt tolerance among the various rootstocks and these differences are attributed to salt exclusion, particularly to chloride exclusion (Walker and Douglas, 1983). The maximum Cl^- concentrations permissible in soil water that do not cause leaf injury to most of the rootstocks are shown in Table 4.12. The tolerance of many of the fruit trees and vine crops can be significantly improved by selecting rootstocks that restrict Cl^- and/or Na^+ accumulation.

Table 4.12. Chloride tolerance limits of some fruit crop cultivars and rootstocks (Maas, 1990)

Crop	Rootstock or cultivar	Maximum permissible Cl^- in soil water without leaf injury[1] mol m^{-3}
Rootstock	West Indian	15
Avocado (*Persea*	Guatemalan	12
americana)	Mexican	10
Citrus	Sunki mandarin, grapefruit	50
(*Vitis* sp)	Cleopatra mandarin, Rangpur lime	50
	Sampson tangelo, rough lemon[2]	30
	Sour orange, Ponkan mandarin	30
	Cirumelo 4475, trifoliate orange	20
	Cuban shaddock, Calamondan	20
	Sweet orange, Savage citrange	20
	Rusk citrange. Troyer citrange	20
Grape	Salt Creek, 1613.3	80
(*Vitis* sp.)	Dog ridge	60
Stone fruit	Marianna	50
(*Prunus* sp.)	Lovell, Shalil	20
	Yunan	15
Cultivars	Boysenberry	20
Berries[3]	Olallie blackberry	20
(*Rubus* sp.)	Indian Summer raspberry	10
Grape	Thompson seedless, Perlette	40
(*Vitis* sp.)	Cardinal, black rose	20
Strawberry	Lassen	15
(*Fragaria* sp.)	Shasta	10

1. For some crops, these concentrations may exceed the osmotic threshold and cause some yield reduction.
2. Data from Australia indicate that rough lemon is more sensitive to Cl^- than sweet orange.
3. Data available for one variety of each species only.

CRITERIA FOR RATING CROP SALT TOLERANCE

The ability of crops to tolerate salinity has been determined on the basis of one or more of the following three criteria :

(1) Plant survival.

(2) Absolute plant growth or yield.

(3) Relative growth or yield.

Plant survival on saline soils has been widely used by ecologists to judge the tolerance of native species on uncultivated lands. However, survival without an economic yield has little value to the farmer. Indirectly, this criterion is potentially useful because plant breeders can use it as the basis for selecting salt tolerant lines and cultivars.

Absolute salt tolerance is directly related to the maximum growth or yield potential of given plant under high salinity regardless of its growth rate and yield behaviour under nonsaline conditions. A farmer is mostly interested in absolute yield. Salinity may considerably decrease plant growth without affecting its absolute tolerance. A vigorously growing plant which gets markedly affected by salinity is likely to have low relative salt tolerance, if, however, it still yields more than a slow growing plant (lower yields) with high relative tolerance, it would be judged as having higher absolute salt tolerance.

Although the absolute yields enable to calculate economic returns under given salinity conditions, but also reflect crop responses to many other environmental factors. Even when other conditions like climate, fertility, soil moisture, plant protection etc. are optimum, absolute yield data do not permit comparison between crops because yields for different crops are not expressed in comparable terms.

Relative yield is the yield of a crop grown on a saline soil expressed as a fraction of what would be obtained otherwise under nonsaline conditions with other factors remaining the same. This criterion permits a comparison of crops whose yields may be expressed in quite different units. The reliability of relative yield data depends upon whether yield reductions are unaffected by the level of other essential factors. If relative yield losses due to salinity are independent of differences in absolute yield caused by irrigation, climate, soil and other environmental variables, the relative yield v/s salinity relationship permits a useful expression of salt tolerance data. Although absolute yield under salinity is of greater importance than relative yield, but both must be considered. If a cultivar is relatively more tolerant but grows slowly or has poor quality characteristics, its high relative tolerance should not be ignored if that characteristics can be transferred to high yielding cultivars. Relative tolerance may also contribute to the overall fitness of the species to produce satisfactorily in a range of saline and nonsaline environments - a character referred to as environmental plasticity. Such considerations should inspire and promote significant research efforts on the genetic mechanisms of salt tolerance.

In addition to these criteria attempts have also been made to evolve a criterion based on metabolic parameters. The sodium and potassium contents in plant tissues and particularly their ratio have been found to be suitable parameters to adjudge the relative alkali tolerance.

The only agronomically important plant response measurement for establishing salt tolerance of a crop is the commercial yield. For crops that produce either vegetative or seed yields eg. alfalfa, corn, pearl millet etc., it is advisable to measure both responses to salinity. However, vegetative responses are generally unreliable for predicting salt effects on fruit or seed production (Maas and Hoffman, 1977). Salinity may decrease grain yields of corn and rice without appreciably affecting straw yields. With some other crops, eg. barley, wheat, cotton and some tolerant grasses, seed or fibre production are decreased much less than vegetative growth. For root crops, storage root yields may be decreased much more than that of tops or fibrous roots (Hoffman and Rawlins, 1971).

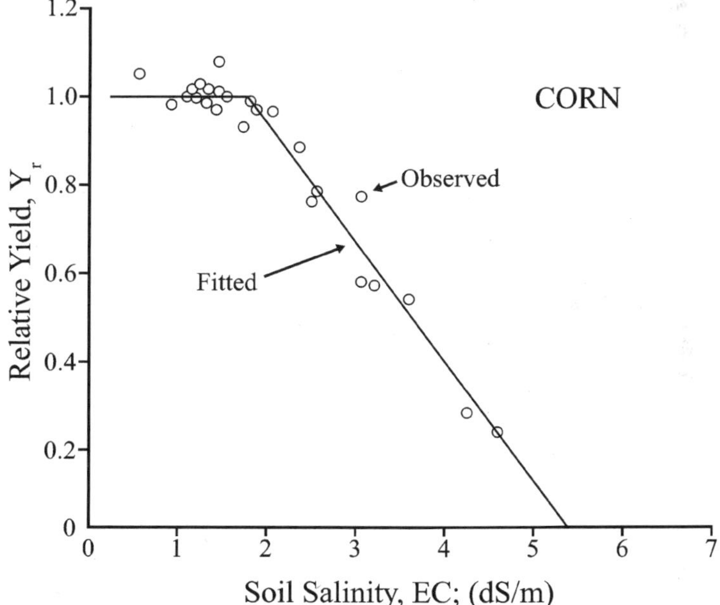

Fig. 4.3. The piece-wise linear response function fitted to actual yield data obtained from corn. Data from Hoffman *et al.* (1982)

Maas and Hoffman (1977) proposed that the yield response curve for agricultural crops could be represented by two linear lines, one a horizontal line depicting no response to increasing salinity at low concentrations, and the second, a concentration dependent line whose slope indicates the yield reduction per unit increase in salinity at higher concentrations. The point at which the two lines intersect designates the ``threshold", i.e. the maximum soil salinity that does not reduce yield below that obtained under nonsaline conditions. Figure 4.3 shows the two-piece model fitted to actual grain yields obtained in a salt tolerance experiment on corn (*Zea mays* L.) This two-piece linear response function provides a reasonably good fit for commercially acceptable yield when plotted against time-and depth averaged salinity in the rootzone. For soil salinities exceeding the threshold of any given crop, relative yield (Y_r) can be estimated with the following equation :

$$Y_r = 100 - b\ (EC_e - a)$$

where, a = the salinity threshold expressed in dS/m (1 dS/m = 1 mmho/cm); b = the yield reduction, or slope, expressed in % per dS/m; and EC = the mean electrical conductivity of saturated extracts taken from the root-zone.

RELATIVE SALT TOLERANCE RATINGS

Maas and Hoffman (1977) reviewed the available salt tolerance data, selecting only those studies which related some known soil salinity parameters to the yield. Subsequently Maas (1986, 1990) and Francois and Maas (1994) incorporated the data obtained in the last 15 years and updated the information (Table 4.13). Crops are listed in different groups such as grain and fibre, grasses and forage and vegetables and fruits crops.

Table 4.13. Salt tolerance of herbaceous crops[1] (Maas, 1986)

Crop		Electrical conductivity of saturated soil extract (dSm^{-1})		Rating[4]
Common name	Botanical name[2]	Threshold[3]	Slope % per dSm^{-1}	
Fibre, grain and special crops				
Barley[5]	*Hordeum vulgare*	8.0	5.0	T
Bean	*Phaseolus vulgaris*	1.0	19.0	S
Broadbean	*Vicia faba*	1.6	9.6	MS
Cotton	*Gossypium hirsutum*	7.7	5.2	T
Cowpea	*Vigna unguiculta*	4.9	12.0	MT
Flax	*Linum usitatissimum*	1.7	12.0	MS
Groundnut	*Arachis hypogaea*	3.2	29.0	MS
Guar	*Cyamopsistetragonola*	8.8	17.0	T
Kenaf	*Hibiscus cannabinus*			MT
Maize[6]	*Zea mays*	1.7	12.0	MS

Crop		Electrical conductivity of saturated soil extract (dSm^{-1})		Rating[4]
Common name	Botanical name[2]	Threshold[3]	Slope % per dSm^{-1}	
Millet, foxtail	*Setaria italica*			MS
Oats	*Avena sativa*			MT*
Rice, paddy	*Oryza sativa*	3.0[7]	12.0[7]	S
Rye	*Secale cereale*	11.4	10.8	T
Safflower	*Carthamus tinctorius*			MT
Sesame[8]	*Sesamum indicum*			S
Sorghum	*Sorghum bicolor*	6.8	16.0	MT
Soybean	*Glycine max*	5.0	20.0	MT
Sugarbeet	*Beta vulgaris*	7.0	5.9	T
Sugarcane	*Saccnarum officinarum*	1.7	5.9	MS
Sunflower	*Helianthus annuus*			MS*
Triticale	*X. Triticosecale*	6.1	2.5	T
Wheat	*Triticum aestivum*	6.0	7.1	MT
Wheat (semidwarf)[10]	*T. aestivum*	8.6	3.0	T
Wheat, Durum	*T. turgidum*	5.9	3.8	T
Grasses and forage crops				
Alfalfa	*Medicago sativa*	2.0	7.3	MS
Alkali grass, Nuttall	*Puccinellia airoides*			T*
Alkli sacation	*Sporobolus airoides*			T*
Barley (forage)[5]	*Hordeum vulgare*	6.0	7.1	MT
Bent grass	*A. stolonifera palustris*			MS
Bermuda grass[11]	*Cynodon dactylon*	6.9	6.4	T
Bluestem, Angleton	*Dichanthium aristaum*			MS*
Brome, smooth	*Bromus marginatus*			MT*
Buffelgrass	*B. inermis*			MS
Bunet	*Cenchrus ciliaris*			MS*
Canary grass, reed	*Phalaris arundinaceae*			MT
Clover, alsike	*Trifolium hybridium*	1.5	12.0	MS
Clover, Berseem	*T. alexandrinum*	1.5	5.7	MS
Clover, Hubam	*Meliotus alba*			MT*
Clover, Ladina	*Trifolium repens*	1.5	12.0	MS
Clover, red	*T. pratense*	1.5	12.0	MS
Clover strawberry	*T. fragiferum*	1.5	12.0	MS
Clover sweet	*Melilotus*			MT*
Clover, white Dutch	*Trifolium repens*			MS*
Cowpea (forage)	*Vigna unguiculata*	2.5	11.0	MS

Crop		Electrical conductivity of saturated soil extract (dSm^{-1})		Rating[4]
Common name	Botanical name[2]	Threshold[3]	Slope % per dSm^{-1}	
Dallisgrass	*Paspalum dilatatum*			MS*
Fescus, tall	*Festuca elatior*	3.9	5.3	MT
Fescue, meadow	*F. pratensis*			MT*
Foxtail, meadow	*Alopecurus pratensis*	1.5	9.6	MS
Grama, blue	*Blouteloua gracilis*			MS*
Harding grass	*Phalaris tuberosa*	4.6	7.6	MT
Kallar grass	*Diplachne fusca*			T*
Lovegrass[12]	*Eragrostis* sp.	2.0	8.4	MS
Maize (forage)[6]	*Zea mays*	1.8	7.4	MS
Milkvetch, Cicer	*Astragalus cicer*			MS*
Oat grass, tall	*Arrhenatherum, Danthonia*			MS*
Oats (forage)	*Avena sativa*			MS*
Orchard grass	*Dactylis glomerata*	1.5	6.2	MS
Panic grass, blue	*Panicum antidotae*			MT*
Rape	*Brassica napus*			MT*
Rescue grass, blue	*Bromus unioloides*			MT*
Rhodes grass	*Chloris gayana*			MT
Rey (forage)	*Secale cereale*			MS*
Rye grass, Italian	*Lolium italicum multflorus*			MT*
Rye grass, perennial	*L. perenne*	5.6	7.6	MT
Salt grass, desert	*Distichlis stricta*			T*
Sesbania	*Sesbania exaltata*	2.3	7.0	MS
Sirato	*Macroptilium atropurpureum*			MS
Sphaerophyse	*Sphaerophysa salsula*	2.2	7.0	MS
Sudan grass	*Sorghum sudanense*	2.8	4.3	MT
Timothy	*Phleum pratense*			MS*
Trefoll, big	*Lotus uliginosus*	2.3	19.0	MS
Trefoil, narrowleaf birdsfoot	*L. corniculatus tenuifolium*	5.0	10.0	MT
Trefoil, broadleaf birdsfoot[13]	*L. corniculatus arvenis*			MT
Vetch common	*Vicia angustifolia*	3.0	11.0	MS
Wheat (forage)[10]	*Triticum aestivum*	4.5	2.6	MT
Wheat, Durum (forage)	*T. turgidum*	2.1	2.5	MT
Wheatgrass, stand, crested	*Agropyron sibiricum*	3.5	4.0	MT

Crop		Electrical conductivity of saturated soil extract (dSm^{-1})		Rating[4]
Common name	Botanical name[2]	Threshold[3]	Slope % per dSm^{-1}	
Wheatgrass, m fairway created	*A. cristatum*	7.5	6.8	T
Wheatgrass intermediate	*A. intermedium*			MT*
Wheat grass, slender	*A. trachycaulum*			MT
Wheatgrass, tall	*A. elongatum*	7.5	4.2	T
Wheatgrass, western	*A. smithii*			MT*
Wild ryle, Altai	*Elymus angustus*			T
Wild rye, beardless	*E. triticoides*	2.7	6.0	MT
Wild rye Canadian	*E. canadensis*			MT*
Wild rye Russian	*E. junceus*			T
Vegetables and fruit crops				
Artichoke	*Helianthus tuberosus*			MT*
Asparagus	*Asparagus officinalis*	4.1	2.0	T
Bean	*Phaseolus vulgaris*	1.0	19.0	S
Beet, red[8]	*Beta vulgaris*	4.0	9.0	MT
Broccoli	*Brassica oleraceae botrytis*	2.8	9.2	MS
Brussel sprouts	*B. oleracea germifera*	1.8	9.7	MS*
Cabbage	*B. oleracea capitat*	1.0	14.0	MS
Carrot	*Daucus carota*			S
Cauliflower	*Brassica oleracea botrytis*	1.8	6.2	MS*
Celery	*Apium graveolens*	2.5	13.0	MS
Cucumber	*Cucumis sativus*	1.1	6.9	MS
Eggplant	*Solanum melongena esculentum*			MS
Kale	*Brassica oleracea acephala*			MS*
Kohlrabi	*B. oleracea gongyoide*	1.3	13.0	MS*
Lettuce	*Lactuca sativa*	1.7	12.0	MS
Maize, sweet	*Zea mays*			MS
Muskmelon	*Cucumis melo*			MS
Okra	*Ablemoschus esculentus*	1.2	16.0	S
Onion	*Allium cepa*			S
Parsnip	*Pastinaca sativa*			S*
Pea	*Pisum sativum*	1.5	14.0	S*
Pepper	*Capsicum annuum*	1.7	12.0	MS
Potato	*Solanum tuberosum*			MS
Pumpkin	*Cucurbita pepo pepo*	1.2	13.0	MS*
Radish	*Raphanus sativus*	2.0	7.6	MS

Crop		Electrical conductivity of saturated soil extract (dSm^{-1})		Rating[4]
Common name	Botanical name[2]	Threshold[3]	Slope % per dSm^{-1}	
Spinach	*Spinacia oleracea*	3.2	16.0	MS
Squash, scallop	*Cucurbita pepo melopepo*	4.7	9.4	MS
Squash, zucchini	*C. pepo melopepo*	1	33	MT
Strawberry	*Fragaria* sp.	1.5	11	S
Sweet potato	*Ipomoa batatas*	12.5	9.9	MS
Tomato	*Lycopersicon lycopersicum*	0.9	9	MS
Turnip	*Brassica rapa*			MS
Watermelon	*Citrullus lanatus*			MS*

1. These data serve only as a guideline to relative tolerance among crops. Absolute tolerances vary, depending upon climate, soil conditions and cultural practices.
2. Botanical and common names follow the convention of Hortus Third where possible.
3. In gypsiferous soils, plants will tolerate EC_e about 2 dSm^{-1} higher than indicated.
4. T = Tolerant, MT = Moderately Tolerant, MS = Moderately Sensitive and S = Sensitive. Ratings with an * are estimates.
5. Less tolerant during seedling stage, EC_e at this stage should not exceed 4 or 5 dS/m.
6. Grain and forage yields of DeKalb XL-75 growm on an organic muck soil decreased about 26% per dS/m above a threshold of 1.9 dSm^{-1}.
7. Because paddy (rice) is grown under flooded conditions, values refer to the electrical conductivity of the soil water while the plants are submerged. Less tolerant during seedling stage.
8. Sesame cultivars, Sesaco 7 and 8, may be more tolerant than indicated by the S rating.
9. Sensitive during germination and emergence, ECe should not exceed 3 dSm^{-1}.
10. Data from one cultivar, "Probred"
11. Average of several varieties. Suwannee and Coastal are about 20% more tolerant, and common and greenfield are about 20% less tolerant than the average.
12. Average for Boer, Wilman, Sand and Weeping cultivars. Lehman seems about 50% more tolerant.
13. Broadleaf birdsfoot trefoil seems less tolerant than narrowleaf.

Salt tolerance data, where available, are given in terms of two key parameters :

(1) Threshold salinity i.e. the maximum soil salinity, that does not reduce yield below that produced under non-saline conditions; and

(2) The slope i.e. the percentage yield decrease per unit increase in salinity beyond the threshold.

The data presented for the relative salt tolerance of various crops are in terms of the electrical conductivity of the saturated soil extract (ECe) with units of decisiemens per metre (1 dSm^{-1} = 1 mmhos cm^{-1}). However, these data serve only as a guideline to assess relative tolerance among crops. Absolute tolerance may change according to climate, soil conditions and cultural practices.

The threshold and slope obtained from the above model can be used to calculate relative yield (Yr) for any given soil salinity exceeding the threshold by using the equation :

$$Yr = 100 - B\ (ECe - A)$$

Where A = the threshold salinity expressed in dSm^{-1}

B = the slope expressed in % per dSm^{-1} and ECe is the mean electrical conductivity of the saturated soil extract of the root zone.

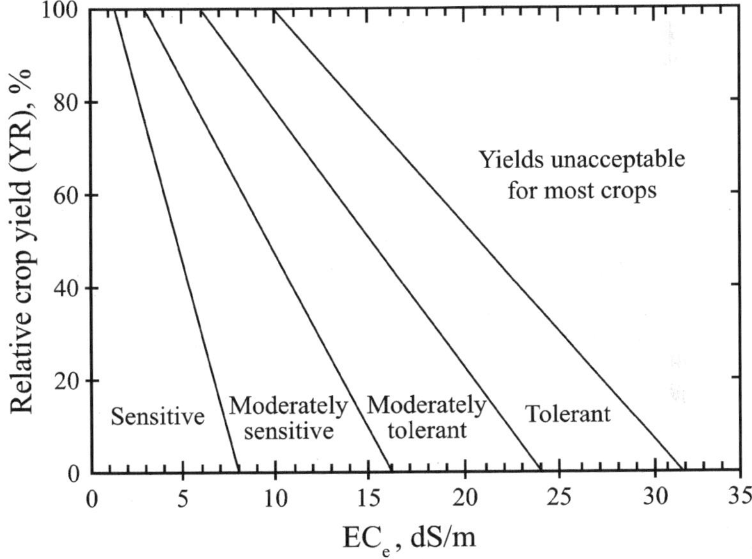

Fig. 4.4. Divisions for classifying crop tolerance to salinity.

The four qualitative salt tolerance ratings are defined by the division boundaries shown in Figure 4.4 for quick comparison among crops. Generally, the threshold and linear slope for a crop remain within one class. Where the linear curve for a crop crossed division boundaries, the crop was classified based on the tolerance at lower salinity levels at which yields are commercially acceptable.

Significant differences among varieties have been observed for many of these crops. It is not always clear, however, whether varietal differences reflect differences in salt tolerance or differences in adaptation to climate or nutritional conditions under which the plants were tested. Consequently, the data (Table 4.10) represent the average tolerance of the varieties studied.

Salt Tolerance of Cereal Crops

Most of the major cereal crops exhibit high tolerance to soil salinity. Sorghum, wheat, triticale, rye, oats, barley and pearl millet, come in this group. The only exceptions are corn and rice (Maas, 1990). Regardless of the overall salt tolerance, all cereals tend to follow the same sensitivity or tolerance pattern in relation to their stage of growth. The seedling or early vegetative stage appears to be the most sensitive, with subsequent stages showing increased tolerance. This phenomenon has been reported for sorghum (Maas *et al.,* 1986), wheat (Maas and Poss, 1989) (Fig. 4.5), barley (Ayers *et al.,* 1952), corn (Maas, 1990) and rice (Pearson and Bernstein, 1959).

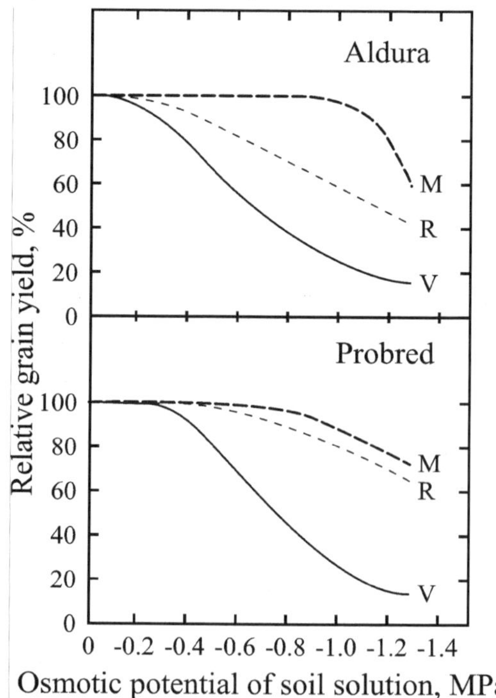

Fig. 4.5. Relative grain yields of Probred and Aldura wheat as a function of the osmotic potential of the soil solution imposed during three growth stages- vegetative (V), reproductive (R) and maturation (M)

Salinity can have a significant effect upon the development process occurring at a particular time. The sequence of events has been separated into three distinct but continuous developmental phases (Maas and Grieve, 1992). In the first phase which encompasses the early vegetative growth stage, leaf and spikelet primordia are initiated, leaf growth occurs and the tiller buds are produced in the axils of the leaves. High soil salinity at this time reduces the number of leaves per culm, the number of spikelets per spike, and the number of tillers per plant (Maas and Grieve, 1992). Differentiation of the terminal spikelet signals completion of this phase.

During phase II, the tillers grow, main stem and tiller culms elongate and the final number of florets is set (Kirby, 1988). Salinity stress during this phase may affect tiller survival and reduce the number of functional florets per spikelet. This phase ends with anthesis. Carpel fertilization and grain filling occur during the final phase (Kirby, 1988). At this time salinity affects number and seed size.

The effect of salinity on spikelet and tiller number established during phase I has a greater influence on final seed yield than the effects exerted on yield components in the later two phases (Maas and Poss, 1989).

Salt Tolerance of Legumes

Legumes are used as grain legumes (soybean, pea, beans, peanuts, chickpea, cowpea), forage crops (alfalfa) or components of pastures (clovers) and extend from cool temperate regions to the tropics. Legumes have long been recognised to be either sensitive or moderately resistant to salinity (Maas and Hoffman, 1977). Considering legumes, two features are relevant : there is variation in salt resistance among legumes and most of them respond to saline conditions by salt exclusion (Lauchli, 1984), that is exclusion of sodium and/or chloride from the leaves.

Except for *Lupins lutens*, a legume of limited agricultural significance in Australia, all the legumes respond to 50 mM NaCl salinity in solution culture by growth inhibition (Lauchli, 1984). However, at 100 mM NaCl, *Lupins lutens* was also adversely affected. Peanut, chickpea, soybean, *Phaseolus vulgaris* and pea are most sensitive to salinity. More salt resistant are particularly the clovers and alfalfa. Significant cultivar differences have been reported in species like *Lupins, Trifolium, glycine* and *Phaseolus.* More recently, varietal differences have also been studied in peanut, chickpea and clusterbean.

Salt Tolerance of Forage Crops

Forage crops fall into two broad salt tolerance categories. Most grasses belong to the tolerant group, with the majority of legumes in the sensitive group. Exceptions to this generalization are, meadow foxtail (*Alopecurus pratensis*) and orchard grass (*Dactylis glomerata*) which are moderately sensitive to salt stress, and birdshoot trefoil (*Lotus corniculatus* var. *tenuifolium*) and the sweet clovers (*Meliotus* sp.) which are moderately tolerant. Many of the forage grasses possess the same growth habit as the cereal grasses and like the cereals, are more sensitive to salinity during the early seedling of growth (Marcar, 1987). Once they are well established beyond the early seedling stage, they are less sensitive to soil salinity.

Clovers are the predominant legumes of pastures but they are generally salt sensitive. The salt tolerance of clovers and alfalfa is highly dependent upon the stage of growth at which salinity is first imposed (West and Taylor, 1981; Smith *et al.,* 1981). There is evidence that the genetic variability exists among the grass and legume species and cultivars which offers the possibility of developing strains with higher salt tolerance (Lauchli, 1984; Francois and Maas, 1994).

Salt Tolerance of Vegetable Crops

Vegetable crops fall more into salt sensitive than salt tolerant categories. The only notable exceptions are asparagus, red beet and zucchini squash. The growth of many vegetables is stunted under marginal conditions of salinity without other visible injury symptoms. At high salinity levels, some vegetables exhibit pronounced injury symptoms in the later stages of growth. Bean leaves develop a marginal chlorosis-necrosis with an upward cupping of the leaves while onions also develop leaf necrosis. Some lettuce cultivars develop calcium deficiency symptoms when sulphate level in the soil is too high. With most vegetable crops, however, the osmotic effect predominates and nutritional effects are either absent or are of secondary importance.

Most vegetable crops produced on saline soils are not of prime market quality. This is seen in such diverse ways as smaller fruit size of tomatoes or peppers, reduced petiole length of celery and misshapen potatoes (Bernstein, 1969). On the other hand the flavor of carrots and asparagus is enhanced by a measurable increase in sugar content when grown under saline conditions. Likewise a number of studies have shown that total soluble solids in tomatoes are significantly increased as salt stress is increased (Jobes *et al.,* 1981; Adams and Ho, 1989). Unfortunately this gain in quality is more than offset by lower yields.

Salt Tolerance of Fruit Trees and Vine Crops

With the exception of date palm and a few others believed to be salt tolerant, most of the fruit trees are sensitive to salinity (Table 4.14). Stone fruits, citrus and avocado have all shown specific sensitivity to foliar accumulation of chloride and sodium. Some of the more sensitive fruit crops may accumulate toxic levels of Na^+ and Cl^- over a period of years from soils that would be classified as nonsaline and nonsodic (Bernstein, 1980). Chloride toxicity in woody species is generally more severe and is observed on a wide range of species than sodium toxicity. However, growth and yield reduction may occur with woody fruit species in the absence of specific ion toxicity. Francois and Clark (1980) working with valencia orange, reported a 50% reduction in fruit yield from salinity with no visible leaf injury symptoms. Once salts have accumulated to toxic levels, however, growth and yield are suppressed by additive effects of osmotic stress and specific ion toxicities.

Table 4.14. Salt tolerance of fruit crops[1] (Maas, 1986).

Crop		Electrical counductivity of saturated soil extract		Rating[4]
Common name	Botanical name[2]	Threshold[3] dSm^{-1}	Slope % per dSm^{-1}	
Almond[5]	*Prunus duclis*	1.5	19.0	S
Apple	*Malus sylvestris*			S
Apricot[5]	*Prunus armeniac*	1.6	24.0	S
Avocado[5]	*Persea americana*			S
Blackberry	*Rubus* sp.	1.5	22.0	S
Boysenberry	*Rubus ursinus*	1.5	22.0	S
Castor bean	*Ricinus communis*			MS*
Cherimoya	*Annona cherimola*			S*
Cherry sweet	*Prunus avium*			S*
Cherry sand	*P. besseyi*			S*
Currant	*Ribes* sp.			S*
Date palm	*P. dactylifera*	4.0	3.6	T
Fig	*Ficus carica*			MT*
Gooseberry	*Ribes* sp.			S*
Grape[5]	*Vitis* sp.	1.5	9.6	MS
Grapefruit	*Citrus paradisi*	1.8	16.0	S
Guayule	*P. argentatum*	15.0	13.0	T
Jojoba[5]	*S. chinensis*			T
Jujube	*Ziziphus jujuba*			MT*

Crop		Electrical counductivity of saturated soil extract		Rating[4]
Common name	Botanical name[2]	Threshold[3] dSm^{-1}	Slope % per dSm^{-1}	
Lemon[5]	*Citrus limon*			S
Lime	*C. aurantiifolia*			S*
Loquat	*Eriobotrya japonica*			S*
Mango	*Mangifera indica*			S*
Olive	*Olea europaea*			MT
Orange	*Citrus sinensis*	1.7	16.0	S
Papaya+5	*Carica papaya*			MT
Passion fruit	*Passiflora edulis*			S*
Peach	*Prunus persica*	1.7	21.0	S
Pear	*Pyrus communis*			S*
Persimmon	*Diospyros virginiana*			S*
Pineapple	*Ananaas comosus*			MT*
Plume; prun	*Prunus domestica*	1.5	18.0	S
Pomegranat	*Punica granatum*			MT*
Pummelo	*Citrus maxima*			S*
Raspberry	*Rubus idaeus*			S
Rose apple	*Syzygium jambos*			S*
Sapote, white	*Casimiroa edulis*			S*
Tangerine	*Citrus reticulata*			S*

1. These data are applicable when roostock are used that do not accumulate Na^+ or Cl^- rapidly or when these ions do not predominate in the soil.
2. Botanical and common names follow the convention of Hortus Third where possible.
3. In gypsiferous soils, plants will tolerate ECe above about dSm^{-1} higher than indicated.
4. T = Tolerant, MT = Moderately Tolerant, MS = moderately Sensitive and S = Sensitive. Ratings with an* are estimates.
5. Tolerance is based on growth rather than yield.

Limitations of Salt Tolerance Ratings

Salt tolerance ratings cannot provide an accurate estimate of the expected yield as it depends not only upon salinity levels but upon many other cultural and environmental factors. The interactions between salinity and various soil, water, and climatic factors change the plant's ability to tolerate salinity. Some of these factors have been discussed else where in greater detail (Chapter 8).

Crop salt tolerance data must also be considered in light of irrigation management. Plants can tolerate a moderately saline water better when irrigations are frequent. As they extract water from the soil, the remaining water becomes more concentrated, consequently plant must consume increasing saline water as the time between irrigations increase. Crops subjected to foliar injury caused by foliar salt absorption from sprinkling waters are less tolerant under sprinkler irrigation than with various soil methods. Therefore, it should be clear that salt tolerance ratings serve only as guidelines to the relative tolerance among crops. They cannot nor they are intended to be used for estimating economic crop losses accurately and precisely.

RESPONSES OF CROPS TO SOIL SODICITY

While lists of crops tolerance to soil salinity are common (Maas and Hoffman, 1977; Maas, 1990; Maas and Grieve, 1992; Francois and Maas, 1994), only qualitative data are available for tolerance to soil sodicity. Rhodes (1982) and Maas (1986) indicated that sodium toxicity is apparently real for woody plants which too show sodium toxicity symptoms after sufficient accumulation from the soil. The adverse effects of exchangeable sodium on crop growth may be due to the following factors :

1. Direct toxic effects
2. Structural deterioration of the soil (adverse physical condition) restricting moisture transmission and aeration. It may physically impede root elongation and seedling emergence.
3. Nutritional imbalance.

In general, the adverse effect appears as a result of the combination of the above three factors. For example; an excess of specific ion may be toxic to various physiological process. At the same time or in some other cases it may also cause nutritional disorders. In alkali soils, specific ions, which may be accumulated by plants are Na^+ or HCO_3^-. When ion imbalance affects the plants, the situation can often be diagonsed and ameliorated.

Relative Salt Tolerance

Crops differ widely in their tolerance to alkali environment of the root zone. Even crop varieties respond differently to alkali conditions. Therefore, selection of crops and cropping pattern is of practical and economic importance to exploit alkali lands. The relative tolerance of different crops to soil sodicity is given in Table 4.15 and 4.16.

Table 4.15. Tolerance of crops to various levels of ESP

Range of ESP	Crops
10-15	Safflower, mash, peas, lentil, pigeon pea, urd bean
15-20	Bengal gram, soybean, maize
20-25	Groundnut, cow pea, onion, pear millet
25-30	Linseed, garlic, guar, lemon grass, palmarosa, sugarcane, cotton
30-50	Wheat, raya, sorghum, sunflower, berseem, shaftal, senji, blue panic, gutton panic
50-60	Barley, sesbania, para grass, Rhodes grass, matricaria
60-70	Rice, sugarbeet, Karnal grass

Note : Relative yields are only 50% of the maximum in the indicated ESP range

Table 4.16. Critical limits of soil ESP at three yield reduction levels for different crops

	Critical ESP for yield reduction		
	10%	25%	50%
Rice	38.0	54.5	72.5
Wheat	22.5	33.3	46.0
Pearl millet	14.5	10.5	27.5
Barley	29.5	42.0	52.2
Cowpea (fodder)	11.0	15.5	24.3
Castor	14.5	22.5	37.0
Safflower	9.5	13.0	15.0
Sunflower	22.5	36.8	52.5
Linseed	14.2	22.0	34.5
Soybean	9.7	14.5	22.3
Garlic	15.0	23.0	35.5
Onion	14.3	22.5	32.3
Pea	9.8	12.5	18.5
Raya (Mustard)	19.5	37.5	68.0
Rye	14.0	18.5	28.0
Dhaincha	15.5	38.5	80.5
Bermuda grass	60.0	85.0	nd
Karnal grass	12.5	nd	nd
Hybrid Napier	21.0	33.0	52.5
Para grass	30.0	52.0	89.5
Sataria grass	24.5	37.5	60.0
Marvel grass	23.3	49.0	92.0
Rhodes grass	14.5	76.5	87.0

nd : determined. Source : Gupta and Gupta (1987).

Gupta and Sharma (1990) tested the model proposed by Maas and Hoffman (1977) to examine its adequacy in describing salinity tolerance of crops to present sodicity tolerance indices for commonly grown crops. In this condition ECe was changed to the level of ESP and the following equation was proposed :

$$Y = \begin{cases} 100 & 0 < ESP \leq ESP_t \\ 100 - S\,(ESP - ESP_t) & ESP < ESP_t \leq ESP_0 \\ 0 & ESP > ESP_0 \end{cases}$$

Where Y is the relative yield, ESP is the average root zone sodicity, ESP_t is the threshold ESP (maximum ESP without yield reduction) and S is the absolute value of the slope (decline in yield per unit increase in the ESP beyond the threshold) of the response function between ESP_t and ESP_0 and ESP_o is the ESP beyond which yield is zero.

The salt tolerance indices in terms of the threshold ESP, slope and ESP_0 as related to above equation for various crops are listed in Table 4.17. The yield of all crops declines with increased ESP beyond a threshold.

Table 4.17. Threshold, slope and ESP_{50} of crops for medium textured soils (Sandy loam)

Crop		Piecewise linear model		
		ESP_t (%)	Slope	ESP_{50} (%)
Barley	*Hordeum vulgare* (P)	8.5	3.5	22.3
Cowpea	*Vigna unguiculata*	13.5	9.1	19.0
Sesbania	*Sesbania aculeata*	46.9	2.4	67.7
Gram	*Cicer arietinum*	7.7	5.0	17.7
Groundnut	*Aracis hypogaea*	8.0	2.2	29.7
Guar	*Cyamopsis psoraloides*	11.9	3.2	27.5
Lentil	*Lens esculentum*	4.9	5.5	14.0
Linseed	*Liunum usitatissinum*	13.3	2.3	25.0
Pearl millet	*Pennisetum typhoideu*	13.6	2.6	32.8
Raya (Indian mustard)	*Brassica juncea* L.	7.6	0.8	70.1
Rice	*Oryza sativa*	24.4	0.9	80.0
	(P)	32.7	1.3	71.2
Rye	*Secale cereale*	11.0	2.6	30.2
Sesame	*Sessamum indicum*	9.0	3.6	22.8
Soybean	*Glycine max*	8.0	3.5	22.3
Sunflower	*Helianthus annuus* L.	11.3	1.1	56.8
Safflower	*Carthamus tinctorius*	7.6	5.2	17.2

Crop		Piecewise linear model		
		ESP_t (%)	Slope	ESP_{50} (%)
Wheat	*Triticum aestivum*	16.4	2.1	40.2
	(P)	20.4	2.6	39.6
Pea	*Pisum sativum*	7.7	4.1	19.9
Onion	*Allium cepa*	9.8	2.2	32.5
Garlic	*Allium sativum*	9.5	1.8	37.3

ESP values indicates average soil ESP of 0-15 cm layer.

ESP_t are the lowest ESP values at which trials were conducted

P = The studies have been conducted in pots.

Response curves for tolerant (with either ESP_t is high or slope is less so that $ESP_{50} \geq 50$). Moderately tolerant and sensitive crops ($ESP_{50} \leq 25$) are presented in Fig. 4.6 which show that crop plants cover a fairly wide spectrum from sensitive to tolerant. It must be, however, kept in mind that crop varieties differ in their salt tolerance widely. Therefore, not only crop selection is important, but also selection of crop varieties is equally important.

Table 4.18. Sodicity tolerance of rice genotypes (Gupta and Sharma, 1990).

Genotypes	Grain yield of rice		
	Threshold ESP_t	Slope	ESP_{50}
CSR-1 Mutant	20.36	2.03	45.00
MI-48	8.02	2.22	30.60
Basmati-370	15.92	1.84	43.10
CSR-1	20.63	0.82	81.60
CSR-3	26.16	0.73	94.65

Table 4.19. Sodicity tolerance of wheat genotypes (Gupta and Sharma, 1990).

Genotypes	Grain yield rice		
	Threshold ESP_t	Slope	ESP_{50}
HD-4502	14.00*	2.83	31.7
HD-2009	14.00*	2.76	32.1
KRL-3-23	14.00*	2.45	34.4
KRL-1-4	14.00*	2.77	32.1
KRL-4-14	14.00*	2.10	37.8
Kharchia-65	17.84	1.85	44.9

*The threshold is less than 14.00. The assumption is made that threshold is 14.00, the lowest ESP at which the experiment has been conducted.

In this regard, significant genotypic differences to alkali conditions have been reported for rice and wheat crops (Gupta and Sharma, 1990). Data presented in Table 4.17 and 4.18 indicate that CSR-3-a natural selection among rice genotypes and Kharichia-65 among wheat genotypes were found to be most tolerant. However, as very limited information is available on this aspect, further studies are warranted in this area.

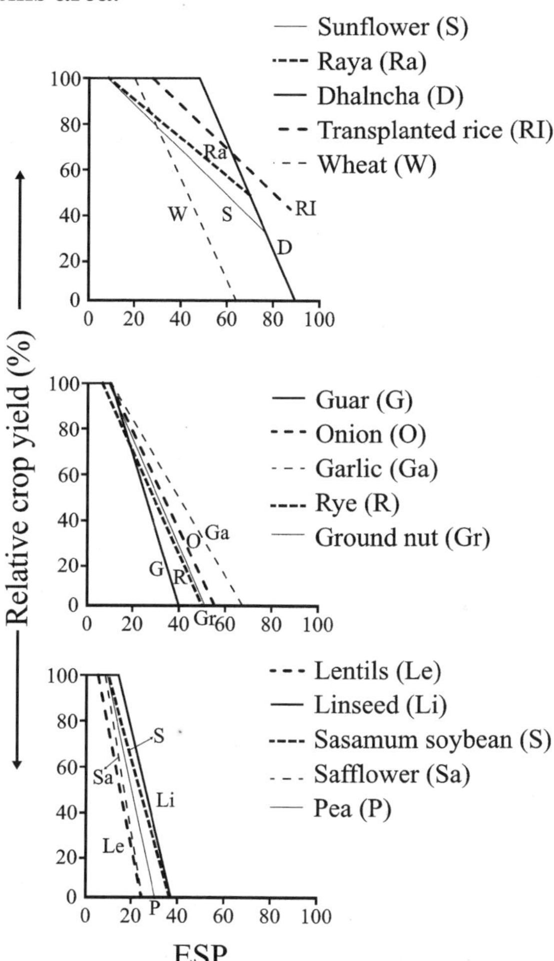

Fig. 4.6. Crops tolerant, moderately tolerant and sensitive to soil ESP

Although genotypic variations in ionic composition and metabolic changes in tolerant and sensitive genotypes of Indian mustard have been reported under salinity (Sharma and Gill, 1994, Garg *et al.*, 1999) but such information is lacking under sodicity stress. Therefore, Burman *et al.* (2001) studied the responses of five genotypes (Varuna C2M-2, Pusa Bold, PCR-7 and BIO-902) of Indian mustard under varying levels of SAR of irrigation water (0.7, 15 and 30). Genotypes Pusa Bold and Bio-902 displayed higher sodicity tolerance whereas genotypes CZM-2 and Varuna were most sensitive with 25% reduction in seed yield at SAR level of 30 (Table 4.20). Sensitive genotypes had higher concentrations of Na and lower K : Na ratio at all SAR levels as compared to tolerant genotype at various growth stages. Increasing levels of SAR progressively decreased net photosynthetic rates and increased leaf diffusive resistance in all genotypes but more so in the sensitive genotypes (CZM-2 and Varuna). Levels of leaf metabolites were also affected more adversely in sensitive than tolerant genotypes under sodicity stress.

Table 4.20. Seed yield (g plant^{-1}) of Indian mustard genotypes as influenced by irrigation with sodic waters

Genotypes	Seed yield at indicated SAR levels			
	0.7	15.0	30.0	Mean
Varuna	2.58	2.28	2.04	2.30
CZM-2	2.54	2.17	1.88	2.19
PCR-7	3.39	3.04	2.75	3.06
Pusa Bold	2.87	2.56	2.42	2.61
BIO-902	2.99	2.73	2.70	2.80
Mean	2.87	2.55	2.35	–
LSD (0.05)	Genotypes 0.32	Sodicity 0.25	GxS NS	

PLANT RESPONSES TO SALINE SPRINKLING WATERS

The salt tolerance data in Table 4.12 or 4.13 apply to crops irrigated with surface methods, such as furrow or basin-type flooding. Sprinker-irrigated crops are subject to additional damage from salt spray on the foliage (Maas, 1985). Salts may be directly absorbed by the leaves, resulting in injury and loss of leaves. In crops that normally restrict salt movement from the roots to the leaves foliar salt absorption can cause serious problems not normally encountered with surface irrigation systems. For example, compared to nonsaline water (EC = 0.6 dS/m), water with an EC = 4.5 dS/m reduced pepper (*Capsicum annuum* L.) yields by over 50% when applied by sprinkler, but only 16% when applied to the soil surface (Bernstein and Francois, 1973).

Crops irrigated with sprinkler are subject to injury not only from salts in the soil but also from salts absorbed through wetted leaf surfaces (Maas, 1985). Some species become severely injured and loose their leaves. The degree of injury is related to the concentration of the salt in the leaves but the weather and water stress can influence the onset of injury (Maas, 1990). However, no guidelines exist for correlating injury of leaves with concentration of salts in the leaves (Garg and Gupta, 1995).

The foliar injury in irrigation waters that are sprinkled on plants depends on the concentration of individual ions in the water, the sensitivity of the crops, the frequency of sprinkling and environmental factors such as temperature, relative humidity, water stress of the plants before irrigation and wind (Maas, 1990).

Table 4.21. Relative susceptibility of crops to foliar injury from saline sprinkling water[1] (Maas, 1990)

Na or Cl conc. (mM/L) causing foliar injury[2]			
< 5	5-10	10-20	>20
Almond	Grape	Alfalfa	Cauliflower
Apricot	Pepper	Barley	Cotton
Citrus	Potato	Cucumber	Sugarbeet
Plum	Tomato	Maize	Sunflower
		Safflower	
		Sesame	
		Sorghum	

1. Susceptibility based on direct accumulation of salts through the leaves.

2. Foliar injury is influenced by cultural and environmental conditions. These data are presented only as general guidelines for day-time sprinkling.

The leaves of many plants readily absorb sodium, calcium and chloride when sprinkled with saline waters containing these elements (Maas, 1985). Ehlig and Bernstein (1959) proved that the ions were directly absorbed by the leaves in sprinkling experiment in which the root medium was protected from salt contamination. However, the susceptibility to foliar injury, varies considerably among species. Several fruit trees (almond, apricot, plum) were partially or completely defoliated when they were sprinkled for over 50 hr with either 10 mM NaCl or $CaCl_2$ (Ehlig and Bernstein, 1959). Citrus leaves absorbed these ions more slowly but still fast enough to cause severe leaf burn. Avocado, in contrast, absorbs salt so slowly that foliar absorption is a negligible factor in sprinkling. Similar results wee obtained for strawberry, which is very sensitive to soil salinity (Ehlig, 1961). Table 4.21 provides the relative susceptibility of various crops to foliar injury due to saline sprinkling waters. The sensitivity of plants to foliar salt injury depends on leaf characteristics affecting rate of absorption and is not generally correlated with tolerance to soil salinity.

Strategies to Minimize Sprinkling Injury

Saline irrigation water is obviously best distributed through surface distribution systems. If sprinkling irrigation with marginally saline water cannot be avoided, several precautions should be taken. When possible, irrigate susceptible crops below the plant canopy to avoid, or at least reduce, wetting the foliage. Hot, dry winds are a major factor in the concentration, absorption and deposition of salts. Avoiding these periods for overhead sprinkling minimises the problem and avoids leaf burn caused by drift to downwind crop areas.

Night sprinkling is very effective in reducing or eliminating both sodium and chloride toxicity due to foliar absorption. Likewise sprinkling during early morning, late evenings and night hours when the winds are likely to be less than in the middle of the day are helpful in avoiding sprinkling drift.

Slowly rotating sprinkler allow appreciate drying on the leaves between sprinkler rotations. Therefore, more frequent or continuous wetting of foliage allows less drying of leaves and less absorption than intermittent wetting and drying. Similarly, a higher rate of water application may reduce the damage by reducing the total period of crop wetting. Increasing the application rate can be accomplished by enlarging the sprinkler orifices, increasing the pressure of reducing the spacing on the sprinkling system.

In extreme cases it may be necessary to change from the more sensitive to tolerant crops, if they can no longer be economically produced. Therefore, crops tolerant to sprinkler irrigation must be identified and evaluated under varying climatic condition. Planting crops during the cooler part of the growing season reduces total water use and the hazards from sprinkler applied water. Therefore, winter crops should be preferred under spinkler irrigation (Garg and Gupta, 1995).

REFERENCES

Abdullah, Z., Khan, M.A. and Flowers, T.J. 2001. Causes of sterility in seed set of rice under salinity stress. J. Agron. Crop Sci 167: 25-32.

Adams, P. and Ho, L.C. 1989. Effects of constant and flunctuating salinity on the yield, quality and calcium status of tomatoes. J. Hort. Sci. 64 : 725-732.

Ahmad, P., Hakeem, K.R., Kumar, A., Ashraf, M. and Akram, N.A. 2012. Salt induced changes in photosynthetic activity and oxidative defence system of three cultivars of mustard (*Brassica juncea* L.) Afr. J. Biotechnol. 11: 2694-2703.

Ahmad, P., Jaleel, C.A., Salem, M.A., Nabi, G. and Sharma, S. 2010. Roles of enzymatic and non-enzymatic antioxidants in plants during abiotic stress, Crit. Rev. Biotechnol. 30: 161-175.

Akbar, M. and Yabuno, S.S. 1977. Breeding for saline resistant varieties of rice. IV Inheritance of delayed type panicle sterility induced by salinity. Japan J. Breed. 27: 237-240.

Akbarimoghaddam, H., Galavi, M., Ghambari, A. and Panjehkeh, N. 2011. Salinity effects on seed germination and seedling growth of bread wheat cultivars. Trakia J. Sci. 9: 43-50.

Asha and Dhingra, H. 2007. Salinity mediated changes in yield and nutritive value of chickpea (*Cicer arietinum* L.) seeds. J. Plant Physiol. 12: 271-275.

Ayers, A.D. and Hayward, H.E. 1948. A method for measuring the effects of soil salinity on seed germination with several crop plants. Soil Sci. Soc. Am. Proc. 13 : 224-226.

Ayers, A.D., Brown, J.W. and Wadleigh, C.H. 1952. Salt tolerance of barley and wheat in soil plots receiving several salinization regimes. Agron. J. 44 : 307-310.

Bernstein, L. 1964. Effect of salinity on mineral composition and growth of plants. Plant Anal. Fert. Prob. 4 : 25-44.

Bernstein, L. 1969. Salinity factors and their limits for citrus culture. Proc. First Int. Citrus Symp. 3 : 1779-1782.

Bernstein, L. 1974. Crop growth and salinity. In : Drainage for Agriculture (ed J. Van Schilf-gaarde) Agronomy 17 : 39-54.

Bernstein, L. 1975. Effects of salinity and sodicity on plant growth. Ann. Rev. Photopathol. 13 : 295-312.

Bernstein, L. 1980. Salt tolerance of fruit crops. USDA Information Bull. 292.

Bernstein, L. and Francois, L.E. 1975. Effect of frequency of sprinkling with saline waters compared with daily drip irrigation. Agron. J. 67: 185-191.

Bernstein, L. and Harward, H.E. 1958. Physiology of salt tolerance. Ann. Rev. Plant Physiol. 9 : 25-46.

Bernstein, L., Ehlig, C.F. and Clark, R.A. 1969. Effect of grape root stocks on chloride accumulation in leaves. J. Am. Soc. Hort. Sci. 94 : 584-590.

Bishnoi, N.R., Laura, J.S., Sharma, K.D. and Singh, N. 1990. Effect of salinity salinization and desalinization on flowering and various yield parametrs in pea (*Pisum sativum* L.) and chickpea (*Cicer ariatum* L.). Inter.. J. Tropical Agric. 8: 143-153.

Boursier, P. and Lauchli, A. 1990. Growth responses and mineral relation of salt stressed sorghum. Crop Sci. 30: 1226-1230.

Burman, U., Garg, B.K. and Kathju, S. 2001. Genotypic variations in growth, mineral composition, photsynthesis and leaf metabolism of Indian mustard under sodicity stress. Indian J. Plant Physiol. 6(N.S.): 374-380.

Chapin, F.S. 1990. Integrated responses of plants to stress - a centralized system of physiological responses. Bioscience 41 : 29-37.

Chapman, V.J. 1966. Vegetation and salinity. In : Salinity and Aridity : new approaches to old problems (ed. H. Boyko) pp. 23-42. Junk, The Hague.

Cooper, W.C., Gorton, B.S. and Edwards, C. 1951. Salt tolerance of various citrus root stocks. Proc. Rio. Grande Valley Hort. Soc. 5 : 46-52.

Cramer, G.R. and Bowman, D.C. 1991. Short term leaf elongation kinetics of maize in response to salinity are independent of the root. Plant Physiol. 95 : 965-967.

Cramer, G.R., Abdel-Basset, R. and Seemann, J.R. 1990. Salinity-calcium interactions of root growth and osmotic adjustment of two corn cultivars differing in salt tolerance. J. Plant Nutr. 13 : 1453-1462.

Cramer, G.R., Epstein, E. and Lauchli, A. 1988. Kinetics of root elongation of maize in response to short term exposure to NaCl and elevated calcium concentration. J. Exp. Bot. 39 : 1513-1522.

Dang, Y.P., Dalal, R.C., Mayer, D.G., McDonald, M., Routely R., Schwenke, G.D., Buck, S.R., Daniells, I.G., Singh, D.K., Manning, W. and Ferguson, N. 2008. High subsoil chloride concentrations reduce soil water extraction and crop yield on Vertisols in north-eastern Australia. Aust. J. Agric. Res. 59: 321-330.

Dantas, B.F., De Sa, R.L. and Aragao, C.A. 2007. Germination, initial growth and cotyledon protein content of bean cultivars under salinity stress. Rev. Bras. de Sementes 29: 106-110.

Deo, R. and Ruhal, D.V.S. 1971. Effect of salinity on the yield and quality of Indian rape (*Brassica campestris* L. var. *sarson*) and linseed (*Linum usitatissium*). Indian J. Agric. Sci. 41 : 134-136.

Dhingra, H.R. and Varghese, T.M. 1997. Flowering and sexual reproduction with salt stress. In : Strategies in Improving Salt Tolerance in Higher Plants (Eds. P.K. Jaiwal, R.P. Singh and A. Gulati) pp. 221-245. Oxford and IBH Publishing Co., New Delhi.

Dolatabadian, A., Modarressanavy, S.A.M. and Ghanati, F. 2011. Effect of salinity on growth, xylem structure and anatomical characteristics of soyabean. Not Sci. Biol. 3: 41-45.

Ehlig, C.F. 1961. Salt tolerance of strawberries with sprinkler irrigation. Proc. Am. Soc. Hort. Sci. 77: 376-379.

Ehlig, C.F. and Bernstein, L. 1959. Foliar absorption of sodium and chloride as a factor in sprinkler irrigation. Proc. Am. Soc. Hort. Sci. 74: 661-670.

Evlagon, D., Ravina, I. and Neumann, P.M. 1990. Interactive effects of salinity and calcium on hydraulic conductivity, osmotic adjustment and growth in primary roots of maize seedlings. Israel J. Bot. 39 : 239-247.

Fernandez-Torquemada, Y. and Sanchez-Lizaso, J.L. 2013. Effects of salinity on seed germination and early seedling growth of the Mediterranean seagrass *Posidonia oceanica* (L.) Delile. Estuarine, Coastal Shelf Sci. 119: 64-70.

Flowers, T.J. 1985. Physiology of halophytes. Plant Soil 89 : 41-56.

Flowers, T.J., Hajibagheri, M.A. and Clipson, N.J.W. 1986. Halophytes. Quart. Rev. Biol. 61: 313-337.

Flowers, T.J., Troke, P.F. and Yeo, A.R. 1977. The mechanisms of salt tolerance in halophytes. Ann. Rev. Plant Physiol. 28 : 89-121.

Francois, L.E. and Clark, R.A. 1980. Salinity effects on yield and fruit quality of Valencia orange. J. Am. Soc. Hort. Sci. 105 : 199-202.

Francois, L.E. and Maas, E.V. 1994. Crop response and management in salt affected soils. In : Handbook of Plant and Crop Stress (ed. M. Pessarakli) pp. 149-181, Marcel Dekker, Inc. New York.

Francois, L.E., Donovan, T.J. and Maas, E.V. 1989. Salinity effect on rye grain yield, quality, vegetative growth and emergence. Agron. J. 81: 707-712.

Francois, L.E., Donovan, T.J. and Maas, E.V. 1989. Salinity effects on seed yield, growth and germination of grain sorghum. Agron. J. 76: 741-744.

Francois, L.E., Donovan, T.J. and Maas, E.V. 1990. Salinity effects on emergence, vegetative growth and seed yield of guar. Agron. J. 82: 587-592.

Francois, L.E., Maas, E.V., Donovan, T.J. and Young, V.I. 1986. Effect of salinity on grain yield and quality, vegetative growth and germination of semi-dwarf and durum wheat. Agron. J. 78: 1053-1058.

Garg, B.K. and Garg, O.P. 1980. Effect of sodium carbonate and bicarbonate on the growth and absorption of essential macro-nutrients and sodium in pea (*Pisum sativum* L.). Proc. Indian natn. Sci. Acad. B46 : 694-698.

Garg, B.K. and Garg, O.P. 1981. Effect of sodium carbonate and bicarbonoate on seed germination. Geobios 8 : 31-33.

Garg, B.K. and Gupta, I.C. 1995. Plant responses to saline spinkling waters. Curr. Agric. 19: 1-11.

Garg, B.K. and Gupta, I.C. 2001. Salt sensitivity of crops at various growth stages. Curr. Agric. 25: 1-12.

Garg, B.K. and Lahiri, A.N. 1986. Problems of salt stress in arid zone crops. In : Proc. natn. symp. on physiological, biochemical and genetic aspects of crop plants in relation to environmetal stresses (eds. R. Singh. I.S. Sheoran and M.R. Saharan) 63-68., H.A.U. Hissar, India.

Garg, B.K., Kathju, S., Vyas, S.P. and Lahiri, A.N. 1997. Sensitivity of clusterbean to salt stress at varius growth stages. Indian J. Plant Physiol. 2: 49-55.

Garg, B.K., Kathju, S., Vyas, S.P. and Lahiri, A.N. 1999. Genotypic differences to salt stress in Indian mustard. In: A.S. Faroda, N.L. Joshi, S. Kathju and Amal Kar (eds.) Management of Arid Ecosystems pp. 295-302. Arid Zone Research Association of India and Scientific Publishers (India), Jodhpur.

Garg, N. and Manchanda, G. 2008. Salinity and its effects on the functional biology of legumes. Acta Physiol. Plant. 30: 595-618.

Glenn, E.P., Brown, J.J. and Khan, M.D. 1997. Mechanisms of salt tolerance in higher plants.

Gopalchari, N.C., Sriamamurthy, C.H. and A. Hanumanta Rao 1968. Effect of forms and levels of nitrogen on the uptake of chloride by flue-cured tobacco from a saline soil. Indian J. Plant Physiol. 11 : 27-37.

Gowing, D.J., Davies, W.J. and Jones, H.G. 1990. A positive root-sourced signal as an indicator of soil drying in apple. *Mauz* x *domestica* Borkh. J. Exp. Bot. 41 : 1535-1540.

Grattan, S.R., Shannon, C., May, D.M., Mitchell, J.P. and Burns, R.G. 1987. Use of drainage water for irrigation of melons and tomatoes. Calif. Agric. 41: 27-28.

Greenway, H. and Munns, R. 1980. Mechanisms of salt tolerance in non-halophytes. Ann. Rev. Plant Physiol. 31 : 149-190.

Gupta, I.C. and Gupta, S.K. 1987. Management of Saline Soils and Waters. Oxford and IBH Publishing Co. New Delhi, 339 p.

Gupta, S.K. 1986. Agricultural drainage methods for salt affected soils. Sinchan 4 : 91.

Gupta, S.K. and Sharma, S.K. 1990. Responses of crops to high exchangeable sodium percentage. Irrig. Sci. 11 : 173-177.

Hasanuzzaman, M., Fujita M., Islam, M.N., Ahamed, K.U. and Nahar, K. 2009. Performance of four irrigated rice varieties under different levels of salinity stress. Int. J. Integr. Biol. 6: 85-90.

Hasson, E. and Poljakoff-Mayber, A. 1981. Does salinity induce early ageing of pea tissue? Oceologia 50 : 94-97.

Hoffman, G.J. and Rawlins, S.L. 1971. Growth and water potential of root crops as influenced by salinity and relative humidity. Agron. J. 63:877-880.

Jennings, D.H. 1976. The effect of sodium chloride on higher plants. Biol. Rev. 51 : 453-486.

Kaddah, M.T. 1963. Salinity effects on growth of rice at the seedling and inflorescence stages of development. Soil Sci. 96: 105-111.

Kaddah, M.T., Lehman, W.F., Merk, B.D. and Robinson, F.E. 1975. Salinity effects on rice after the boot stage. Agron. J. 67: 436-459.

Kaveh, H., Nemati, H., Farsi, M. and Jartoodeh, S.V. 2011. How salinity affect germination and emergence of tomato lines. J. Biol. Environ. Sci. 5: 159-163.

Khan, M.A. and Weber, D.J. 2008. Ecophysiology of high salinity tolerant plants (tasks for vegetation science), 1st edn. Springer, Amsterdam.

Khodarahmpour, Z., Ifar, M. and Motamedi, M. 2012. Effects of NaCl salinity on maize (*Zea mays* L.) at germination and early seedling stage. Afr. J. Biotechnol. 11: 298-304.

Kim, C.M. 1958. Effect of saline and alkali salts on the growth and internal compounds of selected vegetable plants. Physiol. Plant. 441-450.

Kirby, E.J.M. 1988. Analysis of leaf, stem and ear growth in wheat from terminal spikelet stage to anthesis. Field Crops Res. 18 : 127-140.

Kumar, D. 1984. The value of certain plant parameters as an index for salt tolerance in Indian mustard (*Brassica juncea* L). Plant Soil 79 : 261-272.

Lauchi, A. and Grattan, S.R. 2007. Plant growth and development under salinity stress. In: Jenks, M.A., Hasegawa, P.M. and Mohan, J.S. (eds.) Advances in Molecular Breeding towards Drought and Salt Tolerant Crops. Springer, Berlin, pp. 1-32.

Lauchli, A. 1984. Salt exclusion: an adaptation of legumes and pastures crops under saline conditions. In: Salinity Tolerance in Plants : Strategies for Crop Improvements (eds. R.C. Staples and G.H. Toenniessen) pp. 171-187, John Wiley and Sons, New York.

Lunin, J., Gallatin, M.H. and Batchelder, A.R. 1963. Salinity irrigation of several vegetable crops at various growth stages. I. Effect on yields. Agron. J. 55 : 107-110.

Maas, E.V. 1985. Crop tolerance to saline sprinking water. Plant Soil. 89: 273-284.

Maas, E.V. 1986. Crop tolerance to saline soil and water. Proc. US Pak Biosaline Res. Workshop, Karachi, Pakistan, 205.

Maas, E.V. 1990. Crop salt tolerance. In : Agricultural Salinity Assessment and Management. (ed. K.K. Tanji). pp. 262-304, ASCE New York.

Maas, E.V. 1993. Plant growth response to salt stress. In: Towards the Rational Use of High Salinity Tolerant Plants (Eds. H. Lieth and A. All-Masoon) Vol. 1: 279-281. Kluwer Academic Publisher, Netherlands.

Maas, E.V. and Hoffman, G.J. 1977. Crop salt tolerance-current assessment. J. Irrig. Drain. Div. ASCE 103 : 115-134.

Maas, E.V. and Poss, A. 1989. Salt sensitivity of wheat at various growth stages. Irrig. Sci. 10 : 29-40.

Maas, E.V., Hoffman, G.J., Chaba, G.D., Poss, J.A. and Shannon, M.C. 1983. Salt sensitivity of corn at various growth stages. Irrig. Sci. 4 : 45-57.

Maas, E.V., Poss, J.A. and Hoffman, G.J. 1986. Salinity sensitivity of sorghum at three growth stages. Irrig. Sci. 7 : 1-11.

Marcar, N.E. 1987. Salt tolerance in the genus *Lolium* (rye grass) during germination and growth. Aust. J. Agric. Res. 38 : 297-307.

McCree, K.J. 1986. Whole plant carbon balance during osmotic adjustment to drought and salinity stress. Aust. J. Plant Physiol. 13 : 33-45.

Mizarhi, Y. 1982. Effect of salinity on tomato fruit ripening. Plant Physiol. 69 : 966-970.

Mizrahi, Y. and Pasternak, D. 1985. Effect of salinity on quality of various agricultural crops. Plant soil 89 : 301-307.

Munns, R. 1993. Physiological processes limiting plant growth in saline soil: some dogmas and hypotheses. Plant Cell Environ. 16: 15-24.

Munns, R. 2005. Genes and salt tolerance: bringing them together. New Phytol. 167: 645-663.

Murty, K.S. and Rao, C.N. 1965. Studies on salt tolerance in rice 1. Effect of salt concentration on yield and chemical composition of rice. Oryza 2 : 87-91.

Nahar, K. and Hasanuzzaman, M. 2009. Germination, growth, nodulation and yield performance of three mung bean varieties under different levels of salinity stress. Green Farming 2: 825-829.

Orcutt, D.M. and Nilsen, T. 2000. Physiology of Plants under Stress-Soil and Biotic Factors. John Wiley and Sons, Inc. New York.

Osawa, T. 1957. Effect of various concentrations of sodium chloride on the growth, flowering, fruit bearing and chemical composition of egg-plants in sand culture. Hort. Assoc. Japan 26 : 9-14.

Osawa, T. 1960. Studies on the salt tolerance of vegetable crops in sand culture I. On fruit vegetables. Hort. Assoc. Japan. 29 : 294-304.

Osawa, T. 1961. Studies on the salt tolerance of vegetable crops in sand culture. II. On leaf vegetables. Jap. Soc. Hort. J. 30 : 48-56.

Othman, Y., Al-Karaki, G., Al-Tawaha, A.R. and Al-Horani, A. 2006. Variation in germination and ion uptake in barley genotypes under salinity conditions. World J. Agric. Sci. 2: 11-15.

Paliwal, K.V. 1972. Irrigation with Saline Water. IARI Monograph No. 2 (New Series) WTC/IARI, New Delhi.

Paliwal, K.V. and Maliwal, G.L. 1972. Effects of salts on the growth and chemical composition of Okra (*Abelmaschus esculentus)* and sponge gourd *(Luffa cylindrica*). J. Hort. Sci. 47 : 517-524.

Pang, Q., Chen, S., Dai, S., Wang, Y. Chen, Y. and Yan X. 2010. Comparative proteomics of salt tolerane in *Arabidopsis thaliana* and *Thellungiella halophila*. J. Proteome Res. 9: 2584-2599.

Pasternak, D., Twersky, M. and De Malach, Y. 1979. Salt resistance in agricultural crops. In: Stress Physiology in Crop Plants. (Eds.: H.W. Mussel and R.C. Staples) pp. 127-142. Wiley, New York.

Pearson, G.A. 1959. Factors influencing salinity of submerged soils and growth of Calor Rice. Soil Sci. 87 : 198-206.

Pearson, G.A. and Bernstein, L. 1959. Salinity effects on several growth stages of rice. Agron. J. 51 : 654-657.

Pearson, G.A., Ayers, A.D. and Eberhard, D.L. 1966. Relative salt tolerance of rice during germination and early seedling development. Soil Sci. 102 : 151-156.

Poljakoff-Mayber, A. and Lerner, H.R. 1994. Plants in saline environments. In : Handbook of Plant and Crop Stress (Ed. M. Pessarkali) pp. 65-96. Marcel Dekker, Inc. New York.

Raju, K.V. and Ranganayakulu, C. 1978. Effect of saline water irrigtion on salt distribution in profile and performance of sunflower crop. Indian J. Agric. Res. 12 :183-186.

Ramayya, B., Raju, K., Rao, R.V. and Samra, D.A. 1976. Effect of application of saline water at different growth stages on the yield of rice varieties of Sona, IR-8 and Jaya. Andhra Agric. J. 23: 228-234.

Rana, R.S. 1983. Ann. Rep. CSSRI, Karnal, pp. 93-95.

Rao, T.S., Purnapragnachar, H. and Hadimani, A.S. 1969. Effect of soil salinity on the germination of paddy varieties. J. Indian Soc. Soil Sci. 17 : 431-435.

Rhodes, J.D. 1982. Reclamation and management of salt affected soils after drainage. Proc. 1st Annual Water Provincial Conf. Lethbridge, Canada, 123 p.

Rogers, M.E. and Noble, C.Z. 1991. The effect of NaCl on the establishment and growth of Balansa clover (*Trifolium micholinum* var. *balansae*). Aust. J. Agric. Res. 42 : 847-857.

Semiz, G.D., Unulukara, A., Yurtseven, E., Suarez, D.L. and Telei, I. 2012. Salinity impact on yield, water use, mineral and essential oil content of fennel (*Foeniculum vilgare* Mill.). J. Agric. Sci. 18: 177-186.

Shalhevet, J. 1994. Using water of marginal quality for crop production : Major issues. Agric. Water Mgmt. 25 : 233-269.

Shalhevet, J., Yaron, B. and Horowitz, V. 1974. Salinity and citrus yield - an analysis of results from a salinity survey. J. Hort. Sci. 49 : 15-27.

Shanker, A.K. and Venkateswarlu, B. 2011. Abiotic stress in plants - mechanisms and adaptations. In: Tech Janeza Trdine 9, 51000 Rijeka, Croatia.

Shannon, M.C. 1997. Adaptation of plants to salinity. Adv. Agron. 60: 75-120.

Sharma, P.C. and Gill K.S. 1994. Salinity induced effects on biomass yield, yield attributing characters and ionic constituents in genotypes of Indian musard. Indian J. Agric. Sci. 64: 785-790.

Smith, D., Dobrenz, A.K. and Schonhorst, M.H. 1981. Response of alfalfa seedlings to high levels of chloride salts. J. Plant Nutr. 4 : 143-174.

Solomon, M., Gedulovich, E., Mayer, A.M. and Poljakoff-Mayber 1986. Changes induced by salinity to the anatomy and morphology of excised pea roots in culture. Ann. Bot. 57 : 811-818.

Somani, L.L. 1991. Crop Production with Saline Water. Agro Botanical Publishers. Bikaner, India, 308 p.

Storey, R. and Wyn Jones, R.G. 1977. Quarterly ammonium compounds in plants in relation to salt resistance. Phytochemistry 16 : 447-453.

Strogonov, B.P. 1964. Physiological Basis of Salt Tolerance in Plants. IPST, Jerusalem.

Szaboles, I. 1974. Salt affected soils in Europe. Martinus Nijhoff, The Hague, p. 63.

Tanji, K.K. 1990. Agricultural Salinity Assessment and Management, ASCE, New York, 619 p.

Ulfat, M., Athar, H., Ashraf, M., Akram, N.A. and Jamil, A. 2007. Appraisal of physiological and biochemical selection criteria for evaluation of salt tolerance in canola (*Brassica napus* L.) Pak. J. Bot. 39: 1593-1608.

Vorasoot, N., Songsri, P., Akkasaeng, C., Jogloy, S. and Patanothai, A. 2003. Effect of water stress on yield and agronomic characters of peanut. Songklanakarin J. Sci. Technol. 25: 283-288.

Wahhab, A., Muhammad, F. and Ahmad, M. 1957. Soil salinity conditions and growth of crops. PWC. Pakistan Sci. Conf. 9 : 2-4.

Waisel, Y. 1972. Biology of Halophytes. Academic Press. New York, 395 p.

Walker, R.R. and Douglas, T.J. 1983. Effect of salinity on uptake and distribution of chloride sodium and potassium ions in citrus plants. Aust. J. Agric. Res. 34 : 143-153.

West, D.W. and Francois, L.E. 1982. Effect of salinity on germination, growth and yield of cowpea. Irrig. Sci. 3 : 169-175.

White, P.J. and Broadley, M.R. 2001. Chloride in soils and its uptake and movement within the plant: a review. Ann. Bot. 88: 967-988.

Xu, C., Sibicky, T. and Huang, B. 2010. Protein profile analysis of salt-responsive proteins in leaves and roots in two cultivars of creeping bent grass differing in salinity tolerance. Plant Cell Rep. 29: 595-615.

Xu, S., Hu, B., He, Z., Ma, F., Feng, J., Shen, W. and Yan, J. 2011. Enhancement of salinity tolerance during rice seed germination by presoaking with hemoglobin. Int. J. Mol. Sci. 11: 2481-2501.

Zidan, I., Shaviv, A., Ravina, I. and Neumann, P.M. 1992. Does salinity inhibit maize leaf growth by reducing tissue concentrations of essential mineral nutrients? J. Plant Nutr. 15 : 1407-1419.

■■■■

Chapter - 5

MECHANISMS OF SALT INJURY IN PLANTS

INTRODUCTION

Salinity is a potential problem for semi-arid and arid regions of the world. Though all soils contain some amount of soluble salts which are essential for crop production, but problem develops only when soluble salt accumulation in the soil reaches a level harmful to the growth of plants. Saline conditions drastically change the environment of the root aeration, osmotic potential of the soil solution and the normal equilibrium of the dissolved ions. Although salinity is not incompatible with plant life yet only a few crops are able to grow in the presence of high concentration of salts. The most common effect of salinity on plants is suppression of growth which is associated with reduction in crop yields.

Salinity has several deleterious effects on plants such as: Osmotic stress (resulted in loss of cell turgor), Ion toxicity (mainly due to the Na^+ and Cl^- and $SO^{2-}{}_4$) mineral deficiency (specially Potassium and Zinc), Ion imbalance, oxidative stress, stomatal blockade (resulted in reduction of carbon dioxide availability), photosynthesis prohibition, cell division prohibition (interaction of salinity with plants cellular components including DNA, proteins lipids and pigments), increase of sensitivity to diseases and a combination of these (Fahramand *et al.,* 2014; Rani *et al.,* 2012; Aslam *et al.,* 2011; Tester and Davenport, 2003). Also decrease in PSII Photochemistry and chlorophyll control because of salinity and subsequently reduction in photosynthesis and yield production in plants is well reported (Afzal *et al.,* 2014; Jamil *et al.,* 2014; Ashraf and Harris, 2013). Genotype, environment and genotype by environmental (GxE) interactions are major factors that evolutionary influence features of plants, so due to adaptation to the local environments, different genotypes show genetic variation which may impact on the traits of plants including salinity resistance (Bahmani *et al.,* 2012a; Izadi *et al.,* 2011).

SALT INJURY

Levitt (1980) has enumerated 3 types of salt injuries in plants. These are: Primary direct salt injury. Primary indirect salt injury and Secondary salt induced injury.

Primary direct salt injury: This type of injury is very sudden and occurs if there is inundation of plant under very high saline conditions. The plant is under salt shock. This injury is identified by their *speed of action*. Even most halophytes, for instance are sensitive to direct salt effects produced by a sudden and instantaneous salt shock (Repp, 1958). Only species like *Salicornia* can survive high and sudden increase in salt concentration in their root medium. These injuries become too sudden to be osmotic or secondary salt injury. This could be produced even under laboratory conditions. It induces following effects.

(i) This produce, vacuole contraction and protoplasm swelling which may lead to death in few hours.

(ii) It is a salt induced increase in impermeability and therefore, affects the plasma membrane i.e. both *Tonoplast* and *plasmalemma.*

(iii) There is change in the protein and liquid composition of the membranes.

(iv) More direct evidence of primary effect is the strong efflux of K^+ from the leaves as has been observed in case of bean and barley which can only be alleviated by Ca. This efflux is concerned with increase in permeability.

(v) There is irreversible changes in structural and confirmational changes in the ultra structure of the membrances. It affects the selectivity, structure and permeability of the membranes.

(vi) This also leads to high electroconductivity resulting in shunting of ions, leading to interference with membrane transport mechanisms and ultimately leakiness of the membranes, diruption and rupturing of the membranes and finally the death of plant.

Primary indirect salt injury: The salt injury which is induced by disturbances in the metabolic machinery of the plant is called primary indirect salt injury. The inhibition of growth and development including decrease in plant height, tiller number, leaf area, leaf elongation, leaf number, and other yield attributes like ear length and number, grain weight and 1000-grain weight are considered due to drainage and disruption of the normal equilibrium of life processes.

There are the results of aberrations of alterations in normal metabolic functions of plant.

The common analogy is that the plant must maintain a state of equilibrium with its environment in order to survive. If it is growing under normal conditions of soil with usual low concentrations of ions, it must increase its ion concentration well balance in its protoplasm so as to have normal functioning of the cell. However, under salt stress, the maintenance of this unbalanced, steady state requires the continuous expenditure of energy on the part of plant. It must also concentrate to balance the excessive ions like Na, Cl, SO_4 and other toxic ions. So in order to balance the ionic concentrations, the plant shall have to expend energy to support the normal functioning which otherwise could be utilized for maintaining normal growth processes.

Most of the other observed changes in plants under salt stress are reduced cell elongation, cell division (leading to decrease in leaf area), delay in flower emergence, early ripening of fruits (as in peach), enhanced senescence (as in beans and cotton) are ascribed to metabolic disturbances in plants through osmotically induced loss of turgor. Some of these shall be discussed in subsequent chapters.

Secondary Salt Induced Injury

1. Osmotic stress injury (injury due to osmotic effects)

The reduced growth of plants under salinity was ascribed to lack of water. This concept means a shortage of water within the plants even when growing under moist soil solutions, soil conditions or in saline culture solution. There is lowering of osmotic potential of soil water, resulting from high concentrations of soluble salts, was thought to prevent uptake of water by the plant. This decreases the water potential of the soil water and hence water availability further reduces the turgor. The reduced turgor inhibits growth, development, elongation and enlargement of the plant. Hence less height accrues due to osmotic effect.

2. Specific Ion effects (Ion excess stress)

From experimentations on iso-osmotic concentration of PEG and NaCl or Na_2SO_4 shall induce both osmotic as well as ionic effect at equimolar concentrations. However, the effect on (iso-osmotic concentrations) some legumes and cereals happened to be more due to NaCl or Na_2SO_4 than with PEG. So this showed that the more effect of NaCl and Na_2SO_4 than PEG is due to more accumulation of Cl^- and SO_4^{-2}

ions. This is called the specific ion effect. This higher salt stress may led to stomatal closure and therefore, decreased transpiration and it also leads to defoliation and finally reduction in growth, yield and dry matter accumulation. For example HD-4502 accumulates more Na^+ and Cl^- than KH-65 (Kharchia 65) or (Tolerant) KRL-1-4 wheat.

3. Deficiency stress or nutritional imbalance injury

Saline soils are characterized by low nutrient ion activities and extreme ratio of Na^+/Ca^{2+}, Na^+/K^+, Ca^{2+}/Mg^{2+} and $Cl^-/\ NO_3^-$ in the soil solution. The glycophytes which encompass most cultivated crops are exposed to saline conditions, nutritional disorders may develop. These disorders vary in their intensity and can differ among species as well as among cultivars within a species.

The nutritional imbalance is due to:

(i) Effect of salinity on availability of nutrients. Enhanced concentration of salt ions decreased micro- and macro-nutrients.

(ii) Salinity affects the uptake and partioning of nutrients within the plant.

(iii) Physiological inactivation of a given nutrient resulting in deficiency and hence requirement of the plant for that nutrient is increased.

(iv) Salt induces growth inhibition, which then affects the respiration of roots and hence uptake of essential nutrients is decreased.

The three major potential limitations which modulate the growth and development of plants in saline habitats are discussed in detail here.

(1) Water stress arising from the more negative water potential (elevated osmotic potential) of the rooting medium.

(2) Specific ion toxicity usually associated with excessive intake of chloride, sodium and other ions.

(3) Nutrient imbalance or ion imbalance stress.

At times, these hazards operate in an overlapping manner and it becomes difficult to differentiate between their individual effects. It is particularly difficult to distinguish between osmotic and ionic effects of salts, because their relative effects change with the nature of salinity, salt content of the medium and salt accumulation in the plants (Greenway and Munns, 1980; Wyn Jones, 1981; Lauchli and Epstein, 1990).

There is no concensus about the relative importance of these potentially detrimental stresses and indeed reports differ widely in their assessments. Although these stresses may be formally recognized as separate entities, in physiological and biochemical terms yet they may not be so clearly delineated and may closely be interrelated. It will be however, more convenient to discuss the aforesaid effects of salinity separately, although they may frequently operate simultaneously.

Interrelations between Osmotic and Specific Ion Effects

A general introduction to the interrelations between osmotic and specific ion effects, is shown in Fig. 5.1. Salt accumulation in the root environment lowers its water potential with an almost inevitable decrease in water potential of the plant tissues. When there is no salt uptake by the plant, the consequences will be the same as for plants grown on dry soils i.e. in environment with low matric potentials. Such plants may be of low turgor and hydration (A1). Alternatively, organic solute accumulation by the plant cells may restore turgor and hydration, a process called osmotic adjustment (A2). If this happens, considerable amount of metabolic products would be required, particularly in the case of sugars and amino acids. At external concentration of 100 mmol NaCl the cell sap would have to accumulate about 30 g^{-1} of hexose sugars or 60 g l^{-1} of disaccharides. For many species this presents about 20 and 40% sugars on a dry weight basis.

Osmotic adjustment might also be induced by salt uptake (mostly Na^+ and/or Cl^-) into the cells B1); this would avoid a considerable drain on metabolic products. However, salt uptake by the plant introduces the danger of toxicities (B2). The obvious case is the adverse effects of ions on the metabolic machinery (B2b). A more subtle alternative arises from possible high salt concentrations in the cell walls. These would adversely affect the water relation of individual cells i.e. cause their partial dehydration (B2a). However, if the salts are compartmentalized in the vacuole and the cytoplasm is relatively salt free, the toxic effects of the ions may be greatly overcome (B2c) and plant will display a higher salt tolerance as in the case of many halophytes.

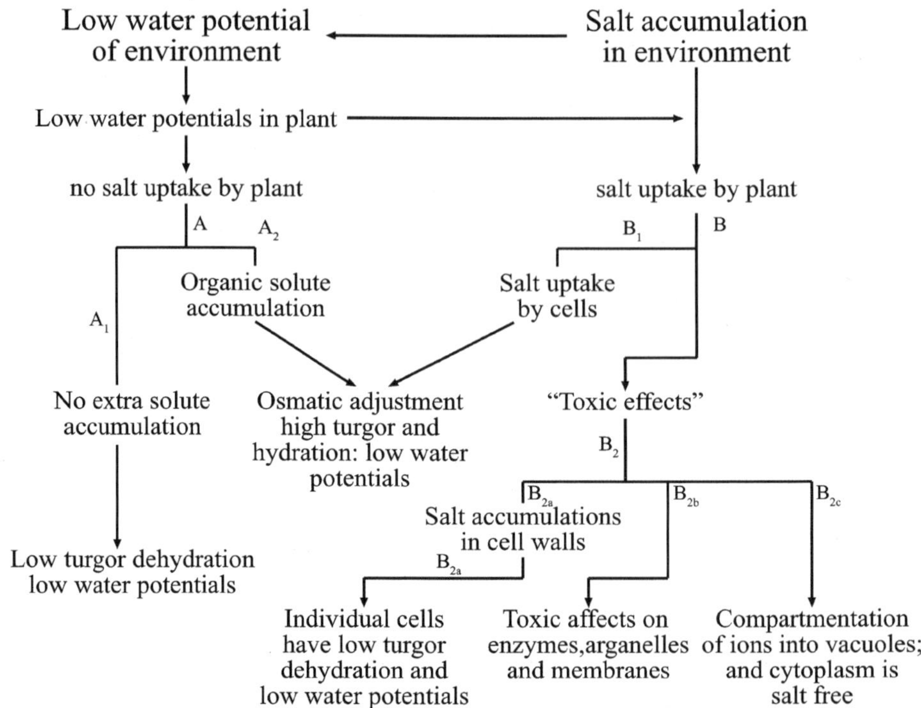

Fig. 5.1. Interrelations between factors involved in the physiology of salt tolerance.

OSMOTIC EFFECTS

Towards the end of 19th century, the reduced growth of plants under salinity was ascribed to lack of water. Schimper (1898) introduced the concept of "physiological drought" which dominated ideas concerning the mechanism of salinity damage in plants for almost half a century. This concept means a shortage of water within the plant even when growing under moist but saline soil conditions or in saline culture solutions. The lowered osmotic potential of soil water, resulting from high concentrations of soluble salts, was thought to prevent uptake of water by the plant. Disturbed water balance was thought to be the main factor of salinity damage, although specific ion effects and response of certain plant species were also recognized. A corollary to the physiological drought hypothesis was that equimolar concentrations of different salts would have the same effect on plant growth.

The physiology drought hypothesis prevailed for some sixty years despite the contrary results of Osterhault (1906) who showed that diluted sea water was less damaging to plant growth than equimolar concentrations of single salts. This hypothesis was further challenged by Bernstein (1961, 1963) and others (Greenway, 1973) who showed that most plants do adapt osmotically under saline conditions and gradient for water flow into the plant is maintained. Concommitantly, resistance of leaves to water loss increases with increasing salinity. This tends to overcome the increased resistance to water uptake in the roots, which is observed to different degrees in crop plants under saline conditions.

Evidence for Osmotic Effects

Water deficits would be indicated by reduced turgor or stomotal opening. Since the water potential of a saline growth medium of soil is lowered by the osmotic potential of the dissolved solutes, the intracellular water potential must decrease either increasing the number of solute species (increasing π sap) or by increasing the water binding capacity (ψ matrix). In the absence of such responses the tissue will loose turgor and dehydrate. Since cell elongation is turgor dependent and contributes massively to vegetative growth, growth will quickly be inhibited. Five evidences in this regard are prescribed below:

(1) The major evidence that water stress is a significant factor in the inhibition of growth by external salts comes from the early work which shows that iso-osmotic concentrations of different salt solution produce a similar reduction in the growth of sensitive glycophytes (Bernstein and Hayward, 1958). Thus, the osmotic potential and not the ionic composition of the medium was taken to be crucial; but these data have not been universally reproduced. In many later experiments different growth responses to salts of various compositions have been established (Lagerwerff, 1969; Greenway and Munns, 1980).

(2) The most persuasive evidence in favor of adverse water relations at high external NaCl is the smaller growth inhibition at high as compared to low relative humidity. This interaction may be greater for salt sensitive than for tolerant species (Hoffman and Rawlins, 1971; O'Leary, 1975) as shown below:

Relative humidity (%)	II ext (bar) causing 50% yield reduction				
	Onion	Radish	Bean	Cotton	Beet
40-45	2.2	2.9	2.2	5	6
85-90	4.3	4.0	3.5	5	7

However, even this evidence is equivocal because net ion uptake to the shoots is often greater at high than low rates of transpiration, particularly when both internal and external ion concentrations are high. It has been demonstrated that net salt transport, especially the transport of potentially toxic ions Na^+ and Cl^- may vary with water flow rates and consequently with relative humidity (Wyn Jones, 1981).

Likewise, Na^+ concentration in leaves of rice increased at high rates of transpiration. Furthermore, ion concentrations in leaves of bean and cotton were higher at low than high relative humidity, the increase being 20% for Cl^- and 45-60% for cations (Na^+, K^+, Mg^{++} and Ca^{++}) at 72 mM NaCl (Nieman and Poulsen, 1967).

(3) A further influence of relative humidity on salt tolerance has been observed in *Atriplex halimus* (Gale *et al.,* 1970) in which at low relative humidity, but not at high RH, moderate salinity (NaCl) promoted growth (Fig. 5.2). This suggests that high tissue salt levels and therefore, high π sap are required for growth under conditions of high transpirational water loss.

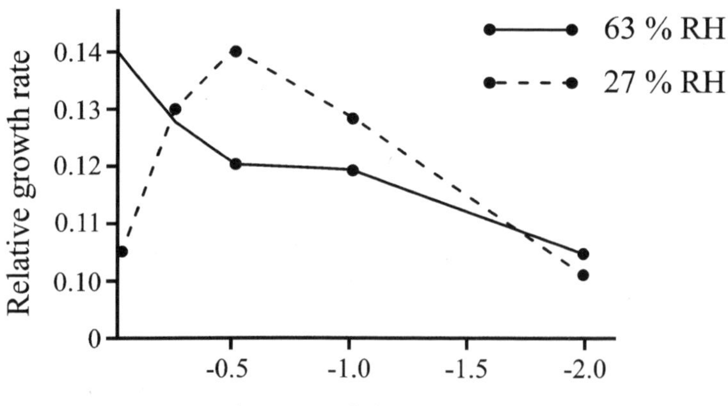

Fig. 5.2. Influence of atmospheric relative humidity and salinity (NaCl) on the relative growth rate of *Atriplex halimus*

(4) In some halophytes organic solutes make a major contribution of π sap which indicate that some measure of osmotic components is the normal response of plants to salinity. This may imply some degree of turgor homeostasis and a constant turgor has been recorded in few instances (Hoffman and Rawlins, 1971). However, growth has been widely found to be inhibited despite the maintenance of a constant p between leaf sap and external solution at high salinity.

(5) Crassulacean acid metabolism (CAM) is probably a metabolic adaptation to semi-arid habitats. In *Mesembryanthemum crystallinum* shift from conventional C_3 metabolism to CAM is induced by NaCl stress (Wyn Jones, 1980) and by other treatments likely to promote water stress, which again implies that water stress is a major factor in salt toxicity. However, Bloom (1979) failed to induce CAM in *M. crystallinum* by a water stress treatment (PEG 6000) and suggested that salt accumulation in the tissue is a requirement for the induction of CAM. Once again, the relationship between water and specific ion stresses is equivocal.

It has been concluded by Munns and Termaat (1986) that rate of leaf area expansion is short term response i.e. it is set within a day of exposure to salinity. They have suggested that it is governed by the response of the root to the low water potential in the soil, and that this response is communicated to the shoot via growth regulators. Understanding these processes might enable us to increase leaf expansion in saline soils.

Shalhevet and Hsiao (1986) showed a clear distinction between the responses of plants to water stress (WS) and salt stress (SS) in terms of incomplete osmotic adjustment under water stress. There were additive effects of WS and SS in depressing growth but at equivalent levels, the relative magnitude of low soil matric potential (WS) on plant growth was twice as great as that of low solute potential (SS). McCree (1986) suggested that realistic physiological studies on salt and water stress should deal with long term rather than short term effects. Sharma (1993) demonstrated that exposure to salinity causes immediate reduction in growth of wheat adjustment through ion accumulation in the tissues, growth showed recovery which was attributed to both decrease in Na as well as to regain of leaf turgor and water relations.

SPECIFIC ION EFFECTS

The second problem facing plants exposed to salinity is the influence of toxic ions on the physiological and biochemical processes. Certain ions like Na^+ and Cl^- dirupt basic features of protein molecules. The disruptive forces of these ions also affect the membranes whose maintenance is basic to biological reactions. Ion excess or toxicity is defined as the condition where high internal ion concentration reduces growth.

Salinity and Ionic Toxicity

The presence of excessive soluble salts in the soil competes with the uptake and metabolism of mineral nutrients that are essential to plants. The appropriate ion ratios provide a tool to the physiological response of a plant in relation to its optimum growth and development (Wang *et al.,* 2003). However, increased salt uptake induces specific ion toxicities like that of high Na^+, Cl^-, or sulfate (SO_4^{-2}) which decrease the uptake of essential nutrients like phosphorus (P), potassium (K^+), nitrogen (N), and calcium (Ca^{2+}). Salinity enhances the Na^+ content in *Vicia faba* while the Na^+/K^+ ratio was increased (Gadallah, 1999) thus suggesting a negative relationship between Na^+ and K^+. In addition, many of the deleterous effects of Na^+ seem to be related to the structural and functional integrity of membranes. Salinity stress causes an increase in the levels of Na^+ and Cl^- in *Atriplex griffithii* in root, stem, as well as in leaves, and the highest ion accumulation was found in leaves followed by stem and root suggesting a positive relationship between Na^+ and Cl^- concentration. The Ca^{2+} content was reduced in shoots and leaves of *A. griffithii* plants grown at high salinity; however, being stable in roots and the K^+ content was reduced with increased levels of salinity, particularly in leaves. On the other hand, Mg^{2+} concentration was not much affected in stems and roots but the decrease in leaf was more prominent (Khan *et al.,* 2000). Decrease in Ca^{2+} and Mg^{2+} content of leaves under salinity stress suggests increased membrane stability and decreased chlorophyll content, respectively (Parida *et al.,* 2004). Despite the fact that most plants accumulate both Na^+ and Cl^- ions in high concentrations in their shoot tissues when grown in saline soils, Cl^- toxicity is also an important cause of growth reduction. Tavakkoli *et al.* (2011) studied the extent to which specific ion toxicity of Na^+ and Cl^- reduces the growth of four barley genotypes grown in saline soils under varying salinity treatments. High Na^+, Cl^- and NaCl separately reduced the growth of barley; however, the reductions in growth and photosynthesis were greatest under NaCl stress and were mainly additives of the effects of Na^+ and Cl^- stress. They also reported that Na^+ and Cl^- exclusion among barley genotypes are independent mechanism and different genotypes expressed different combinations of the two mechanisms. High concentrations of Na^+ reduced K^+ and Ca^{2+} uptake and reduced photosynthesis mainly by reducing stomatal conductance.

Evidence for Ionic Effects

There is some direct evidence that high internal Cl^- and/or Na^+ concentrations reduce growth. This comes from following two approaches which separate effects of ion excess from water deficits.

(1) Growth or yield of some species is reduced at such low external Cl^- (5.10 mM) that adverse effects of low water potential are implausible eg. avocado (Bernstein, 1975), soybean (Lauchli and Wieneke, 1979) and grapevine (Edelbauer, 1978).

After 14 days in 10 mM NaCl (– 0.43 bar) the dry weight of 24 day old plants of sensitive soybean variety was 40% lower than of control without NaCl.

(2) At high osmotic potential growth of beans and maize (salt sensitive) and of barley (salt tolerant) was substantially better in polyethlene glycol (PEG) than in iso-osmotic salt solutions; wheat germination and early seedling growth proved to be more affected by PEG induced water stress than iso-osmotic levels of NaCl whereas later growth stages were more affected by NaCl. Similarly in pearl millet NaCl proved more detrimental to growth and yield than PEG at iso-osmotic concentrations (Garg *et al.,* 1984).

Poorer growth in salt than PEG is convincing evidence for ion excess because some PEG may be transported to the shoot and exert there its own adverse effect (Greenway and Munns, 1980).

(3) Comparison between iso-osmotic concentrations of different salts has been used to distinguish between adverse water relations or ion excess. Various salts often inhibit growth to different degrees (Lagerwerff, 1969; Greenway and Munns, 1980; Garg *et al.,* 1983). The variable growth inhibition by different salts suggests the existence of specific ion effects (Table 5.1 and 5.2). Garg and Garg (1980) demonstrated that $NaHCO_3$ was more detrimental than Na_2CO_3 at equal concentrations for growth and metabolism of pea.

(4) There are many examples of glycophytes and even halophytes in which their tolerance to NaCl has been correlated with their ability to exclude Cl^- and/or Na^+ from the shoots (Table 5.3). Various physiochemical mechanisms are involved in the exclusion of these ions from the shoots, including active Na^+ efflux from roots and absorption by specialized xylem parenchyma cells. Such observations strongly suggest that specific ion toxicity in the shoot is crucial factor in salt stress.

Table 5.1. Effects of different salts on dry weight of various species (Decrease as % of growth in absence of salt) (Lagerwerff, 1969).

Species	Osmotic potential (due to salt)			
	Bar	NaCl	Na_2SO_4	$CaCl_2$
Kidney bean	– 2	– 28	– 31	– 38
	– 3	– 42	– 51	– 49
	– 4	– 58	– 65	– 59
Guyaule				
	– 1	– 37	– 23	– 3
	– 2	– 60	– 60	– 10
	– 3	– 67	– 96	– 40

Table 5.2. Relative effects of different salts of similar EC (10 dSm^{-1}) on grain yield and dry matter (g $plant^{-1}$) of wheat (Garg *et al.*, 1983)

Salts	Grain yield	Dry matter
Control	2.84	8.83
Na_2SO_4	2.39	9.17
KCl	2.34	8.63
$CaCl_2$	2.31	7.13
$MgCl_2$	2.11	6.12
NaCl	1.94	7.14
$NaHCO_3$	1.65	6.52
Na_2CO_3	1.19	4.71
LSD (0.05)	0.51	0.91

Table 5.3. Examples of species where sodium chloride tolerance has been related to ion exclusion (Wyn Jones, 1981)

Family	Species	Exclusion	Reference
Gramineae	Barley	Cl^-, Na^+	Greenway (1973)
	Wheat	Cl^-, Na^+	Storey and Wyn Jones (1978) Torres and Bingham (1973)
Leguminoseae	*Agropyron elongatum*	Cl^-, Na^+	Shannon (1978)
	Soybean	Cl^-	Abel (1969)
	Chickpea	Cl^-, Na^+	Lauter *et al.* (1981)

Family	Species	Exclusion	Reference
	Berseem	Cl^-, Na^+	Winter and Lauchii (1984)
	Beams	Na^+	Jacoby (1965)
Root stocks	Avocado	Cl^-	Downton (1978)
	Grapevine	Cl^-	Ehlig (1960) Downton (1978)
	Citrus species	Cl^-, Na^+	Maas and Hoffman (1977)
	Stone-fruit crps	Cl^-, Na^+	Maas and Hoffman (1977)

(5) Large differences in ion concentrations occur between species and even varieties. There are also large differences in ion concentrations between different plant parts of the same plant. Other leaves of non-halophytes growing at high salinity usually have higher Cl^- concentrations than younger leaves e.g. beans, soybean and many others (Greenway and Munns, 1980). For example, after beans were exposed for 16 days to 50 mM NaCl. Cl^- concentrations in the leaves (mmol) were as below:

Primary	Fourth trifoliate	Youngest trifoliate	Buds
218	139	68	58

Similar results occur for Na^+ in those species which show large net Na^+ transport into leaves e.g. barley (Greenway, 1973) and rice (Yeo and Flowers, 1985). In sharp contrast, young leaves of halophytes have very similar ion concentration to old leaves (Flowers *et al.,* 1977). The equal distribution of ion may reduce both the changes of water deficits in expanding leaves and of ion excess in expanded leaves.

(6) Further evidence for specific ion concentrations and salt tolerance is based on relationship between internal ion concentrations and salt tolerance. Leaf injury occurs in sensitive varieties of wheat, rice and soybean which have very high Na^+ and Cl^- concentrations respectively in their leaves. Garg *et al.* (1990) showed that sensitive wheat variety HD-4502 had much higher Na^+ and Cl^- concentrations in their shoots as compared to salt tolerant variety Kharchia-65 at 10 dSm^{-1} salinity of irrigation water (Table 5.4). Similarly Lahiri *et al.* (1995) reported that sensitive and tolerant genotypes of clusterbean exhibited significant difference in Cl^- and Na^+ concentrations at 10 dSm^{-1} soil salinity (Table 5.5).

Table 5.4. Influence of salt stress on the concentrations of sodium and chloride in the shoot and grain of wheat varieties differing in salt tolerance (Garg *et al.*, 1990)

Varieties	Salinity (dSm^{-1})	Sodium				Chloride			
		55 DAS	75 DAS	Final shoot	Grain	55 DAS	75 DAS	Final shoot	Grain
Kharchia-65	0.2	0.23	0.13	0.25	0.08	0.33	0.20	0.37	0.12
(Tolerant)	10	1.45	1.00	0.69	0.10	1.74	0.20	0.90	0.16
HD-4502	0.2	0.93	0.44	0.50	0.12	1.00	0.52	0.55	0.18
(Sensitive)	10	2.50	1.75	1.32	0.18	2.55	1.68	1.30	0.28

Table 5.5. Influence of soil salinity (10 dSm^{-1}) on the concentrations (%) of Na, K and Cl in shoot tissue of clusterbean genotypes at the flowering stage (Lahiri *et al.*, 1996).

Genotype	Sodium		Chloride		Potassium	
	C	S	C	S	C	S
Tolerant						
Malosen	0.20	0.33	0.28	0.50	2.18	2.07
HFG-182	0.24	0.48	0.31	0.56	2.37	2.25
Sensitive						
FS-277	0.43	0.78	0.49	0.97	2.08	1.87
Durga Jai	0.42	0.85	0.46	1.06	1.90	1.70

C = Control, S = 10 dSm^{-1}

(7) Chloride sensitive crops usually show injury symptoms when their leaves accumulate 0.3 to 1% Cl^- on dry leaf-weight basis, depending on the crop. Typical injury symptoms are leaf burn on dying, occurring initially at the extreme tip of older leaves and progressing along leaf edges towards the stem. In extreme cases chloride toxicity can cause defoliation. Woody plants are especially susceptible to leaf injury caused by the accumulation of toxic levels of Cl^- or Na^+ in the tissue. Characteristic leaf injury symptoms are most frequent in fruit and nut crops such as berries, grapes, stone fruits, citrus, avocado and woody ornamental plants.

(8) *In vitro* studies have shown that soluble enzymes from halophytes and non-halophytes have similar sensitivity to electrolytes (Flowers *et al.*, 1977; Greenway and Munns, 1980). However, growth of many salt tolerant species is much less sensitive to high initial concentration of electrolytes than

would be expected from the response of isolated enzymes. This discrepancy between *in vivo* and *in vitro* responses may arise from lower ion concentrations in the cytoplasm than in the tissues as a whole. However, for most glycophytes there is no discrepancy between the salt sensitivity of enzyme activities *in vivo* and the growth responses. Thus, the reduced growth may well be the inhibition of enzymes by salt, assuming that there are no adaptations as found in halophytes (Flowers *et al.,* 1977).

(9) To evaluate whether there were specific effects on long term growth, apart from an osmotic effect of the soil solution, Munns and Termaat (1986) grew wheat, barley, white clover and Egyptian clover at iso-osmotic concentrations of NaCl and concentrated macro-nutrients for 30 days. Initially there was no difference in growth between plants with and without NaCl at a given osmotic potieal (-0.24 and -0.48 MPa) but by 11 days it was clear that the plants in NaCl were growing less than those in the concentrated nutrients. This indicated that there was a specific and adverse effect of NaCl mainly due to excessive uptake of Na^+ and Cl^-. However, it could also be due partly to insufficient or excessive uptake of other ions as NaCl appears to affect phosphate or nitrate uptake in certain species. Munns and Termaat (1986) ruled out N deficiency as limiting factor and concluded that in the long run there are toxic effect of Na^+ or Cl^- in the leaves, on the top of any osmotic effect of external NaCl on the roots.

In conclusion, it is evident that salt concentrations in individual leaves usually increase with time. Glycophytes show no signs of regulation of the salt concentrations in their leaves unlike many halophytes which can maintain constant salt concentrations. Thus, for non-halophytes it is inevitable that ion concentrations will eventually build up in the older leaves and they will die. The concentrations at which this occurs will depend upon the ability of the species to compartment the salts in the vacuole, and the time taken will depend mainly on the salinity level, the ability of the roots to exclude the salt and the ambient conditions affected the rate of transpiration. Salt build up in cytoplasm interferes with metabolism and in cell wall causes loss of turgor and then excessive loss of water. Death of many older leaves limits productivity due to decreased photosynthetic area. This results in the decline in the production of the carbohydrate and ultimately production per unit plant falls below the levels that plants are not able to sustain growth (Munns and Termaat, 1986). Therefore, the integration of responses in the whole plant is critical for the survival, growth and development of plants under saline conditions.

TWO PHASE GROWTH RESPONSE TO SALINITY (BIPHASIC MODEL)

Munns and Termaat (1986) proposed Biphasic model for inhibition of growth by salinity. This is very important when screening plants for salt tolerance. The first phase of growth reduction in the model is due to the osmotic effect of soil salinity (Fig. 5.3) and is due to the salt outside the roots. It is essentially a water stress or osmotic phase for which there is little genotypic variation. The growth reduction is persumably regulated by hormonal signals coming from the roots (Munns, 1993). Then there is second phase of growth reduction which takes time to develop and results from internal injury. It is due to the salts accumulating in transpiring leaves to excessive levels exceeding the ability of the cells to compartmentalise salts. This will inhibit growth of the young leaves by reducing the supply of carbohydates to the growing cells. The biphasic growth inhibition model also predicts that differences in growth. A response to salinity and varieties of a crop should only be evident during, the second phase of inhibition as was found for wheat, barley and maize genotypes.

The two-phase growth response has been shown clearly for maize and wheat cultivars. Two maize cultivars with two-fold differences in rates of Na^+ accumulation in leaves had the same growth reduction for 15 d in 80 mM NaCl (Cramer *et al.,* 1994). Furthermore, another two maize cultivars, again with two-fold differences in Na^+ accumulation, had the same growth reduction for 4 weeks in 100 mM NaCl, and it was not until 8 weeks that a growth difference was clearly seen (Fortmeier & Schubert, 1995). Similar results were found in wheat (Munns *et al.,* 1995). Two closely related wheat genotypes that differed in rates of Na^+ accumulation had the same growth reduction for the first 4 weeks in 150 mM NaCl, and only in the fifth and sixth week were there differences in dry weight production between the genotypes (Fig. 5.3). However, visual differences appeared long before that; the genotype with the higher rate of Na^+ uptake showed faster leaf senescence, with injury appearing in the oldest leaf after 2 weeks (Fig. 5.3). Later, when the proportion of dead leaves increased above about 20% of the total, the rate of new leaf production showed down dramatically, and some individuals died. These data illustrate the principle that the initial growth reduction is due to the osmotic effect of the salt outside the roots, and that what distinguishes a salt-sensitive plant from a more tolerant one is the inability to prevent salt from reaching toxic levels in the transpiring leaves.

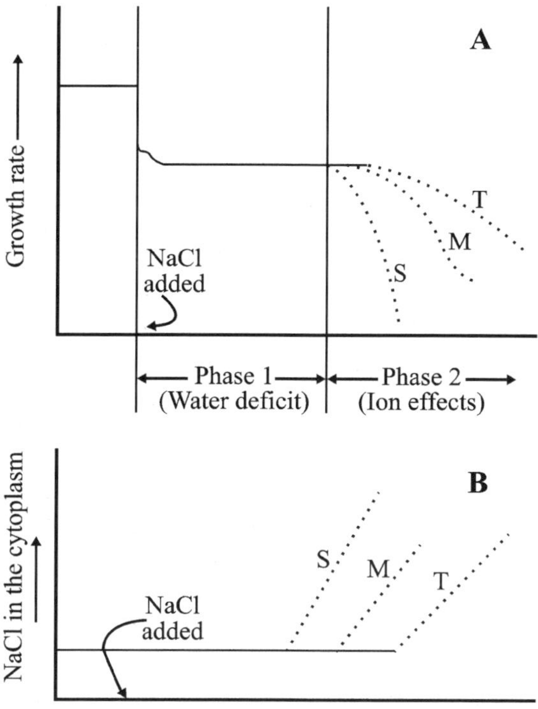

Fig. 5.3. (A) the two phase model for salt inhibition of plant performance. T = a tolerant genotype, M = a moderately tolerant genotypes, S = a sensitive genotype. The initial decrease in growth is due to a water stress effect, and the second decrease in growth is due to ion effects. (B) Cytoplasmic salt concentration increases sooner (early phase 2) in sensitive genotypes while tolerant genotypes can retain a lower cytoplasmic salinity for a longer period.

NUTRIENT IMBALANCE

Saline soils may be characterized by low nutrient ion activities and extreme ratios of Na^+/Ca^{2+}, Na^+/K^+, Ca^{2+}/Mg^{2+} and Cl^-/NO_3 in the soil solution. When glycophytes, which encompass most cultivated crops are imposed to saline conditions, nutritional disorders may develop. These disorders vary in their intensity and can differ among species as well as among cultivars within a species. Nutrient imbalance may result from the effect of salinity on nutrient availability, uptake or partitioning within the plant or may be caused by physiological inactivation of a given nutrient, resulting in an increase in the plant's internal requirement for that essential element (Feigin, 1985; Grattan and Grieve, 1994).

Salinity has been found to reduce N and P accumulation in plants, although this effect may not always be growth limiting (Bernstein *et al.,* 1974; Champagnaol, 1979; Munns and Termaat, 1986; Grattan and Grieve, 1994). The decreased N and P concentrations under salt stress have been attributed to increased Cl^- uptake (Feigin, 1985; Garg *et al.,* 1990). Plants grown in N^- or P^- deficient environments respond positively to addition of these elements provided the plant is not experiencing severe salt stress (Bernstein *et al.,* 1974; Feigin, 1985; Garg *et al.,* 1990, 1993). Levitt (1980) has concluded that reduced uptake of a nutrient is insufficient evidence, by itself, that salt injury is due to a secondary nutrient deficiency. It may, for instance, result from an osmotically produced decrease in root growth or a reduced uptake of a nutrient will occur as a result of salt induced decrease in growth. Therefore, evidence of the reduced uptake must be followed by an elimination of the salt stress injury when the nutrient uptake is returned to normal, for instance by increasing the nutrient supply.

In certain cases, however, P toxicities can develop in salt sensitive plants, if the P concentration in the substrate is too high (Bernstein *et al.,* 1974; Nieman and Clark, 1976; Treeby *et al.,* 1988). This indicates that salt injury is not due to salt induced P deficiency in such crops. Maas and Nieman (1978) suggested that salinity damages the mechanism that controls the inter-cellular phosphate concentration and the cell's energy charge.

Numerous studies have shown that K^+ concentration in plant tissues is reduced as the Na^+ salinity or the Na^+/K^+ ratio in the root medium is increased (Janzen and Chang, 1987; Subbarao *et al.,* 1990). The effect of Na^+ on K^+ uptake is two fold. At low concentrations, Na^+ may actually increase K^+ uptake, though decreases it at higher concentrations (Levitt, 1980). There is evidence that Na^+ can partially substitute for K^+ in many glycophytic species without affecting growth which, however, depend upon the crop species. Despite the overwheming data that show reduced uptake and translocation of K^+ by plants grown in high Na^+ substrate, there are few data which show that the addition of K^+ to sodium dominated soils improved plant growth or yield (Bernstein *et al.,* 1974; Grattan and Grieve, 1994).

Sodium-induced Ca^{2+} deficiencies have notorious growth distorting effects in developing leaves. The uptake of Ca^{2+} from the soil solution may decrease because of ion interactions, precipitation and increases in the ionic strength that reduce the activity of Ca^{2+}. These combined effects are mainly responsible for reduced yield under saline or sodic conditions (Bernstein, 1975; Janzen and Change, 1987). Nutritional imbalances in salt stressed cereals under different molar ratios of Na^+ and Ca^{2+} have demonstrated striking intergeneric differences in corn, rice, wheat, sorghum and pearl millet species (Grattan and Grieve, 1994). Garg and Garg (1980) have reported deficiencies of Ca^{2+} and Mg^{2+} as the major causes of reduction in growth and yield of pea and maize under sodic conditions.

Besides macro-nutrients, salinity or sodicity may also lead to deficiencies or disorders in the levels of micro-nutrients such as zinc, manganese, iron and copper (Hassan *et al.,* 1970; Doering *et al.,* 1984; Suhayda *et al.,* 1992). In saline and sodic soils, the solubility of micro-nutrients is particularly low and plants grown in these soils often exhibit deficiencies in these elements (Page *et al.,* 1990). However, the micro-nutrient concentration in plant shoots may increase, decrease, or have no effect, depending upon the type of plant, tissue, salinity, micronutrient concentration and environmental conditions. For instance salinity increased the concentrations of Zn, Mn and Fe in barley (Suhayada *et al.,* 1992) but decreased in corn (Hassan *et al.,* 1970). It has been reported that saline solutions rich in divalent cations increase shoot Mn concentration whereas a saline environment dominated by monovalent cations reduces the same.

Thus, the available literature indicates that a mineral deficiency or toxicity may sometimes contribute to salt injury but is rarely the sole cause of the injury or inhibition (Levitt, 1980; Grattan and Grieve, 1994).

DETRIMENTAL EFFECTS OF SALINITY

A. Direct effects

Among all plant organs and developmental stages, salinity most directly affects seeds and shallow roots because salts are commonly concentrated in the upper layers of soils. In some cases, aboveground parts are directly affected by salt spray or by inadequate filtration of water moving from the soil into the xylem. Xylem water normally contains few ions (5-30 mM), even in halophytes that concentrates salts in leaves (Glenn *et al.,* 1994). In glycophytes, small increases in soil salinity (50 mM) can rapidly lead to damaging salt buildup in leaf tissues, reaching 500 mM in a few days. The direct effects of excessive ions on plants vary with the species of ion such that sodium ions are more toxic than calcium, and sulfate ions are more toxic than chloride ions. The direct effects of ions can be observed at the level of enzyme activity, membrane function, and several important metabolic processes.

1. Enzyme Activity: As a general rule, enzymes are inhibited by salt *in vitro* equally in glycophytes and halophytes (Greenway and Osmond, 1972; Greenway and Munns, 1980). In fact, salinization of an enzyme extraction is an effective method of precipitating proteins.

Only extremely halophilic bacteria have enzymes that are tolerant of high salinity. Although the enzymes of glycophytes and halophytes are equally sensitive to salinity, the effects of soil solution salts on plant metabolism vary among ionic species and between halophytes and glycophytes. In glycophytes, Na^+ salts are more toxic than K^+ salts, while in halophytes, toxicity follow, the general pattern: $K_2SO_4 > KCl > Na_2SO_4 = NaCl$ (Egan and Ungar, 1998).

Enzymes associated with basic energetic pathways, such as glycolysis, in all plant species studied are sensitive to inhibition by salinity (Blum, 1988). This inhibition of enzyme activity by salinity is not due to osmotic influences, because equimolar concentrations of compounds such as glycine-betaine or proline are much less toxic than NaCl. It has been hypothesized that *in vitro* inhibition of enzyme activity by salt most likely infers that elevated salt concentration (greater than 150 mM) *in vivo* would result in similar inhibition (Greenway and Munns, 1980).

2. Nutrient Imbalance: Many studies have shown that limitations in potassium, calcium, magnesium, and phosphorus can occur in glycophytes exposed to high salinity (Gorham, 1992). Most of these deficiencies and toxicities are discussed in another chapter. Here direct effects of Na^+ and Cl^- on nutrient transport proteins in the root membrane are presented.

It is well-established that crop performance may be adversely affected by salinity-induced nutritional disorders. However, the relations between salinity and mineral nutrition of crops are very complex (Grattan and Grieve, 1999). The nutritional disorders may result from the effect of salinity on nutrient availability, competitive uptake, transport, or distribution within the plant. Numerous reports indicate that salinity reduces nutrient uptake and accumulation of nutrients into the plants (Rogers *et al.,* 2003). The availability of micronutrients in saline soils is dependent on the solubility of micronutrients, the pH of soil solution, redox potential of the soil solution, and the nature of binding sites on the organic and inorganic particle surfaces. In addition, salinity can differently affect the micronutrient concentrations in plants depending upon crop species and salinity levels. Micronurient deficiencies are very common under salt stress because of high pH (Zhu *et al.,* 2004).

Numerous studies have demonstrated that salinity could reduce nitrogen accumulation in plants. Decreased N uptake under saline conditions occurs due to interaction between Na^+ and NH^+_4 and/or between Cl^- and NO^-_3 that ultimately reduce the growth and yield of crop. This reduction in NO^-_3 uptake is associated with Cl^- antagonism or reduced water uptake under saline conditions. The availability of phosphorous is also reduced in saline soil due to (a) ionic strength effects that reduced the activity of PO_4^{3-}, (b) phosphate concentrations in soil solution was tightly controlled by sorpiton processes, and (c) low solubility of Ca-P minerals. Hence, it is noteworthy that phosphate concentration in agronomic crops decreases as salinity increases (Qadir and Schubert, 2002). Sodium concentration in plant tissues increases in the high NaCl treatment and leaf Ca^{2+}, K^+, and N decreases (Tuna *et al.,* 2007). Elevated sodium chloride levels in the root medium reduce the nutrient assimilation, especially of K and Ca, resulting in ion imbalances of K, Ca and Mg (Keutgen and Pawelzik, 2009). In another study, it has been reported that Ca^{2+} and Mg^{2+} concentrations of all plant organs transiently declined in response to external NaCl salinity (Hussain *et al.,* 2013).

Sodium is transported across membranes by a protein complex that moves Na^+ in one direction and H^+ in the other direction. The energy source for this antiport system is a H^+ gradient established across the membrane by H^+-ATPase pumps. ATPase-dependent Na^+/H^+ antiport transporters are inhibited by NaCl. For example, in the salt-tolerant *Plantago* species, high concentrations of Ca^{2+} or Mg^{2+} stimulate H^+-ATPase activity, but H^+-ATPase is inhbited by an equamolar concentration of Na^+. However, long-term exposure to external NaCl can inhibit internal Na^+ stimulation of H^+-ATPase activity, which indicates that long-term exposure to salinity causes a conformational change in the H^+-ATPase complex, or the Na^+ transporter protein in the cell membrane. Because many root membrane transport proteins are dependent upon the electrochemical gradient established by H^+-ATPase, those transporters will be inhibited by the effect of NaCl on H^+-ATPase activity, leading to Na^+-dependent nutrient limitation.

3. Membrane Dysfunction: Some basic structural components of membranes are affected by salinity. Commonly associated with increased salinity is an increase in vesicularization of the plasma membrane (Flowers *et al.,* 1972). There can also be a significant increase in endoplasmic reticulum. The increase is vesicles and endoplasmic reticulum may be a mechanism of compartmentalizing of exporting Na^+ ions.

Under saline conditions, plasma membrane leakage increases in glycophytes. In fact, there is a linear relationship between external salinity and membrane leakage rate. It is clear that Na^+ replaces Ca^{2+} in membranes, and this ionic change disrupts the differential permeability of membranes (Renegal, 1992).

Alongwith the changes in structure and permeability, membrane chemical composition is affected by salinity. Under hypersaline conditions, the total lipids of cells decrease (Kuiper, 1984). In some glycophytes, this ratio increases when exposed to salinity. The saturation level of phospholipids increases with increasing salinity in glycophytes, which reduces membrane fluidity and increases leakage. In addition, salt stress inhibits linoleic acid desaturase, which leads to a lower quantity of linoleic acid in chloroplast membranes and reduces chloro-plast membrane function.

4. General Metabolic Processes: As described above, basic metabolic pathways such as respiration and photosynthesis are inhibited by salinity. Inhibition of respiration is primarily associated with the direct effect of salinity of enzyme function. However, elevated salt content in tissues directly influences photosynthetic enzymes and secondarily influences gas exchange and light reactions.

B. Indirect effects

1. Water Relations: The most observable indirect effect of salinity on plant performance is that of reduced soil water availability, because as salinity increases, soil water potential decreases. Over 100 years ago, Schimper (1898) realized that the osmotic problem of salinity is nearly equivalent to that of water stress. In fact, it has been difficult to separate physiological responses to low soil water potential from those in response to salinity. Only two types of experiments can be used to separate the individual effects of salinity and water stress on plant performance. One is to compare the results between equimolar concentrations of salts and organic osmotica (osmotically active substances). In this case, the water stress effect is the same, but the effect of salinity does not occur in the solution with organic osmotica. Another example is experiments that pressurize the soil solution containing high salinity in comparison to experiments that pressurize the soil solution containing high salinity in comparison to unpressurized, equimolar solutions. the pressurized solution has an osmotic effect but no water stress effect, while the unpressurized solution has both the salinity and water stress effects.

Water potential is an important physiological parameter for determining the water status of the plants (Parida and Das, 2005). According to Romero-Aranda *et al.* (2001), an increase of salt in the root medium can lead to a decrease in leaf water potential and, hence, may affect many plant processes. At very low soil water potentials, this condition interferes with plant's ability to extract water from the soil and maintain turgor. However, at low or moderate salt concentration (higher soil water potential), plants adjust osmotically (accumulate solutes) and maintain a potential gradient for the influx of water.

Stomatoal closure is often a rapid response to the initiation of hypersalinity (Dejong, 1978, Seeman and Sharkey, 1985, Plaut *et al.*, 1989). This rapid closure could be due to a reduction in tissue turgor pressure due to the low water potential of the medium, a specific effect of sodium on guard cell wall plasticity, or an increase in a root hormonal signal (most likely abscisic acid, or ABA). The evidence favors a specific ion effect or a hormonal signal rather than the turgor response. When epidermal strips of *Aster trifolium* are exposed to KCI solutions, stomata open, but when exposed to equimolar NaCl solutions, stomata close (Perera *et al.*, 1994). Termaat *et al.* (1985) showed that there is no association between turgor pressure and leaf growth under hypersaline conditions.

2. *Collateral Nutrient Limitation.* Plants can experience limitations in other nutrients due to the effects of salinity, even in situations where the nutrients are not limiting in the soil solution. Trace elements are not particularly limiting in saline soils. Those most likely to be limiting are copper, iron, manganese, and zinc because these elements have low solubility in saline or alkaline soils (Tanji, 1990). By contrast, boron, molybdenum and selenium have higher solubility in alkaline soils; therefore, plants may accumulate these elements to a higher level in saline soils. An increase in boron concentration in plant tissue could lead to phytotoxicity because plants have a limited tolerance to this trace element. This does not usually occur for Se^{2+} or Mo^{2+} because plants have a relatively high tolerance for these trace elements (compared with other heavy metals).

Macronutrients can become deficient in saline soils because of interference by sodium (Na^+) ions. In barley, the transport of Ca^{2+}, Mg^{2+}, K^+, and N from the root to shoot is reduced when plants are treated with NaCl (Termaat, 1986). This inhibition of macronutrient accumulation is particularly true for potassium (K^+) ions because salinity induces a reduction of K^+ accumulation by roots (Wu *et al.*, 1996). In a saline environment where Na^+ predominates over K^+, the paramount nutritional requirement for plants is K^+ (Rains and Epstein, 1967). Even when the transport mechanisms are highly selective for K^+ over Na^+ (high-affinity transporters), K^+ can become deficient in saline conditions (Ball *et al.*, 1987). In fact, the ability to maintain metabolically significant concentrations of K^+ may be essential for salt tolerance in glycophytes (Zhu *et al.*, 1998).

3. Oxidative Stress : Besides direct impact of salinity on plants, a common consequence of salinity is induction of excessive accumulation of reactive oxygen species (ROS) which can cause peroxidation of lipids, oxidation of proteins, inactivation of enzymes, DNA damage, and/or interact with other vital constituents of plant cells. Salt stress can lead to stomatal closure, which reduces carbon dioxide availability in the leaves and inhibits carbon fixation, exposing chloroplasts to excessive excitation energy which in turn increase the generation of ROS such as superoxide (O^{-}_{2}), hydrogen peroxide (H_2O_2), hydroxyl radical (OH^+), and singlet oxygen ($^1O_2$; Parida and Das, 2005). On the other hand as salt stress is complex and imposes a water deficit because of osmotic effects on a wide variety of metabolic activities (Greenway and Munns, 1980; Cheeseman, 1988). This water deficit leads to the formation of ROS. ROS are highly reactive and may cause cellular damage through oxidation of lipids, proteins, and nucleic acids (Ahmad *et al.,* 2010a,b). In many studies, it was observed that production of ROS is increased under saline conditions and ROS-mediated membrane damage has been demonstrated to be a major cause of the cellular toxicity by salinity in different crop plants such as rice, tomato, citrus, pea, and mustard (Gueta-Dahan *et al.,* 1997; Dionisio-Sese and Tobita, 1998; Mittova *et al.,* 2004; Ahmad *et al.,* 2009, 2010b). In one of the studies, it has been shown that long-term salinity treatments (EC 5.4 and 10.6 dSm^{-1}, 60 days) causes significant increase in H_2O_2 and lipid peroxidation in wheat seedlings, which was higher in salt-sensitive cultivar than salt tolerant cultivar (Sairam *et al.,* 2002). In another study, increased lipid peroxidation and levels of H_2O_2 were observed with increased salinity in *Brassica napus* (Hasan-uzzaman and Fujita, 2010a) and *Triticum aestivum* (Hasanuzzaman and Fujita, 2011b). It has been shown that the production of ROS during environmental stresses such as salinity is one of the main causes for decreases in crop productivity. Therefore, regulation of ROS is a crucial process to avoid unwanted cellular cytotoxicity and oxidative damage.

Several environmental characteristics common in saline habitats can lead to an increase in reactive oxygen species (ROS) and an inhibition of photosynthetic performance. Soil salinity rises to its highest value during warm and relatively dry periods. At these times, soil water potential is low, because of both the high salinity and the low water availability that causes a reduction in stomatal aperture. Saline habitats are also characterized by relatively high light intensity, which in combination with low stomatal aperture can result in the production of ROS in chloroplasts and mitochondria. This increase in ROS for plants experiencing high salt conditions is most likely ubiquitous among plant species (Smirnoff, 1993).

ROS accumulate in the chloroplast because light is stimulating electron flow, in the presence of a limited supply of NADP, to accept the electrons flowing through the photochemical reactions. Therefore, electrons are diverted into ROS through the Mahler reaction (Alscher *et al.,* 1997). An initial oxyradical product, superoxide (O_2^-), cannot cross biological membranes, which restricts it to the site of production (chloroplast and mitochondria). ROS such as superoxide can severely disrupt enzyme activity and membranes associated with photosynthesis. Free-radical scavenging systems such as superoxide dismutase (SOD), and the peroxidase cycle, can be a critical component of salinity tolerance (Olmos *et al.,* 1994, Olmos and Helin 1996, Bohnert and Jensen, 1996) due to their protection of chloroplast function during high light and high salinity.

COMPARATIVE RESPONSES OF PLANTS TO SALT AND WATER STRESS

Plant responses to salt and water stress have much in common. Salinity reduces the ability of plants to take up water, and this quickly causes reductions in growth rate, along with a suite of metabolic changes identical to those caused by water stress. The initial reduction in shoot growth is probably due to hormonal signals generated by the roots. There may be salt-specific effects that later have an impact on growth; if excessive amounts of salt enter the plant, salt will eventually rise to toxic levels in the older transpiring leaves, causing premature senescence, and reduce the photosynthetic leaf area of the plant to a level that cannot sustain growth. These effects take time to develop. Salt-tolerant plants differ from salt sensitive ones in having a low rate of Na^+ and Cl^- transport to leaves, and the ability to compartmentalize these ions in vacuoles to prevent their build-up in cytoplasm or cell walls and thus avoid salt toxicity. In order to understand the processes that give rise to tolerance of salt, as distinct from tolerance of osmotic stress, and to identify genes that control the transport of salt across membranes, it is important to avoid treatments that induce cell plasmolysis, and to design experiments that distinguish between tolerance of salt and tolerance of water stress (Munns, 2002).

Studies on cell water relations and growth, show that the early responses to water and salt stress are essentially identical. Salt-specific effects occur mainly in old leaves where salt brought in with the transpiration stream accumulates to high levels over time.

The similarity between water and salt stress also apply to most metabolic processes; all processes apart from those relating to ion transport. Hormonal responses are similar; for instance, abscisic acid levels rise within 1 h of a imposition of water stress (Bensen *et al.*, 1988) and salt stress (He and Cramer, 1996). Photosynthesis also decreases in both water and salt stress, the responses being over the time scale of days rather than minutes. Although there can be strong correlations between increases in leaf ion concentrations and reductions in photosynthesis or stomatal conductance, there is as yet no unequivocal evidence for casual relationships. Correlations can disappear when considering different leaves, or different salinities (Rawson *et al.*, 1988). Experiments using different genotypes differing in rates of Na^+ or Cl^- accumulation may be able to distinguish between the effects of salt in the leaf, and salt in the soil. Alternatively, stress-relief experiments can show whether the salt in the leaf is causing the decrease in stomatal conductance or photosynthesis, as leaching the soil will quickly restore the water relations of the plant but not affect salt levels in leaves. Such experiments with olive trees have indicated that the decreases in photosynthesis were due to the osmotic effect of the salt outside the roots, not a specific effect of the salt in the leaves (Tattini *et al.*, 1995).

Combined Effects of Salt and Water Stress

The cummulative effects of water and salt stress assume importance when drought affected crops are given a life saving irrigation even with saline waters under semi-arid conditions. In a number of studies with pearl millet and clusterbean (Garg *et al.*, 1984, 1986, 1996) it was found that life saving irrigation with saline water to water stressed plants can protect the crop from total failure and loss of full yields. However, such an irrigation should be provided in the begining of a dry spell rather than at a belated stage when cumulative effects of both the stresses may lead to severe detrimental effects on plant productivity.

Data presented in Table 5.6 indicate that application of saline water upto 8 dSm^{-1} to drought affected clusterbean was beneficial (Garg *et al.*, 1986). The seed yield obtained in undroughted plants even with application of 12 dSm^{-1} salinity water significantly exceeded the yield obtained in droughted plants irrigated with normal tap water. Since drought and salt stress appear to have cumulative adverse effects it was proposed that life saving irrigation at the critical stage of flowering should the provided in the beginning of the dry spell as it is most rewarding. The beneficial gain of such an irrigation may be limited when the plants have already suffered severe water stress.

Table 5.6. Effect of saline waters on seed yield of clusterbean plants with and without water stress

Salinity levels (dSm^{-1})	Seed yield ($g\ plant^{-1}$)		
	Control plant	Water stressed plants	Mean
0	3.25	2.30	2.78
4	3.05	2.25	2.65
8	2.89	1.92	2.40
12	2.69	1.83	2.26
Mean	2.96	2.07	–
LSD (0.05)	Salinity 0.26	Drought 0.18	SxD 0.36

Understanding the mechanisms operating at the whole plant level has implications for screening techniques to distinguish plants that are tolerant of salinity as distinct from soil drying. Salinity can affect growth in a number of ways. The first phase of the growth response is due to the osmotic effect of the salt in the soil solution, and produces a suite of effects identical to those of water stress caused by drought. Later, there may be an additional effect on growth; if excessive amounts of salt enter the plant they will eventually rise to toxic levels in the older transpiring leaves, causing premature senescence. This will reduce the amount of assimilate that the plant can produce, and a reduction in the assimilate transported to the growing tissues may further limit growth. This is the second phase of the growth response, and is the phase that clearly separates species and genotypes that differ in the ability to tolerate saline soil.

Thus, growth is limited predominantly by osmotic stress, but in species that have a high rate of salt uptake, or cannot compartmentalize salt effectively in vacuoles, salt-specific effects develop with time, impose an additional stress on the plant through failing capacity to produce photoassimilate, and give rise to the categories of `salt-sensitive' and `salt-tolerant'. This implies that any improvement in drought resistance would make a plant more adapted to saline soil. However, the processes that adapt a plant specifically to saline soil involve the regulation of the uptake and compartmentation of salt, to delay as long as possible the time when it accumulates to toxic levels in leaves that are actively photosynthesizing. Breeding or genetic engineering of plants better adapted to saline soil should focus on these processes.

The above discussion has not led to a clear definition of the primary source of injury in salt stress because in addition to specific ion, osmotic and nutritional stresses, interactions between these factors also seems to operate. A close relationship between osmotic stress and ion toxicity can be readily visualized even without involvement of possible hormonal interactions. It seems that at a molecular or biochemical level, the rigorous distinction between specific ion and water stresses would appear to have little meaning. At present no discrete universal source of injury component can be recognised. In fact, the scope for interaction and complexity of the mechanisms involved in tolerance is so great, that there is no *prima facie* case for expecting the single universal mechanism of salt injury. These problems are not independent and in most terrestrial plants the solutions to each other are mutually exclusive. The bare fact is that, in a saline environment, a plant needs to adapt osmotically; it is therefore, liable to suffer from ion excess if it accumulates inorganic ions and from water deficits if it not. Furthermore, if a salt resistant glycophyte can osmotically adjust to a saline medium, the increased rates of ion uptake and transport and particularly the synthesis of organic solutes requires additional expenditure of energy that would otherwise support growth processes. Although the energy costs of osmotic adjustment are still poorly understood, the drain of energy for ion compartmentalisation may contribute substantially to the observed growth reduction. The allocation of carbon to the synthesis of organic solutes for osmotic regulation will also involve a cost in terms of reduced growth.

REFERENCES

Afzal, M., Ahmad, A., Alderfasi, A.A., Ghoneim, A. and Saqub, M. 2014. Physiological tolerance and cation accumulation of different genotypes of *Capscium annum* under varying salinity stress. P. Int. Acad. Ecol. Environ. Sci. 4(1): 39-49.

Ahmad, P. and Sharma, S. 2008. Salt stress and phyto-biochemical responses of plants. Plant Soil Environ. 54: 89-99.

Ahmad, P., Hakeem, K.R., Kumar, A. Ashraf, M. and Akram, N.A. 2012. Salt-induced changes in photosynthetic activity and oxidative defense system of three cultivars of mustard (*Brassica juncea* L.) Afr. J. Biotechnol. 11: 2694-2703.

Ahmad, P., Jaleel, C.A. and Sharma, S. 2010b. Antioxidative defence system, lipid peroxidation, proline metabolizing enzymes and biochemical activity in two genotypes of *Morus alba* L. subjected to NaCl stress. Russ. J. Plant Physiol. 57: 509-517.

Ahmad, P., Jeleel, C.A., Azooz, M.M. and Nabi, G. 2009. Generation of ROS and non-enzymatic antioxidants during abiotic stress in Plants. Bot. Res. Intern. 2: 11-20.

Ahmad, P., Nabi, G. and Ashraf, M. 2011. Cadmium-induced oxidative damage in mustard (*Brassica juncea* (L.) Czem. & Coss.) plants can be alleviated by salicyclic acid. South Afr. J. Bot. 77: 36-44.

Akbarimoghaddam, H., Galavi, M., Ghanbari, A. and Panjehkeh, N. 2011. Salinity effects on seed germination and seedling growth of bread wheat cultivars. *Trakia J. Sci.* 9: 43-50.

Apel, K. and Hirt, H. 2004. Reactive oxygen species : metabolism, oxidative stress and signal transudction. Annu. Rev. Plant Biol. 55: 373-399.

Asada, K. 1994. Production and action of active oxygen species in photosynthetic tissue. In: Foyer, C.H., Mullineaux, P.M. (eds.) Causes of Photoxidative Stress and Amelioration of Defense System in Plants. CRC Press, Boca Raton, pp. 77-104.

Ashraf, M. and Harris, P.J.C. 2013. Photosynthesis under stressful environments: An overview. Photosynthetica 51: 163-190.

Aslam, R., Bostan, N. Amen, N., Maria, M. and Safdar, W. 2011. A critical review on halophytes: Salt tolerant plants. J. Med. Plant Res. 5(33): 7108-7118.

Bahmani, K., Izadi Darbandi, A., Jafari, A.A., Sadat Noori, S.A. and Farajpour, M. 2012b. Assessment of Genetic Diversity in Iranian Fennels using ISSR Markers. J. Agr. Sc. 4(9): 79-84.

Bahmani, K., Izadi Darbandi, A., Sadat Noori, S.A. and Jafari, A.A. 2013. Assessment of the Genetic Diversity in Iranian Fennels by RAPD Markers. J. Herbs Spices Med. Plants. 19: 275-285.

Bahmani, K., Izadi Darbandi, A., Sadat Noori, S.A., Jafari, A.A. and Moradi, N. 2012a. Determination of interrelationships among phenotypic traits of Iranian fennel (*Foeniculum vulgare* Mill.) using correlation, stepwise regression and path analyses. J. Essent Oil Bear Pl. 15(3): 424-444.

Bar, Y., Apelbaum, A., fi Kafka, U. and Goren, R. 1997. Relationship between chloride and nitrate and its effect on growth and mineral composition of avocado and citrus plants. J. Plant Nutr. 20: 715-731.

Bensen, R.J., Boyer, J.S. and Muller, J.F. 1988. Water deficit induced changes in absasic and growth, polysomes and transable RNA in soybean hypoctotyls. Plant Physiol. 88: 288-294.

Bernstein, L. 1961. Osmotic adjustment of plants to saline media. I. Steady state. Am. J. Bot. 48 : 909-918.

Bernstein, L. 1963. Osmotic adjustment of plants to saline media. II. Dynamic phase. Amer. J. Bot. 50 : 360-370.

Bernstein, L. 1975. Effects of salinity and sodicity on plant growth. Ann. Rev. Phytopathol. 13 : 295-312.

Bernstein, L. and Hayward, H.E. 1958. Physiology of salt tolerance. Ann. Rev. Plant Physiol. 9 : 25-46.

Bernstein, L., Francois, L.E. and Clark, R.A. 1974. Interactive effects of salinity and fertility on yields of grains and vegetables. Agron. J. 66 : 412-421.

Bloom, A.J. 1979. Salt requirement for Crassulacean acid metabolism in the annual succulent *Mesembryanthemum crystalinum*. Plant Physiol. 63 : 749-753.

Blum, A. 1988. Plant Breeding in Stress Environments. CRC Press, Boca Raton,, FL. 223 p.

Champagnol, F. 1979. Relationship between phosphate nutrition of plant and salt toxicity. Phosphorus in Agriculture 76 : 35-44.

Cheeseman, J.M. 1988. Mechanism of salinity tolerance in plants, Plant Physiol. 87: 547-550.

Cramer, G.R., Alferico, G.J. and Schidt, C. 1994. Salt tolerance is not associated with the sodium accumulation of two maize hybrids. Aust. J. Plant Physiol. 21: 675-692.

Dionisio-Sese, M.L. and Tobita, S. 1998. Antioxidant responses of rice seedlings to salinity stress. Plant Sci. 135: 1-9.

Doering, H.W.. Schulze, G. and Roschder, P. 1984. Salinity effects on the micronutrient supply of plants differing in salt resistance. Proc. 6th Inter. Collog. Optimization Plant Nutrition. pp. 165-172. Montpellier France.

Elstner, E.F. 1987. Metabolism of activated oxygen species. In: Davies D.D. (ed.) The biochemistry of plants, vol. II, Biochemistry of metabolism. Academic, San Diego, pp. 252-315.

Fahramand, M., Mahmoody, M., Keykha, A., Noori, M. and Rigi, K. 2014. Influence of abiotic stress on proline, photosynthetic enzymes and growth. Int. Res. J. Appl. Basic Sci. 8(3): 257-265.

Feigin, A. 1985. Fertilization management of crops irrigated with saline water. Plant Soil 89 : 285-299.

Gadallah, M.A.A. 1999. Effects of proline and glycine betaine on *Vicia faba* response to salt stress. Biol. Plant. 42: 249-257.

Garg, B.K. and Garg, O.P. 1980. Sodium carbonate and bicarbonate induced changes in growth, chlorophyll, nucleic acids and protein contents in leaves of *Pisum sativum*. Photosynthtica 14 : 594-598.

Garg, B.K. Vyas, S.P., Kathju, S., Lahiri, A.N., Mali, P.C. and Sharma, P.C. 1993. Salinity-fertility interaction on growth, mineral composition and nitrogen metabolism of Indian mustard. J. Plant Nutr. 16 : 1637-1650.

Garg, B.K., Kathju, S., Vyas, S.P. and Lahiri, A.N. 1990. Effect of saline water irrigation on tolerant and sensitive wheat varieties under disparate soil fertility conditions. Ann. Arid Zone. 29 : 179-189.

Garg, B.K., Vyas, S.P., Kathju, S. and Lahiri, A.N. 1984. Study of relative consequences of water and salt stresses in pearl millet. Sci. Cult. 50 : 96-97.

Glenn, E.P., Olsen, W.E., Frge, R.J. and Moore, D.W. 1994. How much soldium accumulation is necessary for salt tolerance of subspecies of the halophyte *Atriplex canescens?* Plant Cell Environ. 17: 711-719.

Grattan, S.R. and Grieve, C.M. 1994. Mineral nutrient acquision and response by plants grown in saline environments. In : Handbook of Plant and Crop Stress (ed. M Pessarakli) pp. 203-226. Marcel Dekker, Inc. New York.

Grattan, S.R. and Grieve, C.M. 1999. Salinity-mineral nutrient relations in horticultural crops. Sci. Hortic. 78: 127-157.

Grattan, S.R. and Maas, E.V. 1988. Effect of salinity on phosphate accumulation and injury in soybean. Influence of $CaCl_2$/NaCl ratios. Plant Soil 105 : 25-32.

Greenway, H. 1973. Salinity, plant growth and metabolism. J. Aust. Inst. Agric. Sci. 39 : 24-34.

Greenway, H. and Munns, R. 1980. Mechanisms of salt tolerance in non-halophytes. Ann. Rev. Plant Physiol. 31 : 149-190.

Greenway, H. and Osmond, C.B. 1972. Salt resonses of enzymes from species differing in salt tolerance. Plant Physiol. 49: 256-259.

Gueta-Dahan, Y., Yaniv, Z., Zilinskas, B.A. and Ben-Hayyim, G. 1997. Salt and oxidative stress: similar and specific responses and their relation to salt tolerance in citrus. Planta 204: 460-469.

Halliwell, B. and Gutteridge, J.M.C. 1985. Free radicals in biology and medicine, Clarendon, Oxford.

Halliwell, B. and Gutteridge, J.M.C. 1989. Free radicals in biology and medicine, 2nd edn. Clarendon, Oxford.

Hasanuzzaman, M. and Fujita, M. 2011a. Selenium pretreatment upregulates the antioxidant defense and methylglyoxal detoxification system and confers enhanced tolerance to drought stress in rapeseed seedlings. Biol. Trace Elem. Res. 143: 1758-1776.

Hasanuzzaman, M. and Fujita, M. 2011b. Exogenous silicon treatment alleviates salinity-induced damage in *Brassica napus* L. seedlings by upregulating the antioxidant defense and methylglyoxal detoxification system. Abstract of Plant Biology 2011, American Society of Plant Biology, http://abstracts-aspb.org/pb2011/public/P10/P10001.html

Hassan, N.A.K., Drew, J.V., Knudsen, D. and Olson, R.A. 1970. Influence of soil salinity on production of dry matter and uptake and distribution of nutrients in barley and corn I. barley (*H. vulgare* L.). Agron. J. 62 : 43-45.

He, T. and Cramer, G.R. 1996. Abscisic acid concentrations are correlated with leaf reduction in two salt stressed rapid cycling *Brassica* species. Plant Soil 179: 25-33.

Heywood, V.H. 2002. The conservation of genetic and chemical diversity in medicinal and aromatic plants. In: S. Ener, B. (ed.) Biodiversity: Biomolecular aspects of Biodiversity and Innovative Utilization. Kluwer Academic, Plenum Publishers, New York, pp. 13-22.

Hoffman, G.J. and Rawlins, S.L. 1971. Growth and water potential of root crops as influenced by salinity and relative humidity. Agron. J. 63 : 822-826.

Hu, Y. and Schmidhalter, U. 2005. Drought and salinity: a comparison of their effects on mineral nutrition of plants. J. Plant Nutr. Soil Sci. 168: 541-549.

Hussin, S, Geissler, N. and Koryo H.W. 2013. Effect of NaCl salinity on *Atriplex nummularia* (L.) with special emphasis on carbon and nitrogen metabolism. Acta Physiol. 35: 1025-1038.

Jamil, M., Rehman, S. and Rha E.S. 2014. Response of growth, PSII photochemistry and chlorophyll content to salt stress in four *Brassica* species. Life Sci. J. 11(3): 139-145.

Janzen, H.H. and Chang, C. 1987. Cation nutrition of barley as influenced by soil solution composition in a saline soil. Can. J. Soil Sci. 67 : 619-629.

Keutgen, A.J. and Pawelzik, E. 2009. Impacts of NaCl stress on plant growth and mineral nutrient assimilation in two cultivars of strawberry. Environ. Exp. Bot 65: 170-176.

Khan, M.A., Ungar, I.A. and Showalter, A.M. 2000. Effects of salinity on growth water relations and ion accumulation of the subtropical perennial halophyte. *Atriplex griffithii* var. *stocksii.* Ann. Bot. 85: 225-232.

Kuiper, P.C. 1984. Functioning of plant cell memberanes under saline conditions: membrane lipid composition and ATPase. In: Salinity Tolerance in Plants Strategies for Crop Improvement (eds.) R.C. Staples and G.H. Toenniessen. pp. 71-91. John Willey, New York.

Kurth, E., Cramer, G.R., Lauchli, A. and Epstein, E. 1986. Effect of NaCl and $CaCl_2$ on cell enlargement and cell production in cotton roots. Plant Physiol. 82: 1102-1106.

Lagerwerff, J.V. 1969. Osmotic growth inhibition and electrometric salt tolerance evaluation of plants. Plant Soil 31 : 77-96.

Lahiri, A.N., Garg, B.K., Vyas, S.P., Kathju, S. and Mali, P.C. 1996. Genotypic differences to soil salinity in clusterbean. Arid Soil Res. Rehabilit. 10 : 333-345.

Lauchli, A. and Epstein, E. 1990. Plant responses to saline and sodic conditions. In : Agricultural Salinity Assessment and Management (Ed. K.K. Tanji). pp. 113-137. ASCE, New York.

Lauchli, A. and Wieneke, J. 1979. Studies on growth and distribution of Na^+, K^+ and Cl^- in soybean varieties differing in salt tolerance. Z. Pflanzenernachr. Bodenkd. 142 : 3-13.

Lauter, D.J., Munns, D.N. and Clarkin, K.L. 1981. Salt response of chickpea as influenced by N supply. Agron. J. 73 : 961-966.

Lea-Cox, J.D. and Syvertsen, J.P. 1993. Salinity reduces water use and nitrate N-use efficiency of citrus. Ann. Bot. 72: 47-54.

Levitt, J. 1980. Responses of Plants to Environmental Stresses. Vol. II Water, Radiation, Salt and other Stresses. Academic Press Inc. New York, 607 p.

Maas, E.V. and Hoffman, G.J. 1977. Crop salt tolerance - current assessment. J. Irrig. Drain. Div. ASCE 103 : 115-134.

Maas, E.V. and Nieman, R.H. 1978. Physiology of plant tolerance to salinity. In: Crop Tolerance to Suboptimal Land Conditions (Ed. G.A. Jung) Chap. 13 Am. Soc. Agron. Spec. Publ. 32 : 277-289.

McCree, K.V. 1986. Whole plant carbon balance during osmotic adjustment to drought and salinity stress. Aust. J. Plant Physiol. 13 : 33-45.

Mittova, V., Guy, M., Tal, M. and Volokita, M. 2004. Salinity upregulates the antioxidative system in root mitochondria and peroxisomes of the wild salt-tolerant tomato species *Lycopersicon pennellii.* J. Exp. Bot. 55: 1105-1113.

Munns, R. 1994. Physiological processes limiting plant growth in saline soil: some dogmas and hypothesis. Plant Cell Environ. 16: 15-24.

Munns, R. 2002. Comparative physiology of salt and water stress. Plant Cell Environ. 25: 239-250.

Munns, R. and Termaat, A. 1986. Whole plant responses to salinity. Aust. J. Plant Physiol. 13 : 143-160.

Munns, R., Schachtman, D.P. and Condon, A.G. 1995. The significance of a two-phase growth response to salinity in wheat and barley. Aust. J. Plant Physiol. 22: 561-569.

Nieman, R.H. and Clark, R.A. 1976. Interactive effects of salinity and phosphorus nutrition on the concentrations of phosphate and phosphate esters in mature photosynthesizing corn leaves. Plant Physiol. 57 : 157-161.

Nieman, R.H. and Poulsen, L.L. 1967. Interactive effects of salinity and atmospheric humidity on the growth of bean and cotton plants. Bot. Gaz. 128 : 69-73.

O'Leary, J.W. 1975. Salinity induced changes in hydraulic conductivity of roots. Proc. Symp. Tertranika Lomnica Czechoslovakia 1971 (Ed. J. Kolek) pp. 309-314, Veda Publ. House Slovak Acad. Sci. Bratislava.

Orcutt, D.M. and Nilsen, E.T. 2000. Salinity stress In: Physiology of Plants under Stress - Soil and Biotic Factors. John Willey and Sons, Inc. New York.

Page, A.L., Chang, A.C. and Afriana, D.C. 1990. Deficiencies and toxicities of trace elements. In : Agricultural Salinity Assessment and Management (Ed. K.K. Tanji) pp. 138-160. ASCE Manuals and Reports on Eng. Practice No. 71, ASCE New York.

Parida, A.K., Das, A.B. and Mitra, B. 2004. Effects of salt on growth, ion accumulation, photosynthesis and leaf anatomy of the mangrove, *Brugulera parviflora.* Trees-Struct Funct. 18: 167-174.

Perera, L.K.R., Mansfield, T.A. and Malloch, A.C. 1997. Stomatal responses to sodium ions in *Aster trifolium*. Plant Cell Environ. 17: 335-340.

Rani, C.R., Reema, C., Alka, S. and Singh, P.K. 2012. Salt tolerance of *sorghum bicolor* cultivars during germination and seedling growth. Res. J. Rec. Sci. 1(3): 1-10.

Romero-Aranda, R., Soria, T. and Cuartero, S. 2001. Tomato plant water uptake and plant-water relationships under saline growth conditions. Plant Sci. 160: 265-272.

Seaman, J.R. and Critichley, C. 1985. Effects of salt stress on the growth, ion content, stomatal behaviour and photosynthetic capacity of a salt sensitive species. *P. vulgaris* L. Planta 164: 151-162.

Shalhevet, J. and Hsiao, T.C. 1986. Salinity and drought - a comparison of their effects on osmotic adjustment, assimilation, transpiration and growth. Irrig. Sci. 7 : 249-264.

Sharma, S.K. 1993. Physiological aspects of crop response to saline/sodic irrigation II. Plant water and ionic relations. In : Advances in Research on Sustainable Irrigation in Saline Environment (Eds. N.K. Tyagi, S.K. Kamra, P.S. Minhas and N.T. Singh). pp. 84-95. CSSRI, Karnal, India.

Smirnoff, N. 1998. Plant resistance to environmental stress. Curr. Opin. Biotechnol. 9: 14-19.

Subbarao, G.V., Johansen, C., Jana, M.K. and Kumar Rao, J.F.D.K. 1990. Effects of sodium/calcium ratio in modifying salinity response of pigeonpea (*Cajanus cajan*). J. Plant Physiol. 136 : 439-443.

Suhayda, C.G., Wang, X and Redmann, R.E. 1992. Trace metal composition of *Hordeum* species as altered by salinity. In : Salinity and Sustainable Agriculture (Ed. H. Steppuhn and D. Curtin). Swift Current Research Station. Saskatchewn.

Tanji, K.K. 1990. Agricultural Salinity Assessment and Management, ASCE, New York, 619 p.

Tattini, M., Gucci, R. Coradeicli, M.A., Ponzio, C. and Everard, J.D. 1999. Growth, gas exchange and ion content in *Olea europaea* plants during salinity and subsequent relief. Physiol. Plant. 95: 117-124.

Tavakkoli, E., Fatehi, F., Coventry, S., Rengasamy, P. and McDonad, G.K. 2011. Additive effects of Na^+ and Cl^- ions on barley growth under salinity stress. J. Exp. Bot. 62: 2189-2203.

Termaat, A. and Munns, R. 1996. Use of concentrated macronutrient solutions to separate osmotic from NaCl-specific effects on plant growth. Aust. J. Plant Physiol. 13:509-522.

Termaat, A., Passioura, J.B. and Munns, R. 1985. Shoot turgor does not limit shoot growth of NaCl-affected wheat and barley. Plant Physiol. 77: 819-872.

Tester, M. and Davenport, R. 2003. Na tolerance and Na transport in higher plants. Ann. Bot. 91: 503-527.

Treeby, M.T. and Van Steveninck, R.F.M. 1988. Effects of salinity and phosphate on ion distribution in lupin leaflets. Physiol. Plant. 73 : 317-322.

Tuna, L.A., Kaya, C., Ashraf, M., Altunlu, H., Yokas, I. and Yagmur, B. 2007. The effects of calcium sulphate on growth, membrane stability and nutrient uptake of tomato plants grown under salt stress. Environ. Exp. Bot. 59: 173-178.

Volkmar, K.M., Hu, Y. and Steppuhn, H. 1998. Physiological responses of plants to salinity: A review. Can. J. Plant Sci. 78: 19-27.

Wang, W., Vinocur, B. and Altman, A. 2003. Plant responses to drought, salinity and extreme temperatures: towards genetic engineering for stress tolerance. Planta 218: 1-14.

Wu, S., Ding, L. and Zhu, J. 1991. SOSI- a genetic locus essential for salt tolerance and potassium acquisition. Plant Cell. 8: 617-627.

Wyn Jones, R. 1981. Salt tolerance. In: Physiological Processes Limiting Plant Productivity (Ed. C.B. Johnson) pp. 271-292. Butterworths, London.

Yeo, A.R. and Flowers, T.J. 1986. Salinity resistance in rice (*Oryza sativa* L.) and a pyramiding approach to breeding varieties for saline soils. Aust. J. Plant Physiol. 13 : 161-173.

Zhu, B.C., Su, J., Chen, M.C., Verma, S.P.S., Fan, Y.I. and Wu, R. 1998. Over expression of Δ1-pyrroline-S carboxylate synthetase gene and analysis of tolerance to water stress and salt stress in transgenic rice. Plant Sci. 139: 41-48.

Zhu, J.K. 2002. Salt and drought stress signal transduction in plants. Ann. Rev. Plant Biol. 5: 247-273.

Zhu, Z.J., Wei, G.Q., Li, J., Qian, Q.Q. and Yu, J.Q. 2004. Silicon alleviates salt stress and increases antioxidant enzymes activity in leaves of salt stressed cucumber (*Cucumis sativus* L.) Plant Sci. 167: 527-533.

■■■■

Chapter - 6

SALINITY AND METABOLIC PROCESSES

INTRODUCTION

Salinity affects all the aspects of plant metabolism and induces changes in plant anatomy and morphology. Some of these changes are often considered to be adaptations which help the plants to endure the stress imposed by salinity. However, most of the observed changes seem to be symptom of salinity damage. The primary observable effect of salinity on all types of plants is reduced growth. Extreme salinities result in death, preceded by loss of ionic control in the root, chlorosis, wilting and necrosis. At sub-lethal salinities growth processes although slowed, are not prevented. Cells are not visibly damaged and development follows the normal pattern (Maas and Nieman, 1978).

Crop productivity is severelly affected by salinity stress. This occurs directly due to the impact of salinity on photosynthesis, respiration, nutrient assimilation, hormonal imbalance, etc. The indirect adverse effect of salinity is enhanced generation of reactive oxygen species in stressed plants which subsequently cause damage to macromolecules such as lipids, proteins, and nucleic acids, and thus, constrain crop productivity.

The salt induced growth inhibition is accompanied by one or many metabolic disturbances which are expected to inhibit growth or may be a factor in cell injury. Most of these disturbances are mediated via adverse effects of salts on the enzymatic pathways. However, such effects depend on the type and degree of salinity, extent of the salt resistance and susceptibility of the plant species, stage of growth and environmental conditions. Some of the major metabolic processes, as affected by salinity, are described below.

PHOTOSYNTHESIS

Photosynthesis is one of the most important biochemical pathways by which plants convert solar energy into chemical energy and grow. The reduction in photosynthetic rates in plants under salt stress is mainly due to the reduction in water potential.

Photosynthesis is also inhibited when high concentrations of Na^+ and/or Cl^- are accumulated in the chloroplasts and chlorophyll being important content of photosynthesis is adversely affected.

The effects of salt stress on photosynthesis are either direct or indirect. Direct effects are the diffusion limitations through the stomata and the mesophyll and alteration in the photosynthetic metabolism. The secondary or indirect effects are imposition of the multiple stress arising due to osmotic and specific ion effects of salts (Chaves *et al.,* 2009).

There are numerous reports of a general reduction in photosynthesis in crop plants by various types of salinity e.g. cotton, onion, beans, tomato, wheat, barley, pearl millet, maize, pea, sugarbeet, rice and many others (Boyer, 1965; Gale, 1975; Downton, 1977; Papp *et al.,* 1983; McCree, 1986). An exception to the general rule can be seen in halophytes, where low concentrations of salt do not always reduce, and may even enhance photosynthesis.

Photosynthesis involves a long chain of mechanisms, enzyme activities and generation of intermediate products, which are regulated by large number of internal and external parameters. Salinity may affect the overall process of photosynthesis at different points, either by inhibiting enzyme activity or by altering it. The effects of salinity on photosynthesis can be both stomatal (decreased intracellular CO_2 due to stomatal closure, lowering of stomatal conductance (Brugnoli and Lauteri, 1991)) or non-stomatal (decrease in RuBPase activity). The other effects include decrease in the quantum efficiency of CO_2 uptake (Seeman and Chritchley, 1985) and change in the ionic relations of the chloroplast (Long and Baker, 1986) or change in photochemical reactions (Reddy *et al.,* 1992).

Soil salinity decreases the rates of photosynthesis and plant growth to various degrees in crop plants. The reduction in gas exchange has been attributed to salt damage of the photosynthetic tissue (chloroplasts), to stomatal effects and consequent restriction of the availability of CO_2 for carboxylation, to non stomatal effects and to acceleration of senescence (Passarakli, 1994, Garg and Gupta, 1997). However, in wheat only limited work has been carried out on photosynthesis under saline conditions. Rates of photosynthesis are usually lower in NaCl-treated plants and may decrease with time of exposure to a given salt stress level (Munns and Termaat, 1986). Sharma (1996) observed that net photosynthetic rate, transpiration rate and stomatal conductance were the highest in flag leaf, declined in the middle and fully expanded leaves and were minimum in the oldest leaves. These processes were reduced by salinity with similar leaf-to-leaf gradients. Intercellular CO_2 concentration in the older leaves was higher, than in the flag leaf in non-saline plants and increased similarly with salinity. Leaf age was the major factor in reducing net photosynthesis and senescence processes were promoted by salinity. In this regard Munns *et al.* (1995) reported that salt within the plant reduces growth by causing premature senescence of old leaves and hence a reduced supply of assimilates to the growing regions.

The reports on the influence of salinity on photosynthesis in barley are rather conflicting. According to several workers salt stress itself does not exert adverse effect on the primary photosynthetic process (Sharma and Hall, 1991; Dunn and Neales, 1993; Belkhodja *et al.,* 1955). However, other investigators reported decrease in photosynthesis under salinity stress (Rawson, 1986; Kalaji and Nalfoezyk, 1991; Cho *et al.,* 1998). NaCl salinity (0, 75 and 150 mM) increased the leaf chloride contents and reduced peak photosynthesis, the initial slope of the light response curve and dark respiration of young leaves at ligule emergence but dark respiration of slightly older leaves increased by salinity as were the compensation points (Rawson, 1986). Dunn and Neales (1993) found that 25 mM NaCl reduced barley growth by 44% without affecting gas exchange but at 100 mM NaCl growth was reduced by 62% and net photosynthesis and leaf conductance were also reduced. Kalaji and Nalborezyk (1991) also observed that with increasing NaCl concentrations (30-170 mM) the growth rate, leaf area, total DM, photosynthesis rate and transpiration rate of 8 barley cultivars kept in a growth chamber decreased. Cho *et al.* (1998) found that photosynthesis activity decreased only in sensitive variety of barley at varying concentrations of NaCl (0, 50, 100 and 150 mM) after 10 days exposure to the salt. There was little effect in the tolerant genotype Bunong. In sorghum only limited work has been carried out on photosynthesis under saline conditions. Patil *et al.* (1983) observed that soil salinity affected photosynthetic area as well as rate of carbon assimilation in sorghum. The lowered photosynthetic ability under salt stress conditions was due to stomatal closure, inhibition of chlorophyll synthesis and decreased activities of PEP and RUBP carboxylases. The reduction in yield was attributed to the low rate of photosynthesis and reduced synthesis of sugars.

Sharma and Hall (1992a) investigated the effects of salinity, high light intensity and their interactions in leaves and isolated chloroplasts of barley (C_3) and sorghum (C_4) plants. The light saturated rate of CO_2 uptake and maximum quantum yield decreased with increasing salt concentrations in the range of 0-20 mM. Stomatal conductance was more sensitive to salt stress than photoinhibition. It was found that salt stress itself does not exert adverse effects on the primary photosynthetic process since the Fv:Fm ratio and electron transport activity did not decrease with higher salt concentrations. Further studies on changes in carotenoid composition and photosynthesis indicated that salt stress resulted in the formation of zeaxanthin but it was significantly less than that observed in the high light stressed plants (Sharma and Hall, 1992b). Formation of zeaxanthin and degradation of beta-carotene are apparently involved in protection against photo inhibition.

The concentration at which salinity causes severe reduction in photosynthesis varies greatly in different plant species and varieties. The step of photosynthesis affected may also vary depending upon the type and level of salinity and plant species. The reduction in photosynthesis in glycophytes may occur in the following ways :

(1) Low diffusion of CO_2 into the chloroplasts.

(2) Alternations in structure and function of organelles responsible for photosynthesis.

(3) Changes in photosynthetic reactions i.e. light and dark reactions.

(4) Effects on transport of assimilated products and intermediary compounds.

Stomatal Inhibition of Photosynthesis

Originally, scientists assumed that salinity was inhibiting photosynthesis by inhibiting biochemical reactions (Long and Baker, 1986). However, with the an increasing number of studies, it became clear that photosynthesis is inhibited by both stomatal and nonstomatal factors. Elevated salinity reduced photosynthesis in *Spartina* primarily by stomatal mechanisms (Longstreth and Strain, 1977). In a study by Robinson *et al.* (1983), photosynthesis was inhibited 65% by salinity, and stomatal conductance was inhibited by a similar amount, while there was no change in chlorophyll concentrations. Zisca *et al.* (1990), showed that photosynthesis was inhibited by salinity through effects on mesophyll conductance only. These and other studies led many researchers to believe that salinity inhibits photosynthesis primarily by its influence on stomata.

However, many other studies found different results. Photosynthesis in *Atriplex paula* was inhibited by a decrease in both mesophyll and stomatal conductance (Longstreath and Nobel, 1979), as was that of the gray mangrove (Ball and Farquhar, 1984). Moreover, Pearcy and Ustin (1984) demonstrated that when exposed to high salinity, photosynthesis was primarily inhibited by nonstomatal factors in several salt marsh species. When all these experiments are considered together, the relative control of mesophyll and stomatal conductance by salinity varies among species without a general pattern among halophytes or glycophytes.

In many instances, the factors responsible for the reduction in photosynthesis have been investigated but no clear mechanism of inhibition has emerged. In several investigations the primary effect of salinity is increased stomatal resistance to CO_2 diffusion as known under water stress (Gale *et al.*, 1967; Walker *et al.*, 1979). This causes a reduction of photosynthesis whenever CO_2 is a limiting factor eg. under high light intensity or due to the failure of guard cells to adjust osmotically, ionic interferences with stomatal function and increased indigenous content of abscisic acid.

Non-stomatal Inhibition of Photosynthesis

Boyer (1965) noted a 25% reduction in photosynthesis of cotton grown at -0.85 MPa NaCl, even though stomatal resistance was unaffected by salinity. Gale *et al.* (1967) detected stomatal closure in salt stressed cotton, but photosynthesis was also inhibited at low light intensities and at elevated CO_2 concentrations indicating salt damage to biochemical or photochemical reactions. Similarly 2/3rd reduction in photosynthesis in bean plants under salinity has been attributed to non-stomatal causes. Downton (1977) working with grape vines also attributed the major decrease to non-stomatal factors like liquid phase transport resistances between the mesophyll wall and carboxylating sites, excitation resistance and carboxylaction resistance (Table 6.1). Carbohydrate concentrations (sugars and starch) decreased by 20-40% in leaves at 75 mM NaCl, showing that the reduced growth was due to the reduction in photosynthesis. A primary effect in mesophyll resistance to CO_2 exchange could account for two-third of the 50% reduction in photosynthesis, the rest being due to stomatal resistance. This was accompanied by accumulation of radioactive carbon in intermediates of the photorespiratory pathway and could be related to the chloride status of leaves. Lloyd *et al.* (1987) reported that NaCl salinisation reduced CO_2 assimilation rates of valencia orange leaves even though the leaves maintained turgor; lower assimilation rates were therefore attributed to ion toxicity.

Table 6.1. Influence of increased stomatal resistance on decline of photosynthesis in grapevine leaves containing different levels of chloride (Downton, 1977).

Salinity (mM NaCl)	Leaf Cl^- % of dry wt.	Photosynthesis % of control	Percentage reduction in photosynthesis from control due to increase in stomatal resistance
Control	0.02	–	–
15	0.42	92	0.0
75	2.04	53.5	11.2
125	2.70	21.5	13.2

Photochemical reactions of isolated thylakoids are known to be inhibited by high concentrations of NaCl *in vitro*. In the sensitive species like beans both NaCl and Na_2SO_4 reduced the cyclic and non-cyclic photophosphorylation but in moderately tolerant maize and sunflower, NaCl stimulated both types of photophosphorylation whereas Na_2SO_4 inhibited it. Robinson *et al.* (1983) showed that NaCl upto 200 mM did not result in any major decrease in the photosynthesis potential of spinach leaves but photosynthesis on per plant basis decreased because of reduction in leaf area. Papp *et al.* (1983) also found that leaf extension growth in sugarbeet was sensitive to even low levels of NaCl, while leaf photosynthetic capacity was reduced only at higher salinity levels. Curtis and Lauchli (1986) concluded that growth in kneaf (*Hibiscus cannabinus*) under moderate salt stress is affected primarily through a reduction in expansive growth and leaf area development rather than a decline in photosynthetic capacity.

Photosystem II (PS II) is a relatively sensitive component of the photosynthetic system with respect to salt stress (Allakhverdiev *et al.*, 2000). A considerable decrease in the efficiency of PSII, electron transport chain (ETC), and assimilation rate of CO_2 under the influence of salinity has been noticed (Piotr and Grazyna, 2005). Demetriou *et al.* (2007) noticed alterations in photosynthetic characteristics of *Scenedesmus obliqus* that result in declined biomass accumulation. In citrus, salinity stress decreased growth by reducing of net photosynthetic rate, stomatal conductance, performance of PSII, and photosynthetic efficiency (Lopez-Climent *et al.*, 2008). Kalaji *et al.* (2011) reported that salinity stress affects growth of barley by altering chlorophyll fluorescence (PSII) and function of oxygen evolving complex. Furthermore, Mittal *et al.* (2012) observed that salt stress affects growth of *Brassica juncea* by affecting photosynthetic (PS II) and electron transport rates, and D1 protein. There are some other factors that reduce photosynthetic rates under salt stress: dehydration of cell membranes which reduce their permeability to carbon dioxide, salt toxicity, enhanced senescence, changes in enzyme activity induced by alterations in cytoplasmic structure, and negative feedback by reduced sink activity (Iyengar and Reddy, 1996).

Yeo *et al.* (1985) studied the effect of NaCl salinity upon net photosynthesis and transpiration by individual leaves of rice, by gas exchange measurements in seedlings at five to six leaf stage. Salinity did not initially reduce net photosynthesis in the whole plant but only in the older leaves in which sodium accumulated (Table 6.2). Analysis of the course of events in fourth leaf following salinization showed that net photosynthesis was inversely correlated with sodium concentration in the leaf tissue (Fig. 6.1). There was no evidence of a threshold effect, net photosynthesis declined linearly with increasing leaf sodium concentrations and was reduced by 50% at only 0.5 m mol sodium per gram dry weight. It has been suggested that the inhibition of net photosynthesis in rice by salinity is mediated by water deficit in the leaf cells due to accumulation of salt in the apoplast.

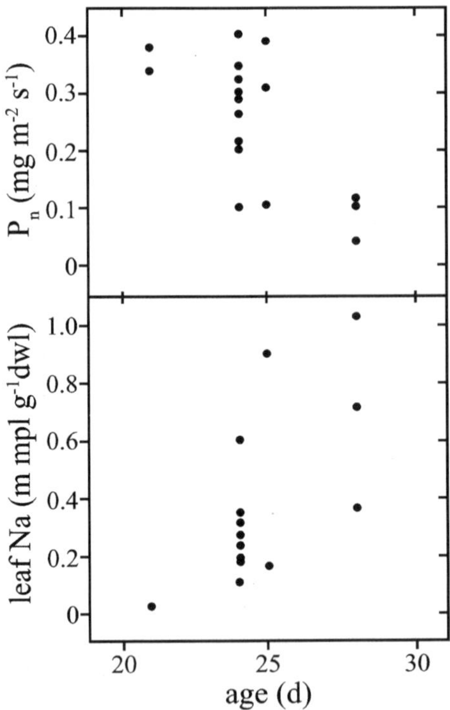

Fig. 6.1. Net photosynthesis and Na concentration as a function of time in the fourth leaf of plants of rice cv. IR2B which has been salinized (25 mol m^{-3} NaCl) at 17 d.

Table 6.2. The effect of NaCl upon net photosynthesis (pn) and leaf Na concentration in different leaves of *Oryza sativa* L. elite breeding line IR 2153-26-3-5-2 (Yeo *et al.*, 1985).

Leaf	Control pn (mg m^{-2} S^{-1})	+ Salt Pn (mg m^{-2} S^{-1})	%	Leaf Na mmol g^{-1} dry wt.
3	0.312 (0.029)*	0.170 (0.032)	54.4	0.513 (0.081)
4	0.383 (0.010)	0.260 (0.049)	67.5	0.355 (0.116)
5	0.334 (0.016)	0.330 (0.041)	98.9	0.070 (0.013)

* Leaf Na concentration is for salinized plants, value in parenthesis is s.e.m.

Thus, on the one hand, leaf elongation and hence the development of photosynthetic surface area is adversely affected by salinity while on the other hand decreases in photosynthetic capacity, as such, are influenced by salinization leading to reduced crop growth and yield.

Effects on Chloroplasts

Chloroplasts seem to be the organelles most affected by salinity. Salinity resulted in swelling within the granal loculi and forests, accumulation of lipid droplets and disappearance of grana in the chloroplasts of tomatoes and spinach. However, these changes occurred at higher concentrations and longer exposure to NaCl in case of spinach as compared to tomatoes. Similar changes have been observed in chloroplasts in leaves of tomato and pea plants exposed to NaCl salinity. NaCl induced shrinkage of the chloroplasts, the rate of photosynthetic phosphorylation increased and the rate of Hill reaction decreased with decreasing osmotic potential. However, Na_2SO_4 was less detrimental than NaCl and it was suggested that the effect is due to the anion and that the cation has virtually no effect on these reactions. The results of many studies reveal that although the photosynthetic system (i.e. Chloroplasts) is resistant to a considerable degree of dehydration, it may become damaged if dehydration is accompanied by an increase in electrolyte concentrations i.e. by the ionic effects (Strogonov, 1974).

Other research has shown that light reactions may be component of photosynthesis most sensitive to salinity. For example, photosystem II is quickly damaged in salt-sensitive species when exposed to 500 mM NaCl (Ball and Anderson, 1986). This idea leads into a consideration of the interaction between light intensity and salt inhibition of photosynthesis. Salinity is more toxic to the photosynthesis of plants adapted to low light conditions than for plants adapted to high light conditions. High light-adapted plants have a relatively large capacity to generate ATP and reducing power through electron transport processes. Accordingly, a small inhibition of electron transport processes by salinity in high light-adapted plant may not be as important as the same inhibition in low light-adapted plants. Furthermore, C_4 plants are normally high light-adapted, and they have a greater tolerance of salinity than C_3 plants (DeJong, 1978). This may be another indication that a high ability to utilize adenolates by photosynthesis may protect the photochemical reactions from damage by the secondary effects of salinity.

Flowers *et al.* (1985) found varietal differences on the effects of NaCl on the chloroplast structure in rice leaves. It was found that similar concentrations of salt in the leaf tissue had a greater effect in bringing about a disorganization of cellular structure and of decreasing net photosynthesis in the land race `Amber' than the elite breeding line `IR 2153'. The main damage to the ultra structure of chloroplasts appeared to be the loss of the envelope and disorganization of the grana in `Amber'. In contrast there were few signs of any effect on `IR 2153' even at a salt concentration of 549 m mol kg^{-1} d wt (Table 6.3).

Table 6.3. Percentage of intact chloroplasts and mitochondria per cell section in the leaf tissues of two varieties of rice (*Oryza sativa* L.) at different tissue salt concentrations (mmol kg^{-1} d. wt.) (Flowers *et al.*, 1985)

Amber			IR 2153		
Na	Mitochondria	Chloroplasts	Na	Mitochondria	Chloroplasts
15	100	100	15	100	100
137	100	94	157	100	100
443	17	8	413	100	100
513	12	7	549	100	97

NB : Sodium was measured on adjacent tissue. Chloroplasts were counted in mesophyll cells and mitochondria in cells of a minor vein.

Effects on Leaf Pigments

Increases as well as decreases in content of chlorophyll in different species under salinity have been reported by various investigators. According to several workers salinity decreases the chlorophyll content of plants (Garg and Garg, 1980; Garg and Lahiri, 1986) whereas several other studies reveal that salinization increases the chlorophyll content of plants (Strogonov, 1974). Increased chlorophyll content in saline habitats has been found in tomatoes, cotton and other glycophytes. Chlorophyll `a' and `b' showed a decrease at the beginning, followed by return to and even above the control values (Sivtsev, 1973). Most striking data are of Strognov (1974) who found a marked increase in both chlorophyll `a' and `b' under saline conditions in a number of plants.

The decrease in chlorophyll content under salt stress is a commonly reported phenomenon, and in various studies, chlorophyll concentration has been used as a sensitive indicator of the cellular metabolic state (Chutipaijit *et al.,* 2011). In *O. sativa* leaves, the reduction of chlorophyll *a* and *b* contents of leaves was observed after NaCl treatment (200 mM NaCl, 14 days) where chlrophyll *b* content of leaves (41%) was affected more than the chlorophyll *a* content (33%) (Amirjani, 2011). In another study, *O. sativa* exposed to 100 mM NaCl showed 30, 45 and 36% reduction in chlorophyll *a,* chlorophyll *b*, and carotenoids contents as compared to the control (Chutipaijit *et al.,* 2011). Saha *et al.* (2010) observed a linear decrease in the levels of total chlorophyll, chlorophyll *a,* chlorophyll *b,* carotenoids, and xanthophylls as well as the intensity of chlorophyll fluorescence in *Vigna radiata* under increasing concentrations of NaCl treatments. Compared to control, the pigment contents decreased on an average by 31% for total chlorophyll, 22% for chlorophyll *a,* 45% for chlorophyll *b,* 14% for carotene, and 19% for xanthophylls (Saha *et al.,* 2010). In one of the studies in cucumber, it has been shown that total leaf chlorophyll contents significantly decreased with an increasing NaCl levels. The decrease in total chlorophyll contents was 12, 21, and 30% at 2, 3 and 5 dSm^{-1} of salt stress, respectively, compared to non-treated plants (Khan *et al.,* 2013). Associated with the decline in pigment levels, there was an average 16% loss of the intensity of chlorphyll fluorescence as well. Usually, there is dominance of chlorophyll `*a*' over chlorophyll `*b*' in plants but their values become closer with increasing salinity (Mane *et al.,* 2010). The decrease in chlorophyll content under stress is a commonly reported phenomenon, and in various studies, this may be due to different reasons, one of them is related to membrane deterioration (Mane *et al.,* 2010).

Chlorophyll content of the leaves decreased under salt stress conditions (Boyer, 1965; Bhambota and Kanwar, 1970; Lapina and Popov, 1971; Ponomarova *et al.,* 1971; Prisco and O'Leary, 1972; Kumar, 1973; Sharma, 1974; Benzioni *et al.,* 1974; Joshi and Naik, 1977; Garg and Garg, 1982). However, on the contrary, Passera and Albuzio (1978) reported an increase in the chlorophyll content of wheat leaves under salt stress. Similar results showing an increase in chlorophyll content under salt stress was also reported by Nieman way back in 1962 in pea leaves. Weimberg (1975) reported that the rate of chlorophyll formation in etiolated pea seedlings exposed to light, in salinised media, was slower than the unsalinised plants. The commonly observed effect of salinity, as induced by various salts, by numerous workers, has been a decrease in photosynthesis per unit leaf area in many of the crops plants (Gale *et al.*, 1967; Udovenko *et al.*, 1971; Lapina and Bikmukhemetova, 1972; Kennedy, 1977; and Downton, 1977).

Table 6.4. Influence of salt stress (10 dSm^{-1}) on total chlorophyll content of salt tolerant (T) and sensitive (S) varieties of certain crop plants (Garg and Lahiri, 1986)

Crops	Variety	Chlorophyll (mg g^{-1} dw)		
		Control	Salt stress	Percent reduction
Wheat	Kh-65 (T)	4.40	3.66	16.8
	HD-4502 (S)	4.05	3.24	20.0
Pearlmillet	Babapuri	9.16	8.43	8.0
	BJ-104	9.33	7.07	24.2
Mustard	DIRA-337 (T)	12.21	11.17	8.2
	RYS-80 (S)	11.57	9.65	16.6
Clusterbean	Malosen (T)	6.52	6.51	0.0
	Durga Jai (S)	7.81	5.39	31.0

In contrast, Ahmed *et al.* (1983) reported that in castorbean, sunflower and flax the leaf pigment content per unit dry weight decreased and hence the total photosynthetic activity also decreased with increase in the level of salinity. Garg and Garg (1980) found that both Na_2CO_3 and $NaHCO_3$ depressed chlorophyll content of pea at different growth stages and the depression was more under Na_2CO_3. Garg and Lahiri, (1986) observed that salinity generally reduced chlorophyll content in a number of arid zone crops such as pearl millet, clusterbean, mung bean, wheat and Indian mustard (Table 6.4). Decreased chlorophyll content has also been reported in pigeon-pea exposed to salinity but photosynthesis on unit chlorophyll basis rather increased under saline conditions.

Treatment of soybean with salinity levels of upto 20 dSm^{-1} inhibited the total chlorophyll content, chlorophyll a/b ratios, net assimilation rates (NAR) and RGR after 25 and 50 days of sowing (Zaidi and Singh, 1995). Similar decline in total chlorophyll content has been reported by Garg *et al.* (1996) in *Cyamopsis tetragonoloba* plants under salinity levels of up to 12 dSm^{-1}. 50 mM NaCl showed a marked reduction in the chlorophyll content in green gram (Muthukumarasamy and Panneerselvam, 1997). 2-six week old chickpea genotypes when treated with 4 levels of salinity showed a reduction in the rate of photosynthesis as well as CO_2 assimilation (Sharma, 1997).

As different crop genotypes show differential response to salinity regarding the influence on chlorophyll content, the chlorophyll stability index has been used as an important criterion for salt tolerance. Recent studies have shown that chlorophyll fluorescence could be an excellent tool for screening since it is easy to measure and may allow for the screening of large number of genotypes in a short time. Balkhodja *et al.* (1994) suggested that chlorophyll fluorescence could be effectively used for the screening of barley genotypes for salinity tolerance.

Photosynthetic Enzymes

The precise effects of salt stress on the *in situ* activities of photosynthetic enzymes are little known and more information is warranted. Results reported are variable and enzyme responses have been found to depend on the species, age of the leaf and type and concentration of salt. Osmond and Greenway (1971) reported that PEP carboxylase from a salt tolerant species, *Atriplex* was more salt sensitive than the same enzyme taken from a salt sensitive species *Zea mays*. Sankhla and Huber (1974) found decreased activities of RUBP carboxylase and PEP carboxylase under saline conditions. However, Weimberg (1975) did not find any difference in eleven enzymes of photosynthetic apparatus in pea seedlings grown under salt stress. Seeman and Critchley (1985) reported that salinity (100 mM NaCl) reduced photosynthetic efficiency independent of stomatal closure in *Phaseolus vulgaris*. This reduction was shown to be a consequence of a reduction in the efficiency of RUBP carboxylase rather than a reduction in the leaf content of photosynthetic machinery. Salinity reduced photosynthesis by reducing both the RUBP pool size by an effect on RUBP regeneration capacity and RUBP carboxylase activity by an unknown mechanism when RUBP was limiting.

Passera and Albuzio (1978) showed that $^{14}CO_2$ fixation by wheat leaves declined with 50 mM NaCl concentration with a decrease in the ratio of RuBP to PEP carboxylase activities as well as with a rise in photorespiration under saline conditions. Lapina *et al.* (1976) observed that iso-osmotic (2-3 atm) concentrations of sodium chloride and sodium sulphate reduced the cyclic and non-cyclic photophosphorylation in *Phaseolus vulgaris*. In the moderately tolerant maize and sunflower, NaCl stimulated both types of photophosphorylation, whereas Na_2SO_4 inhibited the same. Shone and Gale (1983) observed that NaCl salinity diminished both gross and net photosynthesis in lucerne (*Medicago sativa*). As apparent from the foregoing, salt stress in general, reduced the rate of photosynthesis but this could be partially overcome by supplementing an extra dose of CO_2 as was evident from the experiments of Schwarz and Gale (1984), which were conducted in both C_3 (*Phaseolus vulgaris* and *Xanthium strumarium*) as well as C_4 (*Zea mays* and *Atriplex halimus*) plant species.

Shift in the Pathway of CO_2 Fixation in Response to Salinity

A change in the pathway of CO_2 fixation in response to salinity stress has been documented, especially in halophytic plants. There are evidences for a shift from the C_3 pathway of CO_2 fixation to C_4 and crassulacean acid type (CAM) metabolism. The shift to the CAM pathway may be an activity adaptive response conferring ecological advantages to the concerned halophytic plants. This may be due to the reduction in turnover of water and subsequent reduction in passive uptake of salts in these plants. However, this system suffers due to its relatively low CO_2 fixation capacity. It has also been found that NaCl can shift a balance from C_4 to C_3 photosynthesis in *Aleuropus litoralis* leaves (Shomer Ilan and Waisel, 1975). Mature leaves from *Mesembryanthemum crystallium* plant receiving saline treatment for a number of days shift from C_3 photosynthesis to CAM. The development of CAM is characterized by diurnal fluctuation in malate production, the ability to fix CO_2 at night through nocturnal opening of stomata and increased activity of PEP and other CAM enzymes.

Most C_4 plants seem to perform better than C_3 plants with respect to net photosynthesis rates. Under stress conditions, this can be exemplified by the relative independence of quantum yield (mol CO_2 fixed per absorbed einstein) of C_4 plants, found over a wide temperature range, as compared to that of C_3 plants. In C_3 plants, quantum yield decreases to values below those of C_4 plants at temperatures above 30°C.

Karmarkar and Joshi (1969) showed that organic acid synthesis is stimulated in the CAM plant *Bryophyllum pinnatum* by the presence of small amounts (0.04 M) of NaCl. This was shown due to the Cl^{-1} ion and not to Na^+ or $SO^{-2}{}_4$. Winter and Willert (1972) demonstrated activation of the dark fixation pathway and increased malate content in response to NaCl in *Mesembryanthemum crystallinum*. Neither CAM nor malate accumulation occurred in the absence of NaCl. The activity of MDH (enzyme closely related to C_4 metabolism) was found to be activated by NaCl in the leaves of high salt halophytes.

The induction of CAM (and nocturnal stomatal opening) may be stimulated by the effect of water or salinity stress on stomatal conductance. It is generally held that an increase in salinity (or drought) causes a midday depression in photosynthesis. The depression can extend to include most of the day as the stress continues. This time period (of low water potential and low midday conductance) corresponds to a transient period of lower turgor potential. When this period extends for long enough, a genetic response is induced to initiate CAM (about 10 days under stress), stomata stay open only at night, and turgor returns to a value greater than the prestress level (Winter and Gadermann, 1991). The induction of CAM in field situations has been elegantly demonstrated using carbon isotope and malate accumulation information indicating that CAM induction is an important adaptation to salinity and water stress in the field.

SALINITY AND RESPIRATION

Salinity stress in plants can be characterized by a slowing of shoot and root growth, reduced photosynthesis, and the reallocation of respiration from growth to maintenance (Hsiao, 2012; Jacoby *et al.*, 2010; Yang *et al.*, 2013). These decreases in growth and photosynthesis are thought to be caused by a reduction of stomatal conductance, which decreases CO_2 availability (Flexas *et al.*, 2007) or by alterations of photosynthetic metabolism (Lawlor and Cornic, 2002). The respiratory responses to salinity are complex, because there are many inconsistencies in observations between various plant species regarding the rates of respiration in salt-tolerant and salt-sensitive plants (Jacoby *et al.*, 2011). These variable observations are thought to be due to the differing strategies plants may adopt to modulate carbon balance and energy allocation for tissue and osmotic tolerance when it is exposed to salt. Because plant salinity tolerance greatly depends on signalling processes as well as adaptation mechanisms and the efficiency of energy consumption of the plant species exposed to salt, each of these can interact to provide great complexity in the response and the resulting levels of tolerance.

Previous research has revealed the important role that mitochondria play in the salinity responses of plants. Review of this literature shows that two biochemical processes required for respiratory functions are affected under salinity stress: (1) the tricarboxylic acid cycle (TCA) and (2) The transport of metabolites across the inner mitochondrial membrane. However, the mechanisms by which components of these processes are affected or react to salinity are still for from clearly understood (Che-Othman *et al.*, 2017).

The literature survey indicates both increases as well as decreases in respiration rates in response to salinity. Increases in respiration rates have been reported under NaCl and Na_2SO_4 salinity in beans, pea, corn and several other plants (Nieman, 1962; Livne and Levin, 1967; Amthor, 1989). Garg and Garg (1985) found increased rates of respiration in green gram and pea under the influence of Na_2CO_3 and $NaHCO_3$ (Fig. 6.2). On the other hand decreases in respiration rate in response to salinity have also also been found in several crop plants such as wheat, gram, cotton and other glycophytes.

Gale (1975) observed that maintenance respiration (Rm) of several species increases with salinity upto those levels above which apparent signs of toxicity appear. Increased respiration is, however, at the expense of net CO_2 fixation which means reduced over all growth. Lessani and Andrepoulos-Renaud (1969) demonstrated that stimulation or repression of respiration is governed by salt concentration. Respiration of tissue samples taken from lucerne plants grown in culture solution containing 0 to 5 gl^{-1} NaCl was stimulated and the respiration rate increased with increase in NaCl concentration. However, higher levels of salinity decreased respiration and in plants grown at 12 gl^{-1} NaCl it was 10% less than the controls.

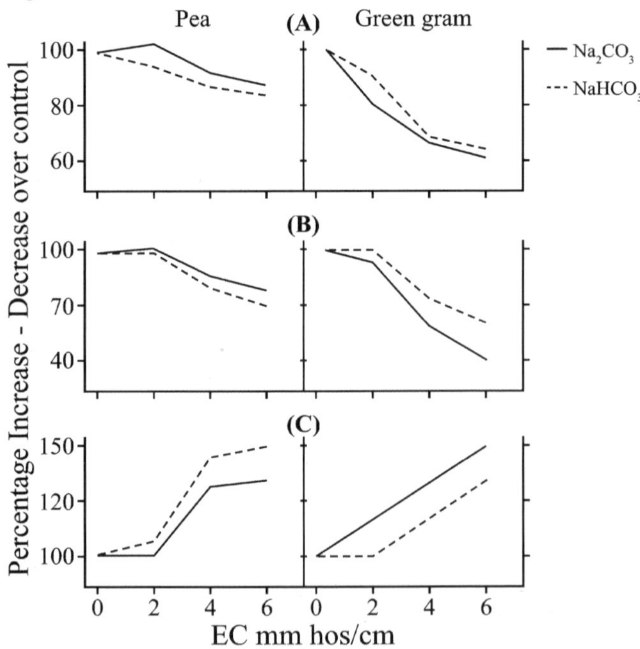

(A) Chlorophyll, (B) Net Photosynthesis, (C) Respiration

Figure 6.2. Effects of Na_2CO_3 and $NaHCO_3$ on chlorophyll, net photosynthesis and respiration in leaves of pea and green gram.

As with other salinity responses, respiration rate also depends on the salt species and level, the plant species, plant organ and the stage of growth. Increased respiration in corn in the presence of either Na_2SO_4 or NaCl and decreases in respiration of tomato plants under NaCl and decreases in respiration of tomato plants under NaCl salinity are well known (Gale, 1985). Similarly, Hoffman and Phene (1971) showed that kidney bean displayed a large increase in respiration at low (–2.0 bars) than at high (-4.4 bars) levels of salinity i.e. 80% and 10%, respectively. However, in cotton respiration was increased by 60% and 85% v/s controls at -6.4 bars and -12.5 bars of salinity, respectively.

MITOCHONDRIAL FUNCTION AND SALT STRESS

Mitochondrial function is one of the major cellular processes that clearly shows a response to salinity exposure in plants. The negative membrane potential of functional mitochondria means that Na^+ in the cytosol will be drawn to accumulate inside the organelle. Mitochondrial adaptation response includes a wide range of metabolic processes and is likely a mixture of direct salt inhibition of enzyme function as well as responses to high-salinity conditions. These inhibitions and adaptations may re-enable efficient respiratory processes that are sustainable or prolong the biochemical function of salt-laden tissues.

Energy is supplied to drive biochemical reactions in the cell in the form of ATP, which is produced from energy stored in sugars. Under normal conditions, this involves glycolysis in cytoplasm and the TCA cycle and ETC in mitochondria (Fernie *et al.,* 2004). Salinity stress leads to enhanced ATP demand to operate a number of energy-consuming adaptive mechanisms, including ion homeostasis, ROS defence, and osmotic adjustment by accumulating inorganic ions and compatible solutes. Ionic toxicity in the cell is a signature of salinity stress, which may affect membrane function and enzyme activities (Flagella *et al.,* 2006; Flowers *et al.,* 2014; Hasegawa, 2013).

TCA CYCLE RESPONSES TO SALINITY STRESS

The TCA cycle is a central component of respiratory ATP synthesis by supplying reductant for the ETC. The proteins that consitite the TCA cycle are well known (Barnes and Weitzman, 1986; Wiskich and Dry, 1985), and the regulation of these proteins has been extensively studied (Broz *et al.,* 2014; Nunes-Nesi *et al.,* 2013). Even so, the role and regulation of these proteins during salinity exposure are still far from understood. The TCA cycle requires the presence of pyruvate, which is transported across the inner membrane or synthesized in the matrix from malate through the action of pyruvate dehydrogenase complex (Jacoby *et al.,* 2011).

Changes in the abundance of the TCA cycle enzymes in plant tissues under salinity stress depend on many factors including growth conditions, developmental stage, the genotype, and the tissue examined. Interestingly, previous studies on changes of TCA cycle protein abundance under salinity stress involving 138 observations did not show any overall salt tolerance specific pattern of total TCA cycle proteins for abundance changes to salinity stress. Over 50% of TCA cycle proteins have been reported to be increased in abundance in salt-sensitive plants and 50% of TCA cycle proteins decreased in abundance in salt-tolerant plants (Che-Othman *et al.,* 2017).

The mitochondrial function, is known to be crucial during salinity stress in plants as it plays roles in signal transduction, energy production, and carbon balance, and it is also known that exposure to salt will directly alter key mitochondrial processes. This occurs due to the flexible activity of the TCA cycle and the need for mitochondria to transport substrates across the mitochondrial membranes. Multiple changes have been observed in the TCA cycle and mitochondrial transport machinery under salinity stress in plants. However, little consistency of response by the same components among different species has been seen, suggesting that different mechanisms may be adopted by different species in response to salinity stress. However, through this combination of the literature, there is clear evidence that in salt-tolerant plants, TCA cycle activity and associated metabolite transport can be altered to maintain ROS levels for signalling and to avoid immature senescence and, at the same time, allow efficient respiration by utilizing appropriate carbon sources and thus provide the energy required for salt tolerance mechanisms in salt-laden tissues. Connecting salt signalling to its organelle impact and this impact to altered energy state may be the key to understanding the diversity of salinity responses in central carbon metabolism and tolerance thresholds of different genotypes (Che-Othman *et al.,* 2017).

Renaud (1969) showed that stimulation or repression of respiration was governed by the salt concentration used and it was generally believed that while lower salt concentration stimulated respiration, higher levels of salinity decreased the same. As with most of the other salinity responses, there appeared to be large differences among various species and the response was dependent upon the salt level, the plant organ and the stage of growth. Lapina and Popov (1970) found decrease in respiration in tomato plants grown in culture solutions containing NaCl, whereas Lapina and Bikmukhametova (1972) found increased rate of respiration in corn grown under Na_2SO_4 or NaCl salinity. Similar increase (Porath and Poljakoff-Mayber, 1964; Udovenko *et al.,* 1974) or decrease (Porath and Poljakoff-Mayber, 1968; Goris, 1969; and Flowers, 1972) in the respiration rate has also been reported under saline conditions which possibly depended upon the plant type, salt concentration and stage of the plant. Increased respiration was also observed in wheat and barley seedlings under NaCl salinity and lucerne and cotton plants under both chloride and sulphate salinity.

Salinity may also alter the structure of mitochondria and thereby affect respiration rates in plants. Buchkolz *et al.* (1971) found that NaCl increased number of mitochondria in root tips of *Zea mays.* Increase in the number of mitochondria in response to salinity has been reported in several other studies and this is considered to be an adaptive mechanism. Gaidamakina (1974) reported that high salinity decreased the amount of mitochondria in sunflower seedlings but increased in the more salt tolerant barley seedlings.

Changes in the pathway of respiration from glycolytic to pentose phosphate pathway (PPP) in response to salinity was reported in cotton and pea roots and barley but there was no inhibition of the O_2 uptake (Porath and Poljakoff Mayber, 1968). The PPP was also increased by NaCl in a number of halophytes. Salinity above 120 mM NaCl increased the percentage of absorbed glucose oxidation via the PPP, but did not affect the glycolytic pathway in the root tips of *Tamarix tetragyna.* At the same time, salinity depressed the glucose-6-phosphate dehydrogenase, pyruvate kinase and oxidative phosphorylation and these effects became more evident at a salinity level of -10 bars (240 mM) concentration, which is rarely exceeded in the root zone of native habitat of the plant.

The route of respiration was changed from glycolytic to pentose phosphate shunt in seedlings of barley and cotton germinated under saline conditions and in pea root tips (Goris, 1969; Razmaev and Azimov, 1969). Chloride salinity depressed CO_2 evolution from C_6 and not from C_1 whereas sulphate salinity depressed both (Porath and Poljakoff-Mayber, 1968). It was further shown that activities of most of the glycolytic enzymes were depressed 10-fold by salinity.

Livne and Levin (1967) reported that while respiration of mitochondria of pea plants exposed to NaCl showed 25-75% increase, the phosphorylation/oxidation ratio (P/O) remained constant. This indicates that respiration remained normal under salinity despite change in its rate. The increased respiration in response to salinity is probably an adaptive and not a pathological reaction. Shone and Gale

(1983) found marked changes in the growth coefficients with changes in the NaCl concentrations in the culture solution in the alfalfa plants (Table 6.5). Amthor (1989) concluded that salinity increased maintenance respiration probably for compartmentation and osmoregulation. However, the role of respiration in regulating partitioning of carbon between growth, maintenance, transport and storage is at present unknown.

Table 6.5. Growth and maintenance respiration coefficients (g and m, respectively) of alfalfa at various NaCl concentrations (Shone and Gale, 1983)

Osmotic potential of NaCl (–MPa)	g (mg CO_2 g^{-1})	m (mg CO_2 g^{-1} h^{-1})
0.00	760	1.06
0.15	900	1.56
0.30	1150	1.83
0.45	1470	1.54
0.60	830	1.21

SALINITY AND CARBOHYDRATE METABOLISM

Both increase in salinity and type of salinity influence carbohydrate metabolism of plants. Gauch and Eaton (1942) reported that in barley grown in sand culture flushed with 104 mM chloride during a diurnal cycle, leaves were generally 20-30% higher in starch and 30-70% higher in sugars than in controls. They suggested that increase in starch implied a lack of utilization and concluded that salinity affected cellular elaboration rather than photosynthesis. The term elaboration clearly refers to cell division and growth. Munns *et al.* (1982) reported that the carbohydrate status of the elongating tissue of the salt tolerant variety of barley supports the idea that its growth at high NaCl was not limited by supply of carbohydrates to the growing regions. Munns and Termaat (1986) have confirmed this view after reviewing the published information.

Rathert (1984) studied the effect of NaCl on total root and leaf sucrose and total foliage starch in crops of differing salt tolerance during the early stages of salinity stress. Leaf sucrose level increased most in bush bean (sensitive) but less in rice (moderately tolerant, whereas it decreased slightly in soybean (moderately tolerant) and more in cotton (tolerant). Initial root sucrose was lowest in bush bean but this sensitive species showed the highest per cent increase after salinisation. Similarly foliage starch increased in sensitive crops but decreased in tolerant crops (Table 6.6). This shows that salinity induced changes in carbohydrates are dependent upon the species and the concentration of salts. Lahiri *et al.* (1987) found that starch content decreased progressively with increasing soil salinity in cluster-bean leaves while reducing sugars increased progressively (Table 6.7). However, at a high salinity level (10 dSm^{-1}) both starch and sugars increased probably due to reduced growth resulting in lesser utilization. A similar observation was made earlier by Downton (1977) in case of grapevine. Several other studies have indicated both decreases as well as increases in starch and sugar contents in response to salinity (Garg and Lahiri, 1986).

Table 6.6. Effect of salinity on sucrose and starch of leaves and roots (Rathert, 1984).

Species	Carbohydrate content (mg g^{-1} dry weight)								
	Sucrose (leaves)			Sucrose (roots)			Starch (leaves)		
	Control	salt	%	Control	salt	%	Control	salt	%
Bushbean	8.9	12.7	143	5.6	12.6	225	18.4	29.7	161
Rice	32.6	35.0	107	9.2	15.8	172	32.9	49.1	149
Soybean	12.3	11.7	95	12.6	14.3	113	11.2	13.4	120
Cotton	12.4	10.9	89	20.6	22.3	108	20.8	12.5	60

*Salt treatment 3.7 dSm^{-1} for all crops except cotton which was 4.9 dSm^{-1} for 7 days.

Table 6.7 Influence of increasing soil salinity on the levels of starch, reducing sugars and chlorophyll in leaves of clusterbean at pre-flowering stage (Lahiri *et al.*, 1987)

Salinity (dSm^{-1})	Starch (mg g^{-1} d wt.)	Reducing sugars (mg g^{-1} dt.)	Total chlorophyll (mg g^{-1} dw)	Leaf dry wt. (g $plant^{-1}$)
0	296.5	10.7	11.0	0.92
2	148.0	11.0	10.9	0.77
4	131.9	14.6	10.9	0.60
6	198.0	17.4	8.9	0.46
8	–	18.0	8.0	0.33
10	272.5	19.6	6.4	0.27

Likewise stimulation of amylase activity under salinity has been reported in several studies (El-Fouly and Jung, 1972; Strongonov, 1974) whereas a decline in amylase activity at high salinity has been found in others (Sarin and Narayanan, 1968).

Zhukovskaya *et al.* (1972) reported that salts exert a significant effect on *de novo* synthesis of enzymes of carbohydrate metabolism in germinating seeds and therefore, inhibit meristematic activity of barley.

Hassan-Porath and Poljakoff-Mayber (1973) showed that absorption of externally supplied glucose was depressed in pea root tips from plants grown in NaCl and Na_2SO_4 types of salinity. Sheoran and Garg (1978) found that salinity (NaCl and Na_2SO_4) resulted in the accumulation of soluble carbohydrates in the cotyledons, whereas a decrease was observed in the embryo-axis. Similar effects were also observed by Garg (1977) under the influence of Na_2CO_3 and $NaHCO_3$ in germinating mung bean and maize seedlings. It was found that amylase activity was inhibited in the cotyledons but stimulated in the embryo-axis under salt stress leading to the aforesaid effects of salts on the carbohydrate level.

Saline conditions resulted in shrivelling of *Vigna radiata* L. Wilczek seeds leading to a significant decrease in the total soluble sugars as well as reducing sugars with concomitant increase in the soluble sugars of healthy seeds grown under non-saline conditions (Dhingra and Sharma, 1993). A similar increase in the soluble sugars and a decrease in starch has also been reported by Dhingra *et al.* (1994) in the seeds of 3 promising genotypes of chickpea, viz., ICCV 88102, H82-2 and C-235.

Singh and Singh (1995) while working on 3 cultivars of *Pisum sativum* namely Rachna, NDP-2 & NDP-7 reported a decrease in the reducing, non-reducing sugars and total saccharides with increasing soil exchangeable sodium percentage in the environment. Siddiqui and Krishnamoorthy (1995) observed that application of salinity decreased the amylase activity and starch content with a concomitant accumulation of soluble carbohydrates in the cotyledons of *Vigna unguiculata.* Sheokand *et al.* (1995) reported an increase in the total soluble carbohydrates in 45 day old chickpea plants subjected to 50 and 100 mM NaCl treatments. Muthukumarasamy and Panneerselvan (1995) observed that 20, 40 and 60 mM salinity levels had an inhibitory effect on the carbohydrate metabolism in the seedlings of soybean. Misra *et al.* (1996) also reported an accumulation of total soluble saccharides in *Vigna radiata* L. seedlings under 0.5, 1, 2 and 3 per cent NaCl. Olmos and Hellin (1996) while working on two lines of *Pisum sativum* reported that the relatively tolerant calli accumulated more reducing sugars indicating osmotic adaptation through reducing sugars.

LIPID METABOLISM

Changes observed in lipid metabolism in salt-treated plants can be associated with adverse effects of salt stress, namely its osmotic components, on membrane integrity and function. In a comparative proteomic study of salt-treated *A. thaliana* and *T. salsuginea* (Pang *et al.,* 2010), an augmented abundance of 3-ketoacyl-acyl carrier protein synthase I and phospholipase/carboxyesterase family protein has been found in salt-treated *A. thaliana.* In *T. salsuginea,* an elevated level of a putative long-chain-fatty acid-CoA ligase involved in fatty acid synthesis and a declined level of a putative glycerophosphodiester phospho-diesterase involved in lipid degradation was found (Pang *et al.,* 2010). Increased relative abundance of dihydrolipoamide dehydrogenase and enoyl-ACP reductase has been found in salt-treated rice panicles (Dooki *et al.,* 2006). The changes observed in lipid metabolism in salinity-exposed plants indicate profound changes in cell membrane integrity, composition, and function under stress. A short-term decreased relative abundance of monogalactosyl diacylglycerol synthase, an enzyme involved in biosynthesis of galactoslyglycerolipids (monogalactosyl diacylglycerol, digalactosyl diacylglycerol), the major components of chloroplast inner envelope and thylakoid system, observed in salt-treated maize chloroplasts (Zorb *et al.,* 2009), and an increased relative abundance of UDP-sulfoquinovose synthase involved in biosynthesis of thylakoid membrane sulfolipids observed in creeping bent grass (Xu *et al.,* 2010) indicate profound changes in chloroplast membrane structure and composition in response to salt and osmotic stress. Chagnes in lipid-metabolism associated lipid transfer proteins and temperature induced lipocalins have been observed in apoplastic fluid of tobacco cells (Dani *et al.,* 2005) and tomato radicals (Chen *et al.,* 2009), respectively, subjected to salinity stress.

SALINITY AND NITROGEN METABOLISM

There are three main steps in the nitrogen metabolism of plants, as listed below :

(1) The conversion of inorganic N into organic N compounds of low molecular weight. This is irreversible step in higher plants.

(2) Synthesis of high molecular weight N compounds from low molecular weight N compounds, particularly amino acids.

(3) Breakdown of the N containing macromolecules by increased activity of hydrolytic enzymes.

Saline conditions influence all these steps in one way or the other and thus have a considerable effect on nitrogen metabolism. Disruption of nitrogen metabolism by salinity has been attributed to : (a) decreased nitrogen uptake, (b) decreased nitrate reductase activity, (c) altered amino acids synthesis, (d) impaired protein synthesis and enhanced hydrolysis of proteins, (e) slowed synthesis of nucleic acids, and (f) increased activity of hydrolysing enzymes such as RNase, DNase, protease and several others leading to degradation of macromolecules. Some of these effects are briefly discussed below.

Nitrogen Uptake

Plants can take up and metabolize both NO_3 and NH^+_4 forms of nitrogen but salinity is reported to decrease the nitrate uptake itself. A number of laboratory and greenhouse studies have shown that salinity reduces N accumulation in plants (Feigin, 1985; Garg *et al.*, 1990; Grattan and Grieve, 1994). This is not surprising since with few exceptions an increase in chloride uptake and accumulation is accompanied by a decrease in shoot nitrate concentration. Aslam *et al.* (1984) reported that Cl^- inhibited NO^-_3 uptake more than SO^2_4 in barley, when these anions were present on equal osmolarity basis. Gorham *et al.* (1986) observed drastic reductions of leaf NO^-_3 concentrations in response to salinity, notwithstanding the fact that other nitrogen containing fractions either increased or were marginally reduced. Many other studies have indicated similar effects of Cl^- accumulation on NO^-_3 uptake (Feigin, 1985; Grattan and Grieve, 1994).

In majority of the crop plants studied, saline conditions reduce nitrate level in shoot (Grattan and Grieve, 1994). However, in response to salinity the levels of soluble nitrogenous compounds such as proline, glycine-betaine, total soluble amino acids, polyamines, etc. increase (Gorham *et al.*, 1986). Munns and Termaat (1986) suggested that salt stressed plants may have less N than non-stressed plants and thus they concluded that decreased tissue N level under saline conditions might not be growth limiting. According to these workers, N deficiency in salt stressed plants does not seem to be the mechanism for salt injury.

Under saline conditions, the form of N which is applied to the soil also becomes important. Maize and melon plants grown with NH^+_4 as N source become more sensitive to salinity than the plants grown with NO_3^{-2} as N source (Lewis *et al.* 1989; Feigin, 1990). It was observed by Lewis and coworkers (1989) that the growth of wheat and maize plants was improved due to addition of Ca^{2+} when plants were treated with NO_3^{-2} compared to the plants grown with NH_4 under saline conditions. In cucumber plants Cl^- uptake was enhanced in salinized medium when half of the NO_3^{-2} was replaced with NH^+_4 (Martinez and Cerda, 1989). Salt stressed melon plants also showed more accumulation of Na^+ and Cl^- and less Ca^{2+} and K^+ in leaves as NH_4 was increased and NO_3^{-2} was reduced in the growth medium (Feigin, 1990). For wheat plants grown in salinized soil it has been advocated by many workers that NO_3^{-2} is a better N source than NH^+_4 (Leidi *et al.,* 1991, Silberbush and Lips, 1991a). A decrease in the weight of wheat grain was observed with increase in NH^+_4/NO_3^{-2} ratio under saline conditions in sand culture (Silberbush and Lips, 1991b).

Nitrogen Assimilation Process

The reports of the effect of salinity on N assimilation are varied. No studies have been performed which could measure all the processes of N assimilation simultaneously. NR has been extensively studied by various groups of investigators for its behaviour in different plant species under salinization (Dubey and Pessarakli, 1995).

Nitrate reductase enzyme, which is involved in the reduction of nitrate to nitrite is known to be highly sensitive to all types of stresses, particularly salt and osmotic stresses. Nitrate reductase activity (NRA) is generally reported to be inhibited by salinity in a number of crop plants such as wheat, pearl millet, sorghum, mustard, pea, clusterbean and tomato (Plaut, 1974; Garg *et al.,* 1990, 1993; Lahiri *et al.,* 1996) as well as in several halophytes (Heimer, 1973; Austenfeld, 1974). However, stimulation of NRA, particularly under low salinity levels has also been reported (Austenfeld, 1974; Aslam *et al.,* 1984). The reduction in NRA has been attributed to the inhibition of NO^-_3 uptake by Cl^- in many investigations. At iso-osmotic concentrations of salts, suppression of NRA was greater with NaCl than with Na_2SO_4. Garg *et al.,* (1983) compared the effects of different salts at equal conductivity (10 dSm^{-1}) in wheat at various growth stages. The general inhibition of NRA by all salts and particularly by Na_2CO_3 and $NaHCO_3$ at the advanced stages, suggested the possibility of performance alteration through decreased nitrogen metabolism (Table 6.8). Likewise a progressive decline in NRA with increasing salinity points towards the disturbed N metabolism in a number of other studies (Lahiri *et al.,* 1987; Garg *et al.,* 1993). Gouia *et al.* (1984) found that the greater sensitivity of nitrate reductase activity to NaCl in bean leaves compared to cotton leaves was due to a decreased compartmentalization of ions rather than due to a difference in salt tolerance of the enzyme itself. Lahiri *et al.* (1996) found significant varietal differences in the NR activity of clusterbean and higher reduction in NR activity in sensitive genotypes have been attributed to higher uptake of Na and Cl in the shoot and lower K/Na ratios.

Table 6.8. Effect of various salts at similar EC (10 dSm^{-1}) on leaf nitrate reductase activity and level of soluble protein in wheat on 50th and 70th day of sowing (Garg *et al.*, 1983)

Salts (10 dSm^{-1})	Nitrate reductase (μg NO_2 g^{-1} d.wt. h^{-1})		Soluble protein (mg g^{-1} d. wt.)	
	50 d	70 d	50 d	70d
Control	77.4	70.5	76.0	42.1
Na_2SO_4	45.1	42.4	56.7	55.1
$NaHCO_3$	53.4	29.3	57.1	31.4
Na_2CO_3	55.0	17.8	65.7	32.8
NaCl	57.7	53.9	60.3	35.9
KCl	48.3	51.3	64.5	39.4
$CaCl_2$	19.3	55.8	54.1	45.7
$MgCl_2$	52.9	60.3	53.2	40.5

According to Katiyar and Dubey (1992) the decrease *in vivo* NR activity due to salinity in salt sensitive rice cultivars is possibly due to the decreased rate of synthesis of the enzyme in seedlings of these cultivars under salinization, whereas in the tolerant cultivars, salinity possibly promotes more synthesis of the enzyme in salt stressed seedlings.

Katiyar and Dubey (1992) observed different behavious of NR and NIR in salt sensitive and tolerant genotypes of rice. Salt sensitive rice cvs. Ratna and Jaya showed a decrease in *in vivo* NR activity in shoot and root, whereas tolerant cvs. CSR-1 and CSR-3 showed an increase *in vivo* NR activity in both these tissues under NaCl salinization. However, roots maintained higher NR activity under normal as well as NaCl stress conditions. Similar decrease and increase in NR activity due to salinity in salt sensitive and tolerant rice cultivars, respectively was observed by others (Krishnamurthy *et al.,* 1987; Pandey and Srivastava, 1989). Salt tolerant crop cultivars, thus appear to have better adaptability to saline stress by exhibiting efficient $NO^{-2}{}_3$ reduction under salinization (Katiyar and Dubey, 1992).

The activity of enzyme NiR either gets inhibited or remains almost unaffected under saline stress. In leaves of *Sorghum vulgare* plants growing in presence of NaCl in the medium, the activity of NiR decreased (Rao and Gnanam, 1990). The extent of decrease was a function of NaCl concentration in the growth medium It is suggested that NiR is less sensitive than NR in sorghum plants (Rao and Gnanam, 1990). Similar decrease in NiR activity has been observed in wheat, lentil and *Phaseolus aureus* plants growing under saline stress. In seedlings of 7-day-old horsegram (*Macrotyloma uniflorum*) subjected to NaCl salinity in the growth medium, NiR activity remained unaffected compared to control plants (Reddy *et al.,* 1993). Rice plants differing in salt tolerance show varying levels of NiR activity (Katiyar and Dubey, 1992). The decreased NiR activity under salt stress as observed in many plant species might be helpful to plants by reducing the toxic levels of NH_4 inside the tissues. It is suggested that salinity stress favours accumulation of ammonia in plant parts (Reddy *et al.,* 1993; Ramanjulu *et al.,* 1994).

Ammonia either directly absorbed by plant roots or as a result of reduction of NO^-_3 is further assimilated and incorporated into the amide group of glutamine by the action of the enzyme glutamine synthetase (GS) and subsequently into glutamic acid by glutamine synthase (GOGAT). Salinity has been reported to affect GS/GOGAT pathway of ammonia assimilation and leads to either decrease or increase in their activities depending upon the plant species as well as the tissues studies (Dubey *et al.,* 1991; Garg *et al.,* 1993; Singh and Dubey, 1994). Total activity as well as specific activity of GS and GDH (glutamate dehydrogenase) were inhibited in the roots of pea (salt sensitive) while in corn (more resistant) inhibition was recorded at the stage of glutamine acid synthesis. Huber and Sankhla (1973) reported an inhibitory effect on growth and GDH activity of pearl millet seedlings by 0.5% NaCl.

The enzymes glutamine synthetase (GS) and glutamate synthase (GOGAT) which form the major route of NH_4 assimilation in plants are markedly affected under saline stress (Miranda -Ham and Loyola-Vargas, 1988; Katiyar, 1990; Chaillou *et al.,* 1992; Ramanjulu *et al.,* 1994; Singh and Dubey, 1994). The effects of salinity on GS activity are varied and depend on the nature of plant species as well as the type of tissues studied. Rice plants differing in salt tolerance showed increased level of GS activity in both the sets of cultivars under salinization, however, the extent of increase in enzyme activity was more in sensitive cultivars than the tolerant ones (Singh and Dubey, 1994). In roots of germinating *Phaseolus aureus* seeds (Misra and

Dwivedi, 1990) and in *in vitro* plant shoots of *Lycopersicon esculentum* (Chaillou and Guerrier, 1992), NaCl in the medium caused increase in GS activity in a concentration dependent manner. In leaves of mulberry (*Morus alba*) plants subjected to salinity stress increased activity of GS was observed, which paralleled with the elevated levels of glutamine in the tissues (Ramanjulu *et al.,* 1994). However, contrary to these findings *Canavalia ensiformis* plants (Miranda-Ham and Loyola-Vargas, 1988) or callus tissues from *Lycopersicon esculentum* (Chaillou and Guerrier, 1992; Alvarez and Guerrier, 1994) when grown under saline stress, showed a marked decrease in GS activity in all tissues. In *Lycopersicon esculentum* plants, organ specific response of NaCl on GS activity was observed; 50 mM NaCl caused increase in enzyme activity in shoots, whereas decreased activity was observed in roots (Chaillou and Guerrier, 1992). Similarly, in pea and soybean plants grown under salinized conditions, decreased GS activity was observed in roots (Sharma and Gupta, 1986; Chaillou *et al.,* 1962). In horsegram seedlings GS remained insensitive to salinity (Reddy *et al.,* 1993). It is suggested that higher GS activity in plants growing under saline conditions might be a useful adaptation for such plants in saline environments in detoxifying and assimilating more ammonia by this pathway, leading to increased accumulation of soluble nitrogenous compounds, which act as compatible cytoplasmic solutes under saline stress (Dubey and Rani, 1989). Increased GS activity might be helpful in providing more glutamate required for the biosynthetic reactions under saline conditions (Yamaya and Oaks, 1988).

Lahiri *et al.* (1996) found a general decline in GS activity of clusterbean genotypes at 10 dSm^{-1} soil salinity but the decrease was significantly more in sensitive as compared to tolerant genotypes. Singh and Dubey (1994) also found that high GS activity in salt stressed seedlings is associated with the salt tolerance of rice genotypes.

The activity of the enzyme GOGAT decreases in many salt sensitive plant species and cell lines grown in presence of NaCl (Katiyar, 1990; Gulati and Jaiwal, 1996). It was observed by Katiyar (1990) that GOGAT from salt tolerant rice plants is comparatively more tolerant to NaCl than the enzyme from salt sensitive species. When seedlings of rice cultivars of differing salt tolerance were grown under increasing levels of NaCl salinity, a moderate salinity level of 7 dSm^{-1} caused increase in enzyme activity, whereas the activity of the enzyme was inhibited at a higher *in situ* salinity level of 14 dSm^{-1} (Katiyar, 1990). When NaCl-non-adapted as well as NaCl-adapted callus cultures of *Vigna radiata* were grown in presence of NaCl, GOGAT activity decreased throughout 28-day growth period in non-adapted lines, whereas in NaCl-adapted lines enzyme activity increased up to 14 day and thereafter a decrease was observed till 28-day compared to their controls (Gulati and Jaiwal, 1996). However, in *Canavalia ensiformis* plants (Miranda-Ham and Loyala-Vargas, 1988) and in germinating seeds of *Phaseolus aureus* (Misra and Dwivedi, 1990) increased GOGAT activity was observed under saline stress. In salt-treated (140 mM NaCl) calli derived from cotyledons of the domestic salt sensitive tomato (*Lycopersicon esculentum*), decrease in both NADH-GOGAT and NADPH-GOGAT activity was observed (Rus-Alvarez and Guerrier, 1994). The decrease in the activities of both the enzymes GS and GOGAT under salinization in certain plant species suggests that N assimilation by this pathway is possibly impaired due to salinity in such plants (Dubey and Pessarakli, 1995).

The mitochondral enzyme glutamate dehydrogenase (GDH) has been extensively studied by various groups of investigators for its behaviour in crop plants grown under saline stress. In many crop species examined, the activity of GDH increases showing possible role of this enzyme in detoxifying ammonia or in nitrogen assimilation under salt stress conditions (Srivastava and Singh, 1987). However, in certain cases GDH activity remains maintained as controls or gets decreased under NaCl salinity (Dubey and Pessarakli, 1995).

It is believed that under saline conditions toxic levels of ammonia and certain amino acids accumulate in the tissues resulting from derangement of protein synthesis (Pulich, 1986) and also possibly by efficient NO_3^2 reduction (Gulati and Jaiwal, 1996). Under such conditions, when GS/GOGAT pathway of ammonia assimilation gets impaired, increased level of GDH provides adaptive value to the plants by detoxifying and assimilating more ammonia (Srivastava and Singh, 1987; Dubey and Pessarakli, 1995). Higher endogenous level of ammonia is harmful to metabolism and thus increased GDH activity under stress may play a vital role by protecting the cells from the stress damage (Gulati and Jaiwal, 1996). It is suggested that the increased activity of NADH-GDH under saline stress may be due to acceleration in *de novo* synthesis of the enzyme protein and/or due to modulation of the activity of the enzyme due to NaCl stress (Gulati and Jaiwal, 1996).

The shift of emphasis from organic acids to amino acids may result from an inhibition of malic dehydrogenase (MDH) activity and a stimulation of transamination reactions under saline conditions as suggested by several studies. Alanine transferase and aspartate amino transferase activities were stimulated by 0.5% NaCl in pearl millet shoots (Huber and Sankhla, 1973). Sharma and Garg (1985) also found that both alanine amino transferase (GPT) and aspartate amino transferase (GOT) showed increased activity in wheat at seedling, ear emergence and grain filling stages under iso-osmotic levels of PEG and NaCl. Stimulation of GPT in moth bean and of GPT, GOT and GDH in peanut have also been reported although the level of salinity imposed in the later case was comparatively low. The increase in amination and transmination activities visualize significance of amino acids under stress conditions. It seems that increase in amino transferases coupled with increased GDH activity forms a vital link in plant protective mechanisms under saline conditions.

SALINITY AND PROTEIN METABOLISM

A number of workers have observed that salinity, by and large, depressed protein synthesis and accelerated their degradation and disturbed the soluble amino acids protein ratio in plants (Strogonov, 1964; Nieman and Poulsn, 1964; Rauser and Hanson, 1966; Strogonov *et al.,* 1970; Udovenko *et al.,* 1973). Kahane and Poljakoff-Mayber (1968) showed that both uptake and incorporation of labelled amino acids into proteins were reduced by the presence of salt in the incubation mixture in pea roots. Sulphate salinity was found to be more deleterious than chloride. Breakdown of polysomes accompanied by an increased level of RNase has been demonstrated during stress (Nir *et al.,* 1970; Marin and Vieira De-Silva, 1972).

Although it is generally agreed that salt stress causes a significant reduction in protein content of plants (Helal and Mengel, 1979; Pessarakli and Tucker, 1985; Garg and Garg, 1980; Garg *et al.,* 1983, 1990) but it is still not clearly understood whether this decrease is the result of protein degradation or lowered aminoacids incorportation into proteins. Impaired protein synthesis was reported as a result of reduction in plant growth and crop yield under salt and water stress conditions (Ben-zioni *et al.,* 1967; Kahane and Poljakoff-Mayber, 1968; Pessarakli and Tucker, 1985). According to these reports, while sodium chloride depressed plant growth and nitrogen incorporation into the proteins, it substantially increased inorganic N accumulation in plant tissues. However, some investigators found that soluble protein and free amino acids content of plants were not consistently affected by soluble salts. On the contrary stimulation of protein synthesis has also been reported in the more resistant barley by salinization. The irregular changes in protein-N fraction and N metabolism in these studies may be due to the difference in the salt tolerance of different plants. The disturbed N metabolism appeared more pronounced among the salt sensitive plant species. Shevyakova and Leonova (1975) found that protein synthesis was stimulated at lower salt concentration but inhibited at higher salt concentrations in horsebean leaves. Similar pattern in cotton and wheat under PEG and NaCl stresses is also known. In many studies NaCl has been found to be more depressive than PEG or Na_2SO_4 in reducing the protein content of plants. The higher toxicity of Na_2CO_3 and $NaHCO_3$ over NaCl and Na_2SO_4 was however, observed in wheat plants at various growth stages (Garg *et al.*, 1983).

Dubey (1994) suggested that salinity promotes the synthesis of specific proteins, causes either decrease or increase in the level of total and/or soluble proteins, depending upon the plant parts studied, and leads to increased activity and synthesis of many enzymes involved in protein metabolism. In various crop species, a decrease in the protein level in salt-stressed plant parts is attributed to a decrease in protein synthesis, the decreased availability of aminoacids and the denaturation of enzymes involved in aminoacids and protein synthesis (Levitt, 1980). On the contrary increased protein level under salinization reported in several crops appears to be due to the increased synthesis of pre-existing as well as certain new sets of proteins (Dubey and Rani, 1989).

Saline conditions resulted in shrivelling of *Vigna radiata* L. Wilczek seeds leading to a significant decrease in proteins, free amino acids as well as proline with a concomitant increase of the same in the healthy seeds grown under non-saline conditions (Dhingra and Sharma, 1993). Singh and Singh (1995) reported a decrease in the protein content in three cultivars of *Pisum sativum*, viz., Rachna, NDP-2 and NDP-7 with increasing soil exchangeable sodium percentage in the environment. Aid El-Samad *et al.* (1997) while working on cultivars of *Glycine max* reported that the relative salt tolerance of the cultivars was due to the accumulation of soluble proteins, amino acids and proline. Saline conditions led to a decrease in the protein content with a concomitant increase of protease activity in the root, shoot and leaf of chickpea seedlings. However, a reversal of the trends was observed in the cotyledons (Muthukumarasamy and Panneerselvam, 1997). Soybean callus cultures treated with 150 mM of NaCl resulted in a significant increase in the protein content (El-Enany, 1997).

The protein hydrolysis or degradation under salinity is also well documented. During the germination of peas, salinity lowered the protein and peptide contents by stimulating protein hydrolysis (Uprety and Sarin, 1975). This hydrolysis was considered to be a primary effect of the salt. The increase in hydrolysis of proteins in salinized glycophytes has been found to be generally associated with an increase in the products of hydrolysis free amino acids (Levitt, 1980). In a number of studies, accumulation of free amino acids under salt stress has been found to be related to decreased soluble protein content (Lahiri *et al.,* 1987). However, in certain cases, the decreased synthesis of proteins may be due to a disturbance in amino acids metabolism, which was observed in pea seedlings and broad bean leaves (Shevykova, 1966). It has been suggested by Strogonov (1964) that increased hydrolysis of existing proteins in the cotton cotyledons during germination associated with decreased synthesis of new proteins in the developing seedlings occur under saline conditions. Similar results were reported by Garg and Garg (1981) in germinating mung bean seedlings under the influence of Na_2CO_3 and $NaHCO_3$.

Accumulation of Amino Acids

Plants subjected to salt stress show increased levels of total free amino acids (Dubey and Rani, 1989; Dubey and Pessarakli, 1995). The amino acid proline accumulates in larger amounts compared to other amino acids. In most of the plant species studied, salt stress leads to a dramatic increases in proline level in the tissues. Irrigation with saline water increased proline content in tomato leaves (Heuer, 1994). Callus cultures of *Cicer arietinum* (Pandey and Ganapathy, 1985), eggplant (Jain *et al.,* 1987), rice (Kishor, 1988), Brassica (Chandler and Thorpe, 1987), *Medicago sativa* (Shah *et al.,* 1990), sugarbeet (Dilly *et al.,* 1991), tomato (Rus-Alvarez and Guerrier, 1994), etc. accumulate proline in response to NaCl. Cultivated crop plants like wheat, barley, sorghum, rice, pearl millet, maize, cotton, sunflower, soybean, pea, chickpea, mustard, alfalfa, tomato, potatoes, jojoba show manifold accumulation of proline when grown under NaCl treatments (Heuer, 1994). Rice seedlings subjected to NaCl stress show about four times increase in proline level compared to nonstre-ssed seedlings (Nyan and Shyon, 1984; Dubey and Rani, 1989).

The most commonly observed effect of salinity on amino acids metablism is free proline accumulation both in glycophytes and halophytes (Greenway and Munns, 1980; Levitt, 1980; Heuer, 1994). NaCl and PEG induced osmotic stress, increased free proline accumulation in wheat and pearl millet (Garg *et al.,* 1984) which was 8-10 times than the control value. In moth bean NaCl treatment increased proline, arginine, serine and glutamic acid content of leaves. Table 6.9 indicates that increasing salinity progressively increased free proline and free amino acids in leaves of clusterbean which was associated with decreased level of soluble protein. It has also been found (Heuer,

1994) that free proline accumulation was higher when tomato plants were subjected to both water and salt stresses as compared with salt stress alone (Table 6.10).

Table 6.9. Influence of increasing soil salinity on levels of soluble protein, free aa and free proline (mg g^{-1} dw) of clusterbean at pre-flowering stage (Lahiri *et al.*, 1987)

Salinity (dSm^{-1})	Soluble protein	Free aa	Free proline
0	265.2	13.7	0.60
2	269.3	12.9	0.55
4	227.7	24.5	1.09
6	213.8	28.2	1.15
8	229.6	29.8	1.15
10	196.8	29.7	1.31

Table 6.10. Accumulation of proline in water or salt-stressed tomato leaves* (Heuer, 1994)

Salinity level	Proline, content (uM g^{-1} fresh weight)	
	High soil moisture	Low soil moisture
Nonsalinized	8.83	13.27
Low salinity	15.64	15.77
High salinity	18.31	21.47

*Plants were irrigated with saline water (4 or 10 dSm^{-1}) or irrigation was withheld for 7 or 14 days.

Plants differ in their capacity to accumulate proline. Salt tolerant species generally show more accumulation of proline than sensitive species when subjected to saline stress (Dubey and Rani, 1989). It is believed that the accumulation of proline in salt stressed plants results from its increased synthesis and decreased utilization (Stewart and Larher, 1980). The major functional role of accumulating proline under salt stress appears to be in osmoregulation. Proline along with other soluble nitrogenous compounds serves as compatible cytoplasmic solute and osmoregulator in plants. Increased proline level provides protection to the enzymes against denaturation due to NaCl. Under *in vitro* condition, proline level of 200-800 mM does not inhibit enzyme activities. Proline also serves as a soluble N sink and as an indicator for plant tolerance to salt stress (Heuer, 1994).

Belkhodja (1996) further observed that the amount of proline differs in the different plant organs depending upon the development stage, type of plants and salt treatment and increases proportionally with the increasing salt concentration in the salt tolerant varieties of *Vicia faba*. Durga Prasad *et al.* (1996) while working on the differential behavious of two cultivars of soybean towards salinity reported an increase in the protease activity, amino acids and proline accumulation under NaCl salinity, which probably served as an osmoticum to protect the cell organelles and enzymes under saline conditions. Gupta and Singhal (1996) worked on *Arachis hypogea* and reported that the negative effect of NaCl could be ameliorated by 0.5 mM of $CaCl_2$, indicating that cells required a specific Ca^{2+}/Na^{+} ratio for their growth. This ratio also signified an increase in proline content, thereby suggesting a link between proline content and increased growth. Garg *et al.* (1996) reported that increasing salt concentrations of up to 10 dSm^{-1} decreased chlorophyll content with a concomitant increase in RNA and free proline in leaves at the flowering stage of clusterbean, mung bean and moth bean, Clusterbean was relatively more tolerant than the other two. Misra *et al.* (1996) also observed 4-5 times higher proline accumulation in the roots of mungbean seedlings when grown under 0.5, 1, 2 and 3 per cent (m/v) NaCl. Similarly, Saradhi *et al.*, (1996) advocated over 10 fold higher proline level in the 5 day old *Vigna radiata* seedlings exposed to 200 mM NaCl. *Medicago sativa* and *Glycine max* plants showed a 5 fold proline and a 2-fold trigonelline increase under salt stress (Tramontano *et al.*, 1997). Abd El-Samad *et al.* (1997) attributed the salt tolerance of 2 cultivars of soybean to the accumulation of amino acids and proline under salt stress with cultivar clark being able to tolerate NaCl salinity upto -1800 kpa osmotic potential while cultivar forest up to -1500 kPa. Garg (1991) and Muthukumarasamy *et al.* (1997) reported that saline conditions enhanced the proline content over control in all parts of chickpea seedlings. An increase in proline, glycine and betaine over control along with a marked reduction in growth and chlorophyll content was observed in root, stem and leaf of green gram at 100 and 150 mM NaCl treatments (Muthukumarasamy and Paneerselvam, 1997).

Many studies have indicated that adaptive role of proline is related to survival rather than maintenance of growth (Greenway and Munns, 1980). There is no clear cut evidence that proline accumulates more in salt tolerant than salt sensitive species. In fact, most of the available data are conflicting and the role of proline in salt tolerance has not been clearly understood at present. Heuer (1994) concluded that in general proline is specific to a genotype and generalization for different varieties of a crop is not always possible. Therefore, additional studies are required to elucidate the role of proline in plant adaptation to stress.

SALINITY AND NUCLEIC ACIDS

A reduction in the synthesis of nucleic acids (RNA and DNA) and their degradation has been reported in plants subjected to salt stress (Nieman, 1965; Rauser and Hanson, 1966; Garg and Garg, 1980). However, few studies also report an increase in the RNA content of plants subjected to salinity (Pathak *et al.,* 1974; Garg *et al.,* 1983). Nieman and Poulsen (1964) found that salt stress imposed by NaCl reduced net synthesis of DNA by 60% and of RNA by 90%. Nieman (1965) concluded that reduced rates of cell division and enlargement, RNA and protein synthesis are the primary causes of added salinity. Rausder and Hanson (1966) found reduced rates of DNA synthesis in two varieties of soybean and Na_2SO_4 was more detrimental than K_2SO_4. Na_2SO_4 drastically reduced the incorporation of P^{32} into DNA in salt sensitive but lesser in salt tolerant variety while K_2SO_4 had a mild effect. Decreases in RNA and DNA synthesis in the leaves at the flowering stage of okra and sponge gourd under salinity are also known.

Salinization has been found to interfere with the synthetic processes following the stages of replication and translation. Kaul (1981) observed decrease in nucleic acids of guava leaves under salt stress and the effect was more pronounced on RNA content. Decrease in nucleic acids was probably due to increased hydrolytic activity, reduced synthesis or increased degradation. Garg and Garg (1980) found that both Na_2CO_3 and $NaHCO_3$ depressed leaf growth and reduced the contents of RNA, DNA and protein in leaves of pea at different growth stages (Table 6.1) but Na_2CO_3 was more detrimental than $NaHCO_3$. They concluded that suppressed RNA and DNA contents may be due to impaired synthesis and/or intensified activity of cytoplasmic RNase and DNase, as well documented under NaCl and Na_2SO_4 type of salinity. While enhanced RNase activity under salt stress is well established (Arad and Richmond, 1976; Levitt, 1980). enhanced DNase activity under saline conditions has also been reported (Sheoran and Garg, 1978).

Garg and Garg (1981) found that the hydrolysis of RNA and DNA was retarded, in the cotyledons of germinating green gram seeds, and their translocation to the growing embryoaxis was impaired by Na_2CO_3 and $NaHCO_3$. At iso-osmotic concentrations, $NaHCO_3$ was more depressive than Na_2CO_3. They further found that the inhibitory effects of these salts were more pronounced in a sensitive than a tolerant variety (Garg and Garg, 1982).

Table 6.11. Effect of Na_2CO_3 and $NaHCO_3$ on RNA, DNA and protein (mg per plant) in leaves of pea at 30, 60 and 90 days after sowing (Garg and Garg, 1980)

Treatment (dSm^{-1})		Days after sowing								
		RNA			DNA			Protein		
		30	60	90	30	60	90	30	60	90
Control	0.2	2.06	4.03	9.38	0.37	0.48	1.05	31.7	57.1	126.8
Na_2CO_3	2	1.69	3.43	8.16	0.32	0.43	0.74	27.9	46.6	82.7
	4	1.56	1.62	5.36	0.23	0.31	0.58	16.3	33.2	64.8
	6	1.24	1.30	2.28	0.15	0.19	0.30	11.6	17.9	37.0
$NaHCO_3$	2	1.90	4.08	10.0	0.33	0.46	0.97	28.2	51.2	122.7
	4	1.45	2.60	6.80	0.26	0.33	0.69	19.2	39.2	75.9
	6	1.30	1.94	3.82	0.23	0.25	0.42	17.5	22.5	47.8
LSD (0.05)	–	0.30	0.40	0.71	0.05	0.05	0.19	34	4.8	13.3

SALINITY AND ENZYME ACTIVITY

Salinity induces changes in the activities of proteolytic, amylolytic, nucleolytic, phosphorolytic and oxidative enzymes in germination seeds and growing plant parts (Dubey and Rani, 1990, Garg *et al.,* 1993). Salinity causes either an increase or a decrease in the activity of enzymes, depending on the nature of the enzymes, the plant parts studied, and the genotypes of plant species differing in salt tolerance (Dubey, 1994). Many enzymes have been reported to increase as well as decrease, as a result of salinisation, for instance amylase, peroxidase, malate dehydrogenase, catalaase, nitrate reductase and glutamate dehydrogenase (Table 6.12). In some cases, the increase in activity occurs at low concentrations of salt and the decrease at higher concentrations. One salt may inhibit the activity while the other may stimulate the activity of the same enzyme in the same plant. The increase or decrease in enzyme activity may also depend upon the plant species, the nature of plant tissue and the stage of development. Many enzymes are inhibited in the cotyldeons and roots of germinating seedlings but are stimulated in the embryo-axis and leaves e.g. amylase, RNase, DNase, protease and peroxidase (Sheoran and Garg, 1978, 1979). Likewise nitrate reductase is generally inhibited by salts in many crop plants but may be stimulated in few others, particularly in the halophytes. In some cases progressive decline with increasing salinity has been found for nitrate reductase (Lahiri *et al.,* 1987; Garg *et al.,* 1993) while peroxidase progressively increased with increasing salinity (Lahiri *et al.,* 1987). Although the list (Table 6.12) is not very exhaustive yet it provides a fair idea about the magnitude of large variation in responses of different enzymes to salinity. In certain cases e.g. MDH and peroxidase, the iso-enzyme pattern changes as a result of salt stress. Some new iso-enzymes may appear which may be adaptation to salinity. Some existing bands may also disappear at high salinity levels (Sheoran and Garg, 1979).

Table 6.12. Effects of salt stress on the activities of different enzymes in higher plants.

Enzyme	Plant and Tissue	Salt treatment	Change in activity	Reference
Amylase	Barley	NaCl	Increased	Dixon and Webb (1957)
Amylase	Wheat (germinating seeds)	NaCl	Decreased	Sarin and Narayanan (1968)
Amylase	Wheat seedlings	NaCl	Increased	El-Fouly and Jung (1972)
Amylase	Pearl millet leaves	NaCl	Decreased	Huber *et al.* (1977)
Amylase	Mungbean germinating seeds	NaCl, Na_2SO_4, KCl, K_2SO_4	Decreased in cotyledons, increased in embryoaxis	Sheoran (1980)
Amylase	Wheat leaves	Saline waters	Increased	Garg *et al.* (1982)
Peroxidase	Pea roots	NaCl	Increased	Strogonov (1964)
Peroxidase	Cotton roots	NaCl and Na_2SO_4	Increased	Rakova *et al.* (1969)
Peroxidase	Pea leaves and roots	NaCl, Na_2SO_4, KCl, K_2SO_4	No change or decreased	Weimberg (1970)
Peroxidase	Mungbean germinating seeds	NaCl, Na_2SO_4, KCl, K_2SO_4	Decreased in cotyledons, roots, increased in embryo-axis and leaves	Sheoran and Garg (1979)
Peroxidase	Pearl millet leaves	NaCl, PEG	Increased	Garg *et al.* (1984)
Peroxidase	Wheat leaves	NaCl, Na_2SO_4, Na_2CO_3, $NaHCO_3$	Increased	Garg *et al.* (1983)
Peroxidase	Clusterbean leaves	Soil salinity (0-2-4-6-8-10 dSm^{-1})	Progressive increase with salinity	Lahiri *et al.* (1987)
Malate dehydrogenase (MDH)	Pea roots	Na_2SO_4, NaCl	Inhibited	Porath and Poljakoff Mayber (1965)

Enzyme	Plant and Tissue	Salt treatment	Change in activity	Reference
MDH	Wheat seedlings	NaCl Na_2SO_4	Increased Inhibited	Nainawatee and Das (1974)
MDH	*Suaeda maritima, Pisum sativum*	NaCl KCl NH_4Cl, $CaCl_2$	Increased	Flowers *et al.* (1976)
MDH isoenzymic pattern	Mung bean germinating seeds	NaCl, Na_2SO_4 KCl, K_2SO_4 (5 and 10 dSm^{-1})	Disappearance of isoenzymers bands at high concen.	Sheoran and Garg (1980)
RNase	Barley leaves	NaCl	Stimulated	Arad and Richmond (1973)
RNase	Pea leaves	NaCl	Stimulated	Kabanov (1973)
RNase	Tomato leaves	NaCl	Stimulated	Tal (1977)
RNase	Mungbean germinating seeds	NaCl, Na_2SO_4, KCl, K_2SO_4	Inhibited in cotyledons, roots increased in E-axis, leaves	Sheoran and Garg (1978)
DNase	Mungbean	NaCl, Na_2SO_4, KCl, K_2SO_4	Na salts stimulated, K salts no effect	Sheorena and Garg (1978)
	Tomato leaves	NaCl	Stimulated	Tal (1977)
Protease	Mungbean germinating seedlings	NaCl, Na_2SO_4, KCl, K_2SO_4	No effect, only NaCl increased the activity n leaves	Sheoran and Garg (1978)
Protease	Clusterbean leaves	Soil salinity (0-10 dSm^{-1})	Progressive increase with salinity	Lahiri *et al.* (1987)
ATPase	Beans	NaCl	Stimulated	Horovitz and Waisel (1970)
ATPase	Wheat leaves	Saline water (0-150 me/l)	Stimulated	Garg *et al.* (1982)
ATPase	Wheat leaves	NaCl, Na_2SO_4 $NaHCO_3$, Na_2CO_3	Stimulated	Garg *et al.* (1983)
Acid pyrophosphates	Wheat leaves	-do-	Inhibited	Garg *et al.* (1983)
Acid pyrophosphate	Clusterbean leaves	Soil salinity	Progressive increase	Lahiri *et al.* (1987)

Enzyme	Plant and Tissue	Salt treatment	Change in activity	Reference
Alkaline pyro-phosphate	Wheat leaves	Saline water (0-150 me/l)	Stimulated	Garg *et al.* (1982)
Alkaline pyro-phosphate	Pearl millet leaves	PEG, NaCl	Inhibited	Garg *et al.* (1984)
Alkaline pyrop-hosphatse	Clusterbean leaves	Soil salinity	Stimulated	Lahiri *et al.* (1987)
Nitrate reductase (NR)	*Suaeda maritima* leaves	NaCl	Increased	Boucard (1972)
NR	Tobacco leaves	NaCl, KCl	Inhibited	Heimer (1973)
NR	Mungbean	Mungbean	Increased	Sankhla and Huber (1975)
NR	Wheat leaves	Saline waters	Inhibited	Garg *et al.* (1982, 90)
NR	Wheat leaves	NaCl, Na_2SO_4 Na_2CO_3, $NaHCO_3$	Inhibited by all most by Na_2CO_3	Garg *et al.* (1983)
NR	Pearl millet leaves	PEG, NaCl	Inhibited	Garg *et al.* (1984)
NR	Wheat leaves	NaCl, Na_2SO_4	Inhibited	Sharma and Garg (1985)
NR	Barley seedlings	NaCl	Stimulated	Aslam *et al.* (1984)
NR	Clusterbean leaves	Soil salinity	Progress decrease with salinity	Lahiri *et al.* (1987), Lahiri *et al.* (1996)
NR	Pea leaves	NaCl	Decreased	Lal and Bhardwaj (1987)
NR	Tomato and cucumber leaves	NaCl	Decreased	Marnez and Cerda (1989)
NR	Pearl millet varieties	Soil salinity	More decrease in sensitive than tolerant varieties	Sharma and Gill (1990)
NR	Indian mustard	Saline waters	Progressive decrease	Garg *et al.* (1993)
Glutamine synthetase (GS)	*Suaeda maritima*	NaCl	Stimulated	Boucard and Billard (1978)
GS	Pea leaves	NaCl	Decreased	Lal and Bhardwaj (1987)

Enzyme	Plant and Tissue	Salt treatment	Change in activity	Reference
GS	Indian mustard	Saline waters	Progressive decrease	Garg *et al.* (1993)
GS	Rice seedlings	NaCl	Increased	Singh and Dubey (1994)
GS	Clusterbean genotypes	Soil salinity	Decreased	Lahiri *et al.* (1996)
Glutamine synthase (GOGAT)	Pea leaves	NaCl	Inhibited	Lal and Bhardwaj (1987)
GOGAT	Indian mustard	Saline waters	Inhibited	Garg *et al.* (1993)
GOGAT	Rice seedlings	NaCl	Inhibited	Dubey *et al.* (1991)
Glutamine dehydrogenase (GDH)	Peanut leaves	NaCl	Inhibited	Rao *et al.* (1981)
GDH	Wheat leaves and roots	NaCl	Promoted at lower and inhibited at higher concentration	Sharma and Garg (1985)
GDH	Indian mustard	Saline waters	Promoted at lower and inhibited at higher	Garg *et al.* (1993)
Aspartate amino transfer (GOT)	Moth bean leaves	NaCl	Increased	Huber and Sankhla (1973)
GOT	Peanut leaves	NaCl	Increased	Rao *et al.* (1981)
GOT	Wheat leaves and roots	NaCl	Increased	Sharma and Garg (1985)
Alanine aminotransferase (GPT)	Moth bean leaves	NaCl	Increased	Humber and Sankhla (1973)
GPT	Peanut leaves	NaCl	Increased	Rao *et al.* (1981)
GPT	Wheat leaves and roots	NaCl	Increased	Sharma and Garg (1985)

Dubey (1994) indicated that salinity cause either an increase or decrease in the activity of enzyme depending on the nature of enzymes, the plant parts studied, the genotype of a species differing in salt tolerance and the concentration and composition of salts. However, the sensitivity of key enzyme of nitrogen metabolism-nitrate reductase is well documented (Garg and Gupta, 1997). In sorghum detrimental effects of salt stress on enzymes of nitrogen metabolism have been reported by several investigations (Khan *et al.,* 1990; Rao and Gnanam, 1990). The activities of both nitrate reductase (NR) and nitrite reductase (NIR) were inhibited, albeit differently, as a function of NaCl concentrations in the growth medium (Rao and Gnanam, 1990). Results indicated that NR is more sensitive than NIR to NaCl salinity both *in vivo* and *in vitro*. Furthermore, anion salinity was more toxic to NR activity than cation salinity. Khan *et al.* (1990) found that shoot NR decreased with increasing NaCl salinity but the reduction was significantly more in sensitive (IS-4807) than tolerant (Giza-10) genotype at all levels of salinity. Such effects are well known in other crops also (Garg and Gupta, 1997).

Besides NR, the activities of many other enzymes have been reported to be adversely affected by salinity in sorghum plants. Patil *et al.* (1983) reported decreased activity of PEP carbophylase and RuBP carboxylate which led to lowered photosynthetic activity under different levels of salinity. Changes in amylase and acid protease activities during seedling growth of sorghum were observed in two genotypes increased salt concentrations (Ogra and Baijal, 1982). However, variety T8B was more salt tolerant than MP chari in relation to enzyme activities and seedling growth. Marambll and Ando (1995) found that amylase activity in the germinating seeds declined in response to salinity but the amylase may be compensated by the appearance of new iso-enzymes of amylase and this may be related to salt tolerance capacity of the sorghum plants. Koyro *et al.,* (1993) found that ATPase activities and membrane fine structures of rhizodermal cells from sorghum and spartina roots were differently affected under mild salt stress. The vanadate-sensitive tissue ATPase activities increased about 5 times in sorghum and about 2 times in Spartina whilst the nitrate and azide sensitive tissue ATPase activities increased about 6 times only in sorghum.

The activities of aspartate aminotransferase (GOT), alanine aminotransferase (GPT) and glutamate dehydrogenase (GDH) were increased in the roots and leaves of wheat at high NaCl salinity (Sharma and Garg, 1985). It seems that increase in aminotransferases coupled with increased GDH activity forms a vital link in plant protective mechanisms under saline conditions.

Salt induced changes in the activities of various enzymes are often due to disturbances in the normal metabolism of plants. In general synthetic enzymes are inhibited while hydrolysing enzymes are stimulated by salt stress. Although, on the basis of whole plant many enzymes from halophytes are affected less adversely than those from glycophytes yet *in vitro* studies have shown that soluble enzymes from halophytes and glycophytes have similar senstivities to electrolytes (Greenway and Munns, 1980). It seems that compartmentalization of electrolytes in halophytes prevents the enzymes from direct injury of salts but the enzymes as such may not be tolerant to salinity.

In order to understand various salt effects, it must be realized that the salt does more than simply existing in solution as charged particles. Salts have two antagonistic effects on proteins : (a) tend to break the electrostatic bonds and (b) cause increase in the hydrophobic interactions. There is, therefore, no basic reason for expecting a general increase or decrease in enzyme activity. On the contrary, both effects may be expected depending on the specific structural characteristics of the protein.

In view of the above two antagonistic effects of salts on proteins, and therefore on enzyme activity, the existence of single, salt induced, master metabolic disturbance, leading to all others, is extremely unlikely if not impossible.

Greenway and Osmond (1972) showed that enzymes malate dehydrogenase, aspartate transaminase, glucose-6-phosphate dehydrogenase and isocitrate dehydrogenase, extracted from salt sensitive *Phaseolus vulgaris* and salt tolerant *Atriplex spongiosa* or *Salicornia australis* and grown in saline cultures, showed no significant changes in specific activity with an increase in salt concentration. Similar response to several other enyzmes from pea leaves, roots and stems were also reported by Weimberg (1970) under NaCl, KCl, Na_2SO_4 and K_2SO_4 salinisation. Sheoran and Garg (1978) reported an inhibition of RNase and DNase activities in the cotyledons and roots but an increased activity in embryo axis and leaves without any effect on the protease activity in any of these organs during the early seedling growth in mung bean under salinised conditions. Similar variable response of different plant parts was also reported by these authors with regard to the peroxidase activity. Dias and Costa (1983) observed direct correlation between nitrate reductase, peroxidase and sugar contents of sugar beet leaves by NaCl salinity.

Guerrier (1987) has shown a negative relationship between IAA-oxidase, catalase and peroxidase activities and Na contents of seedlings in tomato, radish and red cabbage. Garg *et al.* (1988) reported significant increase in peroxidase enzyme activity in the nodules of mung bean with increasing levels of salinity in the rooting medium.

Huang and Chen (1995) while working on soybean advocated an escalation in the activities of glutathione reductase (GR) and its related enzymes like ascorbate peroxidase and dehydro ascorbate reductase in salt acclimated plants. They further observed that GR kept a high ratio of GSH/GSSG in cytosol, which seemed to have a positive effect on the plants to enable them to tolerate the salt stress environment.

Cordovilla *et al.* (1996) reported a decrease in the activities of (GS, NADH-GOGAT activity and NADH-GDH) in nodules of *Vicia faba* by 8 mM KNO_3 and this concentration of KNO_3 increased the salinity tolerance even at 100 mM NaCl treatment. Nitrate reductase (NR) and nitrite reductase (NiR) activities substantially declined in leaves and roots of two soybean genotypes grown under a range of NaCl and Na_2CO_3 salts (0-12 dSm^{-1}) (Khan, 1996). Olmos and Hellin (1996) while working on 2 lines of *Pisum sativum* revealed the higher presence of enzymes related to sugar metabolism (glucokinase, frutokinase and acid invertase) in response to NaCl salinity.

SALINITY AND REACTIVE OXYGEN SPECIES

Salinity limits plant growth and productivity by affecting water deficit, ionic imbalance, osmotic stress and the secondary induced oxidative stress by the production of reactive oxygen species (ROS). Any imbalance in the cellular redox homeastasis can be called oxidative stress which results in the production of ROS, such as superoxide radical (O_2^-), hydrogen peroxide (H_2O_2) hydroxyl radical (OH), alkoxyl radical (RO) and singlet oxygen (O_2) formation via enhanced leakage of electron to oxygen. ROS attack protein, lipids and nucleic acids. ROS inactivate enzymes and damage important cellular components (Mittler, 2002).

Plants have developed a complex antioxidative defence systems to alleviate the damage caused by ROS and the degree of damage depends on the balance between the formation of ROS and its removal by the antioxidative scavenging systems that defend against them. The antioxidative system includes carotenoids, ascorbate, glutalthione and enzymes such as superoxide dismutase (SOD), catalase (CAT), glutathione reductase (GR), peroxidases and enzymes involved in ascrobate glutathione (ASC-GSH) cycle, namely ascorbate peroxidase (APOX), hydro-ascorbate reductase monodehydro ascorbate reducatse and glutathione reductase (Asada, 1992, Hernandez *et al.,* 2000; Apel and Hirt, 2004).

Antioxidant enzymes (viz SOD, APOX, GR and CAT) are important in limiting oxidative damage and in destroying ROS that are produced in excess under various abiotic stresses. Antioxidant enzymes show wide variability in tolerant and susceptible genotypes under drought, salinity and high/low temperature (Hernandez *et al.*, 2000, Sairam *et al.*, 2002). Madan Pal *et al.* (2004) found that maintenance of higher RWC and induction of SOD and APX activities under salt stress in CSR-13 genotype of Rice contributed to its salt tolerant characteristics as compared to sensitive variety MI-48. Sangeeta *et al.* (2007) reported similar findings in tolerant and sensitivie genotype of wheat. Rani *et al.* (2008) reported that antioxidative defence system was more efficient in salt tolerant genotype KRL 1-4 in comparison to HD-2329 of wheat based on their results on activities of enzymes SOD, CAT, POX, APX and GR.

Recently the role of ROS as signaling molecules triggering production of antioxidants enzyme in their own scavenging has also gained importance. It has been shown that ABA and salicyclic acid produced under various abiotic stresses induce ROS via a membrane linked NADPH oxidate which is the generator of redox signal (O_2/H_2O_2) leading to production of various antioxidant enzymes. Foyer and Noctor (2005) suggested that ROS which are involved in oxidative signal transduction what in turn trigger the antioxidant defence system associated with the mechanisms by which plant cell sense the environment and make appropriate adjustment to gene expression, metabolism and physiology. Hung *et al.* (2005) advocated that H_2O_2 seems to serve as a common stress signal in plants. Stress signal is transduced via ABA, Ca^{2+} and H_2O_2 which activates a common transcription factor associated with SOD, APX and Catalase (Agarwal *et al.,* 2005). Devi *et al.* (2008) while studying the role of H_2O_2 on the enzymes of antioxidant defence process in wheat under NaCl salinity demonstrated that an inductive pulse of H_2O_2 treatment could partially alleviate the salt effects on plants.

NODULATION AND NITROGEN FIXATION UNDER SALT STRESS

Symbiotic nitrogen fixation is dependent on the host plant genotype, the *Rhizobium* strain, and the interaction of these symbionts with the pedoclimatic factors and the environmental conditions (Bordeleau and Prevost, 1994). There is general agreement that rhizobia are able to cope with salinity more than their host legumes and there is considerable strain variability in survival and growth under saline conditions (Bhardwaj, 1975). Levels of salinity that inhibit the symbiosis between legumes and rhizobia differ from those that inhibit the growth of individual symbiont. NaCl concentrations that affect symbiosis between *Rhizobium meliloti* and alfalfa are lower than those that affect the growth and survival of individual alfalfa genotypes or *Rhizobium* species (Sprent, 1984). In contrast, to their host legumes, some rhizobia can survive in the presence of extremely high levels of salt both in culture and in soil. *R. meliloti* can survive in soil water salinities greater than that of sea water (46 dSm^{-1}) but most strains of *R. japonicum* grew poorly at salinities of 12 dSm^{-1} (Bernard *et al.,* 1986). Studies comparing various *Rhizobium* species report the salt tolerance as *R. meliloti* > *R. trifoli* > *R. leguminosarum* > *R. japanicum*. The salt effect on rhizobia appeared to be ion specific, Cl^- salts of Na^+, K^+ and Mg^{2+} being more toxic than corresponding $SO^2{}_4$ salts (Elsheikh and Wood, 1989).

Inhibition of nodulation and N_2-fixation by salt stress (mostly NaCl) has been reported in soybean (Singleton and Bohlool, 1984; Elsheikh and Wood, 1995), alfalfa (Lakshmikumari *et al.,* 1981), mung bean (Balasubramanian and Sinha, 1976, Hafeez *et al.,* 1988), clusterbean (Garg *et al.,* 1984), faba bean (Yousuf and Sprent, 1983), green beans (Pessarakli and Zhou, 1990, Cordovilla *et al.,* 1995), pea (Siddiqui *et al.,* 1985), pigeon pea (Subbarao *et al.,* 1990) and several others (Sprent, 1984; Zhou *et al.,* 1995). Retardation of growth as a result of salt stress was much higher in plants depending on symbiotic N_2-fixation than in plants receiving fertilizer nitrogen (Wilson, 1970; Lauter *et al.,* 1981). Several recent studies indicate that inoculation with a salt tolerant *Rhizobium* strain may improve biological N_2-fixation under saline conditions (Subbarao *et al.,* 1990; Elsheikh and Wood, 1995; Zhou *et al.,* 1995).

Salinity and alkalinity inhibit nodulation depending upon the degree of salt stress. Bernstein and Ogata (1966) observed that salinity was inhibitory to growth and nodulation of soybean and alfalfa and that salinity inhibition of nodulation was a specific effect of salinity and not a reflection of restricted root growth. Wilson (1970) found that nodules developed prior to salt treatment appeared remarkably resistant to stress, although the development of new nodules was inhibited in *Glycine wightii*. Balasubramanian and Sinha (1976) found that the reduction due to salinity in total nodule weight in chickpea was a result of decreased number of nodules, but further development of nodules was not affected.

Tu (1981) demonstrated that the reduction in nodulation because of high salinity was attributable to (a) decreased growth and multiplication of rhizobia and (b) decreased availability and susceptibility of root hairs. The former appeared to be more sensitive to the increased salinity than the later because the rapid decrease in rhizobial growth started at 0.2% NaCl but the decrease in number and increase in plasmolysis of root hairs, occurred at 1% NaCl or above. Lakshmi Kumari *et al.* (1974) had also reported that in lucerne, salinity resulted in a root system devoid of root hairs, mucilaginous layer and infection thread formation, inspite of the optimum growth of rhizobia under such conditions. Complete failure of nodulation under high salinity and alkali conditions was found in pea. Sprent (1980) found the *R. meliloti* could tolerate upto 3% NaCl but could form nodules only at concentrations below 0.6% NaCl. It was observed that although crop plants do not like saline conditions, at low salt levels they might compensate by producing larger nodules. Garg *et al.* (1984) observed that salt stress due to NaCl, Na_2SO_4 and $NaHCO_3$ decreased the nodule number and dry weight in both clusterbean and moth bean but Na_2SO_4 did not have significant effect in either crop. However, $NaHCO_3$ showed more toxicity than NaCl and Na_2SO_4.

Hutton (1971) ranked eight legumes in the order of their response to 60 mM NaCl. Lucerne was most resistant and *Neonatonia wightii* and *Desmodium* the most susceptible. Nodulation was not prevented but nitrogen fixation was reduced. Most legumes respond to saline conditions by exclusion of sodium or chloride or both ions from the leaves. It has been argued that organic solutes such as proline and glycine betaine enhance the salt tolerance of legumes and rhizobia. Intercellular accumulation of these solutes have been correlated with salt stress tolerance in *R. melioti* (Botsford and Lewis, 1990). Instead of osmotic adjustment by organic solutes, a more energy efficient way of achieving salt tolerance is to select for subcellular compartmentalisation of inorganic ions coupled with salt sequestration in the cell vacuole of plant tissue (Kramer, 1984).

Nodules with vacuolate cells seem to be more efficient in compartmentalizing and distribution of ions, and thereby avoid salt toxicity (Zahran, 1991). Improved ion compartmentalisation within the cells of nodules and their organelles could promote growth under saline conditions and this appears to be a basic criterion for the selection and breeding of salt tolerant legumes.

There are number of reports on the adverse effects of salinity on the nitrogen fixation and nitrogenase activity in different legumes (Hopmans *et al.,* 1984; Sprent, 1984; Pessarakli and Zhou, 1990). Garg *et al.,* (1984) reported that NaCl, Na_2SO_4 and $NaHCO_3$ (5 and 10 dSm^{-1}) significantly reduced the nitrogenase activity per plant except in case of 5 dSm^{-1} NaCl. However, the specific activity of the nodules did not decline under the salt treatments. On the contrary, it showed an increase at higher concentration of NaCl and Na_2SO_4 in clusterbean. Similarly, N_2-ase activity showed different response to various salts in moth bean (Table 6.13). Here the specific activity of N_2-ase increased under the influence of NaCl and $NaHCO_3$ while it declined slightly under Na_2SO_4. It was concluded that in clusterbean and moth bean, salt stress caused a decrease in N_2-ase activity through a reduction in nodule number and weight. However, the efficiency of nitrogen fixation per unit weight of existing nodules (specific activity) often increased, particularly in moth bean. This suggests the possibility of the existence of an inbuilt compensatory mechanism which enables the salt affected legumes to mitigate partly the imbalance in nitrogen fixation. This compensatory mechanism seems to be more effective in moth bean where nodulation and N_2-ase were affected more adversely.

Table 6.13. Effect of salt stress on nodulation and N_2-fixation in clusterbean and moth bean (Garg *et al.*, 1984).

Treatment (dSm^{-1})		Nodule no. $plant^{-1}$		Dry wt. nodules (mg $plant^{-1}$)		N_2-ase activity (nmoles C_2H_4 $plant^{-1}$)		Specific activity N_2-ase (μ moles C_2H_4 g^{-1} h^{1})	
		Cluster bean	Moth bean	Cluster bean	Moth bean	Cluster bean	Moth bean	Cluste r bean	Mot h bean
Control	0	20.5	18.3	83.3	28.7	471.9	54.7	5.66	1.90
NaCl	5	21.5	9.9	27.7	12.0	462.5	32.3	6.36	2.70
	10	17.0	7.3	47.2	9.2	351.9	31.8	7.45	3.44
Na_2SO_4	5	18.7	10.0	79.8	15.4	373.7	21.4	4.68	1.39
	10	15.3	5.2	58.5	5.8	338.5	8.8	5.79	1.53
$NaHCO_3$	5	19.0	8.1	60.9	8.6	309.9	27.0	5.09	3.14
	10	11.0	3.8	37.9	2.5	239.7	10.9	6.33	0.35
LSD (0.05)		4.6	4.0	16.8	3.4	69.1	6.3	N.S.	0.85

Hopmans *et al.* (1984) found that acetylene reduction activity (ARA) showed a rapid decrease on addition of varying amounts of NaCl (0, 25, 75, 150, 250 and 500 mg) to 200 g sandy loam soil and the maximum effect was observed after 2 days.

A number of studies have shown that solutes have effects on nitrogen fixation which are quite distinct from those of atmospheric stress or water stress. Working with detached nodules of soybean, Sprent (1984) showed that NaCl and KCl (0.1 M) had a very rapid inhibitory effect on acetylene reduction as compared to mannitol at iso-osmotic concentration. Kuhad and Sheoran (1982) compared the effects of iso-osmotic concentrations of NaCl and PEG on N_2-fixation and nodule respiration up to 72 h in clusterbean plants. PEG had a much slower inhibitory effect initially (upto 24 h) on ARA but after 48 h PEG affected N_2-fixation much more adversely than NaCl. N_2-fixation was affected more than the nodule respiration by stress treatments. Similar effects were also observed by Sanch-z-Diaz *et al.* (1982) on N_2-fixation, stomotal response and transpiration of *Medicago sativa, Trifolium repens* and *T. bracycalycinium* when comparing the effects of PEG and NaCl at various osmotic potentials.

Sheokand *et al.* (1995) reported that 45 days old chickpea plants subjected to 50 and 100 mM NaCl treatments for 7 and 14 days, respectively, accelerated nodule greening, decreased the leghemoglobin (Lb) content by 41 per cent and 27 per cent of the control and decreased the acetylene reduction activity (ARA) by 63 and 70 per cent 7 days after treatments. However, 14 DAT 50 mM NaCl treated plants showed recovery in leghemoglobin and total soluble carbohydrates and ARA, suggesting that probably they were playing an osmoregulatory role. *Medicago sativa* L. plants when treated with 250 mM Na_2SO_4, NaCl, KCl, K_2SO_4 and $CaCl_2$ showed a reduction in the nodule number by 70, 65, 22 and 2 per cents respectively (Benet *et al.,* 1996).

Fernandex-Pascual *et al.* (1996) reported a significant decrease in the nitrogenase activity and the leghemoglobin content 150 mol m^{-3} in the nodules of *Lupines alba* Nitrate and nitrite reductase activity substantially declined in leaves and roots of two soybean genotypes grown under NaCl and Na_2CO_3 salts (0-12 dSm^{-1}) and (Khan, 1996). 7-week old *Cyamopsis tetragonoloba* plants treated with 50 and 100 mol m^{-3} NaCl showed a decrease in ARA between 18-38 per cent 3 DAT and 33-54 per cent 6 DAT and Lb content by 6-14 per cent 3 DAT and 19-36 per cent 6 DAT (Bishnoi *et al.,* 1997). Comba *et al.* (1997) worked on two soybean cultivar, one NaCl tolerant (377) and the other relatively NaCl sensitive (411) and reported a 50 per cent decrease in leghemoglobin (Lb) as well as ARA in the latter with no significant change in the tolerant cultivar.

Adverse effects of salinity in legumes are generally mediated through detrimental effects on nodulation and symbiotic nitrogen fixation. Inhibition of nodulation and N_2-fixation by salt stress has been reported in a large number of leguminous crops. However, the retardation of growth is much higher in plants dependent on symbiotic N_2-fixation than in plants receiving fertilizer-nitrogen. There is general agreement that rhizobia are able to cope with salinity more than their host legumes but there is considerable strain variability in their survival and growth under saline conditions. Salinity may differently affect various phases of legume-Rhizobium symbiosis which include *Rhizobium* survival and growth in the bulk soil or the rhizosphere, rhizobial infection of the host, nodule initiation and development, nodule functioning or N_2-fixation process and the growth of the host legume and its ability to maintain a supply of photosynthates and nutrients to the root nodules. It has been suggested that development of salt tolerant *rhizobium* strains suitable to be paired with relatively salt resistant legumes would be an important line for future research (Garg and Gupta, 2000). Furthermore, the selection and development of salt-tolerant cultivars of legumes for effective symbiosis under salt stress hold a key to enhanced nodulation and nitrogen fixation. The interactive studies between salt stress and plant nutrition, particularly calcium and potassium on nodulation and symbiotic nitrogen fixation have received less attention but seem important for alleviation of salinity effects on these processes (Garg and Gupta, 2000).

Notwithstanding the adverse effects of salinity on nodulation and N_2-fixation it is feasible to grow legumes in marginal saline lands by adopting a focused research approach to understand the diversity of salt stress in different legumes and to identify tolerant host genotypes and match them with appropriate tolerant rhizobia. Several studies indicate that inoculation with a salt tolerant *Rhizobium* strain may improve biological N_2-fixation under saline conditions (Subbarao *et al.*, 1990a; Elsheikh and Wood, 1995; Zou *et al.*, 1995). However, introduction of rhizobia into saline soils and their survival and persistance is feasible only if rhiziobial strains tolerant to salinity stress are available for different legumes.

Thus, there appears to be a need to search for salt tolerance in rhizobia, the host cultivars, and the host-*Rhizobia,* combinations for the complete spectrum of salts present in the salinity complex. Breeding legumes tolerant to saline soils must also incorporate a component of breeding for improved N_2-fixation.

SALINITY AND PLANT HORMONES

In general, increasing salinity is associated with a decrease in auxin, cytokinin, and gibberelin levels in plant tissues, and an increase in ABA. Such changes in hormone levels are thought to be a primary process regulating the reduction in growth associated with salinity. In fact, the NaCl-induced reduction in growth can be partly counteracted by the application of gibberellic acid (Prakash and Prathapasenan, 1990). Also, pretreatment of salt-sensitive seed with cytokinins or gibberellins will increase salt tolerance of wheat seedlings (Datta *et al.,* 1998).

Exposure to salt also leads to enhanced production of plant hormones especially ABA and jasmonate (JA; Hamayun *et al.,* 2010; Kumar *et al.,* 2013; Zhang *et al.,* 2006). ABA is often referred to as a stress hormone as it usually accumulates in plant tissues immediately after a plant is exposed to stresses, and it is known to induce a number of biochemical pathways that relate to stress tolerance (Zhang *et al.,* 2006). ABA can be synthesized through the terpenoid pathay from isopentenyl pyrophosphate. Some genes involved in this pathway are found to be rapidly induced after drought or salinity stress is imposed and are corrected with the accumulation of ABA and its intermediates (Ruiz-Sola *et al.,* 2014; Welsch *et al.,* 2008). However, the mechanism of how osmotic and salinity stresses are preceived leading to the induction of ABA biosynthesis remains to be elucidated. ABA biosynthesis and signalling is thought to be induced through a Ca^{2+} -dependent protein phosphorylation cascade, and it may also be mediated by ROS (Ismail *et al.,* 2014; Zhang *et al.,* 2006).

ABA-regulated gene expression is generally thought to be important to confer tolerance during dehydration, and part of the salinity response is known to be linked to its osmotic impact on plants. ABA is responsible for mediating stomatal closure to prevent water loss (Jones and Mansfield, 1970) and maintaining root growth (Sharp *et al.,* 2004). ABA has been known to be involved in maintaining cellular osmotic potential by inducing genes that encode enzymes for compatible solute biosynthesis such as proline (Fujii *et al.,* 2011; Krasensky and Jonak, 2012), betaine (Gao *et al.,* 2004), and also genes that encode dehydrins (Brini *et al.,* 2007). Dehydrins are hydrophilic proteins, which are important in retaining water in the cell under stresses that lead to cellular dehydration induced by low temperature, drought, and salinity (Ingram and Bartels, 1996).

Changes in abundance of several enzymes involved in phytohormone metabolism such as jasmonic acid (JA) biosynthesis (allene oxide cyclase, AOC; lipoxygenase, LOX), gibberellin (GA) biosynthesis (DWARF3), ethylene biosynthesis (SAMS), and ABA biosynthesis (*9-cis*-epoxycarotenoid dioxygenase; NCED) have been detected in salt-treated plants. Increased relative abundance of ABA biosynthesis (increase in NCED level) found in *T. salsuginea* (Taji *et al.,* 2004) corresponds with enhanced ABA level observed in salt-treated plants and with an increased expression of several early (ABA-dependent transcription factors) and delayed (genes induced by ABA-dependent transcription factors, e.g., *Lea/Rab* genes) ABA-responsive genes. An enhanced induction of ethylene receptor ETR1 was found in common wheat cv. Jinan under salinity. Activation of JA biosynthesis (increase in AOC and LOX levels) in salt-treated *A. thaliana* indicates an increased relative abundance of JA-induced signaling under salt stress has also been reported (Pang *et al.,* 2010).

Plant hormones, especially cytokinins and abscisic acid (ABA) appear to play an important role in plant water relations through their effect on stomata. Kinetin promotes and ABA reduces stomotal opening (Livne and Vaadia, 1972). Working with tobacco, Mizrahi *et al.* (1972) demonstrated lowering of the level of indogenous kinetin and an increase in ABA, when the plants were exposed to addition of mannitol or NaCl to the culture solutions. They found a relationship between kinetin, ABA and various plant stresses including salinity. It was shown by bioassay and gas liquid chromatography (GLC) methods that the quantity of bound ABA decreased and that of free ABA increased, in the leaves of tobacco soon after the addition of salt to the root medium (Mizrahi and Richmond, 1971). They had demonstrated earlier the accumulation of ABA in tobacco plants within 4 to 48 h of addition of 0.17 M mannitol or 0.1 M NaCl. It was suggested that increase in ABA content of leaves is not necessarily dependent on reduced leaf turgor and ABA may be involved in a more general aspect of water balance in plants, namely the regulation of adaptive response of plants to osmotic stress.

Downton and Loveys (1981) reported that leaves from grape vines receiving 50 and 100 mM NaCl (–0.2 and –0.4 MPa) showed three fold increase in ABA concentration within 6 h of treatment. Stimulation of ABA accumulation by salinity has also been demonstrated in tomato plants (Walker and Dumbroff, 1981) and this was associated with decreased cytokinin levels. It was, however, observed that even though water content and osmotic potential continued to decline over 8 days of stress and accumulation of ABA declined rapidly after 2 days, reaching normal levels in the next 6 days (Fig. 6.3).

Now, there is considerable evidence that ABA levels increase during salinity stress in both halophytes and glycophytes (Nilsen and Orcutt, 1996), and that increase is positively correlated to the salinity of the growth medium. The activity of enzymes associated with pathways leading to osmolyte accumulation (such as BADH, an enzyme associated with glycine-betaine biosynthesis) are increased by salinity, and the increase correlates with a rise in ABA concentration. Also, exogenous application of ABA at low salinity treatments induces an increase in BADH activity (Ishitani *et al.,* 1995). The reverse is not true, exogenous application of glycine-betaine influences stomatal conductance irrespective of ABA concentration (Makela *et al.,* 1998). In response to increased salinity, ABA levels increase rapidly and then decrease after a few days to values similar to those before treatment. The short-term induction of ABA may constitute a signal for tissues to increase osmolality and induce the synthesis of ion transporters, but a long term increase in ABA is not necessary to keep the new metabolism operating and may inhibit appropriate salinity response. In fact, salt-tolerant species such as *Brassica napus* have lower concentrations of ABA than salt-sensitive species such as *Brassica carinata* (He and Cramer, 1996). The ABA signaling mechanism is most likely related to calcium biochemistry. Although ABA has been implicated in the signal transduction pathway between salinity (also water stress) and gene transcription processes, not all salinity responses are induced by ABA. For example, ABA is not the signular factor inducing accumulation of proline and PEPcase in *Mesembrya-nthemum crystallinum* (Thomas *et al.,*1992).

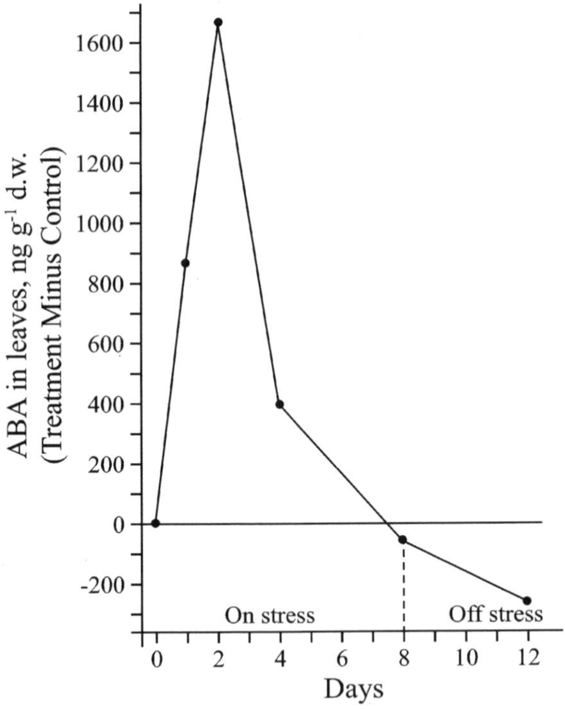

Fig. 6.3. Effect of salt stress on free ABA content of tomato leaves.

Cytokinins are produced in roots and move to plant tops where they are known to have a role in maintenance, especially in maintenance of protein metabolism machinery. Reports have shown that the level of cytokinins in shoot is apparently regulated not only by the rate of supply from the roots but also by their chemical transformation in the leaves. Water and salt stresses have been reported to decrease the transport of cytokinins from the roots to the shoots (Itai and Vaadia, 1971) in several crop plants e.g. sunflower, bean and cotton. This suggests that one reason for metabolic disturbances at the cellular level under saline conditions may be lack of cytokinins, but whether it is the cause or the effect, is not clear. Kahane and Poljakoff-Mayber (1968) reported a decrease in amino acid incorporation into protein of pea roots exposed to NaCl or Na_2SO_4 and the effect of NaCl only was partially counteracted by exogenously applied kinetin. Kinetin application also affected the pattern of distribution of Na^+ and Cl^- in bean and cotton plants.

As salinity is known to affect transport of cytokinins from roots to shoots (Walker and Dumbroff, 1981) they are also known to interact with other growth substances, and therefore, may influence many aspects of shoot physiology particularly under saline conditions. The most important being stomotal behavious, leaf growth, senescence and abscission, all factors that will influence water use by the plants (Davies *et al.,* 1986).

Ethylene (C = C) also increases in the presence of salinity. Several studies with tomato have shown an increase in ethylene by leaves, particularly older leaves (Feng and Barker, 1992). Ethylene is known to induce senescence, and its increase during saline conditions is likely a general stress response. Studies of citrus genotypes showed a biphasic response of ABA, and aminocyclopropane-1-carboxylic acid (ACC) which is the last substrate in ethylene biosynthesis. The initial transitory ACC increase was due to water stress effects, but the subsequent accumulation was due to ionic effects (Gomez-Cadenas *et al.,* 1998). Severe leaf abscission (92%) was correlated with the increase in intercellular Cl and ACC accumulation. Therefore, specific ion effects of salinity may be the ultimate cause of ethylene increase and sensecence induction. Inhibition of ethylene production under saline conditions by salt-tolerant plants could reduce the amount of tissue senescence and allow the accumulation of salt in the tissues before they die. Proline was able to overcome the effect of NaCl on ethylene concentration in *Allenrolfia occidentalis* (Crominski *et al.,* 1988), suggesting that organic osmolytes can inhibit ethylene production during water stress and slow senescence in salt-tolerant plants.

The effects of applied growth regulators which have been reported to protect against salt injury are summarised in Table 6.14. These also include Auxins (IAA) and Gibberellins (GA3) on which only limited information is available under saline conditions. IAA does not affect stomatal aperture directly but it may regulate water balance by modifying the number and size of leaves. GA3 (100 mg l^{-1}) counteracted the negative effects of salinisation on growth, photosynthesis and translocation of bean plants (Strack *et al.,* 1975) as well as on the growth and activities of enzymes of carbohydrate metabolism in pearl millet (Huber and Sankhla, 1973). Ethylene has been reported to reverse the inhibition of germination of lettuce seeds caused by salinity (Verma *et al.,* 1973).

Table 6.14. Effect of applied growth hormones on salt stressed plants (Levitt, 1980)

Hormone	Plant	Effect of applied substance	Reference
Cytokinins (Kinetin)	Pea roots	Promoted amino acids incorporation into proteins of NaCl-stressed tissue	Kahane and Poljakoff-Mayber (1968)
	Wheat	Promoted seed germination and seedling survival in a saline soil	Ramesh and Kumar (1975)
	Tobacco	Prevented salt induced leaf necrosis	Benzioni *et al.,* (1974)
	Barley	Increased RNase and decreased chlorophyll of salinized plants	Arad and Richmond (1976)
Gibberellins (GA_3)	Cotton	Stimulated growth on salinized soils	Agakishiev (1964)
	Bean	Increased growth at low salt levels only	Nieman and Bernstein (1964)
	Bean	Counteracted salt induced inhibition of growth, photosynthesis	Strack *et al.* (1975)
	Pearl millet	Counteracted NaCl effects on growth and enzyme activity	Huber *et al.* (1974)
Auxins (IAA)	Wheat seeds	Overcame Na_2SO_4 induced depression of root growth	Sarin and Rao (1961)
	Maize	Decreased sodium uptake	Darra and Saxena (1973)
Abscisic acid (ABA)	Bean	Inhibited K, Ca, Mg and P uptake in salt stressed plants	Eder and Huber (1977)
	Pearl millet	Inhibited protein synthesis and glucose dehydrogenase activity	Eder and Huber (1977)

On the other hand, application of exogenous ABA inhibited K, Ca, Mg and P uptake from nutrient solution in salt stressed bean plants. Eder and Huber (1977) concluded that since the metabolic changes after application of ABA agree with those following water and salt stress, ABA must play a part in the hardening of plants to stress. Thus, both protein synthesis and the activity of the enzyme glucose dehydrogenase are inhibited by either ABA or NaCl. Kinetin stimulates both and reverses the effects of ABA and NaCl. Kinetin also inhibits protein hydrolysis but has no effect on proline accumulation.

To sum up, in the manifestation of salinity stress, the changes in endogenous levels of hormons are usually implicated (Munns and Termant, 1986, He and Cramer, 1996). The exogenous application of plant growth regulators (PGRs) e.g. auxins, cytokinins, gibberlins, ethylene and even abscisic acid are usually implicated in the alleviation of salinity stress (Amzallag *et al.,* 1992, Dhingra and Varghese, 1997).

However, the limited data available on plant hormones in relation to salinity stress indicate an important role of plant hormones (particularly ABA and cytokinins) in regulation of water balance and other related processes. Therefore, more research efforts are warranted in this area.

REFERENCES

Abd-El-Samad, H.M. and Shaddab, M.A.K. 1997. Salt tolerance of soybean cultivars. Biol. Plant, 39(2) : 263-269.

Abdul Kadir, S.M. and Paulsen, G.M. 1992. Efect of salinity on nitrogen metabolism in wheat. J. Plant Nutr. 5 : 1141-1151.

Agarwal, S., Sairam, R.K., Shrivastava, G.C., Tyagi, A. and Meena, R.C. 2005. Role of ABA, salicyclic acid, calcium and hydrogen peroxide on antioxidatnt enzyme production in wheat seedlings. Plant Sci. 169: 559-570.

Alam, S.M. 1994. Nutrient uptake by plants under stress conditions. In Handbook of Plant and Crop Stress (Ed. Pessarakli, M.) Marcel Dekker New York, pp. 227-246.

Allakhverdiev, S.I., Sakamoto, A., Nishiyama, Y., Inaba, M. and Murata, N. 2010a. Roles of enzymatic and non-enzymaticantioxidants in plants during abiotic stress, Crit. Rev. Bitoechnol. 30: 161-175.

Alva, A.K. and Syvertsen, J.P. 1991. Irrigation water salinity effects, soil nutrient distribution, root density and leaf nutrient levels of citrus under drip fertilizations. J. Plant Nutr. 14 : 715-727.

Amirjani, M.R. 2011. Effect of salinity stress on growth, sugar content, pigments and enzyme activity of rice. Int. J. Bot. 7: 73-81.

Amthor, J.S. 1989. Respiration and Crop Productivity. Springer - Verlag, New York, Berlin, 215 p.

Apel, K. and Hirt, H. 2004. Reactive oxygen species: Metabolism, oxidative stress and signal. Ann. Rev. Plant. Biol. 55: 373-399.

Arad, S. and Richmond, A.E. 1976. Leaf cell water and enzyme activity. Plant Physiol. 57 : 656-658.

Asada, K. 1992. Ascorbate peroxidase - a hydrogen peroxide scavenging enzyme in plants. Physiol. Plant. 55: 225-241.

Aslam, M., Huffaker, R.C. and Rains, D.W. 1984. Early effects of salinity on nitrate assimilation in barley seedlings. Plant Physiol. 76 : 321-325.

Austenfeld, F.A. 1974. The effect of NaCl and other alkaline salts on the nitrate reductase activity of *Salicornia europea* L. Z. Pflanzenphysiol. 71 : 288-296.

Balasubramanian, V. and Sinha, S.K. 1976. Effects of salt stress on growth, nodulation and nitrogen fixation in cowpea and mungbean. Physiol. Plant. 36 : 197-200.

Balkhodja, R., Morales, F., Albadia, A., Gomez, Aprasi, J. and Abadia, J. 1994. Chlorophyll fluorescence as a possible tool for salinity tolerance screening in barley *(Hordeum vulgare* L.). Plant Physiol. 104 : 667-673.

Barner, S.J. and Weitzman, P.D.D. 1986. Organization of citric acid cycle enzymes into a multienzyme cluster. FEBS Letters 201: 267-270.

Belkhodja, R., Medrano, H., Abadia, J. and Mathis, P. 1995. Effects of salinity on photosynthesis parameters of barley (*Hordeum vulgare* L.) grown under a triple line source sprinkler system in the field. Photosynthesis from light to biosphere Vol. IV. IV Proc. Inter Photosynthesis Cong. France 20-25 August 1995 pp. 753-756.

Benzioni, A., Mizahi, Y. and Richmond, A.E. 1974. Effect of kinetin on plant response to salinity. New Phytol. 73 : 315-319.

Bernstein, L. and Ogata, G. 1966. Effects of salinity on nodulation, nitrogen fixation and growth of soybean and alfalfa. Agron. J. 58 : 201-203.

Bhambota, J.R. and Kanwar, J.S. 1970. Effect of different salt concentrations on sweet orange. Ind. J. Agric. Sci. 40 : 485-494.

Bhardwaj, K.K.R. 1975. Survival and symbiotic characteristic of *Rhizobium* in saline alkali soils. Plant Soil 43 : 337-385.

Bishnoi, N.R., Singh, Hanuman and Swaraj, K. 1997. Influence of sodium chloride on nitrogen fixation and enzyme associated with scvenging hydrogen peroxide in cluster bean root nodules. Ind. J. Expt. Biol. 35 (2) : 193-196.

Bordeleau, Z.M. and Prevost, D. 1994. Nodulation and nitrogen fixation in extreme environments. Plant Soil 161 : 115-125.

Botsford, J.L. and Lewis, T.A. 1990. Osmoregulation in *Rhizobium meliloti.* Production of glutamic acid in response to osmotic stress. Appl. Environ. Microbiol. 56 : 488-494.

Boyer, J.S. 1965. Effects of osmotic-water stress on metabolic rates of cotton plants with open stomata. Plant Physiol. 40 : 229-234.

Bradley, P.M. and Morris, J.T. 1991. The influence of salinity on the kinetics of NH^{+}_{4} uptake in *Spartinu alterniflora.* Oceologia 85 : 375-380.

Broz, A.K., Tovar-Mentez, A., Mooney, B.P., Johnston, M.L. and Randel, D.D. 2014. A novel regulating mechanism based upon a dynamic core structure for the mitochondrial pyruvate dehydrogenase complex. Mitochondrion 19(Part B): 144-153.

Bugos, R.C., Hieber, A.D. and Yamamoto, H.Y. 1998. Xanthophyll cycle enzymes are members of the lipocalin family, the first identified from plants. J. Biol. Chem. 273: 15321-15324.

Cannons, A.C. and Pendleton, L.C. 1994. Possible role for mRNA stability in the ammonium controlled regulation of nitrate reductase expression. Biochem. J. 297 : 561-565.

Chaillou, P.B. and Guerrier, G. 1992. Salt responses in *Lycopersicon esculentum* calli and whole plants. J. Plant Physiol. 140 : 494-501.

Chaillou, P.B., Alfocea, F.P. and Guerrier, G. 1992. Comparative effect of N sources on growth and physiological responses of soybean exposed to NaCl stress. J. Exp. Bot. 43 : 1225-1233.

Chen, S., Gollop, N. and Heuer, B. 2009. Proteomic analysis of salt-stressed tomato seedlings : effect of genotype and exogenous application of glycine betaine. J. Exp. Bot. 60: 2005-2019.

Che-Othman, M.H., Millar, A.H. and Taylor, N.L. 2017. Connecting salt stress signalling pathway with salinity induced charges in mitochondrial metabolic processes in C_3 plants. Plant Cell Environ. 40: 2875-2905.

Cho, J; Kim, C; Cho, J.W. and Kim, C.S. 1998. Effect of NaCl concentration on photosynthesis and mineral content of barley seedlings under solution culture. Korean J. Crop Sci. 43 : 152-156.

Chutipaijit, S., Cha-um, S. and Sompornpallin, K. 2011. High contents of proline and anthocyaninicrease protective response to salinity in *Oryza sativa* L. spp. *indica*. Aust. J. Crop Sci. 5: 1191-1198.

Comba, Maria, Maria, E. and Benavides, P. 1997. Relationship between nitrogen fixation and oxidation stress induction in nodules of salt treated soybean plants. Phyton (Buenos Aires), 60 (1-2) : 115-126.

Cordovilla, M.P., Ligero, F. and Lluch, C. 1996. Growth and nitrogen assimilation in nodules in response to nitrate levels in *Vicia faba* under salt stress. J. Exnt. Bot. 47 (295) : 203-210.

Cordovilla, M.P., Ocana, A., Ligero, F. and Leluch, C. 1995. Salinity effects on growth analysis and nutrient composition in four grain legumes - *Rhizobium* symbiosis. J. Plant Nutr. 18 : 377-385.

Curtis, P.S. and Lauchli, A. 1986. The role or leaf area development and photosynthetic capacity in determining growth of kenaf under moderate salt stress. Aust. J. Plant Physiol. 13 : 553-565.

Dani, V., Simon, W.J., Duranti, M. and Croy R.R.D. 2005. Changes in the tobacco leaf apoplast proteome in response to salt stress. Proteomics 5: 737-745.

Davies, W.J., Metcalfe, J., Lodge, T.A. and Costa, A.R. 1986. Plant growth substances and the regulation of growth under drought. Aust. J. Plant Physiol. 13 : 105-125.

Devi, S., Angrish, R. Datta, K.S. and Kumar, B. 2008. Antioxidant defence system in wheat seedlings under sodium chloride stress: An inductive role of hydrogen peroxide. Indian J. Plant Physiol. 13: 118-124.

Dhingra, H.R. and Sharma, P.K. 1993. Biochemical and mineral composition of young healthy and shrivelled mungbean (*Vigna radiata* L. Wilczek) seeds in response to salinity. Ind. J. Plant Physiol. 36 (2) : 115-117.

Dhingra, H.R., Varghese, T.M. and Kajal, N. 1994. Salinity induced changes in carbohydrate metabolism and inorganic constituents in the developing chickpea seeds. Ind. J. Plant Physiol. 37 (1) : 21-24.

Dias, M.A. and Costa, M.M. 1983. Effect of low salt concentrations on nitrate reductase and peroxidase of sugarbeet leaves. J. Expt. Bot. 4(142): 537-543.

Dilly, F.Le, Billard, J.B. and Boucaud, J. 1991. Polyamine levels in relation to growth and NaCl concentrations in normal and habituated sugarbeet callus cultures. Plant Cell Environ. 14 : 327-332.

Dooki, A.D., Mayer-Posner, F.J., Askari, H., Zaiee, A.A. and Salekdeh, G.H. 2006. Proteomic responses of rice young panicles to salinity. Proteomics 6: 6498-6507.

Downton, W.J.S. 1977. Photosynthesis in salt-stressed grapevines. Aust. J. Plant Physiol. 4 : 183-192.

Downton, W.J.S. and Loyeys, B.R. 1978. Compositional changes during grape berry development in relation of abscisic acid and salinity. Aust. J. Plant Physiol. 5 : 415-423.

Dubey, R.S. 1994. Protein synthesis by plants under stressful conditions. In : Handbook of Plant and Crop Stress (Ed. M. Pessarakli) pp. 277-299. Marcel Dekker, Inc. New York.

Dubey, R.S. and Pessarakli, M. 1995. Physiological mechanisms of nitrogen absorption and assimilation in plants under stressful conditions. In Handbook of Plant and Crop Physiology (Ed. M. Pessarakli) pp. 605-625. Marcel Dekker, New York.

Dubey, R.S. and Rani, M. 1989. Influence of NaCl salinity on growth and metabolic status of proteins and amino acids in rice seedlings. J. Agron. Crop Sci. 162 : 97-106.

Dubey, R.S. and Rani, M. 1990. Influence of NaCl salinity on the behaviour of protease, aminopeptidase and carboxyl-peptidase in rice seedlings in relation to salt tolerance. Aust. J. Plant Physiol. 17 : 215-224.

Dubey, R.S., Katiyar, S. and Mittal, R. 1991. Nitrogen assimilation and transamination in developing rice seedlings following NaCl salinity stress. Proc. Inter. Conf. Plant Physiol. Banaras Hindu Univ. pp. 189-194.

Durga Prasad, K.M.R., Muthukumarasamy, M. and Panneerselvam, R. 1996. Changes in protein metabolism induced by NaCl salinity in soybeam seedlings. Indian J. Plant Physiol. 1 (2) : 98-101.

Eder, A. and Huber, W. 1977. Zur wiekung von Abscission saqure and kinetin auf biochemische veranderungen in *Pennisetum typhoides* under stressin wirkungen. Z. Pflanzenphysiol 84 : 303-311.

El-Enany, A.E. 1997. Glutathione metabolism in soybean callus - cultures as affected by salinity. Biol. Plant (Prague). 39 (1) : 35-39.

El-fouly, M.M. and Jung, J. 1972. Enzyme activity in wheat seedlings grown under different sodium chloride salinity levels and their interaction with growth regulators. Biochem. Physiol. Pflanzen. 163 : 492-498.

Elsheikh, E.A. and Wood, M. 1989. Response of chickpea and soybean rhizobia to salt influence of carbon source, temperature and pH. Soil Biol. Biochem. 21 : 883-887.

Elsheikh, E.A. and Wood, M. 1995. Nodulation and N_2-fixation by soybean inoculated with salt-tolerant rhizobia or salt-sensitive Bradyrhizobia in saline soil. Soil Biol. Biochem. 27 : 657-661.

Feigin, A. 1985. Fertilization management of crops irrigated with saline water. Plant Soil 89 : 285-299.

Feigin, A. 1990. Interrelative effects of salinity and ammonium nitrate ratio on growth and chemical composition of melon plants. J. Plant Nutr. 13 : 1257-1269.

Fernandez-Pascual, M., De Lorenzo, C., De Felipe, M.R., Rajalakshmi, S., Gordon, A.J., Thomson, B.J. and Minchin, F.R. 1996. Possible reasons for relative salt stress tolerance in nodules of white Lupin Cv. Multolupa. J. Exp. Bot. 47 (304) : 1709-1716.

Flexas, T.D., Diaz-Espeio, A., Galmes, J. and Ribas-Carbo, M. 2007. Rapid variations of mesophyll conductance in response to changes in CO_2 concentration around leaves. Plant Cell Environ. 30: 1284-1298.

Flowers, T.J. 1972. Salt tolerance in *Suaeda maritima* (L.) Dum. The effect of sodium chloride on growth, respiration and soluble enzymes in comparative study with *Pisum sativum* L. J. Expt. Bot. 23: 301-321.

Flowers, T.J., Duque, E., Hajibagheri, M.A. McGonigle, T.P. and Yeo A.R. 1985. The effect of salinity on leaf ultrastructure and net photosynthesis of two varieties of rice : Further evidence for a cellular component of salt resistance New Phytol. 100 : 37-43.

Foyer, C.H. and Noctor, G. 2005. Redox homeostasis and antioxidant signaling: A metabolic interface between stress perception and physiological responses. Plant Cell. 17: 1866-1875.

Gale, J. 1975. Water balance and gas exchange in plants under saline conditions. In : Plants in Saline Environments. Ecological Studies (Eds. A. Poljakoff-Mayber and J. Gale) pp. 168-185, Springer - Verlag, New York.

Gale, J., Kohl, H.C. and Robert, M.H. 1967. Changes in the water balance and photosynthesis of onion, bean and cotton plants under saline conditions. Physiol. Plant. 20 : 408-420.

Gale, J., Nieman, R. and Poljakoff-Mayber, A. 1970. Growth of *Atriplex halimus* L. in sodium chloride salinated solutions as affected by the relative humdity of the air. Aust. J. Biol. Sci. 23 : 947-952.

Garcia-Olmedo, F., Molina, A., Segura, A. and Moreno M. 1995. The defensive role of nonspecific lipid-transfer proteins in plants. Trends Microbiol. 3: 72-74.

Garg, B.K. 1977. Effects of saline- alkaline conditions on plant growth and metabolism. Ph.D. Thesis. H.A.U., Hissar (Haryana), India.

Garg, B.K. 1998. Role of calcium in plants under salt stress. Ann. Arid Zone 37 : 107-118.

Garg, B.K. and Garg, O.P. 1980. Sodium carbonate and bicarbonate induced changes in growth, chlorophyll, nucleic acid and protein contents in leaves of *Pisum sativum*. Photosynthetica 14 : 594-598.

Garg, B.K. and Garg, O.P. 1981. Metabolism of germinating moong seeds under saline-alkaline conditions due to sodium carbonate and bicarbonate. Indian J. Plant Physiol. 24 : 171-174.

Garg, B.K. and Garg, O.P. 1982. Germination and some metabolic characteristics in green gram seedlings as affected by sodium carbonate. Biol. Plant. 24 : 28-33.

Garg, B.K. and Garg, O.P. 1982. Growth and some metabolic changes in maize leaves as affected by saline - alkaline conditions due to sodium carbonate and sodium bicarbonate. Ind. J. Plant Physiol. 25 (3) : 220-225.

Garg, B.K. and Garg, O.P. 1985. Influence of sodium carbonate and bicarbonate on photosynthesis and respiration of green gram and pea. Curr. Agric. 9 : 39-42.

Garg, B.K. and Gupta, I.C. 1997. Saline Wastelands Environment and Plant Growth. Scientific Publishers (India), Jodhpur 287 p.

Garg, B.K. and Lahiri, A.N. 1986. Problems of salt stress in arid zone crops. In : Proc. Natn. Symp. Physiol. Biochem. Genetic Aspects Crop Plants Environ. Stresses (Eds. R. Singh, I.S. Sheoran and M.R. Saharan) pp. 63-68, H.A.U. Hisar, India.

Garg, B.K., Kathju, S., Vyas, S.P. and Lahiri, A.N. 1983. Influence of salts on growth, yield and metabolism of wheat. Ann. Arid Zone 22 : 329-333.

Garg, B.K., Kathju, S., Vyas, S.P. and Lahiri, A.N. 1990. Effect of saline water irrigation on tolerant and sensitive wheat varieties under disparate soil fertility conditions. Ann. Arid Zone 29 : 179-189.

Garg, B.K., Kathju, S., Vyas, S.P. and Lahiri, A.N. 1996. Relative effects of sodium salts on growth, yield and metabolism of arid legumes. Ind. J. Plant Physiol., 23 (2) : 148-152.

Garg, B.K., Venkateswarlu, B., Vyas, S.P. and Lahiri, A.N. 1984b. Nodulation and nitrogen fixation in clusterbean and moth bean under salt stress. Indian J. Exp. Biol. 22 : 511-512.

Garg, B.K., Vyas, S.P., Kathju, S. and Lahiri, A.N. 1984a. Study of relative consequences of water and salt stresses in pearl millet. Sci. Cult. 50 : 96-97.

Garg, B.K., Vyas, S.P., Kathju, S., Lahiri, A.N., Mali, P.C. and Sharma, P.C. 1993. Salinity-fertility interaction on growth, mineral composition and nitrogen metabolism of Indian mustard. J. Plant Nutr. 16 : 1637-1650.

Garg, N., Dua, I.S., Sharma, S.K. and Garg, O.P. 1988. Nodulation and nitrogen fixing capacity in relation to variable salinity levels in *Vigna radiata* L. (mung). P.U. Res. Bull. (Sci.) 39 (III-IV) : 187-196.

Gauch, H.G. and Eaton, F.M. 1942. Effects of saline substrate on hourly levels of carbohydrates and inorganic constituents of barley plants. Plant Physiol. 17 : 422-434.

Gorham, J., Bodrewics, E., McDonnel, E. and Wyn Jones, R.G. 1986. Salt tolerance in the Triticeae, salinity induced changes in leaf solute composition of some perennial Triticeae. J. Exp. Bot. 37 : 1114-1128.

Goris, I. Ya. 1969. Respiratory metabolism of seeds germinating in salt solution. Sel. Khoz. Biologia. 4 : 246-251.

Gouja, H., Ghorbal, M.H. and Touraine, B. 1994. Effects of NaCl and forms of N and mineral ions on NO_3- reduction rate within whole plants of salt sensitive bean and salt resistant cotton. Plant Physiol. 105 : 1409-1418.

Grattan, S.R. and Grieve, C.M. 1994. Mineral nutrient acquisition and response by plants grown in saline environments. In : Handbook of Plant and Crops Stress (Ed. M. Pessarakli) pp. 203-226. Marcel Dekker, Inc. New York.

Greenway, H. and Munns, R. 1980. Mechanisms of salt tolerance in nonhalophytes. Ann. Rev. Plant Physiol. 31 : 149-190.

Greenway, H. and Osmond, C.B. 1972. Salt responses of enzymes from species differing in salt tolerance. Plant Physiol. 49 : 256-259.

Guerrier, G. 1987. Effects of sodium chloride on oxidase activities during germination of salt sensitive and salt tolerant species. *Biol. Plant.* 29 (4) : 299-306.

Gulati, A. and Jaiwal, P.K. 1996. Effect of NaCl on nitrate reductase, glutamate dehydrogenase and glutamate synthase in *Vigna radiata* Calli. Biol. Plant. 38 : 177-183.

Gupta, A. and Singhal, G.S. 1996. Effect of sodium and calcium chlorides, ABA and Proline on callus cultures of *Arachis hypogea* L. *Biol. Plant.* (Prague), 38 (4) : 525-529.

Hafeez, F.Y., Aslam, Z. and Malik, K. 1988. Effect of salinity and inoculation on growth, nitrogen fixation and nutrient uptake of *Vigna radiata* L. Wilczek. Plant Soil 106 : 3-8.

Hassan-Porath, E. and Poljakoff-Mayber, A. 1973. The effect of salinity on glucose absorption and incorporation by pea roots. Plant Physiol. 14 : 361-368.

Heimer, Y.M. 1973. The effects of sodium chloride, potassium chloride and glycerine on the activity of nitrate reductase of a salt tolerant and two non-tolerant plants. Planta 113 : 279-281.

Helal, H.M. and Mengel, K. 1979. Nitrogen metabolism of young barley plants as affected by NaCl salinity and potassium. Plant Soil 51 : 457-462.

Hermantz, J.A., Jimenz, A., Mullineax, P. and Sevilla, F. 2000. Tolerance of pea to long term salt stress is associated with production of antioxidant defences. Plant Cell Environ. 23: 853-862.

Heuer, B. 1994. Osmoregulatory role of proline in water and salt stressed plants. In : Handbook of Plant and Crop Stress (Ed. M. Pessarakli) pp. 363-381. Marcel Dekker, Inc. New York.

Hoffman, G.J. and Phene, C.J. 1971. Effect of constant salinity levels on water use efficiency of bean and cotton. Trans. Am. Soc. Agril. Engg. 14 : 1103-1106.

Hopmans, P., Dauglas, L.A. and Chalk, P.M. 1984. Aust. J. Agric. Res. 35 : 9-15.

Hsiao, T.C. 2012. In T.C. Hsiao (Ed.), Herbaceous Crops. Rome: FAO.

Huang, Chi-Ying and Yiu-Mei Chen 1995. Role of glutathione reductase and related enzymes in salt tolerant mechanism of soybean plants grown under salt stress conditions. Taiwania 40 (1) : 21-34.

Huber, W. and Sankhla, N. 1973. Eco-physiological studies on Indian arid zone plants. II. Effects of salinity and gibberellin on the activity of the enzymes of amino-acid metabolism in leaves of *Pennisetum typhoides*. Oecologia 13 : 271-277.

Hung, S.H., Yu, C.W. and Hm, C.H. 2005. Hydrogen peroxide functions as a stress signal in plants. Bot. Bull. Acad. Sin. 46: 1-10.

Hutton, E.M. 1971. Variation in salt response between tropical pasture legumes. Sabrao Newsl. 3 : 75-81.

Itai, C. and Vaadia, Y. 1971. Cytokinin activity in water stressed shoots. Plant Physiol. 47 : 87-90.

Iyengar, E.R.R. and Reddy, M.P. 1996. Photosynthesis in highly salt-tolerant plants. In: Pessaraki, M. (ed.) Handbook of Photosynthesis. Marcel Dekker, New York, pp. 897-909.

Jacoby, R.P., Millar, A.H. and Taylor, N.L. 2010. Wheat mitochondrial proteomes provide new links between antioxidant defense and plant salinity balance. J. Proteome Res. 9: 6595-6604.

Jacoby, R.P., Taylor, N.L. and Miller, A.H. 2011. The role of mitochondrial respiration in salinity tolerance. Trends in Plant Science. 16: 614-623.

Jain, R.K., Dhawan, R.S., Sharma, D.R. and Chowdhury, J.B. 1987. Salt tolerance and proline accumulation : a comparative study in salt tolerant and wild type cultured cells of eggplant. Plant Cell Reports 6 : 382-384.

Joshi, G.V. and Naik, G.R. 1977. Salinity effect on ion growth and photosynthetic productivity in sugarcane var. CO. 740. Indian Sugar 27 : 329-332.

Kahane, I. and Poljakoff-Mayber, A. 1968. Effect of substrate salinity on the ability for protein synthesis in pea roots. Plant Physiol. 43 : 1115-1119.

Kalaji, H. and Nalforcyzk, E. 1991. Gas exchange of barley seedlings growing under salinity stress. Photosynthetica 25 : 197-202.

Kalaji, H.M., Govindjee, B.K., Koscielniakd, J. and Zuk-Golaszewka, K. 2011. Effects of salt stress on photosystem II efficiency and CO_2 assimiliation of two Syrian barley landraces. Environ. Exp. Bot. 73: 64-72.

Karmakar, S.M. and Joshi, G.V. 1969. Effect of sand culture and sodium chloride on growth, physical structure and organic acid metabolism in *Bryophyllum pinnatum*. Plant Soil 30 : 41-48.

Katiyar, S. 1990. Studies on polyamines, nitrogen assimination and certain dehydrogenases in ration to salt tolerance in rice. Ph.D. Thesis, Banaras Hindu University, Varanasi, India.

Katiyar, S. and Dubey, R.S. 1990. Salinity induced accumulation of polyamines in germinating rice seeds differing in salt tolerance. Trop. Sci. 30 : 229-240.

Katiyar, S. and Dubey, R.S. 1992. Influence of NaCl salinity on behaviour of nitrate reductase and nitrite reductase in rice seedlings differing in salt tolerance. *J. Agron. Crop. Sci.* 169 : 289-297.

Kennedy, R.A. 1977. The effect of NaCl, PEG and naturally induced water stress on photosynthetic products, photosynthetic rates and carbon dioxide compensation points in C-4 plants. Z. Pflanzenphysiol. 83 : 11-24.

Khan, A.H., Ashraf, M.Y. and Azmi, A.R. 1992. Osmotic adjustment in sorghum under sodium chloride stress. Acta Physiol. Plant. *14* : 159-164.

Khan, A.H., Ashraf, M.Y., Naqvi, S.S.M., Khanzada, B. and Ali, M. 1995. Growth, ion and solute contents of sorghum grown under NaCl and Na_2SO_4 salinity stress. Acta Physiol. Plant 17 : 261-268.

Khan, A.H., Asraf, M.Y. and Azmi, A.R. 1990. Effect of sodium chloride on growth and nitrogen metabolism of sorghum. Acta Physiol. Plant. 12 : 233-238.

Khan, M.G. 1996. Nitrate and nitriate reductase activities in soybean plants raised with saline water. Ind. J. Plant Physiol., 1 (2) : 128-129.

Khan, M.M., Al Masoudi, R.S.M., Al-Said, F. and Khan, I. 2013. Salinity effects on growth electrolyte leakage, chlorophyll content and lipid peroxidation in cucumber (*Cucumis sativus* L.) 2013 International Conference on Food and Agricultural Sciences IPCBEE vol. 55, IACSIT Press, Singapore.

Kishor, K.P.B. 1988. Effect of salt stress on callus cultures of *Oryza sativa* L. J. Exp. Bot. 39 : 235-240.

Koyro, H.U. 1997. Ultrastructural and physiological changes in root cells of sorghum plants (*Sorghum bicolor* x *S. sulanensis* cv. *Sweet Sioux*) induced by NaCl. J. Exp. Bot. 48 : 693-706.

Krishnamurthy, R., Anbazhgan, M. and Bhagwat, K.A. 1987. Salt response of enzymes from rice cultivars differing in salt tolerance. Curr. Sci. 56 : 489-490.

Lahiri, A.N., Garg, B.K., Kathju, S., Vyas, S.P. and Mali, P.C. 1987. Responses of clusterbean to soil salinity. Ann. Arid Zone 26 : 33-42.

Lahiri, A.N., Garg, B.K., Kathju, S., Vyas, S.P. and Mali, P.C. 1996. Genotypic differences to soil salinity in clusterbean. Arid Soil Res. and Rehabilit. 10 : 333-345.

Lakshmi Kumari M., Singh, C.S. and Subba Rao, N.S. 1974. Root hair infection and nodulation in lucerne (*Medicago sativa* L) as influenced by salinity and alkalinity. Plant Soil 40 : 261-268.

Lapina, L.P. and Popov, B.A. 1970. The effect of sodium chloride on photosynthetic apparatus of tomatoes. Sov. Plant Physiol. 17 : 477-481.

Lapina, L.P. and Popov, B.A. 1971. After effect of sodium sulphate and sodium chloride on functional activity and structure of maize chloroplast. Sov. Plant Physiol. 18 : 341-347.

Lapina, L.P., Bikmukhametova, S.A. and Murashov, I.N. 1976. Effect of sodium sulphate and NaCl on activity of photosynthetic phosphorylation in plants of different salt tolerance. Sov. Plant Physiol. 23 : 236-241.

Lauter, D.J., Munns, D.N. and Clarkin, K.L. 1981. Salt response of chickpea as influence by N supply. Agron. J. 73 : 961-966.

Lawlar, D.W. and Cornic, G. 2002. Photosynthetic carbon assimilation and associated metabolism in relation to water deficits in higher plants. Plant Cell Environ. 25: 275-294.

Leidi E.O., Silberbush, M. and Lips, S.H. 1991. Wheat growth as affected by nitrogen type, pH and salinity. I. Biomass production and mineral composition. J. Plant Nutr. 14 : 235-246.

Lessani, M.H. and Andreopoulos Renaud, U. 1969. C.R. Acad. Sci. Paris 269 : 951-953.

Levitt, J. 1980. Responses of Plants to Environmental Stresses. Vol. II Acadmic Press. New York, 607 p.

Lewis, O.A.M., Leidi, E.O. and Lips, S.H. 1989. Effect of nitrogen source on growth and response to salinity stress in maize and wheat. New Phytol. 111 : 155-160.

Livne, A. and Levin, N. 1967. Tissue respiration and mitochondrial oxidative phosphorylation of NaCl-treated pea seedlings. Plant Physiol. 42 : 407-414.

Livne, A. and Vaadia, Y. 1972. Water deficits and hormone relations. In : Water Deficits and Plant Growth. Vol. III (Ed. T.T. Kozlowski) pp. 255-275, Academic Press, New York.

Lopez-Climent, M.F., Arbona, V., Perez-Clemente, R.M. and Gomez-Cadenas, A. 2008. Relationship between salt tolerance and photosynthetic machinery performance in citrus. Environ. Exp. Bot. 62: 176-184.

Maas, E.V. and Nieman, R.H. 1978. Physiology of plant response to salinity. In : Crop Tolerance to Suboptimal Land Conditions (Ed. G.A. Jung) Chap 13. Am. Soc. Agron. Spec. Publ. 32 : 277-299.

Maden Pal, Singh, D.K., Rao, L.S. and Singh, K.P. 2004. Photosyntehtic characteristic and activity of antioxidant enzymes in salinity tolerant and sensitive rice cultivars. Indian J. Plant Physiol. 9: 427-412.

Martinez, V. and Cerda, A. 1989. Nitrate reductase activity in tomato and cucumber leaves as influenced by NaCl and N source. J. Plant Nutr. 12 : 1335-1350.

McCree, K.J. 1986. Whole plant carbon balance during osmotic adjustment to drought and salinity stress. Aust. J. Plant Physiol. 13 : 33-45.

Miranda-Ham, M. and Loyola-Vargas, V.M. 1988. Ammonia assimilation in *Canavalia ensiformis* plants under water and salt stress. Plant Cell Physiol. 29 : 747-753.

Misra, A.N., Murmu, B., Singh, P. and Misra, M. 1996. Growth and accumulation of proline in mungbean seedlings as affected by NaCl. *Biol.* Plant. (Prague). 38 (4) : 531-536.

Misra, N. and Dwivedi, U.N. 1990. Nitrogen assimilation in germinating *Phaseolus aureus* seeds under saline stress. J. Plant Physiol. 135: 719-724.

Mittal, S., Kumari, N. and Sharma, V. 2012. Differential response of salt stress on *Brassica juncea:* Photosynthetic performance, pigment, proline, D1 and antioxidant enzymes. Plant Physiol. Biochem. 54: 17-26.

Mittler, R. 2002. Oxidative stress antioxidants and stress tolerance. Trends Plant Sci. 7: 405-410.

Mizrahi, Y., Blumenfeed, A. and Richmond, A.E. 1972. The role of abscisic acid and salinisation in the adaptive response of plant to reduced root aeration. Plant Cell Physiol. 13 : 15-21.

Moud, A.M. and Maghsoudi, K. 2008. Salt stress effect on respiration and growth of germinated seeds of different wheat cultivars. World J. Agric. Sci. 4: 351-358.

Munns, R. and Termaat, A. 1986. Whole plant responses to salinity. Aust. J. Plant Physiol. 13 : 143-160.

Muthukumarasamy, M. and Panneerselvam, R. 1995. Effect of NaCl salinity on carbohydrate metabolism in germinating soybean seeds. J. Physiol. Res., 8 (2) : 195-199.

Muthukumarasamy, M. and Panneerselvam, R. 1997. Growth and biochemical changes of green gram under NaCl stress. Geobios, 24 (2-3): 107-111.

Nieman, R.H. 1962. Some effects of sodium chloride on growth, photosynthesis and respiration of twelve crop plants. Bot. Gaz. 123 : 279-285.

Nieman, R.H. 1965. Expansion of bean leaves and its suppression by salinity. Plant Physiol. 40 : 151-161.

Nieman, R.H. and Poulsen, L.L. 1964. Separation of salt and water stress effects on plant growth. Plant Physiol. 39 (Suppl.) : 57.

Nir, I., Poljakoff-Mayber, A. and Klein, B. 1970. The effect of water stress on the polysome population and the ability to incorporate amino acid in maize root tips. Israel J. Bot., 19 : 451-462.

Nunes-Nesi, A., Araujo, W.L., Obata, T. and Fernie, A.R. 2013. Regulation of the mitochondrial bicarboxylic acid cycle. Current Opinion Plant Biol. 16: 335-347.

Nyan, V.K. and Shyon, G.K. 1984. Effect of toxic ions on the content of proline in rice seedlings grown under conditions of low pH and high salinity. Soviet Plant Physiol. 31 : 392.

Olmos, Enrique and Eladio Hellin. 1996. Mechanisms of salt tolerance in a cell line of *Pisum sativum* : Biochemical and physiological aspects. Plant Sci., 120 (1) : 37-45.

Osmond, C.B. and Greenway, H. 1972. Salt responses of carboxylation enzymes from species differing in salt tolerance. Plant Physiol. 49 : 260-263.

Pandey, R. and Ganapathy, P.S. 1985. The proline enigma : NaCl tolerant and NaCl-sensitive callus lines of *Cicer arietinum*. Plant Sci. 40 : 13-17.

Pandey, U.K. and Srivastava, R.D.L. 1989. Salinity index in relation to nitrate reductase activity and proline accumulation in paddy genotypes. Indian J. Plant Physiol. 32 : 175-177.

Pang, Q., Chen, S., Dai, S., Wang, Y., Chen, Y. and Yan, X. 2010. Comparative proteomics of salt tolerance in *Arabidopsis thaliana* and *Thellungiella halophilla*. J. Proteome Res. 9: 2584-2599.

Papp, J.C., Ball, M.C. and Terry, N. 1983. A comparative study on the effects of NaCl salinity on respiration, photosynthesis and leaf extension growth in *Beta vulgaris* L. (sugarbeet). Plant Cell Environ. 6 : 675-677.

Passera, C. and Albuzio, A. 1978. Effect of salinity on photosynthesis and photorespiration of two wheat species *(Triticum durum* Cv. PEPE 2122 and *Triticum aestivum* Cv. Marzotto). Can. J. Bot. 56 : 121-126.

Pathak, G.L., Singh, U.V. and Simlot, M.M. 1974. Effect of salinity on RNA and protein biosynthesis in wheat seedlings. Curr. Sci. 43 : 782-783.

Patil, T.M., Mirajkar, P.B., Hedge, B.A. and Joshi, G.V. 1983. Influence of soil salinity on morphology, rate of carbon assimilation, photosynthetic products and enzyme activities in sorghum hybrid CSH-5. Indian J. Plant Physiol. 26 : 155-162.

Pessarakli, M. and Huber, J. 1991. Biomass production and protein synthesis by alfalfa under salt stress. J. Plant Nutr. 14 : 283-293.

Pessarakli, M. and Tucker, T.C. 1985. Uptake of Nitrogen - 15 by cotton under salt stress. Soil Sci. Soc. Am. J. 49 : 149-152.

Pessarakli, M. and Zhou, M. 1990. Effect of salt stress on nitrogen fixation by different cultivars of green beans. J. Plant Nutr. 13 : 611-629.

Pessarakli, M., Huber, J.T. and Tucker, T.C. 1989. Protein synthesis in green beans under salt stress. J. Plant Nutr. 12 : 1105-1121.

Piotr, S. and Grzyna, K. 2005. Antioxidant defense in the leaves of C3 and C4 plants under salinity stress. *Physiol. Plant* 125: 31-40.

Plaut, Z. 1974. Nitrate reductase activity of wheat seedlings during exposure to and recovery from water stress and salinity. Plant Physiol. 30 : 212-217.

Ponomarova, S.A., Procenko, D.F. and Sivcon, M.V. 1971. Effect of sodium chloride on plastid pigments in tomato leaves. Sov. Plant Physiol. 18 : 337-340.

Porath, E. and Poljakoff-Mayber, A. 1963. The effect of salinity in the growth medium on carbohydrate metabolism in pea root tips. Plant Cell Physiol. 9 : 195-203.

Porath, E. and Poljakoff-Mayber, A. 1964. Effect of salinity on metabolic pathways in pea root tips. Israel J. Bot. 13: 115-121.

Ramanjulu, S., Veeranjneyulu, K. and Sudhakar, C. 1994. Short term shifts in nitroten metabolism in mulberry *Morus alba* under salt shock. Phytochemistry, 37 : 991-995.

Rani, C., Sharma, P.K., Kumar, B., Angrish, R. and Datta, K.S. 2008. Alleviation of Boron-salt toxicity by calcium in wheat through associated changes in antioxidant defence system. Indian J. Plant Physiol. 13: 21-28.

Rao, R.K. and Gnanam, A. 1990. Inhibition of nitrate and nitrite reductase activities by salinity stress in *Sorghum vulgare*. Phytochemistry 29 : 1047-1049.

Rathert, G. 1984. Sucrose and starch contents of plant parts as a posible indicator for salt tolerance of crops. Aust. J. Plant Physiol. 11 : 491-495.

Rauser, W.E. and Hanson, J.B. 1966. The metabolic status of ribonucleic acid in soybean roots exposed to saline media. J. Exp. Bot. 27 : 843-852.

Razmaev, I.I. and Azimov, R.A. 1969. Respiration of germination cotton seeds treated with NaCl solutions. Dokl. Akad. Nauk. Uzbek. SSR. 3 : 34.

Reddy, P.S., Sudhakar, C. and Veeranjaneyulu, K. 1993. Ammonia accumulation under salinity stress as a consequence of higher photorespiration. Indian J. Expt. Biol. 31: 485-486.

Robinson, S.P., Downton, W.J.S. and Millhouse, J.A. 1983. Photosynthesis and ion content of leaves and isolated chlroplasts of salt stressed spinach. Plant Physiol. 73 : 238-242.

Rus-Alvarez, A. and Guerrier G. (1994). Proline metabolic pathways in calli from *Lycopersicon esculentum* and *L. pennelli* under salt stress. Biol. Plant. 36 : 277-284.

Saha, P., Chatterjee, P. and Biswas, A.K. 2010. NaCl pretreatment alleviates salt stress by enhancement of antioxidant defense system and osmolyte accumulation in mungbean (*Vigna radiata* L. Wilczek). Indian J. Exp. Biol. 48: 593-600.

Sairam, R.F. and Shrivastava, G.C. 2002. Changes in antioxidant activity in sub-cellular fractions of tolerant and susceptible wheat genotypes in response long term salt stress. Plant Sci. 162: 897-904.

Sangeeta, Seth, P., Sharma, V. and Khandelwal, S.K. 2007. Effect of salinity on antioxidant enzymes in wheat. Indian J. Plant Physiol. 12: 186-188.

Saradhi, P., Pradha, Sandeep, Arara, B. Vani and Jos T. Puthur, 1996. Alteration in NAD^+/NADH ratio regulates salt stress induced proline accumulation in *Vigna radiata* - Proceedings of the National Academy of Sciences India Section B (Biological Sciences) 66 : 89-100.

Sarin, M.N. and Narayanan, A. 1968. Effect of soil salinity and growth regulators on germination and seedling metabolism of wheat. Physiol. Plant. 21 : 1201-1209.

Seeman, J.R. and Critchley, C. 1985. Effects of salt stress on the growth, ion content, stomotal behaviour and photosynthetic capacity of a salt sensitive species, *P. vulgaris* L. Planta 164 : 151-162.

Shah, S.H., Wain, Wright, S.J. and Merrett, M.J. 1990. The interaction of sodium and calcium chlorides and light on growth, potassium nutrition and proline accumulation in callus cultures of Medicago sativa L. New Phytol. 116 : 37-45.

Sharma, S.K. 1996 a. Effects of salinity on uptake and distribution of Na^+, Cl^- and K^+ in two wheat cultivars. Biol. Plant 38 : 261-267.

Sharma, S.K. 1996 b. Soil salinity effects on transpiration and net photosynthetic rates, stomatal conductance and Na^+ and Cl^+ contents in durum wheat. Biol. Plant 38 : 519-523.

Sharma, S.K. 1997. Plant growth, Photosynthesis and ion uptake in chickpea as influenced by salinity Ind. J. Plant Physiol., 2 (2) : 171-173.

Sharma, S.K. and Garg, O.P. 1985. Salinity induced changes in plant growth and activities of glutamate dehydrogenase, aspartate and alanine amino transferases in wheat. Indian J. Plant Physiol. 28 : 407-412.

Sharma, S.K. and Gupta, I.C. 1986. Saline Environment and Plant Growth. Agro Botanical Publishers, Bikaner, India, 172 p.

Sheokand, Sunita, Seema, Dhandi and Kumari Swaraj 1995. Studies on nodule functioning and hydrogen peroxide scavenging enzymes under salt stress in chickpea nodules. Plant Physiol. Biochem. 33 (5) : 561-566.

Sheoran, I.S. and Garg, O.P. 1978. Effect of salinity on the activities of RNase, DNase and protease during germination and early seedling growth of mungbean. Physiol. Plant 44 : 171-174.

Sheoran, I.S. and Garg, O.P. 1979. Quantitative and qualitative changes in peroxidase during germination of mungbean under salt stress. Physiol. Plant. 46 : 147-150.

Shevyakova, N.I. and Leonova, T.G. 1975. Effect of sodium chloride and sodium sulfate on *in vitro* incorporation of carbon 14-leucine into protein. Fiziol. Rast. 22 : 300-305.

Shone, M.G.T. and Gale, J. 1983. Effects of sodium chloride stress and nitrogen source on respiration, growth and photosynthesis in lucerne (*Medicato sativa* L.) J. Exp. Bot. 34 : 1117-1125.

Siddiqui, S. and Krishnamurthy, H.N. 1995. Effect of B9 on the carbohydrate metabolism of cowpea (*Vigna unguiculata* L.) seedlings under saline conditions. Ind. J. Plant Physiol. 38 (2) : 164-167.

Siddiqui, S., Kumar, S. and Sharma, H.R. 1985. Studies on the effects of salinization on nodulation and nitrogen fixation in pea (*Pisum sativum*). Indian J. Plant Physiol. 28 : 369-375.

Silberbush, M. and Lips S.H. 1991a. Potassium, nitrogen, ammonium/ nitrate ratio and sodium chloride effects on wheat growth. I. Shoot and root growth and mineral composition. J. Plant Nutr. 14 : 751-764.

Silberbush, M. and Lips, S.H. 1991b. Potassium, nitrogen, ammonium/nitrate ratio and sodium chloride effects on wheat growth. II. Tillering and grain yield. J. Plant Nutr. 14 : 765-773.

Singh, A.K. and Dubey, R.S. 1994. Salinity induced stimulation of glutamine synthetase activity in rice seedlings differing in salt tolerance. Plant Physiol. Biochem. 21 : 84-90.

Singh, Maharaj and Singh, Sudhanshu, 1995. Sodicity induced changes in saccharides, free proline and protein content of different pea genotypes. Ind. J. Plant Physiol. 38 (2) : pp. 109-113.

Singleton, P.W. and Bohlool, B.B. 1984. Effect of salinity on nodule formation by soybean. Plant Physiol 74 : 72-76.

Sprent, J.I. 1984. Effects of drought and salinity on heterotrophic nitrogen fixing bacteria and on infection of legumes of rhizobia. In : Advances in Nitrogen Fixation Research (Ed. C. Veejer and W.E. Newton) pp. 295-302, Martinus NIghott, The Hague.

Srivastava, H.S. and Singh, R.P. 1987. Role and regulation of L-glutamate dehydrogenase activity in higher plants. Phytochemistry. 26 : 567-610.

Stewart, G.R. and Larher, F. 1980. Accumulation of amino acids and related compounds in relation to environmental stress. In: The Biochemistry of Plants (Ed. Mifin, B.J.). Academic Press, New York, pp. 609-635.

Strogonov, B.P. 1974. Structure and Function of Plant Cells in Saline Habitats. Wiley, New York, IPST, 284 p.

Strogonov, B.P., Kabanov, V.V. and Rakova, N.M. 1970. Feature of protein and nucleic acid metabolism during formative changes of plants under conditions of salinisation. Soviet Plant Physiol. 17 : 324-330.

Subbarao, G.V., Johansen, C., Kumar Rao, J.V.D.K. and Jana, M.K. 1990. Response of the pigeonpea - *Rhizobium* symbiosis to salinity stress : variation among *Rhizobium* strains in symbiotic ability. Biol. Fertil. Soils 9 : 49-53.

Tamman, A.A.E., Ahmad, M.F.A. and Hemeda, M.M. 2008. Study of salt tolerance in wheat (*Triticum aestivum* L.) cultivar Banysolf. Aust. J. Crop Sci. 1: 115-125.

Tramontano, William, A. and Diana Jouve, 1997. Trigonelline accumulation in salt stressed legumes and the role of other osmoregulators as cell cycle control agents. Phytochemistry. 44 (6) : 1037-1040.

Tu, J.C. 1981. Effect of salinity on *Rhizobium* root hair interaction, nodulation and growth of soybean. Can. J. Plant Sci. 61 : 231-239.

Udovenko, G.V., Khazova, G.V. and Kukyanova, N.M. 1971. Phosphate metabolism in plants under conditions of salinisation. Soviet Plant Physiol. 18 : 1003-1009.

Udovenko, G.V., Semushina, L.A., Sakov, V.S., Galkin, V.M., Koshkin, V.A. and Kunchenko, T.A. 1974. Effect of salinisation on the state and activity of photosynthesis apparatus in plants. Soviet Plant Physiol. 21 : 508-513.

Uprety, D.C. and Sarin, M.N. 1976. Physiological studies on salt tolerance in *Pisum sativum.* IV ionic composition and nitrogen metabolism. Acta Agron. Acad. Hurgar, 25 : 455-460.

Verma, C.M., Bohra, S.P. and Sankhla, N. 1973. Lettuce seed germination. Reversal of salinity induced inhibition by ethylene. Curr. Sci. 42:294-295.

Walker, M.A. and Dumbroff, E.B. 1981. Effects of salt stress on abscisic acid and cytokonin levels in tomato. Z. Pflanze. Physiol. 101 : 461-470.

Weimberg, R. 1970. Enzyme levels in pea seedlings grown in highly salinised media. Plant Physiol. 46 : 446-470.

Weimberg, R. 1975. Effect of growth in highly salinized media on the enzymes of the photosynthetic apparatus in pea seedlings. Plant Physiol. 56 : 8-12.

Wilson, J.R. 1970. Response to salinity in Glycine VI. Some effects of a range of short-term salt stresses on the growth, nodulation and nitrogen fixation of *Glycine wightii*. Aust. J. Agric. Res. 21 : 571-582.

Winter, K. and Von Willet, D.J. 1972. Z. Pflanzenphysiol. 67 : 166-170.

Yang, J., Cao, Y., Zhang, W., Sun, L. and Lu, C. 2013. Morphological, physiological and biochemical responses of biofuel plant *Euphorbia lathyris* to salt stress. Acta Agric. Scand. 63: 330-340.

Yang, Q., Chen, Z.Z., Zhoua, X.F., Yina, H.B., Lia, X., Xina, X.F., Honga, X.H., Zhu, J.K. and Gong, Z. 2009. Over-expression of SOS (salt overly sensitive) genes increases salt tolerance in transgenic *Arabidopsis.* Mol Plant 2: 22-31.

Yeo, A.R., Caporn, S.J.M. and Flowers, T.J. 1985. The effect of salinity upon photosynthesis in rice (*Oryza sativa* L.) : Gas exchange by individual leaves in relation to their salt content. J. Exp. Bot. 36 : 1240-1248.

Yousef, A.N. and Sprent, J.I. 1983. Effects of NaCl on growth, nitrogen incorporation and chemical composition of inoculated and NH_4SO_3 fertilized *Vicia faba* (L.) plants. J. Exp. Bot. 34 : 941-950.

Zahran, H.H. 1991. Conditions for successful *Rhizobium* legume symbiosis in saline environmens. Biol. Fert. Soils 12 : 73-80.

Zaidi, P.H. and Singh, B.B. 1995. Effect of growth regulators on IAA-Oxidase and peroxidase activity in soybean under salinity. Ind. J. Plant Physiol. 38 (4) : 337-339.

Zhou, M., Tucker, T.C., Pessarakli, M. and Cepeda, J.A. 1992. Nitrogen fixation by alfalfa with two substrate levels under sodium chloride stress. Soil Sci. Soc. Am. J. 56 : 1500-1504.

Zou, N., Dart, P.J. and Marcar, N.E. 1995. Interaction of salinity and *Rhizobium* strain on growth and N_2-fixation by *Acacia ampliceps*. Soil Biol. Biochem. 27 : 409-413.

■■■■

Chapter - 7

SALINITY AND PLANT MINERAL NUTRITION

INTRODUCTION

The salinity and fertility of soil are determined by the concentration of various ions in the root zone of plants. Salinity develops when the concentration of Na, Ca, Mg, Cl, HCO_3, SO_4 ions and boron in the soil solution is high enough to affect plant growth. As the salt concentration exceeds a threshold level, the growth rate and final size of plants progressively decrease, consequently, the crop yield also declines. The status of soil fertility is determined by the level of available nutrient elements (N, P, K, Ca, Mg, S) and micro nutrients. The availability to N, P and K in soil is often too low for economic yields of agricultural crops. Addition of fertilizers to maintain an adequate fertility level is therefore, a standard practice. The concentration of the other nutrient elements is rarely below the desirable level and they are applied only under specific situations.

The simultaneous presence of salts and nutrient elements in the root zone can influence ion uptake by plants and affect their chemical composition. Synergistic and antagonistic effects can increase or decrease the intensity of these processes. The cation-anion (C-A) balance may also be affected. The level of soil fertility may influence salinity tolerance through its effect on root growth and activity. Since both high salinity and low fertility are important growth limiting factors, moderation or elimination of one factor may affect the crop's response to the other.

Nutrient availability and uptake by plants grown in saline environments is related to :

(1) The activity of the nutrient in the soil solution which depends upon pH, PE, concentration and composition.

(2) The concentration and ratios of the accompanying elements that influence the uptake and transport of this nutrient.

(3) Environmental factors.

The plant growth response in relation to the concentration of an essential nutrient element in the root media under normal conditions is shown in Fig. 7.1. Plant growth is suboptimal when the concentration or activity of the essential nutrient element is less than A and optimal when the concentration is between A and B. Nutrient concentrations that exceed B may inhibit growth as a result of either a toxicity or a nutrient induced deficiency. Available information in the literature indicates that the plant may not exhibit the same response function under saline conditions. In some cases the optimal range may be widened or narrowed, or it may shift to the right or left depending upon the plant species (or cultivar), the particular nutrient, the salinity level, or the environmental conditions. Many studies have demonstrated that an optimal concentration or activity of given nutrient element in nonsaline conditions may be deficient, or in some cases excessive, under saline conditions.

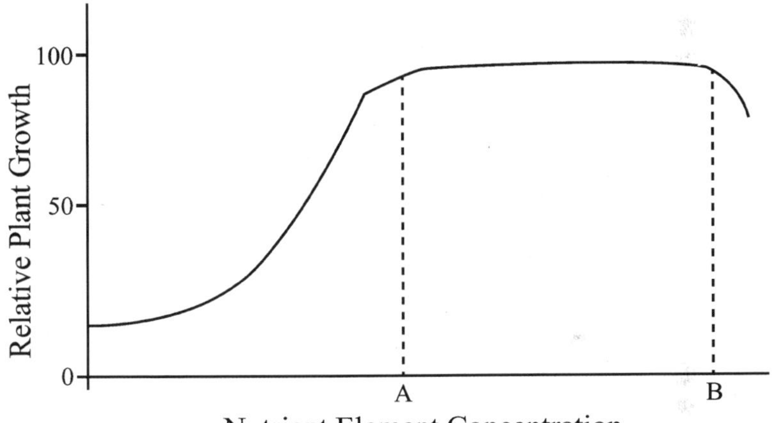

Fig. 7.1. Relative growth of plants in relation to a wide range of concentrations of an essential nutrient element

Most of glycophytes have evolved under non-saline conditions and possess mechanisms for absorbing nutrients in non-saline soils. However, under saline conditions these plants develop certain physiological disorders affecting plant growth, yield and quality.

Nutrient imbalance

Salinity induced nutritional disorders adversely affect crop performance. The relations between salinity and mineral nutrition of plants are very complex (Grattan and Grieve, 1999). These nutritional disorders may result from the effects of salinity on nutrient availability, competitive uptake, transport and distribution within the plant. It is well established that salinity reduces nutrient uptake and accumulation of nutrients into the plant (Rogers, 2003).

Soluble salts in the saline soils compete with the uptake and metabolism of essential mineral nutrients for plants. The desired ion ratios are disturbed affecting physiological processes adversely, affecting plant growth and development. Thus increased salt uptake induces specific ion toxicities like that of Na^+, Cl^-, and SO_4^{2-} while the uptake of essential nutrients like phosphorus, potassium, nitrogen and calcium is decreased. Salinity enhanced the sodium content in *Vicia faba* while Na/K ratio increased (Gadallah, 1999).

Sodium ions cause disturbances in calcium nutrition. Nutritional disorders involving other elements may be linked to the effects of salinity on the transport and metabolism of calcium. Inadequate concentrations of Ca adversely affect membrane functions and growth (Epstein, 1961; Lauchli and Epstein, 1970; Cramer *et al.,* 1988). Some species are particularly susceptible to deficiencies of calcium induced by salinity. Black heart in celery and blossom end rot in tomatoes and pepper occur when excess salt inhibits the uptake of calcium (Geraldson, 1957, Marschner, 1986). Internal browning of lettuce and artichokes buds due to salinity induced calcium deficiency has also been reported (Francois *et al.,* 1991). Similarly in Chinese cabbage an increase in salinity of irrigation water from 1.2 to 10.0 dSm^{-1} increased the percentage of tip burn from zero to 80% (Mizrahi and Pasternak, 1985).

In saline environment where sodium predominates over potassium the plant's paramount nutritional requirement is potassium (Rains and Epstein, 1967). Despite a high degree of selectivity of the potassium transport mechanisms, the salt induced potassium deficiency was reported by Ball *et al.* (1987) in *Avicenia marina* leaf tissue.

Salinity causes leaf injury (Grattan and Maas, 1984, 1988), however, the extent of injury depended upon the concentration of the phosphate, the Ca : Na ratio and the variety. The mechanisms by which moderate salinity accompanied by high phosphate concentrations elicit phosphate toxicity are yet poorly understood (Treeby and Van Steveninck, 1988). Nassery *et al.* (1979) reported calcium induced necrosis in sesame plants due to magnesium deficiency. Sodium and chloride injuries in various fruit crops, scorch symptoms when the levels of chloride in leaves exceeds 1% of dry matter and specific sodium scorch effects in a number of woody species are well documented (Bernstein, 1975). In barley supplied with low concentrations of manganese, high NaCl concentrations in the substrate depressed growth mainly by inhibited manganese uptake and induced manganese deficiency (Cramer and Nowak, 1992).

Salinity also interferes with various metabolic processes related to plant nutrition (Feigin, 1985; Garg *et al.,* 1990, 1993, Grattan and Grieve, 1994). Studies suggest that the major effect of salinity is on protein synthesis (Kahane and Poljakoff-Mayber, 1968; Pessarakli *et al.,* 1989). Helal and Mengal (1979) reported that incorporation of labelled inorganic N into protein fraction was impaired by salinity, even though its concentration in plant tissue was unaffected.

The addition of K enhanced N uptake and its incorporation into proteins indicating that the impairment of protein metabolism under salinity is mediated by disruption of K : Na balance.

SALINITY AND NUTRIENT INTERACTIONS

In salinity and mineral nutrient interaction studies at least two factors operate simultaneously to limit growth and development, (a) the presence of salinity and (b) the imbalance of a particular nutrient element. The salt tolerance of a crop varies depending upon whether salinity or nutrient is the factor more limiting to growth. Bernstein *et al.* (1974) defined three types of idealised salinity and nutrient interactions that could occur (Fig. 7.2) :

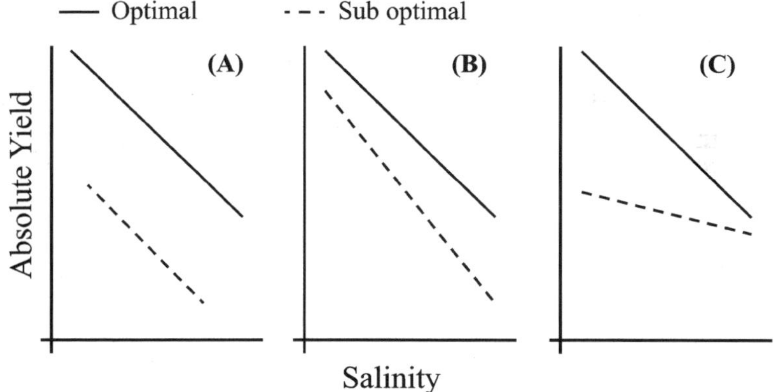

Fig. 7.2. Types of salinity fertility relationship
(A) no effect on salt tolerance; (B) increased salt tolerance and (C) decreased salt tolerance

In the first case (type A), the salinity and fertility effects at the optimum and sub-optimum fertility levels are independent, since the relative yield is the same for the two conditions. The second case (Type B) represents decreased salt tolerance at the suboptimal fertility level, while the third case (Type C) represents increased salt tolerance.

In contrast, Grattan and Grieve (1994) defined the interactions based on plant performance and optimal fertility relative to the performance at suboptimal fertility and this interpretation is shown in Fig. 7.3. Generally, plant growth is promoted more if the more limiting factor is relieved rather than the less limiting factor.

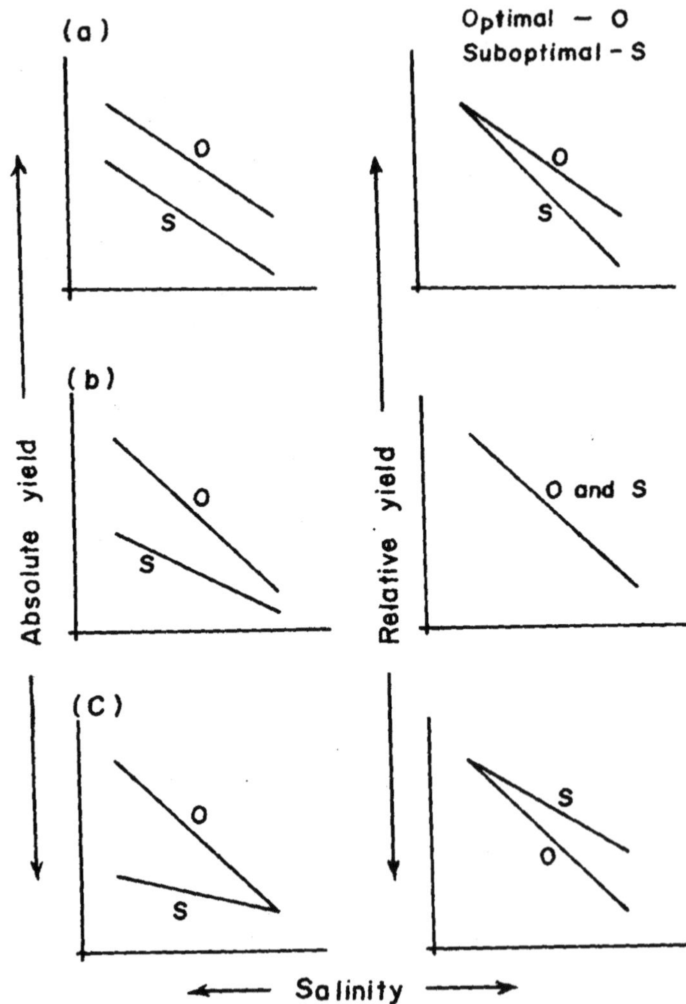

Fig. 7.3. Types of growth responses a plant can exhibit under variable salinity as the nutrient status within the substrate increases from suboptimal to optimal levels.
Salt tolerance (a) increased, (b) unaffected, (c) decreased

Bernstein *et al.* (1974) concluded that the effects of salinity and nutrient on grains and several vegetables are independent and additive when stress imposed upon them by nutrient deficiency and salinity are moderate. When either of these factors severaly limits growth, the other has little influence on yield. It has been pointed out by Grattan and Grieve (1994) that the salinity-fertility interaction described by Bernstein *et al.* (1974) is idealized and therefore may be misleading if used improperly. In many experiments, the nutrient concentration is the most limiting factor in nonsaline or low salinity conditions, yet when the indentical concentration is present in a highly saline environment, salinity is the limiting factor. This point was emphasized by Champagnol (1979) in his literature review on the relationship between salinity and phosphorus nutrition of plants. Therefore, a clear understanding of how this interaction changes from low to high salinity is absolutely essential.

In many cases, a response function similar to that illustrated in Fig. 7.4 is obtained. Under low salinity stress, nutrient deficiency limits plant growth more than salinity and a positive (+) interaction or increased salt tolerance response occurs. Under moderate salinity, nutrient deficiency and salinity stress may equally limit plant growth and no interaction (O) occurs. Under high salinity stress conditions, salinity limits growth more than nutrient deficiency. In fact, plant performance would exhibit a negative (-) interaction or a ``decreased salt tolerance" response if a nutrient element was limiting growth under the nonsaline conditions and the upper salinity treatment was lethal or severely growth limiting.

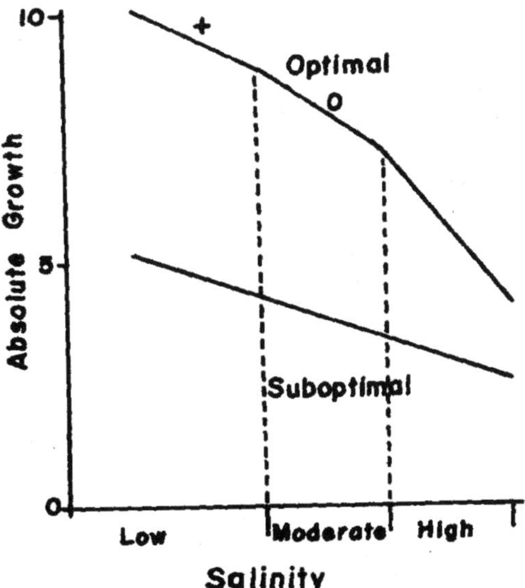

Fig. 7.4. Influence of low, moderate, and high levels of salinity at suboptimal and optimal levels of nutrient supply on plant growth, +, 0 and – signs indicate increase, no effect and decrease in plant tolerance to salinity, respectively

SALINITY AND NITROGEN

In a majority of the plant species, N uptake is generally adversely affected under salt stress conditions, whereas, in the halophytes and in many salt tolerant crop species no significant effect of NaCl on NO_3^{-2} uptake is observed (Aslam *et al.,* 1984; Pessarakli and Tucker, 1985; Alva and Syvertsen, 1991; Alam, 1994).

Salinity reduces N accumulation in plants (Pessarakli and Tucker 1985; Garg *et al.,* 1990, 1993). The possible decrease in N uptake by increasing salinity has been partly attributed to a substituation of chloride for nitrate (Papadopoulos and Rendig, 1983). Examples of this effect are found in barley (Helal *et al.,* 1975, Aslam *et al.,* 1984), cotton (Silverbush and Ben-Asher, 1987), tomato (Kafkafi *et al.,* 1982), melon (Feigin *et al.* (1987), wheat (Torres and Bingham, 1973, Garg *et al.,* 1982, 1990) and Indian mustard (Garg *et al.,* 1993). Sulphate and to a greater extent chloride diminished the rate of nitrate absorption, the degree of inhibition being 80% at 0.2 M NaCl (Aslam *et al.,* 1984). The inhibition of NO_3 absorption by salt was apparent within one minute of imposing the stress. Gorham *et al.* (1986) observed that despite drastic reductions of leaf NO_3 concentrations in response to salinity, other nitrogen containing fractions either increased or only marginally reduced. They suggested that although N-deficiency may occur in NaCl treated plants, there is no evidence to support that this effect is growth limiting.

Many salinity and N interaction studies have been undertaken in soils deficient in N. Therefore, additions of N significantly improved growth and yield of number of crops, such as barley, bean, carrots, cowpea, tomato, corn, clover, beans, millet, spinach, bermuda grass, rice and wheat when the degree of salinity was not severe (Bernstein *et al.,* 1974; Feigin, 1985; Grattan and Grieve, 1994). In several studies, improvement in soil N or soil fertility (N and P_2O_5) increased the relative salt tolerance of crops (Ravikovitch and Yoles, 1971; Selassie and Wagnet, 1981; Garg *et al.,* 1982, 1990, 1993). Increasing salinity of irrigation water progressively decreased grain yield of wheat and Indian mustard but improved soil fertility significantly enhanced the yields as well as imparted higher salt tolerance to fertilized plants (Table 7.1). In another study (Garg *et al.* 1990) improved soil fertility (N_{80} and P_{40}) enhanced the grain yield and DMP of both salt sensitive (HD - 4502) and tolerant (Kharchia-65) wheat varieties and increased their relative salt tolerance (Table 7.2). Improved fertility has been found to be associated with an increase in the concentration and uptake of NPK, particularly of N, decreased tissue concentrations of sodium and chloride and a decrease in the Na: K ratio (Fig. 7.5 and Table 7.3).

Table 7.1. Effect of salt stress on grain yield (g per plant) of wheat and Indian mustard grown under low (LF) and improved (IF) soil fertility conditions (Garg *et al.*, 1982; 1993)

Salinity (mM)	Wheat			Mustard		
	LF	IF	Mean	LF	IF	Mean
0	0.95	2.87	1.91	2.13	2.47	2.30
50	0.80	2.35	1.58	1.81	2.23	2.02
100	0.57	1.92	1.25	1.30	1.94	1.63
150	0.51	1.21	0.86	0.81	1.09	0.95
Mean	0.71	2.09	–	1.51	1.94	–
LSD (0.05)	Salinity	Fertility	SxF	Salinity	Fertility	SxF
	0.26	0.18	0.37	0.24	0.17	0.34

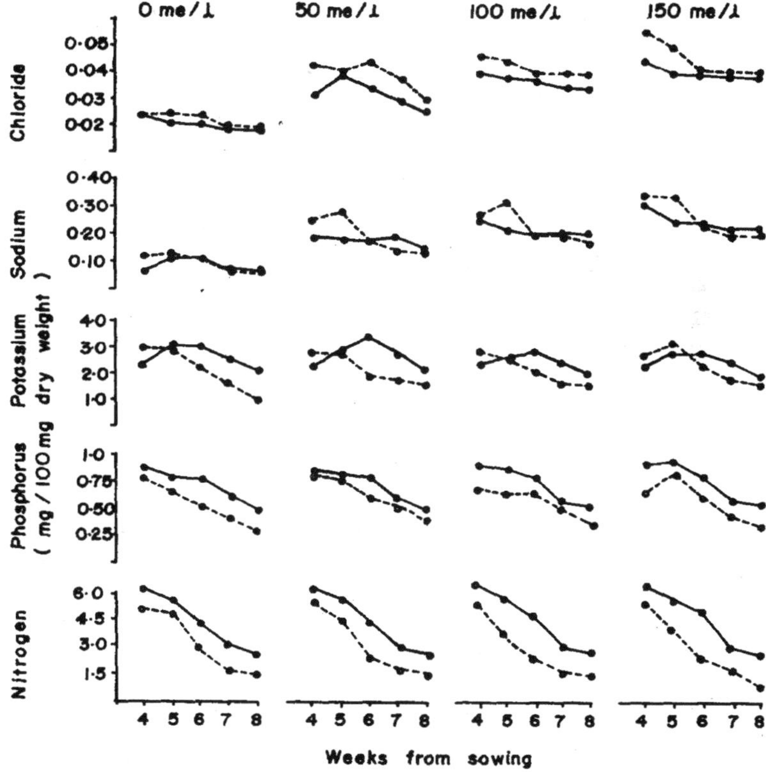

Fig. 7.5 Effects of high (• – •) and low (• – – •) soil fertility on the concentrations of nitrogen, phosphorus, potassium, sodium and chloride in the shoot tissue of wheat grown under different levels of salinity.

The grain yield of barley also improved by 64.6 and 51.0 per cent due to application of 80 kg and 40 kg N ha^{-1} respectively on a moderately saline soil of ECe 7.8 dSm^{-1} (Singh *et al.*, 1969). Mustard, linseed and safflower also responded positively to N application under saline conditions (Panwar and Bhardwaj, 1975). Garg *et al.* (1990) observed that response to improved soil fertility was similar in tolerant as well as sensitive wheat genotypes grown under salt stress. Paliwal (1972) based on a large number of greenhouse and field experiments on the use of fertilizer N in soils irrigated with saline waters concluded that application of normal dose of N and P were beneficial to wheat, barley, maize and peal millet. However, the beneficial effects of fertilizers decreased with increasing salinity and sodicity and the response was better in salt-tolerant than salt-sensitive crops.

There are, however, several studies which report the absence of salinity-N interactions (Bernstein *et al.*, 1974; Feigin, 1985) or a decreased salt tolerance in response to N supply (Khalil *et al.*, 1967; Peters, 1983).

Table 7.2. Influence of soil fertility on the above ground dry matter production and grain yield of wheat varieties grown under normal and saline water irrigation (Garg *et al.*, 1990).

Variety	Salinity (dSm^{-1})	Dry weight shoot (g $plant^{-1}$)			Grain yield (g $plant^{-1}$)		
		HF	LF	Mean	HF	LF	Mean
KH-65	0.2	14.48	7.28	10.88	6.67	3.53	5.10
	10	10.98	4.03	7.49	4.92	1.60	3.26
HD-2009	0.2	12.18	6.39	9.26	5.16	2.71	3.93
	10	8.47	2.08	5.28	3.62	0.95	2.28
HD-4502	0.2	12.97	5.44	9.20	5.44	1.63	3.53
	10	8.11	2.03	5.07	2.87	0.56	1.71
Mean	–	11.18	4.54	–	4.86	1.80	–

LSD	0.05	0.01	0.05	0.01
Variety (V)	0.76	108	0.45	0.64
Salinity (S)	1.55	2.13	0.75	1.04
Fertility (F)	0.52	0.70	0.33	0.45
F x S	0.73	0.99	0.47	0.64
V x S	NS	NS	NS	NS
F x V	NS	NS	NS	NS
F x V x S	NS	NS	NS	NS

Table 7.3. Influence of soil fertility on the sodium potassium ratio (Na : K) in the shoot tissue of wheat varieties grown under normal and saline water irrigation

Variety	Salinity (dSm^{-1})	Na : K ratio at indicated DAS					
		35		55		75	
		HF	LF	HF	LF	HF	LF
KH-65	0.2	0.08	0.08	0.07	0.10	0.09	0.10
	10	0.83	0.91	0.49	1.17	0.48	1.17
HD-2009	0.2	0.10	0.14	0.10	0.09	0.15	0.10
	10	0.45	0.59	0.50	0.93	0.69	0.87
HD-4502	0.2	0.33	0.47	0.38	0.46	0.41	0.49
	10	1.59	2.31	1.24	2.16	2.25	2.50

Table 7.4. Influence of soil fertility on the concentrations of sodium and chloride in the shoot tissue (55 DAS) of wheat varieties grown under normal and saline water irrigation

Variety	Salinity (dSm^{-1})	Sodium (%)			Chloride (%)		
		HF	LF	Mean	HF	LF	Mean
KH-65	0.2	0.20	0.23	0.21	0.31	0.33	0.32
	10	1.00	1.45	1.22	1.25	1.74	1.50
HD-2009	0.2	0.28	0.21	0.25	0.37	0.29	0.33
	10	1.05	1.00	1.02	1.08	0.98	1.03
HD-4502	0.2	0.93	0.93	0.93	1.02	1.00	1.01
	10	2.05	2.50	2.27	2.05	2.55	2.30
Mean		0.92	1.05	–	1.01	1.15	–

LSD	0.05	0.01	0.05	0.01
Variety (V)	0.13	1.21	0.16	0.27
Salinity (S)	0.10	0.16	0.09	0.14
Fertility (F)	ns	ns	0.13	ns
F x S, F x V, V x S	ns	ns	ns	ns
F x V x S	ns	ns	ns	ns

A substantial number of laboratory and greenhouse studies have shown that salinity reduces N accumulation in plants (Feigin, 1985; Pessarakli and Tucker, 1988; Garg *et al.*, 1990, 1993). This is due to the fact that an increase in chloride uptake and accumulation is mostly accompanied by a decrease in shoot nitrate concentration (Torres and Bingham, 1973; Silverbush and Ben-Asher, 1987).

Gorham *et al.,* 1986) observed that despite drastic reductions in leaf NO_3 concentrations in response to salinity, other nitrogen containing fractions either increased or only marginally reduced. These results argue against N deficiency per se as a mechanism of salt injury.

It is noteworthy that reported data indicate that increased NO_3 in the substrates decreased Cl uptake and accumulation (Garg *et al.,* 1982, 1990, Feigin *et al.,* 1987). The decrease in Cl concentration has been found to be associated with decreased Na^+ concentrations also (Table 7.4). The salinity induced decrease in NO_3 concentration is generally accompanied by a decrease in nitrate reductase (NR) actvity which is reponsible for NO_3 reduction. N supply to salt stressed wheat plants enhanced NR activity (Garg *et al.,* 1990) (Fig. 7.6). Similar results have been obtained with Indian mustard (Sharma *et al.,* 1989, Garg, *et al.,* 1993) where improvement in soil fertility significantly enhanced the activity of NR as well as those of ammonia assimilating enzymes such as GS, GOGAT and GDH (Fig. 7.7).

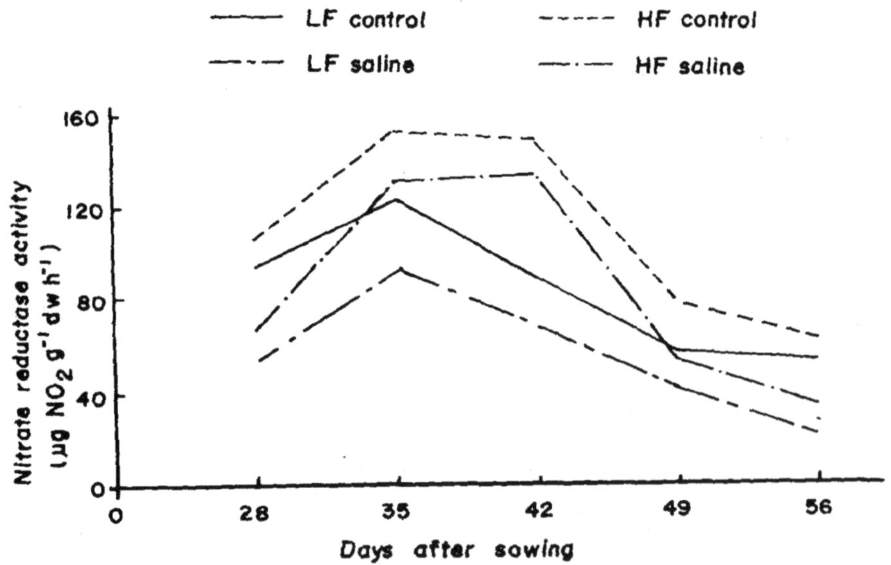

Fig. 7.6. Influence of low (LF) and High (HF) soil fertility on leaf nitrate reductase activity of wheat grown under saline water irrigation (150 mM)

A number of recent studies have indicated positive salinity fertility interactions in isabgol (Burman *et al.,* 2002), pearlmillet (Garg *et al.,* 2001), sesame (Garg *et al.,* 2005) and Indian mustard (Garg *et al.,* 2006). In all these crops, nitrogen fertilized plants displayed higher photosynthesis and more efficient N metabolism under all levels of salinity at different growth stages. This was associated with significant improvement in leaf area development, plant growth and grain yield under salt stress. It was found that seed yield of N fertilized plants even under highest salinity level was comparable with that of non-saline crops plants grown without N application (Garg *et al.*, 2006). Results indicated that improvement of nutrient status through application modulated photosynthetic efficiency, carbohydrate and N metabolism which contributed to better performance of plants ever under saline conditions.

Lewis *et al.* (1989) found that NH^{+}_4 fed maize and wheat plants were more sensitive to salinity than NO_3 fed plants, grown in solution cultures. Leidi *et al.* (1991) concluded that NO_3 is a better source than NH_4 for wheat grown in salt affected areas. However, Shaviv *et al.* (1990) found that wheat grown in soil salinized with NaCl was more tolerant in terms of grain yield under a combination NH_4 and NO_3.

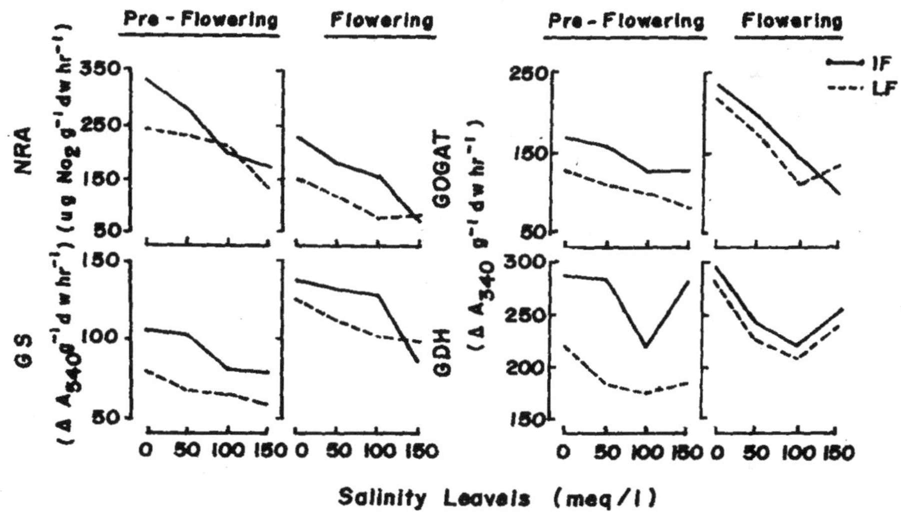

Fig. 7.7. Influence of different levels of salinity on the activity of NR, GS, GOGAT and GDH at the pre-flowering and flowering stages of Indian mustard raised under low and improved soil fertility conditions.

SALINITY AND PHOSPHORUS

The interaction between salinity and phosphorus is complex and highly dependent upon the plant species, plant development stage, the composition and level of salinity and the concentration of P in the substrate.

The solubility and availability of phosphorus in saline water irrigated soils may increase or decrease depending upon the total salinity and nature of the reacting ions. The adverse effect was less with sulfate than the chloride ions because sodium and chloride ions increased the solubility of calcium carbonate more than sulphate ions, which tend to increase phosphate precipitation. Increasing level of P over the recommended dose of 60 kg P_2O_5/ha had a better effect on crop yields when saline waters are used. Application of phosphorus, thus, seems to mitigate the adverse effects of salinity. Minhas and Gupta (1992) observed that magnitude of response of applied P in wheat increased with increase in EC of the irrigation waters. Grain yield of wheat crop increased by 6, 15 and 32% under irrigation with ECiw of 3, 6 and 12 dS/m, respectively when P dose was increased from 30 to 90 kg/ha. (Table 7.6). In another experiment on mustard, the response to higher dose of P was recorded in the year of higher salinity stress (reduction at ECiw 16 dS/m was 97%).

Table 7.5. Effect of P application on wheat yield at different levels of salinity in irrigation waters (Minhas and Gupta, 1992).

ECiw (dS/m)	Grain yield (Mg/ha) at added P (kg/ha)		
	30	60	90
3.0	3.28	3.46	3.47
6.0	2.98	3.26	3.43
12.0	2.17	2.58	2.87

Champagnol (1979) reviewed a number of publications and concluded that P added to saline soils increased crop growth and yield in almost all the crops. However, added P did not necessarily increase crop salt tolerance. After analysing studies with barley, carrot, clover, maize, millet, sorghum, sugarbeet, tomato, vetch and wheat, it was further concluded that added P increased, had no effect, or decreased salt tolerance as salinity increased from low, to moderate and to high levels, respectively. There are many other reports which indicate that addition of P improved the yield of crops under saline conditions (Cerda and Bingham, 1987; Peters, 1983).

Recent evidence indicates that salinity may increase the P requirement of certain plants. Awad *et al.,* (1990) found that when NaCl increased from 10 to 50 and 100 mM, the P concentration in the youngest mature tomato leaf necessary to obtain 50% yield increased from 58 to 77 and 97 mmol kg^{-1} dry weight. Their conclusion was also supported by foliar symptoms of P deficiency that were evident on plants grown at high levels of salt at a given leaf P concentration.

Salinity may decrease, have no effect or increase P accumulation in plants depending upon the plant and environmental conditions (Champagnol, 1979, Feigin, 1985). Although Cl^- and H_2PO_4 ions may not be competitive in terms of plant uptake but there are reports indicating suppressed P uptake and accumulation by increasing Cl^- in tomato shoot (Papadopoulos and Rending, 1983) and by both Cl^- and $SO^2{}_4$ in barley and sunflower (Zhukovskaya, 1973). Garg *et al.* (1982, 1993) also reported decreased P uptake by increasing salinity in wheat and mustard although P concentration was marginally affected. The reduced P uptake may thus be due to decreased plant growth in such cases.

In contrast, in the case of certain grains and vegetables increasing the level of P actually aggravated the salt injury, indicating that the injury is not due to salt-induced P deficiency (Bernstein *et al.,* 1974). Several studies conducted in solution culture have shown that P concentrations that are optimal in non-saline solutions may adversely affect growth or be toxic to corn (Nieman and Clark, 1976), lupin (Treeby and Van Steveninck, 1988) and certain soybean cultivars (Grattan and Maas, 1984) when grown in saline solutions. This indicates that the optimal P range (A to B in Fig. 7.1), narrows under saline conditions.

Type of salinity has also been observed to influence the response to phosphatic fertilizers. When wheat was irrigated with chloride dominated waters, the yield response to phosphate application was higher as compared to sulphate-dominated waters (Table 7.6). Manchanda *et al.* (1982) suggested that application of phosphatic fertilizers most likely will improve the threshold limits of crop to use of chloride dominated saline waters. Mor and Manchanda (1992) observed that salinity induced tolerance depended upon a certain threshold chloride content in the root zone soil beyond which there was no response to any level of applied P. At ECe 4 dS/m, when Cl content was 18 mmol/l under SO_4-dominated salinity (Cl : SO_4 3 :7) and 35 mmol/l Cl-dominated salinity (Cl : SO_4 7 : 3), grain yield of pea was equal at 80 mg/kg of P level under Cl-dominated salinity and at 20 mg/kg of added P level under SO_4 dominated salinity. However, at ECe 8 dS/m when Cl content in the root zone was 37.7 mmol/l under SO_4-dominated and 73.5 mmol/l under Cl-dominated salinity, table pea did not respond to any level of P under Cl-salinity even upto 80 mg P/kg soil and its grain yield decreased by 83 per cent in relation to the non-saline soil. Whereas under SO_4-dominated salinity its grain yield increased with increase in the amount of P in the soil and yield obtained at 80 mg P/kg soil was similar to the yield obtained at ECe 4 dS/m when the added P was only 20 mg/kg soil. The studies of Chauhan *et al.* (1991) also indicated that higher salinity waters can be used for achieving the same level of wheat production (3.5 Mg/ha) if Cl : SO_4 ratio of irrigation waters is lowered or additional doses of P are added. This indicated that P-requirement of wheat is more when irrigated with Cl-dominated than with SO_4-dominated waters having comparable EC values.

Table 7.6. Response of wheat to applied P under saline water irrigation (10 dS/m) with varying Cl : SO_4 ratios (Chauhan, 1998)

Cl : SO_4	Grain yield (q/ha) at the applied P_2O_5 (kg/ha)		
	0	60	90
3 : 1	26.8	34.4	36.8
1 : 1	28.5	37.6	39.5
1 : 3	31.8	39.8	40.2

Maas and Nieman (1978) suggested that salinity damages the mechanism that controls the intercellular phosphate concentration and the cell's sensitive to salinity in the presence of 0.2 mM P which is only 40% of total in the half strength Hoagland solution (Grattan and Maas, 1984). Phosphate susceptible cultivars absorbed and accumulated P at a rate greater than P tolerant cultivars. These studies with soybean show a unique salinity induced nutritional disorder. Direct ion interactions play a minor role. Excessive P accumulates in sensitive cultivars regardless of the Ca^{2+}/Na^{+} ratio (Grattan and Maas, 1988a) and is dependent upon the ionic strength or osmotic potential of the solution regardless of the types of salts used. Combinations of NaCl + $CaCl_2$, KCl + $CaCl_2$ or $NaNO_3$ + $CaNO_3$ all produced similar effects. Plant mortality, on the other hand, is related to high levels of both P and Cl in the leaf tissue.

In most of the agriculture species that have been studied, saline conditions have been found to inhibit P uptake. This effect however, seems to be more severe at low rather than high substrate P concentrations as found in cotton (Martinez and Lauchli, 1994), lupins (Treely and van Steveninck, 1988) and melons (Navarro *et al.*, 2001). However, Rogers *et al.* (2003) found in lucerene that internal concentrations of P in the shoots appeared to be relatively unaffected by external NaCl levels (0-100 mM) across the range of applied P (0.02 to 5.0 mM), whereas root concentrations of P decreased with increasing NaCl salinity, at low P levels (Less than 0.2 mM). A significant NaCl x P effect was found where plants grown at highest P treatment (0.2 M) were more sensitive to increased NaCl. These results indicate the complex relationship between NaCl and P at low P levels.

It needs to be mentioned that such adverse interactions under field conditions are rare as P concentrations used in solution culture in the above studies are many fold higher. In fact, most of the studies conducted in soils show that salinity reduced P concentration in plant tissues. Phosphate availability is reduced in saline soils not only becacuse of the strength effects that reduce the activity of phosphate but also because P concentrations in soil solution are tightly controlled by sorption processes and by the low solubility of Ca-P minerals.

SALINITY AND POTASSIUM

Potassium, which is an essential cytoplasmic element because of its involvement in osmotic regulation and its competitive effect against sodium, is considered very important under saline conditions. Potassium not only participates in osmotic adjustment under saline conditions but also participates in turgor-mediated responses, such as stomatal and leaf movement. The salt tolerance capacity of plants depends on the efficiency of the K uptake mechanism.

Saline soils quite often contain medium to high amounts of available potassium (Sharma *et al.,* 1968), but during reclamation through leaching, losses of K may take place, leading to its low inherent availability (Torres and Bingham, 1973). Generally, plants grown under high salinity show K deficiency due to antagonistic effect of Na and Ca on K absorption and/or disturbed Na : K or Ca : K ratio. Under such conditions, application of potassic fertilisers may increase plant yields.

Sodium toxicity may be less common in saline soils than in alkali soils. However, excess sodium may cause a reduction of K within the plant, leading to low yields. El-Elgabaly (1955) working with resin-sand mixtures observed that increasing Na concentration decreased K content of barley plants and induced thick, short, brown, unbranched roots. Balba (1960) showed that K uptake by plants was depressed by irrigation with water containing NaCl. Finck (1977) found that wheat grown on slightly to moderate saline soils in southern Italy suffered from excess of Na and deficiency of Ca and potassium.

As a result of the similarities in physiochemical properties between Na^+ and K^+, Na^+ could compete with K^+ for major binding sites in key metabolic processes including low and high affinity transporters (HKT) and also disturb plant metabolism (Marschner, 2012; Shabala and Cumin, 2008). Firstly, high levels of Na^+ inhibit K^+ availability. Second Na^+ not only interferes with K^+ translocation from root to shoot but also competes with K^+ for uptake sites at the plasma membrane resulting in lower K^+ uptake. Third salinity leads to plasma membrane disintegration and favour K^+ leaking (Shabala and Cumin, 2008).

Low K^+ status might induce the formation of ROS and related cell damage under saline conditions, which has been attributed to the effects of K^+ deficiency and/or Na^+ toxicity and the imbition of photosynthetic activity that ultimately inhibits plant growth and reduces crop productivity (Wang *et al.*, 2013).

Despite the overwhelming amount of data that shows reduced uptake and translocation of K^+ by plants grown in high Na^+ substrates, there are few data which show that the addition of K^+ to sodium dominated soils improved plant growth or yield. Giorgi *et al.* (1967) found that growth inhibition by the salt stress was overcome by addition of K in *Phaseolus vulgaris, Pisum sativum* and *Citrus aurantium* but not by removal of osmotic stress. On the other hand, Bernstein *et al.* (1974) found that increasing solution K^+ from 0.4 to 2 mM did not affect the leaf K^+ or yield of corn. Bar-Tal *et al.* (1991) found an increase in the yield of corn grown in sandy soil, but the response was proportional at all salinity levels. They concluded that despite its beneficial effects on increasing K^+/Na^+ within the plant, K fertilization did not reduce the deleterious effects of salinity. Muhammed *et al.* (1987) found that shoot and root growth of rice plant grown in 100 mM NaCl solutions was increased when substrate K^+ increased from 1 to 7 mM. Chow *et al.* (1990) showed that difference in the shoot growth of spinach leaves between plants grown at low (50 mM NaCl) and high (250 mM NaCl) salinity at a given level of K^+ can be reduced when K^+ is added to the highest salinity treatment. However, plant growth at the low salinity level only doubled when K in the solution was increased from 0.01 to 10.0 mM.

Many studies have shown that the K^+ concentration in plant tissues is reduced as the Na salinity or the Na^+/Ca^{2+} ratio in the root media is increased (Janzen and Chang, 1987; Lahiri *et al.*, 1987; Subbarao *et al.*, 1990). Reduction in K^+ uptake in plants by Na^+ is a competitive process and occurs regardless of whether the solution is dominated by Na^+ salts of Cl^- or SO_4^{2}. The effect of Na^+ or K^+ uptake is twofold. At low concentrations, Na^+ may actually increase K^+ uptake, though decreasing it at higher concentrations (Nimbalkar and Joshi, 1975). But in most of the studies Na salts decreased dry matter production and the leaf content of K as well as its uptake (Garg and Garg, 1980).

Although plants selectively absorb and translocate K^+ in preference to Na^+, the degree of selectivity varies among species as well as among cultivars within a species. Kafkafi (1984) concluded that the roots of the salt tolerant species (beet) had a higher affinity for K^+, in exchange for Na^+, than the salt sensitive species (bean).

Rathert (1982) found that salinity (Na^+/K^+ = 9) reduced the concentration of K^+ in the leaves of the salt sensitive cotton cultivar (Dandara) more than that in the salt tolerant cultivar (Giza 45). Similarly Lahiri *et al.* (1996) found that soil salinity reduced K^+ concentration much more in salt sensitive genotypes of clusterbean than that of salt tolerant genotypes (Table 7.7) alongwith much lower K^+/Na^+ ratios in the sensitive ones.

Table 7.7. Influence of soil salinity (10 dSm^{-1}) on seed yield. K and Na concentrations and K/Na ratio in salt tolerant (T) and sensitive (S) genotypes of clusterbean (Lahiri *et al.*, 1996).

Genotypes	Seed yield (g $plant^{-1}$)		K (%)		Na (%)		K/Na ratio	
	C	S	C	S	C	S	C	S
Malosen (T)	3.83	3.27	2.18	2.07	0.20	0.33	10.90	6.27
HFG-182 (T)	3.27	2.73	2.37	2.25	0.24	0.48	9.88	4.69
FS-277 (S)	5.63	3.02	2.08	1.87	0.43	0.78	4.84	2.40
Durga Jai (S)	3.88	1.92	1.90	1.70	0.42	0.85	4.52	2.00

* C = control, S = saline

There are evidences that in certain glycophytes Na^+ can partially substitute for K^+ without affecting growth. Marschner (1986) classified many crop species under four groups depending upon the extent by which Na^+ can replace K^+. These groups are :

(1) High proportion of K^+ can be replaced by Na^+ e.g. beets. turnip, Swiss chard.

(2) Moderate replacement of K^+ by Na^+ is possible e.g. cotton, wheat.

(3) Minor substitution of K^+ is possible e.g. rice.

(4) No substitution of K^+ is possible e.g. maize, bean, lettuce.

Despite high selectively of K^+ over Na^+, excessive amounts of K^+ may be detrimental to some plants. Rush and Epstein (1981) found that the wild tomato species could tolerate 200 mM Na^+, but 200 mM K^+ was toxic. On the other hand, the domestic and more salt sensitive tomato species showed the opposite behaviour, it could tolerate K^+ but not Na^+ at the same concentration.

SALINITY, CALCIUM AND MAGNESIUM

Calcium is essential in processes that preserve the structural and functional integrity of plant membranes, stabilize cell wall structures, regulate ion transport and control ion exchange behaviour as well as cell wall enzyme activities (Hanson, 1984). These functions may become seriously impaired by reduced Ca^{2+} availability under saline and particularly sodic conditions. The uptake of Ca^{2+} from soil solutions may decrease because of ion interactions precipitation and increases in ionic strength that reduces the activity of Ca^{2+}. These combined effects are mainly responsible for reduced yields under saline or sodic conditions (Bernstein, 1975; Janzen and Chang, 1987). Therefore the optimum range is shifted to the right (Fig. 7.1) for most crops grown under saline conditions particularly if the solution is dominated by Na^+ salts.

Elevated NaCl levels in the root medium reduce the nutrient assimilation, especially of Ca and K resulting in ion imbalances of Ca, Mg and K (Garg, 1998; Keutgen and Pawel, 2009). The Ca^{2+} content was reduced in shoots and leaves of *A. griffithii* plants grown at high salinity. On the other hand Mg^{2+} concentration was not much affected in shoots and roots but decreased prominently in leaves (Khan *et al.,* 2000). Decrease in calcium and magnesium content of leaves under salinity decreased membrane stability and decreased chloroph-yll content (Parida *et al.,* 2004). Huaain *et al.* (2013) reported decrea-sed Ca^{2+} and Mg^{2+} of all plant organs in response to NaCl salinity.

Calcium has also attracted special attention as a second messe-nger in the signal conduction between environmental factors and plant responses. These functions are seriously impaired by reduced calcium availability under saline and particularly sodic conditions. The uptake of calcium from soil solution may decrease because of ion interactions, precipitation and increase in ionic strength that reduce the activity of Ca^{2+} leading to the reduced growth and yields under saline conditions (Garg, 1998). Salinity induced Ca deficiency has also attracted interest in terms of Ca^{2+} homeostasis at cellular level in the roots and transmitting a signal to the shoots (Rengel, 1992).

Sodium induced calcium deficiency symptoms have been observed in a number of crop plants such as sorghum (Kawasaki and Moritsugu, 1979; Grieve and Maas, 1988), barley (Maas and Grieve, 1988), corn (Maas and Grieve, 1987), rice (Mushammed *et al.,* 1987) and beans (LaHaye and Epstein, 1971; Cachorro *et al.,* 1994). Therefore, when external calcium concentrations are improved, they may mitigate the adverse effects of salinity (LaHaye and Epstein, 1971; Hanson, 1984; Grieve and Fugiyama, 1987; Cramer *et al.,* 1987; Akhavan - Kharazian *et al.,* 1991). Elevated calcium levels may protect the plants from sodium toxicity by reducing displacement of membrane associated calcium (Cramer *et al.,* 1985, 1987), by reducing Na^+ uptake and its translocation to the shoot (LaHaye and Epstein,

1971), by maintaining K/Na selectivity and reducing K^+ efflux in salt stressed plants (Cheeseman, 1988) or by a combination of these effects (Grattan and Grieve, 1994).

Of all the nutrient elements the role of Ca is most critical in plants growing under salt stress (Garg, 1998). Calcium transport to the shoot decreases at high external Na/Ca ratio and in many gramineous species, Ca concentration in the expanding blades may become deficient. Grieve and Maas (1988) investigated the differential susceptibility of sorghum genotypes `NK 265', `NB9040' and `Hegari' to the Na/Ca ratio in solution culture. The tolerance of cv. Hegari to high external Na was attributed to efficient Ca transport to the developing blades. At high external Na, Ca nutrition was adversely affected but the genotypes differed in their ability to acquire, translocate and distribute Ca efficiently.

Calcium application can reduce the adverse effects of NaCl salinity in a number of crop plants (Lahaye and Epstein, 1971; Hanson, 1984; Grieve and Fugiyama, 1987; Cramer *et al.,* 1987; Akhavan-Kharazian *et al.,* 1991). Elevated calcium levels may protect the plants from NaCl toxicity by reducing displacement of membrane associated calcium, by reducing Na^+ uptake and its transport to the shoots or a combination of these effects (Grattan and Grieve, 1994). Calcium also improves potassium uptake under NaCl salinity thereby effectively increasing the K/Na ratio in the tissues. This effect is well known to be important in salt tolerance (Greenway and Munns, 1980). Calcium application enhanced the growth of NaCl stressed clusterbean plants (Garg *et al.,* 1997) (Table 7.8) by reducing Na uptake and increasing K uptake and K/Na ratio. Similar effects have been found in a number of other studies (Grattan and Grieve, 1994; Akavan-Kharzian *et al.,* 1991, Cramer *et al.,* 1987).

Table 7.8. Interactive effects of NaCl and $CaCl_2$ on seed yield, calcium, potassium and sodium contents and K/Na ratio in clusterbean (Garg *et al.,* 1997)

NaCl (mM)	$CaCl_2$ (mM)	Seed yield (g plant^{-1})	Ca (μ mol g^{-1} dw)	K (μ mol g^{-1} dw)	Na (μ mol g^{-1} dw)	K/Na ratio
0	0	11.4	92.1	519.2	83.2	6.24
	2.5	11.5	137.0	549.4	65.7	8.36
	5.0	11.7	169.2	534.6	61.3	8.72
50	0	10.1	91.5	508.9	126.9	4.01
	2.5	10.8	130.2	532.0	105.1	5.06
	5.0	11.0	169.2	532.0	96.3	5.52

100	0	8.6	78.5	475.9	153.2	3.11
	2.5	9.9	123.1	498.7	126.9	3.93
	5.0	10.1	151.0	508.4	113.8	4.47
150	0	7.6	73.0	470.6	205.8	2.29
	2.5	9.0	118.0	485.9	166.4	2.92
	5.0	9.1	139.2	493.6	153.2	3.22

Lahaye and Epstein (1971) observed that plant with poor root systems due to inadequate calcium transferred large quantities of sodium to the upper foliage. However, in the presence of increasing Ca^{2+}, the concentration of Na^+ in *P. vulgaris* leaves progressively declined while the dry matter progressively increased (Fig. 7.8). Similar changes in Na^+ concentrations were reported for stems and roots of beans and leaves, stems and roots of soybean (Wieneke and Lauchli, 1979). There was an interaction between Na^+ and Ca^{2+} on the growth of barley roots, but no interaction was found in the case of beans (Greenway and Munns, 1980). Lynch and Lauchli (1985) proposed that sodium may inhibit the radial movement of Ca^{2+} from the external solution to the root xylem. Cramer *et al.*, (1987) concluded that the primary response to NaCl stress in cotton roots is the displacement of membrane associated Ca^{2+} by Na^+ leading to increased membrane permeability and to loss of K^+/Na^+ selectivity. The addition of 10 mM Ca^{2+} to the saline culture preserved membrane integrity and prevented leakage of K^+.

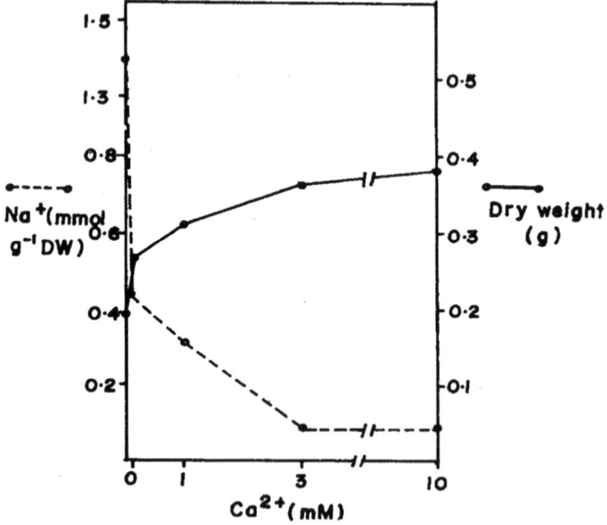

Fig. 7.8. Interaction between calcium and sodium chloride. Effect of Ca^{2+} on dry weight (•——•) and Na^+ concentration in leaves (•——•) of 21-day-old *Phaseolus vulgaris* grown in 50 mM NaCl

Calcium plays a special role in specific ion effects. Cramer *et al.* (1986) observed that wide range of NaCl concentrations does not inhibit the root growth of cotton when supplemental calcium is added (Fig. 7.9).

Shoot growth, however, remained salt sensitive. Calcium addition to the medium apparently favours cell elongation than radial expansion and maintains high rates of cell division (Kurth *et al.,* 1986). At low calcium concentrations, however, the roots became shorter and thicker, and cell production was inhibited. Adding calcium partially restorted the root growth of corn, but adding calcium before adding NaCl completely protected root growth from the inhibitory effects of salt (Cramer *et al.,* 1988).

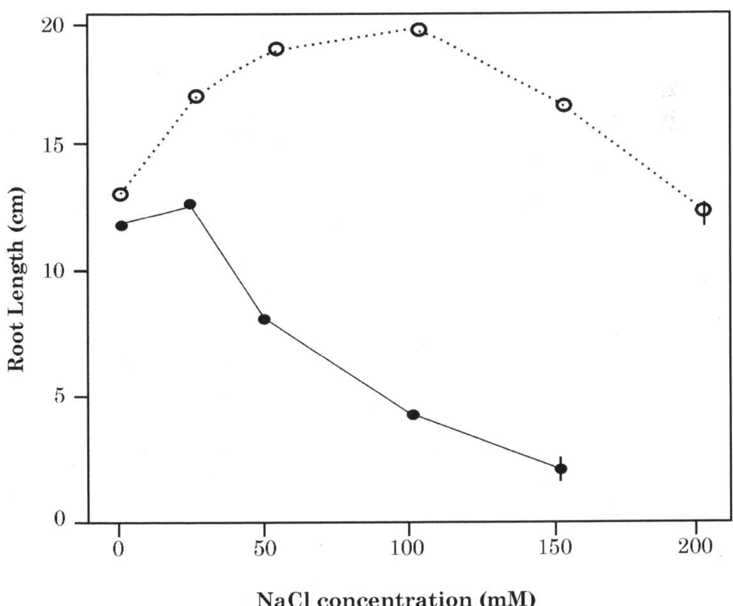

Fig. 7.9. Effect of NaCl and $CaCl_2$ on root length of cotton seedlings. (•) 0.4 mM Ca^{2+}, (o) 10.0 mM Ca^{2+}

Nutritional imbalances in salt stressed cereals under varying molar ratios of Na^+ and Ca^{2+} have demonstrated striking intergeneric differences in corn, rice, wheat, sorghum and pearl millet species (Grattan and Grieve, 1994). Genotypic differences to Na^+/Ca^{2+} ratios have also been reported for sorghum and barley (Grieve and Maas, 1988). The enhanced growth of the tolerant genotypes was attributed to the superior calcium use efficiency and their ability to restrict Na^+ and Mg^{2+} translocation to the leaves.

Therefore, maintenance of an adequate supply of Ca^{2+} in the solution is an important factor in controlling the severity of specific ion toxicities. This is particularly more important for tree and vine crops which are most prone to Na^+ and Cl^- injury than most annual crops. In citrus calcium was found to be effective in reducing the transport of both Na^+ and Cl^- from the roots to leaves, thereby reducing foliar injury (Banuls *et al.,* 1991).

As calcium is strongly competitive with Mg^{2+}, and the binding sites on the root plasma membrane appear to have less affinity for the highly hydrated Mg^{2+} than for Ca^{2+}, therefore high concentrations of substrate Ca^{2+} usually results in increased leaf-Ca alongwith a marked reduction in leaf-Mg (Bernstein, 1975). Increased concentration of $CaSO_4$ in the nutrient solution decreased Mg^{2+} in roots, stems and leaves of *Leucaena leucocephala* (Hansen and Munns, 1988). Bernstein (1975) reported that high levels of $CaCl_2$ in the nutrient medium resulted in a great increase in Ca concentration and a reduction of K and Mg levels in the tissues of bean plants, on the other hand, K addition (as KCl) increased K and decreased Ca and Mg contents in plant tissue. Nassery *et al.,* (1979) have suggested that the detrimental effect of $CaCl_2$ on sesame resulted from the presence of an insufficient level of Mg rather than from the accumulation of toxic levels of Ca in plant tissues.

A simultaneous increase in the concentrations of Na and Ca, as the chlorides, in the nutrient solution resulted in a considerable increase in the levels of these elements in melon shoots and concomitant decrease in K and Mg levels in the plants (Feigin, 1985). The concentration of Ca was almost doubled and that of Na was almost tripled in certain treatments, and the reduction in the K and Mg contents in the melon shoots was of about the same magnitude (Fig. 7.10). In sodic soils, besides a decrease in Ca content, many time Mg and K were also reduced leading to deficiencies and reduced plant growth (Agarwal *et al.,* 1964).

Garg and Garg (1980) found that increasing concentrations of Na_2CO_3 and $NaHCO_3$ progressively reduced the concentrations of both Ca and Mg in pea shoots and roots resulting in their decreased growth. In ground waters of western Rajasthan, in India, magnesium generally dominates over calcium and thus the Mg^{2+}/Ca^{2+} ratios are frequently very high. Garg *et al.* (1983) found that at equal concentration (10 dSm^{-1}) $MgCl_2$ was more detrimental than NaCl and KCl for the growth and yield of wheat. Carter *et al.* (1979) found that barley growth was reduced as the Mg^{2+}/Ca^{2+} ratio increased above 10. Both photosynthetic rate and water use efficiency declined in salt stressed corn (OP = –0.4 MPa), as the external Ca^{2+}/Mg^{2+} ratio increased (Plaut and Grieve, 1988). Excessive leaf Ca concentrations may interfere with CO_2 fixation by inhibition of photosynthetic enzymes, particularly those that are Mg^{2+} activated (Brand and Becker, 1984). Mg salts are also reported to reduce the root growth of *Eucalyptus* more than Na salts and reduction in root growth was associated with low Ca concentrations in the root.

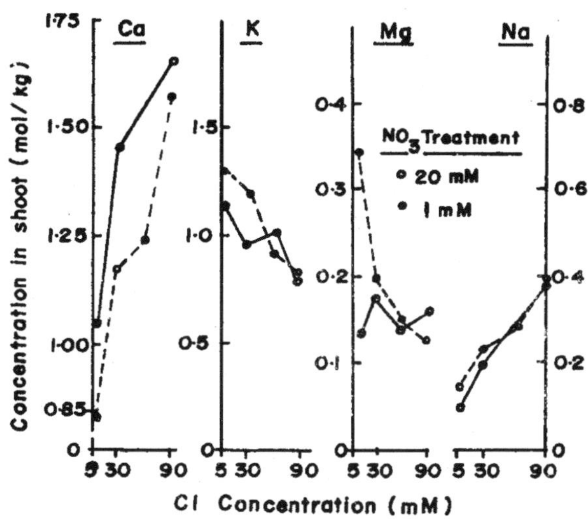

Fig. 7.10. Effect of the chemical composition of the nutrient salt solution on the concentrations of Ca, Mg, K and Na in dry matter in the tops of melon plants

In ground waters of western Rajasthan in India, magnesium generally dominates over calcium and thus Mg : Ca ratios are frequently high. Carter *et al.* (1979) reported that high Mg : Ca ratio in solution may result in Ca deficiencies in barley plants despite high absolute Ca concentrations. Grattan and Grieve (1994) suggested that for plants grown in sea water or dilutions of sea water, it is possible that nutrient disorders can develop because of the high Mg : Ca ratio.

SALINITY AND MICRONUTRIENTS

The solubility of micronutrients (e.g. copper, iron, zinc and manganese) is particularly low in saline and sodic soils. Therefore plants often experience the deficiency of trace elements in such soils. Although the solubility of Zn as zincate ions (ZnO_2) increases with rise in pH (pH > 8.0) but the interaction with Ca between pH 6 to 8 results in the minimum Zn solubility product and therefore Zn deficiency (Totawat and Somani, 1993). Agarwala *et al.* (1964) observed a negative relationship between pH and amount of available Cu in sodic soils. Exchangeable Cu and Mn were found to be low in salt affected soils of Punjab and sodic soils of Rajasthan (Paliwal, 1977). Likewise Fe is often the limiting micronutrient in alkali soils due to high pH and presence of $CaCO_3$. On the other hand some other trace elements such as boron, fluorine and lithium are associated in higher amounts in sodic soils and may prove toxic to crops as well to human beings and animals also who consume products grown on in these soils.

The availability of micronutrients in saline and sodic soils is dependent on the solubility of micronutrients, the pH of soil solution and redox potential of soil solution and the nature of binding sites on the organic and inorganic particle surfaces. In addition salinity can differentially affect the micronutrients concentration in plants depending upon crop species and salinity levels (Oertli, 1991). According to Zhu (2004) deficiency of micronutrients are very common under saline and sodic condition because of high pH.

Consequently, the relationship between salinity and trace element nutrition is complex (Suhayda *et al.,* 1992). The solubility of micronutrients is low in saline and sodic soils. Therefore, plants often experience the deficiency of trace elements in such soils. However, the micro nutrient concentration in plant shoot may increase, decrease or have no effect, depending upon the type of plant, tissues, salinity, micronutrient concentration and environ-mental conditions. The effects of salinity and/or sodicity on the concentration of various micronutrients in different plants is shown in Table 7.9. It may be noted that even in the same plant, an increase or decrease of the same micronutrient element occurs depending upon the experimental conditions and type of salinity. While Hassan *et al.* (1970 a) reported increased manganese (Mn) concentration in barley shoots under salinity, Cramer *et al.* (1991) found a decrease in Mn concentration in barley under the influence of sodium, potassium and calcium salts. Shukla and Mukhi (1985) found that increasing sodicity in maize had a significant reduction in shoot Mn concentration. Agarwal *et al.* (1964) found that in barley increasing ESP increased Mn uptake but decreased the uptake of iron (Fe) and Zinc (Zn).

Table 7.9. Effects of salt stress on micronutrients status in different crop plants

Micro-nutrient	Plant	Effect of salinity or sodicity	Reference
Zinc (Zn)	Barley	Increased concentration	Hassan *et al.,* 1970a Suhayda *et al.,* 1984
	Bean	Increased concentration	Dochiny *et al.,* 1984

Micro-nutrient	Plant	Effect of salinity or sodicity	Reference
	Soybean, Squash and Tomato	Increased concentration	Mass *et al.,* 1972
	Barley	Decreased uptake with increasing ESP	Agarwala *et al.,* 1964
	Sorghum	Decreased uptake with increasing ESP	Bains and Fireman, 1964
	Corn	Decreased concentration	Hassan *et al.,* 1970b
Manganese (Mn)	Barley	Increased uptake with increased ESP	Agarwala *et al.,* 1964
	Barley	Increased concentration with salinity	Hassan *et al.,* 1970 a Suhayda *et al.,* 1992
	Rice	Increased concentration with salinity	Verma and Nene, 1984
	Tomato	Increased concentration with salinity	Maas *et al.,* 1972
	Sorghum	Decreased concentration with ESP increase	Bain and Fireman, 1964
	Pearl millet	Decreased concen. with salt	Ram Deo *et al.,* 1968
	Corn	Decreased concentration with sodicity	Shukla and Mukhi, 1985
Iron (Fe)	Barley	Increased concen. with ESP	Agarwala *et al.,* 1964
	Pea	Increased concentration with ESP	Dahiya and Singh, 1976
	Tomato, Squash and Soybean	Increased concentration with ESP	Maas *et al.,* 1972
	Rice	Increased concentration with ESP	Verma and Nene, 1984
	Barley	Decreased concentration no effect	Hassan *et al.,* 1970 a Suhayda *et al.,* 1992
	Corn	Decreased concentration	Hassan *et al.,* 1972b
	Pearlmillet	Decreased concentration	Ram Deo *et al.,* 1968
Copper (Cu)	Barley	Increased concentration with ESP	Agarwala *et al.,* 1976
	Sorghum	Decreased concentration	Bains and Fireman, 1964

Rao Deo *et al.* (1968) found that $NaHCO_3$ was much more detrimental than NaCl and Na_2SO_4 in decreasing the uptake of Fe and Mn in pearl millet shoots. Bains and Fireman (1964) reported that while molybdenum uptake increased with increasing ESP, the uptake of Mn, Cu and Zn was decreased in sorghum plants.

Long term use of saline water in a soil may decrease the availability of micronutrients such as Zn and Fe and then causes deficiency of these nutrients. This is particularly true in arid and semi-arid regions where soils are generally calcareous and irrigation waters containing high alkalinity/salinity are used for crop production. Chauhan (1998) reported appearance of Zn deficiency in wheat and barley crops in patches when saline waters were continuously used for crop production. This deficiency was corrected by spraying zinc @ 20 kg/ha that improved the crop performance. Minhas and Chhibba (1992) studied Zn transformations in a rice soil irrigated with sodic water. They found that water soluble plus exchangeable Zn fraction increased and insoluble residual Zn fraction decreased with increase in RSC of waters. The increase in water soluble plus exchangeable Zn may be ascribed to the formation of soluble sodium zincate. Bajwa *et al.* (1992) observed that after nine years of sodic irrigations, increasing the sodicity of irrigation waters caused significant decline in DTPA-extractable micronutrients. This decline was more pronounced in rice-wheat than in millet-wheat crop rotation. However, Fe and Mn contents in all the treatments were higher in rice-wheat system. Singh (1999) found that with use of sodic waters (RSC 10 & 20 me/l), zinc application @20 kg Zn/ha improved the grain yield of rice under no gypsum treatment (Table 7.10). In presence of gypsum, increase in yield was non-significant.

Table 7.10. Effect of sodic water irrigations on grain yield of rice (Mg/ha) after two years) (Singh *et al.*, 1999)

RSC (me/l)	Zn levels (kg/ha)			Mean
	0	10	20	
		Without gypsum		
0	5.9	6.2	6.0	6.0
5	5.6	5.8	6.0	5.8
10	5.1	5.3	5.6	5.3
20	4.2	4.4	4.6	4.4
Mean	5.2	5.4	5.5	
CD at 5%	RSC levels - 0.30, Zn levels - 0.26, RSC X Zn - NS			
		With gypsum		
0	5.9	6.2	6.0	6.0
5	6.0	6.0	6.1	6.0
10	5.8	5.9	5.7	5.8
20	5.2	5.6	5.6	5.4
Mean	5.7	5.9	5.9	
CD at 5%	RSC levels - 0.23, Zn levels - NS, RSC X Zn - NS			

Therefore, apart from salinity effects on macronutrients the adverse effects of salt stress on micronutrients may also be responsible for reduced plant growth, particularly under sodic conditions. However, more information is warranted on the trace elements nutrition under saline conditions.

REFERENCES

Agarwala, S.C., Sharma, P.C., Sinha, B.K. and Malhotra, N.K. 1964. Soil and plant relationship with reference to trace elements in usar soils (alkaline) of India. J. Indian Soc. Soil Sci. 12 : 343-354.

Akhavan-Kharazian, M., Campbell, W.F., Jurinak, J.J. and Dudley L.M. 1991. Calcium amelioration of NaCl effects on plant growth, chlorophyll and ion concentration in *Phaseolus vulgaris*. Arid Soil Res. Rehabit. 5: 9-19.

Aslam, M., Huffaker, R.C. and Rains, D.W. 1984. Early effect of salinity on nitrate assimilation in barley seedlings. Plant Physiol. 76: 321-325.

Awad, A.S., Edwards, D.G. and Campbell, L.C. 1990. Phosphorus enhancement of salt tolerance of tomato. Crop Sci. 30 : 123-128.

Bains, S.S. and Fireman, M. 1964. Effect of exchangeable sodium percentage on the growth and absorption of essential nutrients and sodium by five crop plants. Agron. J. 56 : 432-35.

Bajwa, M.S., Sadana, U.S., Nayyar, V.K. and Josan, A.S. 1992. Zinc, iron and manganese status of soils after nine years of sodic irrigation. *J. Res.* (PAU) 29: 37-41.

Banuls, J., Legaz, F. and Plimo-Millo, E. 1991. Salinity calcium interactions on growth and ionic concentration of citrus plants. Plant Soil 133 : 39-46.

Bar-Tal, A., Fiegenbaum, S. and Sparks, D.L. 1991. Potassium-salinity interactions in irrigated corn. Irrig. Sci. 12 : 27-35.

Bernstein, L. 1975. Effects of salinity and sodicity on plant growth. Annu. Rev. Phytopath. 13 : 295-312.

Bernstein, L., Francois, L.E. and Clark, R.A. 1974. Interactive effects of salinity and fertility on yields of grains and vegetables. Agron. J. 66 : 412-421.

Brand, J.J. and Becker, D.W. 1984. Evidence for direct role of calcium in photosynthesis. J. Bioenerg. Biomembr. 16 : 239-249.

Burman, U., Garg, B.K. and Kathju, S. 2002. Interactive effects of saline water irrigation and nitrogen fertilization on growth and metabolism of isabgol (*Plantago ovata* Forsk.). J. Plant Biol. 29: 249-255.

Cachorro, P., Ortz, A. and Cerda, A. 1994. Implications of calcium nutrition on the response of *Phaseolus vulgaris* to salinity. Plant Soil. 159: 205-212.

Carter, M.R., Webster, G.R. and Cairns, R.R. 1979. Calcium deficiency in some solonetzic soils of Alberta. J. Soil Sci. 30 : 161-174.

Cerda, A., Bingham, F.T. and Hoffman, G.J. 1977. Interactive effects of salinity and phosphorus on sesame. Soil Sci. Soc. Am. J. 41 : 915-918.

Champagnol, F. 1979. Relationship between phosphate nutrition of plants and salt toxicity. Phosphorus Agri. 76 : 35-43.

Chauhan, C.P.S. 1998. Efficient management of fertilizer nutrients under saline sodic irrigation. Lead papers and Abstracts. Natl. Sem. Strategies for Mgmt. Poor Quality Irri. Water in Agri. pp. 21-28. CSSRI, Karnal.

Chauhan, C.P.S., Singh, R.B., Minhas, P.S., Agnihotri, A.K. and Gupta, R.K. 1991. Response of wheat to irrigation and phosphorus application. Agric. Water Manag. 20: 223-231.

Cheeseman, J.M. 1988. Mechanisms of salinity tolerance in plants. Plant Physiol. 87: 547-550.

Chow, W.S., Ball, M.C. and Anderson, J.M. 1990. Growth and photosynthetic responses of spinach to salinity : Implications of K^+ nutrition for salt tolerance. Aust. J. Plant Physiol. 17 : 563-578.

Cramer, G.R. and Nowak, R.S. 1992. Supplemental manganese improves the relative growth, net assimilation and photosynthesis of salt stressed barley. Physiol. Plant. 84: 600-605.

Cramer, G.R., Epstein, E. and Lauchi A. 1988. Kinetics of root elongation of maize in response to short term exposure to NaCl and elevated calcium concentration. J. Exp. Bot. 39: 1513-1522.

Cramer, G.R., Epstein, E. and Lauchli, A. 1991. Effects of sodium, potassium and calcium on salt stressed barley. II. Elemental analysis. Physiol. Plant 81 : 197-202.

Cramer, G.R., Lauchi, A. and Epstein, E. 1986. Effects of NaCl and $CaCl_2$ on ion activities in complex nutrient solutions and root growth of cotton. Plant Physiol. 81: 792-797.

Cramer, G.R., Lauchli, A. and Polito, V.S. 1985. Displacement of Ca^{2+} by Na^+ from the plasma lemma of root cells : A primary response to salt stress. Plant Physiol. 79 : 207-211.

Cramer, G.R., Lynch, J., Lauchli, A. and Epstein, E. 1987. Influx of Na^+, K^+ and Ca^{2+} into roots of salt stressed cotton seedlings. Effects of suplemental Ca^{2+}. Plant Physiol. 83 : 510-516.

Dahiya, S.S. and Singh, M. 1976. Effect of Salinity, alkalinity and iron application on the availability of iron, manganese, phosphorus and sodium in pea *(Pisum sativum* L.) crop. Plant Soil 44 : 697-702.

Doering, H.W., Schulze, G. and Roscher, P. 1984. Salinity effects on the micronutrient supply of plants differing in salt resistance Proceedings of the 6th International Colloquim for the Optimization of Plant Nutrition. pp. 165-172. Montpellier, France.

Feigin, A. 1985. Fertilization management of crops irrigated with saline water. Plant Soil 89: 285-299.

Feigin, A. 1990. Interactive effects of salinity and ammonium/nitrate ratio on growth and chemical composition of melon plants. J. Plant Nutr. 13 : 1257-1269.

Feigin, A., Rylski, I., Meiri, A. and Shalhevet, J. 1987. Response of melon and tomato plants to chloride nitrate ratios in saline nutrient solutions. J. Plant Nutr. 10 : 1787-1794.

Francois, L.E., Donovan, T.J. and Maas, E.V. 1991. Calcium deficiency of artichoke buds in relation to salinity. Hort. Sci. 26: 549-553.

Gadallah, M.A.A. 1999. Effects of proline and glycinebetaine on *Vicia faba* response to salt stress. Biol. Plant. 42: 249-257.

Garg, B.K. 1998. Role of calcium in plants under salt stress. Ann. Arid Zone 37: 107-118.

Garg, B.K. and Garg, O.P. 1980. Effect of sodium carbonate and sodium bicarbonate on the growth and absorption of essential macro nutrients and sodium in pea (*Pisum sativum* L.). Proc. Indian Natn. Sci. Acad. B46 : 694-698.

Garg, B.K., Burman, U. and Kathju, S. 2006. Alleviation of salinity effects by nitrogen fertilization on photosynthesis, nitrogen metabolism and yield of Indian mustard. Indian J. Plant Physiol. 11 (N.S.): 145-151.

Garg, B.K., Kathju, S. and Vyas, S.P. 2005. Salinity fertility interaction on growth, photosynthesis and nitrate reductase activity of sesame. Indian J. Plant Physiol. 10 (N.S.) 162-167.

Garg, B.K., Kathju, S., Vyas, S.P. and Lahiri, A.N. 1982. Influence of soil fertility on growth and metabolism of wheat under salt stress. Biol. Plant. 24 : 290-295.

Garg, B.K., Kathju, S., Vyas, S.P. and Lahiri, A.N. 1990. Effect of saline water irrigation on tolerant and sensitive wheat varieties under disparate soil fertility conditions. Ann. Arid. Zone 29 : 179-189.

Garg, B.K., Kathju, S., Vyas, S.P. and Lahiri, A.N. 1997. Alleviation of sodium chloride induced inhibition of growth and nitrogen metabolism of clusterbean by calcium. Biol. Plant. 39: 395-401.

Garg, B.K., Vyas, S.P., Kathju, S., Lahiri, A.N., Mali, P.C. and Sharma, P.C. 1993. Salinity fertility interaction on growth, mineral composition and nitrogen metabolism of Indian mustard. J. Plant Nutr. 16 : 1637-1650.

Giorgi, M.C. Di. Fichera, P. and Tropea, M. 1967. New considerations on the use of saline water III. The influence of the potassium sodium relationship on a saline sensitive culture (*Citrus aurantium*). Agrochemica 11 : 166-175.

Gorham, J., Budrewicz, E., McDonnel, E. and Wyn Jones, A.G. 1986. Salt tolerance in the triticeae: salinity induced changes in leaf solute composition of some perennial Triticeae. J. Exp. Bot. 37: 1114-1128.

Gorham, J., Tomar, O.S. and Wyn Jones, R.G. 1988. Salinity induced changes in the chemical composition of *Leucaena leucocephala* and *Sesbania bispinosa*. J. Plant Physiol. 132 : 678-682.

Grattan, S.R. and Grieve, C.M. 1994. Mineral nutrient acquisition and response of plants grown in saline environments. In : Handbook of Plant and Crop Stress (Ed. M. Pessarakli) pp. 203-226. Marcel Dekker, Inc. New York.

Grattan, S.R. and Grieve, C.M. 1999. Salinity - mineral nutrient relations in horticulutral crops. Sci. Hortic. 78: 127-157.

Grattan, S.R. and Maas, E.V. 1984. Interactive effects of salinity and substrate phosphate on soybean. Agron. J. 76 : 668-676.

Grattan, S.R. and Maas, E.V. 1988a. Effect of salinity on phosphate accumulation and injury in soybean. 1. Influence of $CaCl_2$/NaCl ratios. Plant Soil 105 : 25-32.

Grattan, S.R. and Maas, E.V. 1988b. Effect of salinity on phosphate accumulation and injury in soybean. 2. Role of substrate Cl and Na. Plant Soil 109 : 65-72.

Greenway, H. and Munns, R. 1980. Mechanisms of salt tolerance in non-halophytes. Ann. Rev. Plant Physiol. 31 : 149-190.

Greive, C.M. and Fujiyama, H. 1987. The response of two rice cultivars to external Na/Ca ratio. Plant Soil 103 : 245-250.

Grieve, C.M. and Maas, E.V. 1988. Differential effects of sodium/calcium ratio on sorghum genotypes. Crop Sci. 28 : 659-655.

Hansen, E.H. and Munns, D.N. 1988. Effects of $CaSO_4$ and NaCl on mineral content of *Leucaena leucocephala*. Plant Soil 107 : 101-105.

Hanson, J.B. 1984. The functions of calcium in plant nutrition. In : Advances in Plant Nutrition (Eds. P.B. Tinker and A. Lauchli) Vol. 1, pp. 149-208, Praeger, New York.

Hassan, N.A.K., Drew, J.V., Knudsen, D. and Olson, R.A. 1970a. Influence of soil salinity on production of dry matter and uptake and distribution of nutrients in barley and corn. 1. Barley (*Hordeum vulgare* L.) Agron. J. 62 : 43-45.

Hassan, N.A.K., Drew, J.V., Knudsen, D. and Olson, R.A. 1970b. Influence of soil salinity on production of dry matter and uptake and distribution of nutrients in barley and corn. 2. Corn (*Zea mays* L.). Agron. J. 62 : 46-48.

Helal, H.M. and Mengel, K. 1979. Nitrogen metabolism of young barley plants as affected by NaCl salinity and potassium. Plant Physiol. 51: 457-462.

Helal, H.M., Koch, K. and Mengal, K. 1975. Effect of salinity and potassium on the uptake of nitrogen and on nitrogen metabolism in young barley plants. Physiol. Plant. 35: 310-313.

Hussin, S., Geissler, N. and Koryo, H.W. 2013. Effect of NaCl salinity on *Atriplex nummularia* (L.) with special emphasis on carbon and nitrogen metabolism. Acta Physiol. Plant. 35: 1025-1038.

Janzen, H.H. and Chang, C. 1987. Cation nutrition of barley as influenced by soil solution composition in a saline soil. Can. J. Soil Sci. 67 : 619-629.

Kafkafi, U. 1984. Plant nutrition under saline conditions. In : Soil Salinity under Irrigation Processes and Management (Eds. I. Shainberg and J. Shalhevet) pp. 319-338. Springer Verlag, Berlin.

Kafkafi, U., Valaras N. and Letey, J. 1992. Chloride interaction with nitrate and phosphate nutrition in tomato. J. Plant Nutr. 5: 1369-1385.

Kahne, I. and Poljakoff-Mayber, A. 1968. Effect of substrate salinity on the ability for protein syntehsis in pea roots. Plant Physiol. 43: 1115-1119.

Kawasaki, T. and Mortisugu, M. 1979. A characteristics symptom of calcium deficiency in maize and sorghum. Commun. Soil. Sci. Plant Anal. 10: 41-56.

Keutgen, A.J. and Pawelzik, E. 2009. Impacts of NaCl stress on plant growth and mineral nutrient assimilation in two cultivars of strawberry. Environ. Exp. Bot. 65: 170-176.

Khalil, M.A., Amer, F. and Elgabaly, M.M. 1967. A salinity fertility interaction study in corn and cotton. Soil Sci. Am. Proc. 31: 683-686.

Khan, M.A., Ungar, I.A. and Showalter A.M. 2000. Effect of salinity on growth, water relations and ion accumulation of the subtropical perennial halophyte *Atriplex graffithii* var. *Stocksii.* Ann. Bot. 85: 225-232.

Kurth, E., Cramer, G.R. and Lauchi A. 1986. Effect of NaCl and $CaCl_2$ on cell enlargement and cell production in cotton roots. Plant Physiol. 82: 1102-1106.

Lahaye, P.A. and Epstein, E. 1971. Calcium and salt toleration by bean plants. Physiol. Plant. 25 : 213-218.

Lahiri, A.N., Garg, B.K. Kathju, S., Vyas, S.P. and Mali, P.C. 1987. Responses of clusterbean to soil salinity. Ann. Arid Zone 26 : 33-42.

Lahiri, A.N., Garg, B.K., Vyas, S.P., Kathju, S. and Mali, P.C. 1996. Genotypic differences to soil salinity in clusterbean. Arid Soil Res. Rehabilit. 10 : 333-345.

Leidi, E.O., Silverbush, M. and Lips, S.H. 1991. Wheat growth as affected by nitrogen type, pH and salinity. 1. Biomass production and mineral composition. J. Plant Nutr. 14 : 235-246.

Lewis, O.A.M., Leidi, E.O. and Lips, S.H. 1989. Effect of nitrogen source on growth and response to salinity stress in maize and wheat. New Phytol. 111 : 155-160.

Lynch, J. and Lauchli, A. 1985. Salt stress disturbs the calcium nutrition of barley *(Hordeum vulgare* L.). New Phytol. 99 : 345-354.

Maas, E.V. and Hoffman, G.J. 1977. Crop salt tolerance-current assessment. J. Irrig. Drainage Div. ASCE 103 : 115-134.

Maas, E.V., and Grieve, C.M. 1987. Sodium-induced calcium deficiency in salt stressed corn. Plant Cell Environ. 10 : 559-564.

Maas, E.V., Ogata, G. and Garber, M.J. 1972. Influence of salinity on Fe, Mn, and Zn uptake by plants. Agron. J. 64 : 793-795.

Manchanda, H.R. 1998. Management of saline irrigation. P. 407-431. In: (N.K. Tyagi & P.S. Minhas Eds.) Agric. Salinity Mgmt. In India CSSRI, Karnal, India.

Manchanda, H.R., Verma, S.L. and Khanna, S.S. 1982. Identification of some factors for use of sodic waters with high residual sodium carbonate. J. Ind. Soc. Soil. Sci. 30: 353-360.

Marchner, P. 2012. Mineral Nutrition of Higher Plants. 3rd Ed. Academic Press, London, U.K. pp. 178-189.

Marschner, H. 1986. Mineral Nutrition of Higher Plants. Academic Press, London, 674 p.

Martinez, V. and Lauchli, A. 1994. Salt induced inhibition of phosphate uptake to plants of cotton (*Gossypisum hirsutum* L.) New Phytol. 125: 609-614.

Minhas, Neetu and Chhibba, I.M. 1992. Zinc transformations in a rice soil irrigated with sodic water. Ind. J. Agric. Sci. 69: 251-253.

Minhas, P.S. and Gupta, R.K. 1992. Quality of irrigation water. ICAR Pub. New Delhi.

Mizrahi, Y. and Pastenak, D. 1985. Effect of salinity on quality of various agricultural crops. Plant Soil 104: 57-62.

Mor, R.P. and Manchanda, H.R. 1992. Influence of phosphorus on the tolerance of table pea to chloride and sulphate salinity in a sandy soil. Arid Soil Res. Rehab. 61: 41-52.

Muhammed, S., Akbar, M. and Neue, H.U. 1987. Effect of Na/Ca and Na/K ratios in saline culture solutions on the growth and mineral nutrition of rice (*Oryza sativa* L.) Plant Soil 104 : 57-62.

Munns, R. and Termaat, A. 1986. Whole plant responses to salinity. Aust. J. Plant Physiol. 13 : 143-160.

Nassery, H., Ogata, G. and Maas, E.V. 1979. Sensitivity of sesame to various salts. Agron. J. 71 : 595-597.

Navarro, J.M., Botella, M.A., Cerda A. and Martinez. V. 2003. Phosphorus uptake and translocation in salt-stressed melon plants. J. Plant Physiol. 158: 375-381.

Nieman, R.H. and Clark, R.A. 1976. Interactive effects of salinity and phosphors nutrition on the concentrations of phosphate and phosphate esters in mature photosynthesizing corn leaves. Plant Physiol. 57 : 157-161.

Nimbalkar, J.D. and Joshi, G.V. 1975. Effect of salinity on germination, growth and mineral metabolism of sugarcane var. CO. 740. J. Biol. Sci. 18 : 53-63.

Oertli, J.J. 1991. Nutrient management under water and salinity stress. In: Proceedings of the Symposium on nutrient management for sustained productivity. Dept. Soils, PAU, Ludhiana, India, pp. 138-165.

Paliwal, K.V. 1972. Irrigation with Saline Waters. IARI Monograph No. 2, New Series, New Delhi 190 p.

Papadopoulos, I. and Rendig, V.V. 1983. Interactive effects of salinity and nitrogen on growth and yield of tomato plants. Plant Soil 73 : 47-57.

Parida, A.K, Das, A.B. and Mittra, B. 2004. Effects of salts in growth, ion accumulation, photosynthesis and leaf anatomy of the mangrove *Bruguiera parviflora.* Trees Struct. Funct. 18: 167-174.

Pessaarkali, M. and Tucker, T.C. 1985. Uptake of nitrogen-15 by cotton under salt stress. Soil Sci. Soc. Am. J. 49: 149-152.

Pessaarkali, M., Huber, J.T. and Tucker, T.C. 1989. Protein synthesis in green beans under salt stress. J. Plant Nutr. 12: 1105-1121.

Peters, J.R. 1983. The effect of phosphate fertilizer on the salt tolerance of barley grown on summer fallowed land. Can. J. Soil Sci. 63 : 327-337.

Plaut, Z. and Grieve, C.M. 1988. Photosynthesis of salt stressed maize as influenced by Ca : Na ratios in the nutrient solution. Plant Soil 105 : 283-286.

Rathert, G. 1982. Influence of extreme K : Na ratios and high substrate salinity on plant metabolism of crops, differing in salt tolerance. J. Plant Nutr. 5 : 183-193.

Ravikovitch, S. and Yoles, D. 1971. The influence of phosphorus and nitrogen on millet and clover growing in soil affected by soil salinity. 1. Plant development. Plant Soil 35 : 555-567.

Rogers, M.E., Grieve, C.M. and Shannon, M.C. 2003. Plant growth and ion relations in lucerne (*Medicago sativa* L.) in response to the combined effect of NaCl and P. Plant Soil 253: 187-194.

Rush, D.W. and Epstein, E. 1981. Comparative studies on the sodium, potassium and chloride relations of a wild halophytic and a domestic salt-sensitive tomato species. Plant Physiol. 68 : 1308-1313.

Selassie, T.G. and Wagnet, R.J. 1981. Interactive effects of soil salinity, fertility and irrigation on field corn. Irrig. Sci. 2 : 67-78.

Shabala, S. and Pottosin, I.I. 2010. Potassium and potassium permeable channels in plant salt tolerance. Signal Commun. Plants. 21010: 87-110.

Sharma, P.C., Mali, P.C., Garg, B.K., Vyas, S.P., Kathju, S. and Lahiri, A.N. 1989. Interactive effects of soil fertility and salt stress on the activities of certain enzymes of nitrogen metabolism in mustard. Curr. Sci. 58 : 390-392.

Shaviv, A., Hazan, O., Neumann, P.M. and Hagin, J. 1990. Increasing salt tolerance of wheat by mixed ammonium nitrate nutrition. J. Plant Nutr. 13 : 1227-1234.

Shukla, U.C. and Mukhi, A.K. 1985. Ameliorative role of zinc on maize (*Zea mays*) growth under salt-affected soil conditions. Plant Soil 87 : 423-432.

Silverbush, M. and Ben Asher, J. 1987. The effect of salinity on parameters of potassium and nitrate uptake of cotton. Comm. Soil Sci. Plant Anal. 18 : 65-81.

Singh, B. and Bajwa, M.S. 1990. Effect of poor quality waters on ammonia volatilization losses. J. Ind. Soc. Soil Sci. 30: 779-780.

Singh, Joginder, 1999. Interactive effects of zinc, gypsum and sodic irrigation water on transformations and availability of micronutrient cations to rice and wheat. Ph.D. Dissertation submitted to PAU, Ludhiana.

Singh, K.N., Bains, S.S. and Aggarwal, K.N. 1969. Fertilizer use with saline-cum-high water table conditions. Indian J. Agron. 14: 26-30.

Subbarao, G.V., Johansen, C., Jana, M.K. and Kumar Rao, J.F.D.K. 1990. Effects of sodium/calcium ratios in modifying salinity response of pigeonpea (*Cajanus cajan*). J. Plant Physiol. 136 : 439-443.

Torres, B.C. and Bingham, F.T. 1973. Salt tolerance of Mexixan wheat. I. Effect of NO_3 and NaCl on mineral nutrition, growth and grain production of four wheats. Soil Sci. Soc. Am. Proc. 37 : 711-715.

Treeby, M.T. and Van Steveninck, R.F. 1988. Effects of salinity and phosphate on ion distribution in lupin leaflets. Physiol. Plant. 73 : 317-322.

Verma, T.S. and Neue, H.U. 1984. Effect of soil salinity and zinc application on growth, yield and nutrient composition of rice. Plant Soil 82 : 3 -14.

Wang, M., Zheng, Q. and Gu, S. 2013. The critical role of potassium in plant stress response. Inter. J. Mol. Sci. 14: 7370-7390.

Wieneke, J. and Lauchli, A. 1979. Short-term studies on the uptake and transport of Cl^- by soybean cultivars differing in salt tolerance. Z. Pflanzeneraehr. Botenkd. 142 : 799-814.

Zhu, J.K. 2004. Plant salt tolerance and the SOS pathway In: Proceedings of the XLVIII Italian Society of Agricultural Genetics, Italy, ISBN 88-900622-58.

Zhukovskaya, N.V. 1973. Absorption and accumulation of phosphate by plants under conditions of soil salinization. Soviet Plant Physiol. 20 : 55-61.

■■■■

Chapter - 8

ENVIRONMENTAL FACTORS AND SALT TOLERANCE

INTRODUCTION

The relative effect of salinity is often, if not always, a function of weather. Most crops can bear greater salt stress when the weather is cool and humid than when it is hot and dry. Generally an increase in the severity of weather (heat, wind, dryness) is associated with increased severity of the symptoms of salt damage. Magistad *et al.* (1943) working with identical soil salinities reported that relative yields of alfalfa, strawberry, carrot, cotton onion, squash, and tomato were depressed more in a warm than in cool climates.

Generally, salt tolerance data are only valid for the climatic conditions in which the data were obtained. Temperature, relative humidity, and air pollution all significantly affect plant responses to salinity. Most crops tolerate more salinity stress if the weather is cool and humid than if it is hot and dry. The combined effects of salinity and conditions of high evaporative demand, whether caused by high temperature, low humidity, wind, or drought, are more stressful than salinity alone. Because climate has pronounced effect on plant response to salinity, the time of year when salt tolerance experiments are conducted can also affect the outcome. For example, if the salt tolerance of cool-season vegetable crops was assessed in hot, dry climates, results may underestimate the level of salinity they can tolerate when grown in their normal environment, which is cooler with a lower evaporative demand. Conversely, crops tested in cooler and damper (high humidity) environment than they normally grow in would appear more tolerant than normal (Munns, 1993). Katerji *et al.* (2000) demonstrated that weather conditions affected the threshold value of relative yield of broadbean under salinity, studied under varying growth periods during 1990 (exceptionally warm) and 1998 (exceptionally cold). The relative yield was found to be a of function of difference in temperature of the contrasting weather conditions of 1990 and 1998.

A considerable difficulty in any analysis of weather effects is the non-specific nature of many of the component factors such as temperature, and their mutual interactions. Radiation affects both photosynthesis as well as transpiration and also interacts with air temperature, humidity, wind and turbulance. Another difficulty is the failure of many reports to differentiate between an actual interaction and a mere summation of the effects of salt and other factors on plant growth. Lunt *et al.* (1960) pointed out that the interaction of environment with salt should be analysed on the basis of Baul's principle which states that the resultant of two factors is their product. For example, if extreme heat reduces growth by 20% and excess salt by 30% than together without any special interaction they will reduce growth by 80% x 70% = 56% of the controls (i.e. a reduction of 44%). In this case a reduction of growth equal to or less than 44% would not indicate any interaction of heat and salinity.

Plants exposed to salt in the root environment tend to adjust their internal osmotic concentrations so as to maintain a constant water potential gradient from the rhizosphere to the roots and plant shoots (Gale, 1975). The extent to which they are successful in achieving a complete osmotic adjustment, the level of salt to which they can adjust and the nature of the internal osmoticum are extreme. Plants exposed to salt in the root environment tend to adjust their internal osmotic concentrations so as to maintain a constant water potential gradient from the rhizosphere to the roots and plant shoots (Gale, 1975). The extent to which they are successful in achieving a complete osmotic adjustment, the level of salt to which they can adjust and the nature of the internal osmoticum are extremely variable among plant species. A plant that is unable to achieve a complete water potential adjustment between itself and the root medium becomes very sensitive to the level of evapotranspiration potential (ETP). Even moderate levels of ETP will tend to cause a water imbalance, resulting in a loss of turgor and stomatal closure. As hydraulic conductivity of the root systems tends to decrease under saline conditions, the sensitivity of a plant to ETP increases. Despite full osmotic adjustment, the gas diffusion resistance to the leaves to water vapor loss generally increases under salt stress (Gale *et al.*, 1967) affecting water balance in the opposite direction to the decreased root permeability. As a result, transpiration is reduced to maintain turgor. However, stomatal closure increases the resistance to the diffusion of CO_2 into the leaf, thus reducing photosynthesis and growth. This implies that there are many environmental factors which directly or indirectly modify the transpiration and growth responses to salinity. These are discussed below :

TEMPERATURE AND SALINITY

Temperature is the least specific of all the environmental factors as it affects the movement of salts in the soil, uptake of salts, overall biochemical processes in the plant and transpiration.

For many years it has been observed that high temperatures are associated with lower salt tolerance. Ahi and Power (1938) found that plants grown in diluted sea water provided higher growth in cool than in a hot green house. Bernstein and Ayers (1953) also obtained similar results with vegetable crops. The salinity damage was only apparent when plants were transferred to hot conditions. Ehlig (1940) found that at 90°F or below there were few or no signs of leaf burn due to salinity in table grape vine plants. However, there was extensive damage due to salinity stress above 100°F. Many other reports also indicate more salinity damage during summer months with little or no damage during the winter season.

In some parts of India rice is grown both during the rainy season (summer) and during the dry season (winter). The relative averages of eight rice varieties grown in summer and winter seasons at four salinity levels (Murthy and Janardhan, 1971) indicate that the yield reduction with increasing salinity was much more in the dry than in the wet season (Table 8.1).

Table 8.1. Effect of season on the relative yields* of rice (Murthy and Janardhan, 1971)

Salinity of root zone (dSm^{-1})	Wet season	Dry season
Control	100	100
2-4	93	81
4-8	63	53
10-12	39	11

* Relative yields are comparable only within the same season

Likewise the relative tolerance of crops to soil sodicity is strongly influenced by the prevailing atmosphere evaporative demand. Rice crop grown under submerged condition tolerates higher sodicity in wet years when the rainfall is well distributed and atmospheric humidity is higher during the crop season than in the dry years when the atmospheric evaporation demand is high. This was attributed to the accumulation of sodium in toxic quantities when the evapotranspiration was high (Singh *et al.*, 1979). Acharya *et al.* (1979) observed that the adverse effect of high ESP on plant growth is likely to be accentuated under conditions of high evaporative demands (eg.

summer months) because water supply is limited more in soils of high ESP as compared with that of low ESP under high evapotranspiration.

Lunt *et al.* (1960) reported that under conditions of constant leaf to air humidity gradient either heat or salinity reduced growth of chrysanthemum. However, the combined effect of the two factors was no greater than the arithmetic product, suggesting a lack of some specific interaction.

Francois and Goodin (1972) found that at low temperature, salinity had no effect on sugarbeet germination but germination inhibition, due to salinity, increased between 25° and 40°C. A temperature salinity interaction on the photosynthesis of onion was also reported by Gale *et al.* (1967). These results are in agreement with field observations where onion plants showed lower tolerance to salinity under hot than under cool weather conditions (Bernstein and Ayers, 1953). Several reports in the literature indicate that temperature influences plant salt tolerance to a greater degree than relative humdity (Lunt *et al.,* 1960, Hoffman and Rawlins, 1971).

Temperature has a profound effect on soil properties, particularly in sodic soils. The ESP values are affected in different seasons of the year, being very low in the last summer months (May-June) in arid regions. It seems that the high temperature of about 45° to 50°C during May to June results in top soil temperature of 70° to 80°C that markedly reduces ESP in light and medium textured soils containing excessive $CaCO_3$, specifically having montmorilonite as the dominant clay minerals. In laboratory experiments Gupta and Bandyopadhyay (1975) reported that high temperature caused a remarkable decrease in ESP values of a saline-sodic soils. It is generally known that leaching of salts is better at high than at low temperature.

Climate has a decisive role in determing the suitability of saline waters for irrigation as well as plant growth on saline soils. The most important factor is temperature as high temperature increases evapotranspiration and decreases the salt tolerance of crops. Reduced salt tolerance in hot and dry weather has been reported for alfalfa, clover, bean, beet, cotton and tomato (Bernstein, 1974).

SALINITY AND RAINFALL PATTERN

Gupta and Abichandani (1970) reported that is most cases, the soluble salts and exchangeable sodium which are introduced with saline irrigation waters during the winter season are reduced during the rainy season. However, the leaching of salts depends upon the initial salinity and amount and distribution of rainfall. Normally 350 to 400mm rainfall effectively leaches the salts from the soil profile rendering the top 0 to 40 cm depth in the surface as non saline (EC lesser than 4 dSm^{-1}). The surface soil of Na-Mg-Cl-SO_4 type before rainfall becomes Na-Ca-Cl-HCO_3 type after rainfall (Table 8.2). Similar data from Agra revealed that there is annual build up and leaching of salts following rainfall. Rainfall during the *Kharif* season has a profound effect on leaching.

Table 8.2. Chemical properties of soil before (A) and after (B) rainfall in 4 villages of Jodhpur

Location	Soil depth (cm)	EC_e (dSm^{-1})		ESP		SAR		pH	
		A*	B	A	B	A	B	A	B
Dantiwada $CaCO_3$ 2.0-8.0%	0-20	11.0	2.6	10.9	21.5	26.9	14.6	8.4	8.7
	20-40	8.4	3.8	20.3	34.9	26.1	21.2	8.5	9.0
	40-60	7.2	12.3	56.9	51.1	19.5	29.0	8.5	8.4
	60-100	5.9	11.3	66.2	56.4	23.7	29.9	9.0	8.5
	100-150	5.0	8.3	69.1	39.6	26.8	24.2	9.2	8.7
	150-200	5.0	9.8	40.8	30.2	25.4	19.5	8.9	8.4
Kaparda $CaCO_3$ 4.3-18.5%	0-20	11.8	2.6	59.5	18.0	30.8	8.5	8.3	8.9
	20-40	9.3	3.8	29.7	40.3	31.2	13.8	8.3	8.7
	40-60	9.3	10.8	35.3	32.6	31.1	31.5	8.7	8.5
	60-80	9.3	12.8	29.7	37.6	32.1	34.4	8.7	8.6
Jelwa $CaCO_3$ 0.3-5.2%	0-20	6.7	2.4	16.7	23.9	21.9	16.5	8.3	8.8
	20-40	7.0	3.2	33.5	52.9	37.0	21.0	8.7	9.1
	40-60	5.0	3.1	40.7	4.9	44.6	19.0	9.0	9.0
Shikarpura $CaCO_3$ 9.1-20.8%	0-20	22.2	2.1	34.9	11.0	32.0	3.3	8.2	8.6
	20-40	16.5	2.4	31.6	21.4	30.9	8.8	8.4	9.0
	40-60	13.4	6.6	29.0	9.2	31.1	16.5	8.5	8.7
	60-80	16.0	4.3	36.4	25.6	32.2	8.6	8.4	8.8

Minhas and Gupta (1996) observed that for removal of 80% of salts accumulated during the period preceding the monsoon rainfall would need 1.85, 0.95 and 0.76 cm of rain water per unit depth of fine medium and coarse textured soils in semi-arid regions. In arid regions these values may be somewhat higher because of expected longer lean spells between rain events that may result in resalinization of the soil profile. Sharma and Tyagi (2000) reported that salinity of the soil profile increases with increasing salinity of the irrigation water. Most of the added salts were leaches out of the profile with the monsoon rains. The salinity of the leached soil profiles was within acceptable limits for good germination of the next wheat crop.

LIGHT AND SALINITY

Although there are very few reports on the combined action of radiation intensity and salinity on plant growth but available information does indicate that growth depression due to salinity is generally more under higher than lower light intensities (Gale *et al.*, 1967; Nieman and Poulsen, 1971). In bean plants salinity reduced net photsynthesis (P) only at high radiation levels (30000 lux) and not at all below 8000 lux (Fig. 8.1). However, in case of cotton the same order of reduction of net photosynthesis was found at all light intensities (Fig. 8.2). Gale *et al.* (1967) interpreted this due to the different physiological effect of salinity was probably an interference with CO_2 diffusion into the leaf due to closure of stomata as increasing CO_2 significantly decreased the effect of salinity at high light intensity. However, in cotton, internal biochemical or photochemical processes appeared to have been damaged and in contrast to beans a high concentration of CO_2 could not overcome the effect of salinity at high intensity.

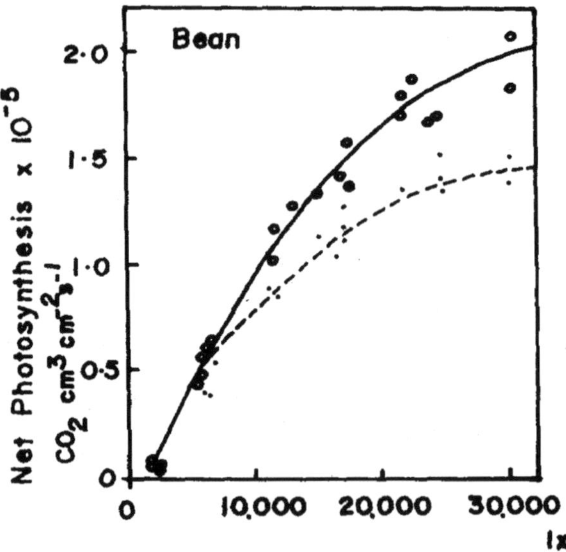

Figure 8.1. Effect of light intensity on the net photosynthesis of control and salinized bean plants. Points are from 5 control and 5 salinized (to - 3.4 bars) plants

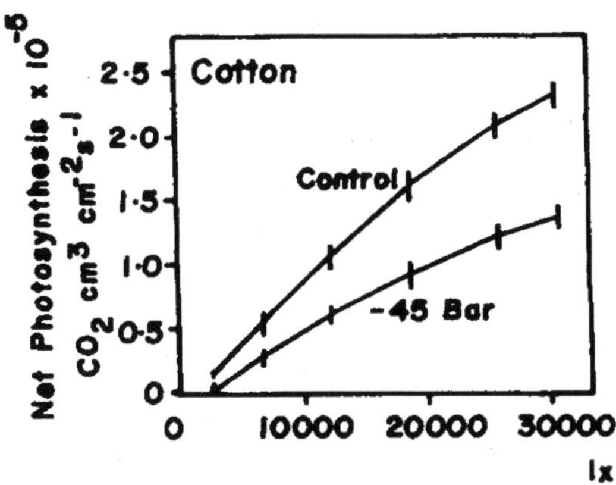

Fig. 8.2 Effect of light intensity on the net photosynthesis of control and salinized (to -4.5 bars) cotton plants

It has been reported by Bernstein (1967) that leaf toxicity symptoms due to salinity were frequently observed on the south side of citrus trees, in response to higher light intensity while leaves on the north side remained symptom free.

Other factors, such as wind or high insolar radiation, that increase the evapotranspirational demands on the plant will likely reduce the plant's tolerance to salinity. However, if light is limiting, increased irradiance may increase salt tolerance. Meiri *et al.* (1982) found that for muskmelon grown in a glasshouse, the salt tolerance of unshaded plants was higher than that of shaded plants. Similar results were reported for broad beans grown under low light intensity in growth chambers (Helal and Mengel, 1981).

Nieman and Poulsen (1971) studied the interactive effects of salinity and light on growth of intact bean seedlings and with excised cotyledonary leaves of radish cultured on control and saline (NaCl) media in the dark and a range of light intensities (0.32, 0.70, 0.97 and 1.72 lux cm^{-2}). In both cases excess NaCl had very little effect on heterotrophic growth in low light intensity and a suppressive effect only in a moderately high light intensities. Excess NaCl was particularly effective in suppressing growth processes requiring high energy light, for example leaf expansion, and the accompanying net synthesis of nucleic acids and protein (Fig. 8.3). These results indicate that light dependent growth processes are particularly vulnerable to salt suppression.

It has been suggested by Gale (1975) that in plants in which one of the primary results of salinity is a partial closure of stomata, photosynthesis and hence overall plant growth will be reduced relatively more at high than at low levels of radiation. In those plant species where osmotic adjustment is incomplete, high levels of radiation will intensify salinity damage by producing water imbalance.

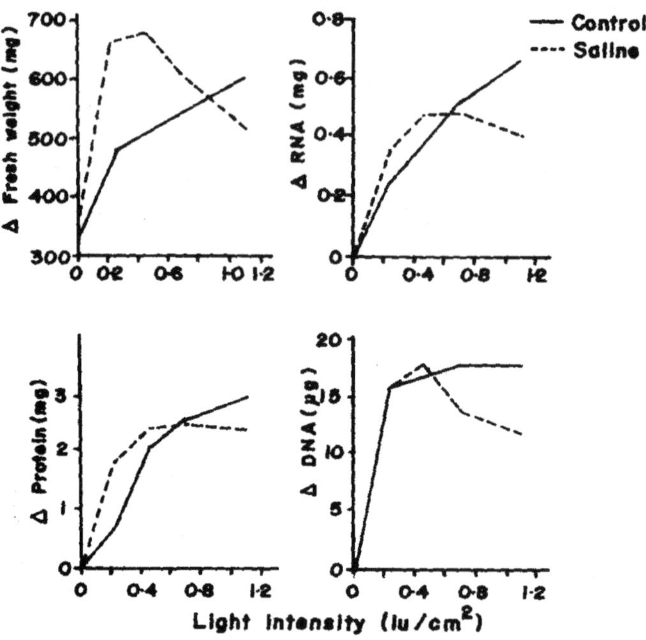

Figure 8.3. Interactive effects of salinity and light on increases in fresh weight, RNA, protein and DNA of excised radish cotyledons incubated for 28 hours. Each point is the average of two replicates. Salinity - 3 bars of NaCl added to the control incubation medium.

RELATIVE HUMIDITY AND SALINITY

High levels of air humidity result in lower rates of transpiration and hence could be expected to alleviate the effects of any water imbalance due to salinity. Furthermore, the fresh : dry weight ratio is increased under humid conditions. This could reduce the concentration of electrolytes and as such was found to be the case for leaves of bean and cotton plants grown under saline condition (Nieman and Poulsen, 1967). High humidity has also been found to ameliorate growth under conditions of salinity by Prisco and O'Leary (1973) in case of beans.

High humidity with temperature held constant has been shown to alleviate salt stress in beans and beets (Hoffman and Rawlins, 1971), barley and corn (Hoffman and Jobes, 1978). However, this effect of high humidity in plants exposed to salinity is not always found as reports for cotton (Hoffman *et al.,* 1971) and wheat (Hoffman and Jobes, 1978) indicate.

Hoffman and coworkers have made extensive studies on the interaction of salinity and atmospheric humidity in a number of crop plants (Table 8.3). Hoffman and Rawlins (1971) found that when the air humidity was raised from 45 to 90%, yield of beet grown in the non-saline media increased by 50% and that of radish by 15% but in onions was unaffected. High RH raised the salinity level at which the growth of onion and radish was reduced by 50% but no interaction was found for beet. However, in case of cotton both air humidity and salinity had a strong independent effect on growth, and humidity did not nullify the effect of salinity (Hoffman *et al.,* 1971). In another study Hoffman and Jobes (1978) found that with a non-saline root medium, increase in RH from 45 to 90% increased the wheat yield by 25%, had no effect on corn yield and reduced barley yield by 16%. High RH increased the salt tolerance of barley and corn but did not affect the tolerance of wheat.

Table 8.3. Growth of plants under saline conditions at two relative humidities (R.H.) (Hoffman and Rawlins, 1971, Hoffman and Jobes, 1978)

Plant species	Root medium salinity (MPa)	Dry weight yield (g/plant^{-1}) at indicated R.H. (%)		Percent increase
		45	90	
Beet	–0.04	6.9	10.2	47.8
	–0.50	4.4 (36.2)*	7.1 (30.4)	61.4
	–1.00	2.4 (65.2)	4.3 (57.8)	79.2
Onion	–0.04	47	53	12.7
	–0.20	33 (29.8)	46 (13.21)	39.4
	–0.40	19 (59.6)	36 (32.1)	89.5
Radish	–0.04	0.56	0.69	23.2
	–0.20	0.46 (17.9)	0.65 (5.8)	41.3
	–0.30	0.22 (60.7)	0.53 (23.2)	140.9
Barley	–0.04	10.2	8.6	–
	–0.50	6.6 (35.3)	8.2 (4.7)	24.2
	–1.00	2.1 (79.4)	3.1 (64.0)	47.6
	–1.50	0.5 (95.1)	0.8 (9.07)	60.0
Wheat	–0.04	7.6	9.4	23.7
	–0.30	6.5 (14.5)	8.3 (117)	27.7
	–0.60	1.9 (75.0)	2.0 (78.7)	5.2
	–0.90	0.4 (94.7)	0.5 (94.7)	25.0
Corn	–0.04	51.3	52.2	1.8
	–0.20	34.2 (33.3)	48.4 (7.3)	41.5
	–0.35	12.9 (74.9)	26.2 (49.8)	103.1
	–0.50	2.1 (95.9)	6.4 (87.6)	204.7

*Figures in parenthesis indicate percent decrease with respect to control.

Salim (1989) found that the relative rates of shoot growth of green gram and red kidney bean and sunflower during a 7-day period of NaCl stress were higher for plants grown at 90% relative humidity than at 30%. No interactive effects of relative humidity and salinity were observed for tomato and *Atriplex spongiosa* because the salt concentrations were too low to affect growth.

According to Pitman (1984) increase in ambient humidity to near saturation has two possible effects on the plant. One is reduction in water flow in the plant and the other is reduction in the water stress. Plant species responded to changing humidity and salinity in different ways.

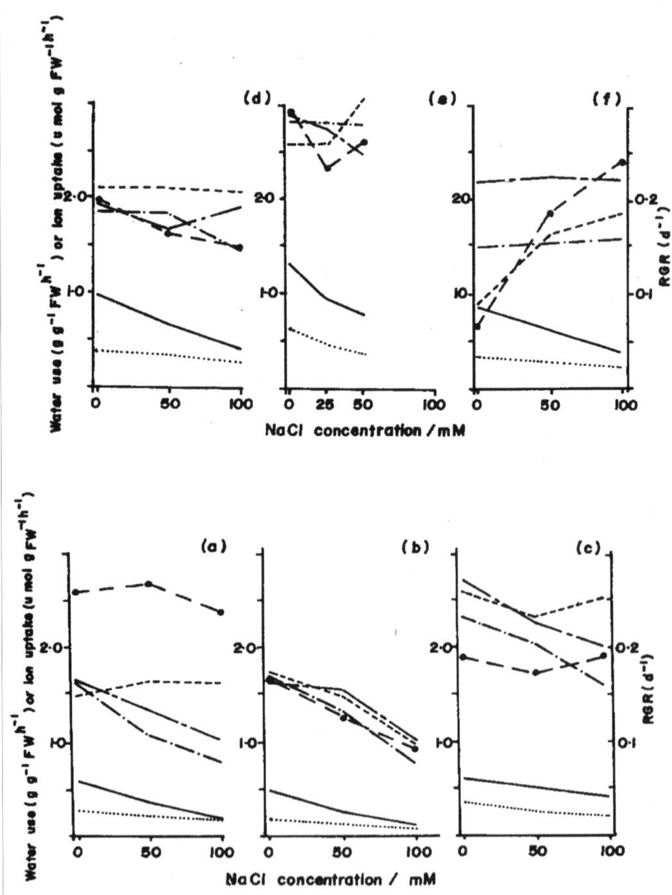

Fig. 8.4. Measurements at higher humidity (closed symbols) and lower humidity (open symbols) of relative growth rates (day-1, 0), water use (g g-1 fresh weight root h^{-1}), and (K + Na) transport from root to shoot (m mol g fresh weight root h-1), for (a) Mung bean, (b) red kidney bean, (c) sunflower (d) *Triticale*, (e) tomato, and (f) *Atriplex*.

Figure 8.4 summarizes the growth responses, water use and rate of transport of K^+ and Na^+ from the root to the shoot in various plant species grown at two different humidities (30-35% and 95%) and on a range of salinities (50 or 100 mM NaCl for all except tomato) for a period upto 20 days. Generally, the relative growth rates were reduced by increased salinity and more so in the lower humidity, though there was less effect of salinity on Atriplex growth. Water use was also reduced with increased salinity and was lower in the high humidity. However, there was much more variability in the rates of transport (K^+ + Na^+) from root to shoot. In mung bean and Atriplex plants the rate of transport was greater at the lower humidity which was associated with higher concentrations in the shoot. In red kidney bean there was little effect of humidity on export from the root, but a reduction with increased salinity proportional to the relative growth rate. For triticale, sunflower and tomato, the rate of transport was more closely related to relative growth rates, being higher at the high humidity than at the low humidity.

It appears therefore that air humidity may interact with salinity in their effect on growth of many plant species. Whether or not high humidity will alleviate the effects of salinity will probably depend upon the ability of the plant to adjust internal osmotic potential and to reduce loss of water thus maintaining its internal water balance.

The influence of NaCl salinity (0 and 80 mM) and relative humidity (30 and 70%) was investigated in two melon cultivars (Revigal C-8, salt sensitive and Galia, Salt tolerant) one month after germination. The growth of the whole plant, leaf, stem and root of cv. Revigal C-8 was increased with increased RH while in cv. Galia, such an increase by RH was only under the NaCl treatment. An increase in RH in the NaCl treatment significantly decreased Na^+ and Cl^- concentrations in leaves of sensitive variety while it had no effect on their concentrations in tolerant variety Galia. Furthermore, under the NaCl treatment cv. Galia had significantly higher crop osmotic potential then cv. Revigal C-8 at both the relative humidities and higher amount of root excudate at 30% RH (Inanaga *et al.,* 2002). Likewise, in tomato, the NaCl treatment decreased the growth of salt sensitive variety Naomi by 34% and 21% at 30% and 70% RH, respectively. However, there was no such decrease in growth of cv. Daniela (salt tolerant) at 30% RH and only a slight decrease at 70% RH. The tolerant variety had significantly lower Na^+ concentration in leaves and stems and a higher root to shoot ratio than cv. Naomi in the NaCl treatment at both the relative humidities. Salt tolerance of cv. Daniela may be related to its ability to exclude Na^+ from the shoot.

AIR POLLUTION AND SALINITY

Ozone, a major air pollutant, decreases the yield of some oxidant sensitive crop more under non-saline than saline conditions (Hoffman *et al.,* 1973, 1975; Maas *et al.,* 1975; Bytnerowicz and Taylor, 1983). This aberration has the tendency to make many crops grown in air polluted regions appear to be more salt tolerant than they really are.

Hoffman *et al.* (1975) found a strong interaction between the effects of ozone and salinity in pintobean, garden beet and alfalfa. At ozone concentrations often prevalent in several agricultural areas, alfalfa yield was increased at moderate salinity levels but not at detrimental salt concentrations. Although salinity reduced ozone damage in pintobean and garden beet but effects were beneficial at salinity and ozone levels too high for economical production (Ogata and Maas, 1973). The salinity ozone interaction may be agronomically important in the air-polluted areas but so far only limited information is available and clearly more research efforts are warranted in this area. It seems at present that the increased ozone tolerance induced by salinity may be more than offset by the detrimental effects of salinity on the harvestable product.

SALINITY AND CLIMATE CHANGE

Increased demand for irrigation in semi-arid climates as a result of both population increase and climate change will tend to increase the extent of secondary salinization. Any increase in the extent of secondary salinization could by offset by positive effects of elevated atmospheric CO_2 on crop yield per unit of water input (Yeo, 1998).

In protected environments, where CO_2 usually enhances plant growth and WUE in the short term. However, for crop production in the open fields, elevated CO_2 is not a factor that can be viewed separately from the climate change that it will bring about. Neither the anticipated CO_2 fertilization nor the WUE benefits to the plant of elevated CO_2 is certain to outweigh the climate effects of elevated CO_2 to water availability and evaporative demand.

Raised temperatures may benefit some crops in some places but disadvantages through increase ET and thermal damage can not be overlooked. Increased WUE may not reduce the salt concentration in a saline environment (Yeo, 1998). Thus, if the benefits of elevated CO_2 are not realized then increased CO_2 will exacerbate rather than moderate the problems of secondary salinity in agriculture.

INCREASED CO_2 CONCENTRATION AND SALINITY

In contrast to ozone, higher CO_2 concentrations in the atmosphere have been shown to increase the salt tolerance of bean, corn and tomato (Enoch *et al.,* 1973; Schwarz and Gale, 1984). Increasing the CO_2 concentration above normal may increase the rate of photosynthesis which in turn may increase the salt tolerance. Enoch *et al.,* (1973) actually showed a marked change in the response of tomato to salinity at 8000 mgl^{-1} (ppm) CO_2 as compared with 350 mgl^{-1} (Fig. 8.5).

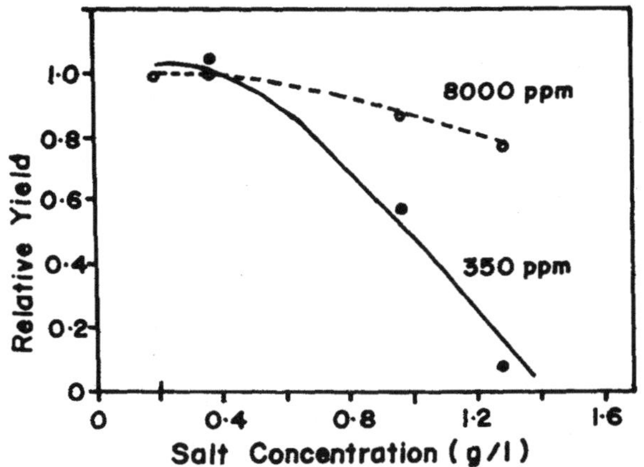

Fig. 8.5. Influence of CO_2 concentration on salt tolerance by tomatoes

Schwarz and Gale (1984) grew four plant species (beans, corn, Xanthium and Atriplex with or without salt in the growth medium at high or low levels of ambient CO_2. In all the four species the relative growth rate increased in response to the addition of CO_2. Furthermore, in all the species the relative response was greater in the salinized than in the control plants, indicating a higher salt tolerance at higher CO_2 concentration.

In another study on the response of sonia roses to salt at different levels of CO_2, Gale and Zerony (1985) found that CO_2 supplementation (600 or 1200 mgl^{-1}) had a very large effect on branch length and flower weight, more so, in the absence of salt. However, at low to medium concentrations of salt (860 and 1700 total soluble salts) the relative response of salt stressed plants was even greater (Table 8.4). In an environment of 1200 mgl^{-1} CO_2 , the roses grew almost as well with 1700 mgl^{-1} TSS as the controls. However, at the highest levels of salt (2600 mgl^{-1}) the roses suffered direct toxicity, which could not be alleviated by supplementary CO_2. In case of alfalfa also high CO_2 concentrations have been reported to alleviate the adverse effects of salt stress. These findings may be of considerable practical importance in controlled environment agriculture in desert areas where non-saline water is not available.

Table 8.4. The effect of CO_2 supplementation on the response of roses to a mixed salt medium (*Rosa hybrida* var. *Sonia*) (Gale and Zerony, 1985)

Salt concentration (mgl^{-1}, TSS)	[CO_2l] ($\mu l\ l^{-1}$)	Branch length (cm)	Flower weight (g)
Normal irrigation water (250)	300±20	26.6±0.6	11.8±0.04
	600±30	42.2±0.2	13.3±0.1
	1200±40	77.2±2.9	23.4±0.9
25% water from brackish well (860)	300±20	21.7±0.9	10.6±0.2
	600±30	45.0±0.8	13.9±0.3
	1200±40	66.6±2.1	28.6±0.7
50% water from brackish well (1700)	300±20	18.4±0.5	8.9±0.1
	600±30	43.9±0.6	12.6±0.3
	1200±40	66.3±1.5	20.7±0.53
75% water from brackish well (2600)	300±20	15.2±0.2	8.6±0.35
	600±30	33.8±0.6	9.6±0.3
	1200±40	54.7±0.8	14.7±0.3

TSS - total soluble salts. N varied from 10 to 28 mg l^{-1}. Figures are means ± S.E. Plants were grown continuously in growth chambers, with conditions simulating the winter greenhouse climate in Israel.

Perez-Lopez *et al.* (2012) made a detailed analysis on the papers published over the last decade relative to physiological parameters affected by salinity and elevated CO_2 interactions. It was demonstrated that the physiological response to this interaction is not only species specific but also cultivar specific. The predicted higher biomass production at elevated CO_2 is not a general trend. Plants adopt different strategies to cope with salt stress including active regulation of plant water potential, stomatal conductance and photosynthetic metabolism as a whole. However, these salt-tolerance mechanisms are energetically expensive, and even if under elevated CO_2 there is generally higher energy supply than under ambient CO_2. In general, biomass production will depend on the trade-off between the anabolic (Photosynthesis) and catabolic (basal respiration and photorespiration) rates as well as the ability of the plants to control the stomatal conductance.

A number of CO_2 enrichment studies have indicated that under NaCl stress high CO_2 improved photosynthetic water use efficiency of leaves whereas NaCl salinity alone has a depensive effect on net phopsyntheic rate and photosynthetic WUE.

WATERLOGGING AND SALINITY

Waterlogging (hypoxia) has a range of effects on plants. Firstly, it rapidly decreases growth, initially of roots, and subsequently of shoots (Barrett-Lennard, 1986), and it increases the senescence of roots, beginning at the tips. Secondly, it affects processes associated with solute movement across membranes, such as the uptake of in-organic nutrietns the regulation of cytoplasmic pH and membrane potentials (Greenway and Gibbs, 2003), and the efflux of internal cell constitutents such as K^+, Cl^-, organic and amino acids, and `basic and acidic metabolites' (Buwalda, 1988). Thirdly, it can decrease stomatal conductance and/or leaf water potentials. Waterlogging interacts with salinity to increase the concentrations of Na^+ and Cl^- in plant shoots. These increases are caused initially by an increase in net rates of transport of Na^+ and Cl^- in the shoots have adverse effects on plant growth and survival the consequences of the interaction. The interaction has major implications for saltland management, and the selection and breeding of plants adapted to saltland.

The literature shows that in general, waterlogging under saline conditions causes increased Na^+ and Cl^- concentrations in the shoot, due initially to increased rates of transport (Barret-Lennard, 2003). These increased concentrations in the shoot have adverse effects on plant growth and survival. Considered as a whole (based on 24 plant species), waterlogging caused average 228% increase in sodium concentration and a 135% increase in chloride concentration in the shoot tissue sampled. Higher ion concentrations in leaves are the result of ion flows into leaves and the diluting effects of growth. In shoots, the concentrations are probably due to decreased rate of shoot growth or increased rates of ion transport under interactive effects of waterlogging and salinity (Barret-Lennard, 2003).

Plants adopt several strategies to improve regulation of salt transport under condition of waterlogging/salinity. Firstly, they avoid hypoxia in the tissues through the formation of arenchyma and maintain the regulation of ion uptake and transport through the formation of an endodermis. Secondly, they protect metabolism through the implementation of salt removal strategies. Glycophytes can protect leaves by absorbing Na^+ and/or Cl^- from the xylem stream into cells by stems petioles or leaf sheaths. On the other hand, halophytes such as

Atriplex amnicula have two mechanisms, for sequestring salt away from the sites of metabolism specialized trichomes (bladder cells) on the leaf surfaces and an ability to sequeater salt into vacuoles.

There is now substantial qualitative and quantitative evidence that waterlogging under saline conditions increases leaf senescence (Barret-Lennard, 2003). In nearly all cases older leaves sencesed before young leaves. Thus interaction between waterlogging and salinity has major implications in management of saline soils and for selection and breeding of plants adapted to waterlogged saline soils.

Thus, it seems that non-edaphic environmental factors may intensify the damage caused by salinity in the following ways : (a) by causing high rates of transpiration, (b) by causing high leaf temperature or both. However, leaf temperature and transpiration rates are determined by the interaction of many plant and environmental factors such as leaf dimensions, stomatal and internal leaf resistances to water loss, optical properties of the leaves, the radiation environment, wind velocity, air temperature and humidity. The effect of these factors on other plant processes such as photosynthesis, respiration and metabolic pathways needs to be investigated.

REFERENCES

Acharya, C.L., Sandhu, S.S. and Abrol, I.P. 1979. Effect of exchangeable sodium on the field drying pattern of soil at two evaporative demands. Soil Sci. 127 : 56-62.

Ahi, S.M. and Powers, W.L. 1938. Salt tolerance of plants at various temperatures. Plant Physiol. 12 : 767-789.

An, P., Inanaga, S., Li, X.J., Enejia, E. and Zhu, N.N. 2005. Interactive effects of salinity and air humidity on two tomato cultivars differing in salt tolerance. J. Plant. Nutr. 28: 459-473.

An, P., Inanaga, S., Lux, A., Li, J., Ali, M. and Sugimato, Y. 2002. Effect of salinity and relative humidity on two melon cultivars differing in salt tolerance. Biol. Plant. 45: 409-415.

Barret-Lennard, E.G. 1986. Effects of waterlogging on the growth and NaCl uptake by vascular plants under saline conditions. Reclam. Revege. Res. 5: 245-261.

Barret-Lennard, E.G. 2003. The interaction between waterlogging and salinity in higher plants: causes, consequences and implications. Plant Soil 253: 35-54.

Bernstein, L. 1969. Salinity factors and their limits for the citrus culture. Proc. First Int. Citrus Symp. 3 : 1774-1782.

Bernstein, L. 1974. Crop growth and salinity. In: Drainage for Agriculture. J. Van. Schilfgaarde (ed.) Agronomy. Am. Soc. Agron. Madison, Wisconsin.

Bernstein, L. and Ayers, A.D. 1952. Salt tolerance of five varieties of onions. Am. Soc. Hort. Sci. Proc. 62 : 367-370.

Bytnerowicz, A. and Taylor, O.C. 1983. Influence of ozone, SO_2 and salinity on leaf injury, stomatal resistance, growth and chemical composition of bean plants. J. Environ. Qual. 12 : 397-405.

Ehlig, C.F. 1940. Effects of salinity on four varieties of table grapes in sand culture. Am Soc. Hort. Sci. Proc. 76 : 323-331.

Enoch, H.Z., Zielsin, N., Biran, Y., Halevy, A.H., Schwartz, M., Kessler, B. and Schimschi, D. 1973. Principles of carbon dioxide nutrition research. Acta. Hot. 32 : 97-110.

Francois, L.E. and Goodin, G.R. 1972. Interaction of temperature and salinity on sugarbeet germination. Agron. J. 64 : 272-273.

Gale, J. 1975. The combined effect of environmental factors and salinity on plant growth. In: Plant in Saline Environments (Eds. A Poljakoff-Mayber and J. Gale) pp. 186-192, Springer Verlag, Berlin.

Gale, J. and Zeroni, M. 1985. The cost of plants of different strategies of adaptation to stress and the alleviation of stress by increasing assimilation. Plant Soil 89 : 57-67.

Gale, J. Kohl, H.C. and Hagan, R.M. 1967. Changes in the water balance and photosynthesis of onion, bean and cotton plants under saline conditions. Physiol. Plant. 20 : 408-420.

Gale, J., Nieman, R. and Poljakoff-Mayber, A. 1970. Growth of *Atriplex halimus* L. in sodium chloride salinated solution as effected by the relative humidity of the air. Aust. J. Biol. Sci. 23 : 947-952.

Greenway, H. and Gibbs, J. 2003. Mechanisms of anoxia tolerance in plants. II. Energy requirement for maintenance and energy distribution to essential processes. Functional Plant Biol. 25: 373-380.

Gupta, I.C. and Abichandani, C.T. 1970. Seasonal variations in the salt composition of some saline water irrigated soils of western Rajasthan. I. Effect of rainfall. J. Indian Soc. Soil Sci.18: 429-445.

Gupta, I.C. and Bandhyopadhyay, A.K. 1975. Effect of heating on the reclamation of a salt affected soil. Symp. Salt Affected Soils, Cairo 1975, 525-528.

Helal, H.M. and Mengel, K. 1981. Interaction between light intensity and NaCl salinity and their effects on growth, CO_2 assimilation and phosphsynthetic conversion in young broad beans. Plant Physiol. 67: 999-1002.

Hoffman, G.J. and Jobes, J.A. 1978. Growth and water relations of cereal crops as influenced by salinity and relative humidity. Agron. J. 70 : 765-769.

Hoffman, G.J. and Rawlins, S.L. 1971. Growth and water potential of root crops as influenced by salinity and relative humidity. Agron. J. 63 : 877-880.

Hoffman, G.J., Gaber, M.J. and Cullen, E.M. 1971. Water relations and growth of cotton as influenced by salinity and relative humidity. Agron. J. 63 : 822-826.

Hoffman, G.J., Maas, E.V. and Rawlins, S.L. 1973. Salinity-ozone interactive effects on yield and water relations of pinto bean. J. Environ. Qual. 2 : 148-152.

Hoffman, G.J., Maas, E.V. and Rawlins, S.L. 1975. Salinity ozone interactive effects on alfalfa yield and water relations. J. Environ. Qual. 4 : 362-371.

Katerji, N., Van Hoorn, J.W., Hamdy, H. and Mastorilli, M. 2000. Salt tolerance classification of crops according to soil salinity and water stress day index. Agric. Water Manag. 45: 99-109.

Lunt, O.R., Oertli, J.J. and Kohl, H.C. 1960. Influence of environmental conditions on the salinity tolerance of several plant species. Trans. 7th Intern. Cong. Soil Sci. 560-570.

Maas, E.V., Rawlins, S.L. and Ogata, G. 1973. Salinity-ozone interactions on pinto-bean : integrated response to ozone concentrations and duration. J. Environ. Qual. 2 : 400-404.

Magistad, O.C., Ayers, A.D., Wadleigh, C.H. and Gauch, H.G. 1943. Effect of salt concentration, kind of salt and climate on plant growth in sand cultures. Plant Physiol. 18 : 151-166.

Meiri, A., Hoffman, G.J., Shannon, M.C. and Poss, J.A. 1982. Salt tolerance of two muskmelon cultivars under two radiation levels. J. Amer. Soc. Hort. Sci. 107: 1168-1172.

Minhas, P.S. and Gupta, R.K. 1992. Quality of Irrigation Water - Assessment and Management. ICAR, New Delhi, 123 p.

Munns, R. 1993. Plant growth response to salt stress in: Towards the Rational Use of High Salinity Tolerant Plants Vol. I. (Eds. H. Heth, and A Al-Masoom) pp. 279-291. Kluwer Academic Publishers. The Netherlands.

Nieman, R.H. and Poulsen, L.L. 1967. Interactive effects of salinity and atmospheric humidity on bean and cotton plants. Bot. Gaz. 128 : 69-73.

Nieman, R.H. and Poulsen, L.L. 1971. Plant growth suppression on saline media : interaction with light. Bot. Gaz. 131: 14-19.

Ogata, G. and Maas, E.V. 1973. Interactive effects on salinity and ozone on growth and yield of garden beet. J. Environ. Qual. 2 : 518-520.

Peter-Lopez, V., Mena-Petite A. and Munz Rueda, A. 2012. Interaction between salinity and elevated CO_2 : A physiological approach. Progress in Botany. 73: 97-126.

Pitman, M.G. 1984. Transport across the root and shoot/root interactions. In : Salinity Tolerance in Plants: Strategies for Crop Improvement (Eds. R.C. Staples at G.H. Toenniessen) pp. 93-123. John Wiley, New York.

Salim, G.R.1989. Effects of salinity and relative humidity on growth and ionic relations of plants. New Phytol. 113: 13-20.

Schwartz, M. and Gale, J. 1984. Growth response to salinity at high levels of carbon dioxide. J. Exp. Bot. 35 : 193-196.

Sharma, D.P. and Tyagi, N.K. 2000. Disposal of drainage water recycling and reuse. Proc. 8th ICID Int. Drainage Workshop, New Delhi, Vol. III : 199-213.

Singh, M.V., Chabbra, R. and Abrol, I.P. 1979. In : Annual Report, Central Soil Salinity Research Institute, Karnal, India.

Wall, R.F. and Hartman, E.L. 1942. Sand culture studies of the effects of various concentrations of added salts upon the composition of tomato plants. Am. Soc. Hort. Sci. Proc. 40 : 460 - 466.

Yeo, A. 1998. Predicting the interactions between the effects of salinity and climate change on crop plants. Scientific Horticulturae 78: 159-174.

■■■■

Chapter - 9

PHYSIOLOGICAL MECHANISMS OF SALT RESISTANCE

INTRODUCTION

Plants differ widely in salt resistance, from the most sensitive ones (e.g. beans) whose growth in affected by very low concentrations of sodium chloride (50 mM) to the most resistant halophytes (*Rhizophora mangle* and *Avicennia germinans*) which thrive only at sea water salinities (500 mM) or even higher. The obligate halophytes which grow only in the presence of sufficient salt include both lower and higher plants, for instance, a blue green algae (*Aphanothece halophytica*), diatoms, a marine yeast (*Rhodotorula glatinis*) and the higher plants (*Suaeda maritima* and *Salicornia europaea*). While obligate halophytes are irreversibly inhibited at low salinity, facultative halophytes (e.g. *Salicornia rubra*) are found at the highest salinity yet are capable of growing normally in low to non-saline environments. Among crop plants, beet grows in the presence of NaCl. The growth response of cultivated species (glycophytes) to salinity overlaps with that of halophytes (Greenway and Munns, 1980) and this is not surprising since some species, such as sugarbeet, have halophytic ancestors.

One of the most detrimental effects of salinity stress is the accumulation of Na^+ and Cl^- ions in tissues of plants exposed to soils with high NaCl concentrations. Entry of both Na^+ and Cl^- into the cells causes severe ion imbalance and excess uptake might cause significant physiological disorders. High Na^+ concentration inhibits uptake of K^+ ions which is an essential element for growth and development that results into lower productivity and may even lead to death (Marschner, 1996; James *et al.,* 2011). In response to salinity stress, the production of ROS, such as singlet oxygen, superoxide, hydroxyl radical, and hydrogen peroxide, is enhanced (Apel and Hirt, 2004; Ahmad and Umar, 2011). Salinity induced ROS formation can lead to oxidative damages in various cellular components such as proteins, lipids, and DNA, interrupting vital cellular functions of plants.

Genetic variations in salt tolerance exist, and the degree of salt tolerance varies with plant species and varieties within a species.

Among major crops, barley (*Hordeum vulgare*) shows a greater degree of salt tolerance than rice (*Oryza sativa*) and wheat *(Triticum aestivum*). The degree of variation is even more pronounced in the case of dicotyledons ranging from *Arabidopsis thaliana,* which is very sensitive towards salinity, to halophytes such as *Mesembryanthemum crystallinum, Atriplex* sp., *Thellungiella salsuginea* (previously known as *T. halophila*) (Munns and Tester, 2008; Pang *et al.,* 2010). In the last two decades sumptuous amount of research has been done in order to understand the mechanism of salt tolerance in model plant *Arabidopsis.* Genetic variations and differential responses to salinity stress in plants differing in stress tolerance enable plant biologists to identify physiological mechanisms, sets of genes, and gene products that are involved in increasing stress tolerance and to incorporate them in suitable species to yield salt tolerant varieties.

RELATIVE CROP SALT TOLERANCE

Although most crop plants are glycophytes, yet there is a rather wide range of salt resistance among them from a maximum in beet roots to minimum in carrots. Among grains barley is more resistant than oats which is more resistant than wheat as judged by the conductivity of the root medium. However, the responses of crop plants vary widely in saline soil, sodic soil or saline water irrigated conditions (Table 9.1 to 9.3). Even within a species there is wide variability among different genotypes. Furthermore, the order of crop salt resistance is not the same in all soils. This is due to the fact that the relative resistance of a given species is not the same for different salts. Most plants are less resistant to NaCl than to Na_2SO_4 but some show the reverse relationship. However, sodium carbonate is almost always more toxic to crop plants than NaCl or Na_2SO_4 (Table 9.4).

Table 9.1. Relative tolerance of crops to salinity

Tolerant (5-10 dSm^{-1})	Medium tolerant (3-5 dSm^{-1})	Sensitive (1.5 - 3 dSm^{-1})
	Crops	
Barley, Sugarbeet, Cotton, Wheat, Safflower, Sorghum, Bajra, Rape, Mustard	Rice, Sugarcane, Groundnut, Maize, Sunflower, Guar, Tobacco, Castor, Soybean, Oat	Linseed, Sesame, Cowpea, Gram, Pea Broad bean, Mung, Pigeon pea, Black gram
	Fruits	
Date Palm	Fig, Apricot, Pomegrante, Grapes, Guava, Ber	Citrus, Apple, Peach

	Vegetables	
Sugarbeet, Amaranthus Spinach, Turnip	Tomatao, Brinjal, Cucumber, Cauliflower, Cabbage, Knol-khol, Potato, Sweet potato, Peas, Carrot, Onion, Lady's finger, Cucurbits	Radish, Beans
	Grasses and Fodders	
Doob, Desmostachya, Para-grass, Sudan grass	Dhaincha Karnal grass	Berseem

Source : Yadav and Gupta (1984)

Table 9.2. Relative tolerance of crops to exchangeable sodium (sodic soils)

Tolerant	Semi-tolerant	Sensitive
Bermuda grass	Wheat	Cowpea
Para grass	Barley	Gram
Rice	Oats	Groundnut
Sugarbeet	Raya	Lentil
	Senji	Black gram
	Berseem	Green gram
	Sugarcane	Pea
	Millets	Maize
	Cotton	Cotton (at gerination)

Table 9.3. Critical Limits of EC_{iw} for different degrees of yield reduction in some crops under different agro-climatic conditions in India

Crop	Soil texture	Location	Critical limit of EC_{iw}		
			10%	25%	50%
Wheat	Sandy loam	Agra	9.0	12.9	17.6
	Loamy sand	Jodhpur	7.3	10.3	13.6
	Sandy loam	Karnal	8.5	11.0	14.2
	Sandy loam	Hisar	7.4	10.2	13.1
	Black clay	Dhanwad	2.7	7.4	18.0
	Black clay	Indore	3.8	8.8	–
	Sand (dune)	Karnal	12.5	16.0	–

Crop	Soil texture	Location	Critical limit of EC_{iw}		
			10%	25%	50%
Barley	Sandy loam	Agra	7.6	16.2	–
Rice	Clay	Bapatala (S)*	1.8	4.0	7.5
		(W)*	1.5	2.6	4.7
	Black clay	Dharwad	3.8	6.5	9.9
Maize	Sandy loam	Agra	2.7	5.5	10.3
	Black clay	Indore	1.2	3.1	9.5
Sorghum	Sandy loam	Agra	8.3	12.8	17.8
	Black clay	Dharwad	2.4	6.5	13.4
Pearl millet	Sandy loam	Agra	7.3	11.6	16.0
Finger millet	Sand	Bapatla	1.6	2.4	8.0
Mustard	Sandy loam	Agra	7.9	10.9	14.6
Sunflower	Sandy loam	Agra	6.0	8.5	11.6
Safflower	Sandy loam	Agra	6.2	11.4	–
	Black clay	Dharwad	4.9	12.2	24.7
Groundnut	Sand	Bapatla	1.7	3.1	5.4
Sesamum	Sandy loam	Pali	5.6	8.5	12.0
Cotton	Black clay	Dharwad	3.3	7.7	15.1
Pigean pea	Sandy loam	Agra	1.8	3.3	6.2
Cowpea	Black clay	Dharwad	2.0	4.6	8.8
	Sandy loam			9.3	12.3
Clusterbean	Sand	Jobner	7.3	9.3	12.3
Onion	Sandy loam	Bapatla	2.5	5.4	10.2
	Sandy loam	Agra	1.7	2.8	5.1
Lady's finger	Sandy loam	Agra	1.4	7.2	13.5
Fenugreek	Sandy loam	Jobner	1.7	8.4	11.8

*S = summer; W = winter Source : Gupta and Yadav (1986)

Table 9.4. The degree of toxicity of various salts for different plants in decreasing order

White lupin	Alfalfa	Wheat	Maize
$MgSO_4$	$MgSO_4$	$MgSO_4$	Na_2CO_3
$MgCl_2$	$MgCl_2$	$MgCl_2$	$NaCl$
Na_2CO_3	Na_2CO_3	Na_2CO_3	$NaHCO_3$
$NaHCO_3$	Na_2SO_4	$NaHCO_3$	Na_2SO_4
Na_2SO_4	$NaCl$	Na_2SO_4	$MgCl_2$
$NaCl$	$NaHCO_3$	$NaCl$	$MgSO_4$

Sorghum	Oats	Cotton	Sugarbeet
$MgCl_2$	$MgSO_4$	$MgSO_4$	$MgSO_4$
$MgSO_4$	$MgCl_2$	$MgCl_2$	$MgCl_2$
Na_2CO_3	Na_2CO_3	$NaCO_3$	Na_2CO_3
$NaHCO_3$	$NaHCO_3$	Na_2SO_4	$NaHCO_3$
Na_2SO_4	Na_2SO_4	NaCl	Na_2SO_4
NaCl	NaCl	$NaHCO_3$	NaCl

Variability in Salt Tolerance Among Higher Plants

Salt tolerance is considered to be a secondary adaptation of angiosperms, since salt tolerance is rare or absent among the lower orders of land plants from which higher plants are thought to have evolved. Higher plants exhibit a wide variety of adaptations to salinity and they vary in salt tolerance from sensitive plants such as beans which is harmed by 0.05 M NaCl in the irrigation supply to halophytes from the euryhaline zone of desert estuaries which grow and complete their life cycles in root zone salinities of 1 M NaCl. Extreme salt tolerance, or halophytism, probably evolved separately in several higher plant families growing in coastal or estuarine locations, then spread to inland habitats (O'Leary and Glenn, 1994). Approximately 2-3,000 species of halophytes, displaying variability in degree of salt tolerance, dominate the world's salt marsh and salt desert floras. Approximately one-third of angiosperm plant families contain halophytic genera.

Table 9.5. Variability of salt tolerance and diversity of tolerance mechanisms among higher plants. Tolerance level is the NaCl salinity that causes 50% reduction in dry matter production. Mechanisms of tolerance are based on literature reports

Plant	Tolerance Level (mM)	Tolerance Mechanisms	References
Bean	40	Na exclusion Na retransport leaf to root	Mass, 1987; Lauchli, 1984
Rice	40	Na exclusion Na storage in older leaves	Mass, 1987, Yeo & Flowers, 1986b
Maize	60	Na exclusion?	Mass, 1987; Cramer *et al.*, 1994

Wheat	140	Partial Na exclusion, storage in leaf vacuoles	Maas, 1987; Schachtman & Munns, 1992
Barley	200	Controlled Na uptake, storage in leaf vacuoles	Mass, 1987; Jeschke & Wolfe, 1993
Diplachne fusa	313	Controlled Na uptake, storage in leaf vacuoles + Salt glands	Warwick & Halloran, 1994
Atriplex conescens spp. *canescens*	350	Controlled Na uptake, storage in leaf vacuoles + Salt bladders	Glenn *et al.*, 1994
Distichlis palmeri	600	Controlled Na uptake, storage in leaf vacuoles + Salt glands	Miyamoto *et al.*, 1995 Glenn, 1987
A. canscens ssp. *linearis*	700	Higher Na uptake than ssp. *canescens*	Glenn *et al.*, 1994
Salicornia bigelovii	>720	High Na uptake succulence	Glenn & O'Leary, 1984 Glenn *et al.*, 1991

Despite their difference, higher plants share some underlying similarities in their mechanisms of salt tolerance, even though some species have developed these mechanisms to operate with far greater efficiency than other species, and are therefore much more salt tolerant. The following sections will document some of the mechanisms of salt tolerance found among higher plants.

SALT RESISTANCE

Salinity can damage the plant through its osmotic effect which is equivalent to a decrease in water activity, through specific toxic effects of ions and by disturbing the uptake of essential nutrients (Chapter 5). These modes of action may operate on the cellular as well as on higher organisation level.

Salt resistance include both avoidance and tolerance mechanisms (Greenway and Munns, 1980; Wyn Jones, 1981; Gorham *et al.*, 1985). Salt avoidance may operate through, (a) passive exclusion of ions due to membrane impermeability, (b) active extrusion of ions by specific pumps or (c) dilution by rapid growth associated with an increase in water content. An unavoidable consequence of growth on a saline soil/solution containing high salt concentrations is the development of osmotic stress which is followed by a loss of turgor.

Tolerance to osmotic stress may operate either through dehydration tolerance, which permits the cells to survive without growing when the turgor decreases, or by avoiding dehydration through osmoregulation, which includes an increase of solute concentration in the cell and consequently rehydration. The solutes may be either organic substances or salt ions which can be sequestered in the vacuole and osmotically balanced by organic solutes in the cytoplasm. The former occurs when salt ions are prevented from entering the cells. These various resistance mechanisms which may be operative in different plants are shown schematically in Fig. 9.1, 9.2 and 9.3.

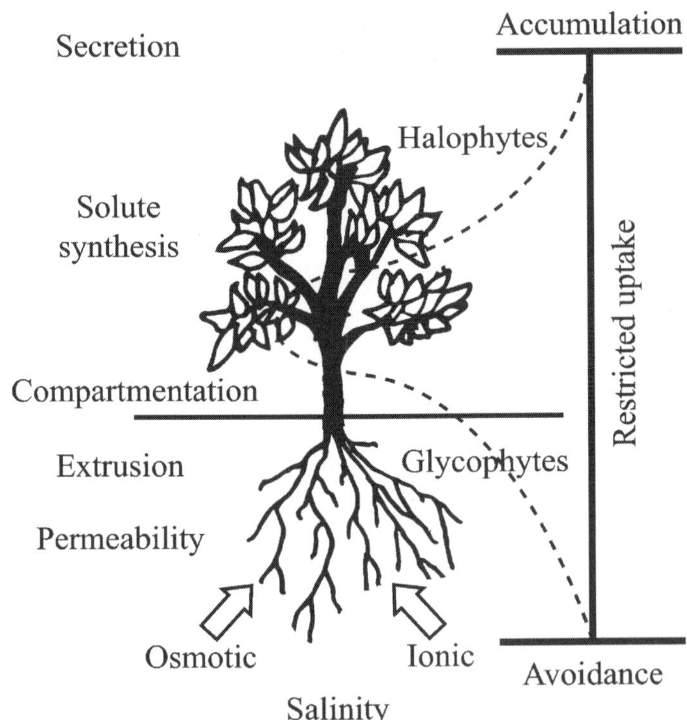

Fig. 9.1. A schematic representation of salinity effects and plant responses

The first strategy (osmoregulation with salt ions) typical to halophytes (Flowers *et al.,* 1977) was also found to characterize the wild relatives of cultivated tomato (Rush and Epstein, 1976) and also in certain genotypes of glycophytes (Alfocea *et al.,* 1993; Ashraf and Waheed, 1993). The second strategy (osmoregulation with organic solutes) seems to characterize the salt resistant varieties of barley, soybean, wheat, grape and several other species (Greenway and Munns, 1980).

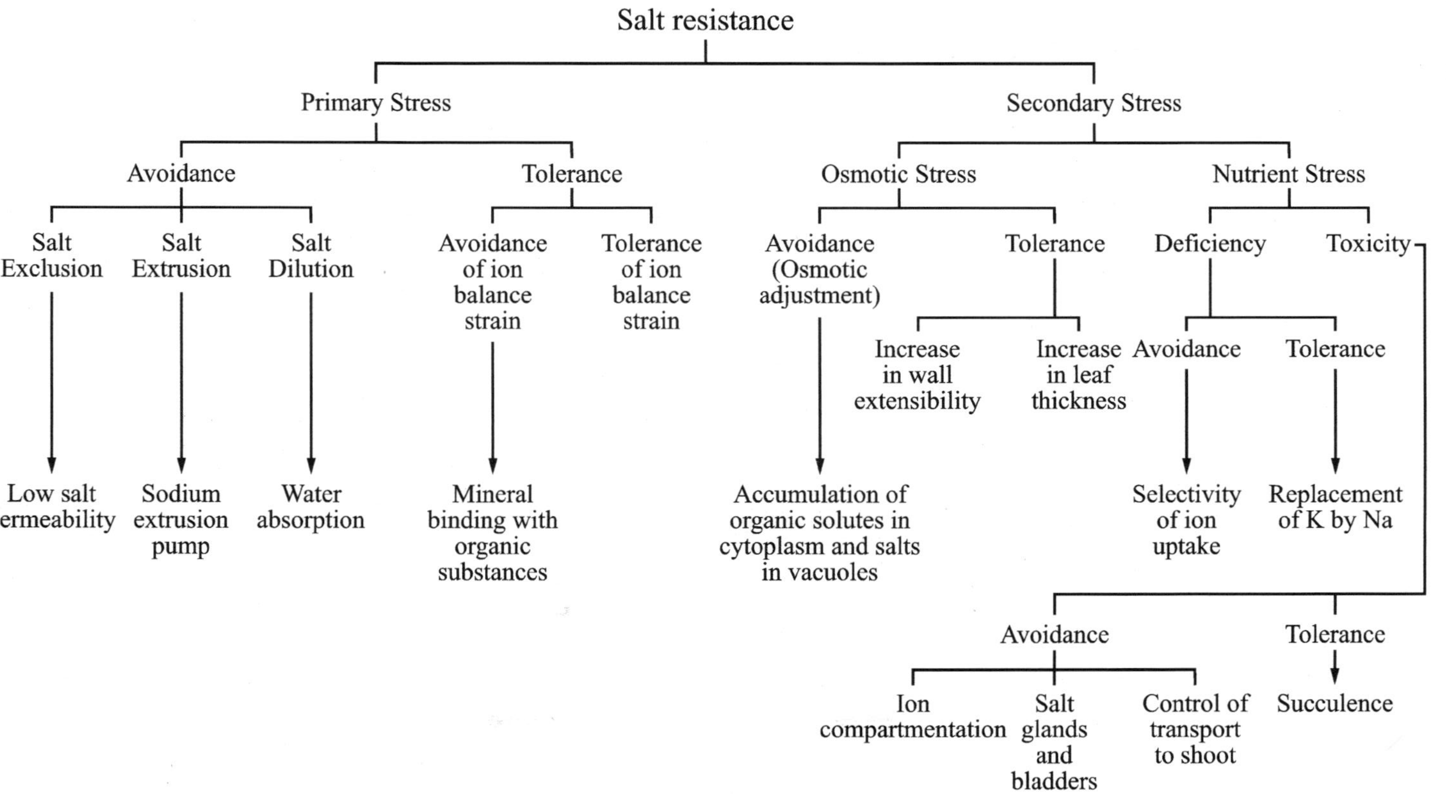

Fig. 9.2. Mechanism of salt resistance in different plants

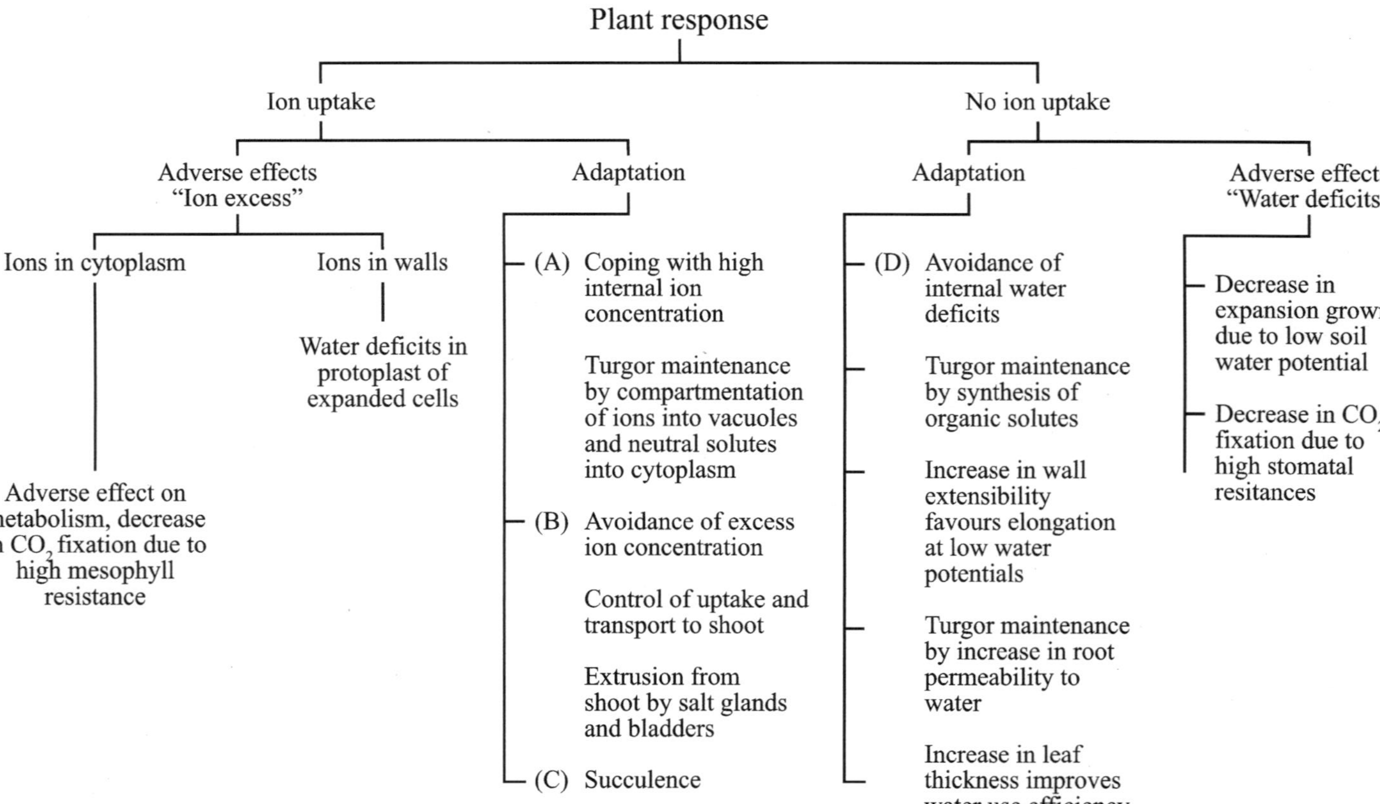

Fig. 9.3. Plant responses to salinity : Adverse effects and possible adaptations.

According to the available information (Greenway and Munns, 1980; Gorham *et al.,* 1985, Poljakoff-Mayber and Lerner, 1994) salt sensitivity of glycophytes may result from the following :

(1) Inability of osmoregulation due to insufficient uptake of salt ions or lack of synthesis of organic solutes being used as osmotica.

(2) Injury caused by inorganic ions which are absorbed by the cells and are not compartmentalized. This situation characterizes the sensitive varieties of crop plants.

Most studies on mechanisms of salt tolerance in higher plants have focussed on osmotic adjustment and ion compartmentation at either the whole plant (roots/shoots) or cellular level (cytoplasm/vacuole). These studies, particularly in the last decade have produced a coherent and widely accepted hypothesis regarding the principal features of tolerance at the cellular level. Three interrelated propositions have been advanced :

(1) Under saline conditions the large quantities of salts (mainly, but not exclusively, sodium chloride) that are absorbed into the leaves and contribute massively to osmotic adjustment are accumulated mainly in the vacuole when tissue concentrations exceed about 200 mM.

(2) The concentration of inorganic ions in the cytoplasm is held in the range of 100 to 200 mM and the cytoplasm shows a strong selectivity for potassium over sodium, magnesium over calcium and phosphate over chloride or nitrate.

(3) Under hyperosmotic conditions (cell osmotic pressure greater than –0.9 MPa) the maintenance of osmotic equilibrium across the tonoplast requires the accumulation in the cytoplasm of nontoxic organic solutes, which are cytoplasm of nontoxic organic solutes, which are usually referred to as compatible solutes. The chemical nature of these compatible solutes varies from one taxonomic group to another, but most are derivatives of polyols or nitrogen dipoles.

Some of these propositions will be described below, however, this subject has been reviewed (Greenway and Munns, 1980; Wyn Jones, 1981; Flowers and Lauchli, 1983; Yeo, 1983; Munns and Termaat, 1986; Flowers and Yeo, 1990, Jacoby, 1994; Garg and Gupta, 1996).

In recent years plant responses to salinity stress have been reviewed with emphasis on molecular mechanisms of signal transduction and on the physiological consequences of altered gene expression that affect biochemical reactions downstream of stress sensing (Glenn *et al.,* 1999, Hasegawa *et al.,* 2000, Orcutt and Nielson, 2000). Among biochemical responses, emphasis has been given to osmolyte biosynthesis and function, water flux control and membrane transport of ions for maintenance of homeostasis (Hasegawa *et al.,* 2000, Sairam and Tyagi, 2004).

SALT AVOIDANCE MECHANISMS

Salt exclusion : Ion Homeostasis

Salt exclusion is one of the most important factors determining the ability to withstand salinity in glycophytes (Levitt, 1980; Lauchli, 1984). Changes in selectivity for K-Na at the plasmalemma, K and Na fluxes and the ability to prevent Na and Cl in the shoot have been implicated in the salt exclusion mechanisms of several species.

Salt tolerance in glycophytes is generally associated with salt exclusion while salt tolerance in halophytes is generally associated with salt accumulation (Gorham, 1994). For example, in glycophytes such as sorghum, beans, wheat, barley and corn, salt tolerance is associated with sodium exclusion. Glycophytes normally maintain a constant or low Na^+/K^+ ratio when challenged with salinity (Greenway and Munns, 1980). In contrast, halophytes may drastically increase their Na^+/K^+ ratio in saline conditions (Glenn and O'Leary, 1984). However, if glycophytes do increase Na^+/K^+, then succulence is required. On the other hand succulence is not needed in halophytes when they increase their Na^+/K^+ ratio during high salinity. Exclusion of salt at the plasma membrane is important in ion homeostasis but cannot be the only mechanism for salt tolerance in halophytes.

In glycophytes there is generally an inverse relationship between salt uptake and salt tolerance, that is, exclusion is the predominant strategy. Differences in the capacity for Na and Cl exclusion also exist among cultivars of species. For example the high salt tolerance of certain cultivars of wheat, barley, pigeonpea and citrus is related to a more effective restriction of shoot transport of both Na and Cl (Marschner, 1986) whereas in soybean cultivars and grapevine root stocks, it is primarily related to the restriction of Cl transport (Downton, 1977; Lauchli and Wieneke, 1979). Restriction of Na in root and restriction of its translocation to the shoot seem to play an important role in the salinity tolerance of wild relatives of pigeonpea (Subbarao *et al.,* 1990).

Classification of species as excluders or includers is helpful in demonstrating the principles of adverse effects (or adaptations) of salinity but in reality very few glycophytes are strict excluders or includers. most are of intermediate types (Marschner, 1986). In tomato, plants of both the types have been described (Jones, 1986; Alfocea *et al.,* 1993).

Results to date give the overwhelming impression that rates of accumulation of Na and/or Cl in the shoot are the critical processes determining varietal differences in salt tolerance. For a set of closely related genotypes, there is often a negative correlation between Na and/or Cl concentration in the leaves and plant growth, that is leaves (or shoots) of the more salt tolerant genotypes usually have lower internal NaCl concentrations for the same time of exposure to NaCl than the salt sensitive ones (Munns, 1990).

Greenway (1965) found that less resistant barley variety accumulated higher contents of Cl^- and Na^+ and lower K^+ content than the two resistant varieties under 125 and 250 mM NaCl. Kingsbury and Epstein (1986) reported that sensitive lines of wheat accumulated more sodium than tolerant lines. They suggested that superior compartmentation of toxic ions, particularly Na^{++} may be a mechanism of salt resistance in this case. Garg *et al.* (1990) found that under saline water irrigation tolerant wheat variety Karchia-65 accumulated less sodium and chloride in their shoots as compared to highly sensitive variety HD-4502. On the other hand potassium concentration was higher in the tolerant as compared to the sensitive genotype resulting in a lower Na/K ratio in the case of former (Table 9.6). Sharma and Gupta (1986) also reported similar results with these two varieties of wheat under NaCl salinity.

Table 9.6. Concentrations of Na, K and Cl and Na/K ratio in the shoot tissue (55 DAS) of wheat varieties grown under normal and saline water irrigation (Garg *et al.,* 1990)

Variety	Salinity level (dSm^{-1})	Sodium	Potassium	Chloride	Na/K ratio
			%		
Kharchia-65	0.2	0.20	2.90	0.31	0.07
(Tolerant)	10	1.00	2.05	1.25	0.49
HD-4502	0.2	0.93	2.48	1.02	0.38
(Sensitive)	10	2.05	1.65	2.05	1.24

In another study under sodicity conditions, Sharma *et al.* (1985) found that resistant wheat varieties protected their reproductive organs against Na by preventing their increase, even when the plants were growing under high sodicity (Fig. 9.4). These organs were also characterized by lower increases in Na/K ratio. Studies on rice varieties with different resistance to soil sodicity also showed that highest Na concentrations were in the sensitive varieties (Fig. 9.5) and its distribution was in the order of sensecising leaves > earheads > stem > younger green leaves. On the other hand in the resistant variety it was sensecising leaves > stem > younger > green leaves > earheads. Very high Na accompanied by lower K appears to be the primary reason of the poor growth and yield of sensitive varieties. Moreover, in the resistant variety Na is mainly localized in the older leaves which are shed and may be serving as a secondary process of resistance. Analysis of different leaves revealed that leaf to leaf gradients in terms of Na, K and Na/K ratio were lowest in the youngest leaf and highest in the oldest surviving leaf and these values were much higher in the sensitive variety (Table 9.7).

Table 9.7. Sodium, potassium (% dry weight) and Na/K ratio in different rice leaves under highest level of sodicity (ESP 45 in case of Basmati-370 and 72 in case of CSR-1) (Sharma, 1986)

Leaf number*	CSR-1			Basmati-370		
	Na (%)	K (%)	Na/K	Na(%)	K (%)	Na/K
1st	1.30	130	1.00	0.80	0.80	1.00
2nd	2.00	1.50	1.33	1.15	0.65	1.77
3rd	2.30	1.50	1.53	2.30	0.65	3.54
4th	2.30	1.15	2.00	2.65	0.50	5.30
5th	2.50	0.80	2.87	–	–	–

*1st is the youngest leaf and 5th is the oldest surviving leaf, 5th leaf could not survive in Basmati - 370.

Lauchli (1984) concluded that most of the legumes respond to saline conditions by salt exclusion, that is, exclusion of sodium and/or chloride from the leaves. Na is excluded from the leaves of many plant species, including legumes (Table 9.8). Jacoby (1965) demonstrated that bean plants exclude Na by retention in the basal parts of the plants but readily translocated Cl to the tops. On the other hand Cl is known to be excluded in certain legumes. Varietal differences in salt resistance of soybean were correlated with exclusion of Cl, the resistant varieties controlling Cl content in the shoots and leaves to a low level (Abel and Mackenize, 1964). Furthermore, Abel (1969) showed that Cl transport to the tops of soybean varieties is under genetic control. In particular, Cl exclusion in soybean cv. Lee was proposed to be controlled by a dominant, single gene pair, whereas Cl accumulation in the leaves of soybean cv. Jackson appeared to be regulated by a recessive single gene pair. It may also be found that some legumes exclude both Na and Cl while others are only Na excluders (Table 9.8).

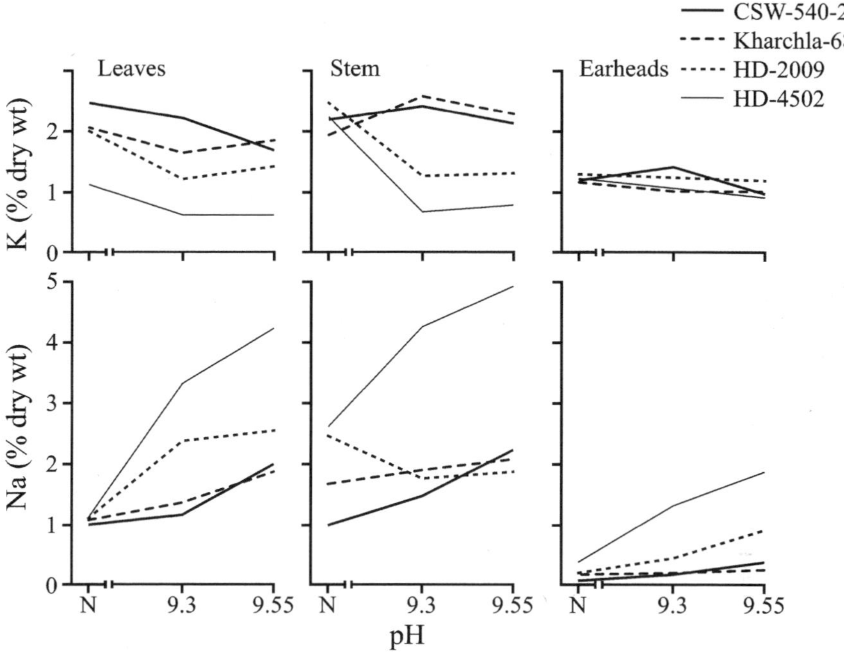

Fig. 9.4 Na and K distribution in different plants parts of wheat genotypes differing in sodicity tolerance at grain filling stage

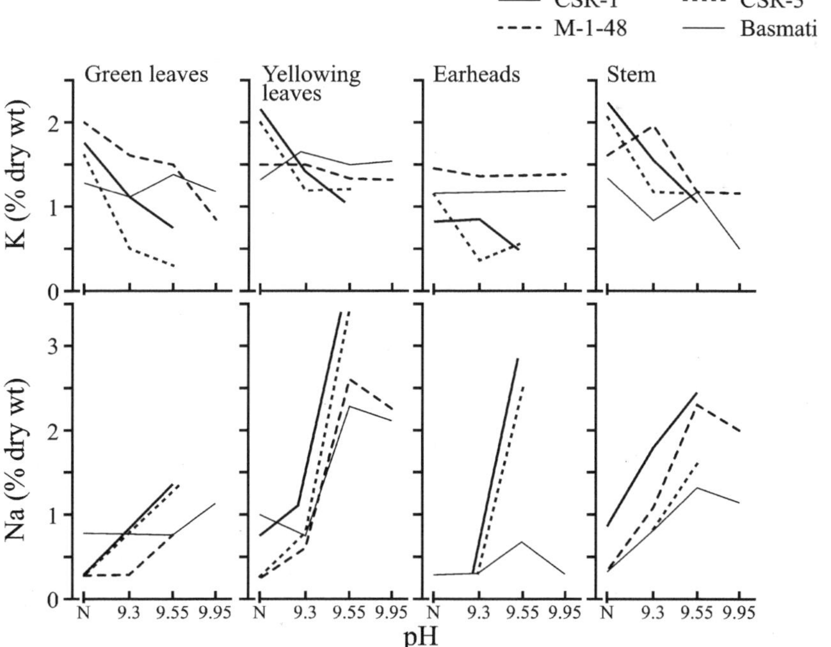

Fig. 9.5. Distribution of Na and K in different plant parts of rice genotypes differing in sodicity tolerance at grain filling stage

Table 9.8. Example of species where salinity tolerance has been related to ion exclusion

Species	Excluded ion	Reference
Barley	Cl^-, Na^+	Greenway (1973) Storey and Wyn Jones (1979)
Wheat	Cl^-, Na^+	Torres and Bigham (1973)
Agropyron elongaturm	Cl^-, Na^+	Shannon (1978)
Soybean	Na^+	Lauchli and Wieneke (1979)
Soybean	Cl^-	Abel and Mackenzie (1964)
Berseem	Cl^-, Na^+	Abel (1969) Winter and Lauchli (1982)
Beans	Na^+	Jacoby (1965)
Chickpea *Root stock*	Na^+	Lauter and Munns (1981)
Avocado	Cl^-	Downton (1978)
Grapes	Cl^-	Downton (1978)
Citrus species	Cl^-, Na^+	Maas and Hoffman (1977)

Subbarao *et al.* (1990) working with salt sensitive pigeonpea (*Cajanus cajan* var. ICP 3783) and its salt tolerant wild relative *Atylosia albicans* and F_1 hybrids under varying salinity levels suggested that the physiological attributes conferring salinity tolerance in *A. albicans* and its F_1 hybrids include: (a) Na and Cl retention in the roots (b) limited translocation to the shoots, (c) high K selectivity and (d) maintenance of transpiration under saline conditions (Fig. 9.6). Likewise Ashraf and Waheed (1993) working with chickpea lines differing in salt tolerance observed that maintenance of high K/Na ratio and K^+ versus Na^+ selectivity were the important factors which contribute to the salt tolerance of genotypes. Similarly salt tolerant genotypes of clusterbean (Malosen and HFG-182) had comparatively lower levels of Na and Cl in their shoots than sensitive genotypes (FS-277 and Durga Jai) at 10 dSm^{-1} soil salinity (Table 9.9). The tolerant varieties had also higher concentrations of K in their shoots and consequently much higher K/Na ratio as compared to sensitive genotypes (Lahiri *et al.,* 1996).

Table 9.9. Concentrations of Na, K and Cl (%) and K/Na ratio in control (C) and salt stressed (S), plants of salt tolerant (T) and sensitive (S), genotypes of clusterbean at pre-flowering and harvest stages of growth (Lahiri *et al.*, 1996)

Genotypes	Na		Cl		K		K/Na Ratio	
	C	S	C	S	C	S	C	S
At pre-flowering stage								
Malosen (T)	0.22	0.35	0.29	0.49	3.03	2.85	13.7	8.1
HFG-182 (T)	0.21	0.32	0.28	0.58	3.03	2.86	14.4	6.8
FS-277 (S)	0.25	0.62	0.28	0.78	2.85	2.45	11.4	4.0
Durga Jai (S)	0.26	0.65	0.33	0.97	2.69	2.22	10.3	3.4
At harvest								
Halosen (T)	0.20	0.33	0.28	0.50	2.18	2.07	10.9	6.3
HFG-182 (T)	0.24	0.48	0.31	0.56	2.37	2.25	9.9	4.7
FS-277 (S)	0.43	0.78	0.49	0.97	2.08	1.87	4.8	2.4
Durga Jai (S)	0.42	0.85	0.46	1.06	0.90	1.70	4.5	2.0

C - 0 dSm^{-1}, S - 10 dSm^{-1}

Varietal differences in salt exclusion have also been observed in a number of other crops such as Indian mustard (Kumar, 1984), *Brassica* species (He and Cramer, 1993), olive (Tattini, 1994) and tomato (Rush and Epstein, 1976; Alfocea *et al.*, 1993; Alarcon *et al.*, 1994).

A hypersaline environment most commonly mediated by high NaCl, results in perturbation of ionic steady state not only for Na^+ and Cl^- but also for K^+ and Ca^{2+} (Niu *et al.*, 1995). External Na^+ negatively impacts intracellular K^+ influx, attenuating acquisition of this essential nutrient by cells. High NaCl causes cytosolic accumulation of Ca^+ and this signals stress responses that are either adaptive or pathological. Ion homeostasis in saline environment is dependent on transmemberane transport proteins that mediate ion fluxes, including H^+ translocating ATPases and pyrophosphatases, Ca^{2+} ATPases, secondary active transporters and channels (Sze *et al.*, 1999). The molecular identity of many transport proteins that putatively mediate Na^+, K^+, Ca^{2+} and Cl^- transport has been determined. It is now clear that many of the transport determinants that mediate ion homeostasis in yeasts and plants are very similar (Serrano *et al.*, 1999). It is likely that the rudiments of salt responsive signal regulatory pathways controlling ion homeostasis in both organisms are analogous (Bressan *et al.*, 1998).

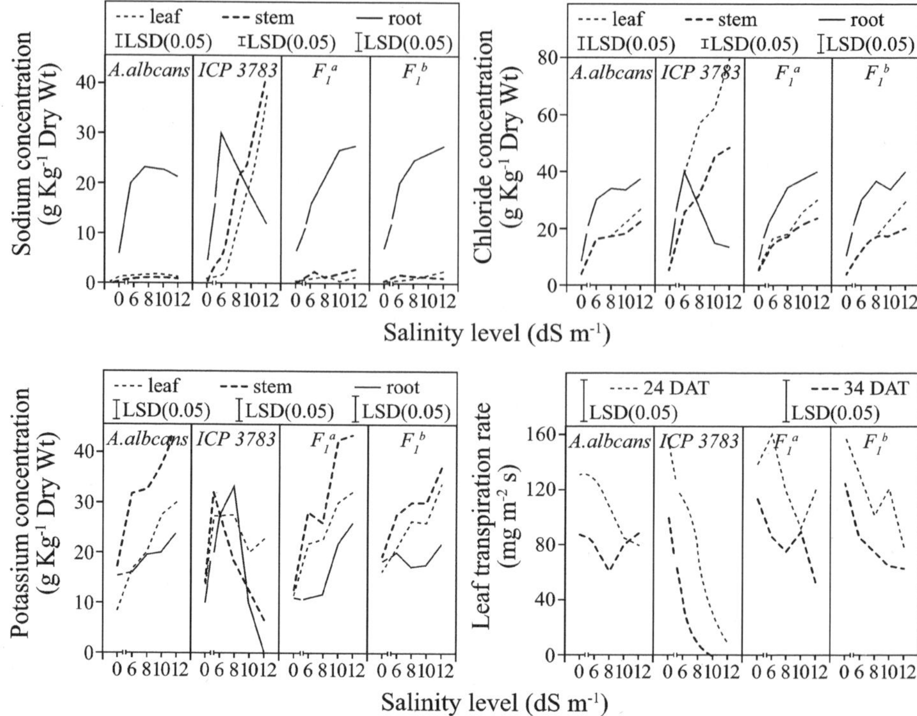

Figure 9.6. Influence of salinity on the tissue concentrations of Na, K and Cl and transpiration rates of *A. albcans, C. cajan* and their F_1 reciprocal hybrids, 75 days after transplanting (DAT)

Salt extrusion, excretion and secretion

In most of the exclusion cases, it is not possible to decide whether the Na salts have been excluded passively or excreted actively. At the cell level, salt exclusion is possible only if the cell is impermeable to the salt or its ions. It is difficult to distinguish between such a passive exclusion and an active extrusion (or excretion) of the salt due to a salt excretion pump. It seems that both the mechanisms may be involved in salt resistant plants. Hansen *et al.* (1976) reported that salt marsh grass actively excretes excess Na^+, K^+ and Cl^- in a dirunal rhythum. Some halophytes avoid an excessive accumulation of ions in some of their tissues by mass excretion. Salt accumulates to excess in some of their leaves which are then shed. Usually, the older leaves are shed and replaced by new, low salt leaves (Ayoub, 1975).

Some halophytes have specialized structures on their leaf and stem surfaces that serve to eliminate salts from their tissues. These include salt glands, salt bladders and salt hairs (Gorham, 1992). Salt glands and salt bladders are found in at least 10 dicotyledonous plant families and one monocotyledonous family. The grass family is particularly rich in species (33 genera) containing salt glands and many members of the chenopodiaceae have salt bladders (e.g. 200 species of *Atriplex* possess salt bladders). Both salt glands and bladders are derived from epidermal cells, but they differ in structure and function. The glands serve to concentrate mostly Na^+, K^+, Ca^{2+} and Cl^- to values 50 times those in the cytoplasm. The first known ion pump in salt glands is the Cl^- pump in *Limonicum vlugare* (Sachar-Hill and Hill, 1970).

Salt bladders are modified leaf hairs that can either concentrate salt at more than 60 x the concentrations found in the leaf, or serve simply as holding tanks for salt at the same concentration as that of the leaf (an extended vacuole). Eventually, bladder cells die and rupture depositing salt on the leaf surface. This is in contrast with salt glands, which continually excrete salt onto the leaf surface. Similar salinity tolerance can occur in halophytes with or without salt glands. Thus, salt glands are not imperative for salt tolerance.

In some highly adapted halophytes, the extrusion mechanism is localized in salt glands which consist of both collecting and excretion cells. Attempts have been made to locate extrusion pumps in these glands. It has been reported that the concentrations of salt secreting species may reach as high as 0.5% NaCl (Schloander *et al.,* 1962). Several Atriplex species possess epidermal bladders with high concentration of ions. Chloride was excreted from the solution on the lamina to the bladders against a concentration gradient. The vacuole of the bladder was highly electronegative with respect to the bathing solution, suggesting an active uptake into the bladders.

Large variation in the salt glands has been reported in the literature. Salt glands of *Tamarix* develop from a single protoderm cell and consist of six secretory cells with granular cytoplasm, large nuclei and two inner, vacuolate collecting cells. The vesiculate hair or bladders of *Atriplex* have a large, highly vacuolate bladder cell attached to a stalk composed of one or more cells which in turn is attached to an epidermal cell.

The first type includes glands of *Tamarix, Frankenia, Limonium* and many mangroves. The characteristic feature of these glands is that they are almost entirely enclosed by a cuticular layer except for small regions, the transfusion zones, between the innermost gland cells and the sub-basal cells. The cuticle is frequently separated from the walls of the secretory cells along the outer surface of the gland creating a large, generally electron transparent cavity or collecting compartment between the secretory cells and cuticle. Characteristically numerous plasmodesmata occur in the transfusion zone wall between the inner secretory cells and the collecting cells. The secretory cells of the gland are also interconnected by plasmodesmata.

The second type of glands are the trichomas of *Atriplex* and the glands are covered by a cuticular layer with the lateral walls of the stalk cell as completely cuticularized and/or suberized. The mature bladder cell has a large central vacuole with a peripheral cytoplasm containing mitochondria, small vesicles, dictyosoms and chloroplasts. Two cells gland of *Spartina* represent the third structural type of glands. The external surface of the gland was covered by a thick cuticle, but the walls of the underlying basal were not cuticularized.

The secretion from some salt glands can be quite extensive, varying from 70 mg of fluid to 100 mg in 24 h or even higher. The secretion from some salt glands is composed of different ions and the percentage of ions varies with the composition of the environment of the roots. The excretion process is highly selective for Na^+ in *Aleuropus litoralis*, K^+ and Ca^{2+} being retained in the leaves to a greater degree than Na^+.

The salt glands function in regulating the salt content of the leaves by secreting excess salt from the tissue. There are many evidences about the role of ATPase in transport of ions in plants. These ion-stimulated ATPase (e.g. sodium pumps), have a key function in halophytes and salt resistance. Several glycophytes also possess an active extrusion system e.g. maize and sorghum. Transport of Na^+ and K^+ by an (Na^+, K^+) activated ATPase has been postulated as a basic mechanism of salt resistance in glycophytes as well as in halophytes. The moderately salt resistant barley appears to possess an outwardly directed Na^+ pump, for the Na^+ efflux which is inhibited by ouabain and stimulated by ATP and pi. The location of the pump was postulated to be at or near the endodermis. Even a more salt resistant wild species of the glycophytic tomato possesses a gland like extrusion system for its guttation fluid has a higher conductivity than that of the cultivated tomato (Tal and Gavish, 1973).

Salt dilution : Succulence

When Na^+ and Cl^- are not excluded, retranslocated, or extruded they can be diluted enough in leaf msophyll to keep cytoplasmic salinity below a toxic level. Succeulence, a high proportion of water to dry weight (or a high ratio of vacuole/cytoplasm), is thought to be an important aspect of halotolerant plants because of the number of halophytes that express some form of succulence (Gorham, 1992). In some extreme halophytes (such as *Sueda maritima*) expansion and growth of succulent tissues requires relatively high salinity, and low salinity limits plant growth (Yeo and Flowers, 1986). Cytoplasmic toxicity rapidly develops when the central vacuole becomes saturated with salt (Cheeseman, 1988). Therefore, plants that have very large vacuoles (high succulence) can store more salt before the vacuole is saturated and toxicity ensues. Some halophytes exhibit a diurnal variation in succulence as measured by leaf thickness, in which succulence is greatest at night and lower during the day (Rozema, 1991). In extreme succulent species, such as *Suaeda maritima* the rate of salt uptake must be very high to support expansion growth and cell division, and under normal conditions growth is limited by ion availability in the leaves (Yeo and Flowers, 1986). Hence, problems of ion excess would not normally occur in extremely succulent leaves or stems.

In many mangrove species, the bulk ion concentration of tissue is maintained at relatively constant levels due to changes in tissue succulence. It may be particularly important for the salt tolerance of developing and expanding tissues.

Succulence is not required for high salinity tolerance in many halophytes, and many nonhalophytes express succulence. Thus, succulence is not highly associated with salinity tolerance in plants.

Rhizophora mucronata does not excrete salt, yet its growing leaves retain a constant concentration though they continue to receive ions. Due to the rapid growth, water is also absorbed in sufficient amounts to prevent an increase in concentration. This dilution of the cell sap due to growth is also found in some moderately resistant non-halophytes. Barley rapidly increases its NaCl concentration during early tillering but shows little further change until grain formation due to rapid growth. Even the varietal salt resistance of twenty two accessions of *Glycine javanica* seemed to be directly related to the growth rate (Gates *et al.,* 1966). This is also true of gram and wheat. The slow growing varieties suffered more than the rapidly growing varieties from toxic concentrations of NaCl during early seedling growth. Similar observations have been made in rice genotypes where the vigour of the tall varieties may enable them to tolerate growth reduction and this alone, rather than any physiological adaptation to salinity may predominate (Yeo and Flowers, 1986; Munns, 1990). It has been suggested that growth has a dilution effect, the tall land race Pokkali had the same net transport of Na^+ through its roots as the semi dwarf IR 22, but had much lower shoot Na^+ concentrations. Garg *et al.* (1990, 1993) observed that wheat and mustard plants grown under improved soil fertility conditions had lower concentrations of sodium and chloride in their leaves than the low fertility plants although the total uptake was higher in the former case. The large growth in case of fertilized plants probably had a dilution effect which resulted in better salt tolerance of fertilized plants besides other factors.

Salt Reabsorption and Retranslocation

If Na^+ and Cl^- ions are not strongly discriminated against at the root membrane, salinity of the xylem stream will increase under saline conditions. Some of the ions in the transpiration stream could be selectively accumulated by parenchyma cells in the xylem and retranslocated back to the roots via the phloem. Particularly in legumes, some parenchyma cells differentiate into transfer cells and reabsorb Na^+ from the transpiration stream (Jeschke and Wolf, 1993). Perhaps the best case of reabsorption of Na^+ from the xylem stream is that of *Phragmites communis*, in which the phloem cells in the shoot base actively reabsorb Na^+ from the xylem stream (Matsushita and Matoh, 1992).

Reabsorption and retranslocation of Na^+ is found only in species that have minimum salt tolerance. In the most salt-tolerant species, most of the salt they are exposed to is accumulated and a large portion is transported to the shoots.

Continued retranslocation of Na^+ back to the roots is not a good strategy for tolerating salinity because it results in an excessive buildup of salt in roots, where the main salt problem already exists (Jeschke and Wolf, 1993).

SALT TOLERANCE MECHANISMS

Plants develop various physiological and biochemical mechanisms in order to survive in soils with high salt concentration. Principle mechanisms include, (1) ion homeostasis and compartmentalization, (2) ion transport and uptake, (3) biosynthesis of osmoprotectants and compatible solutes, (4) activation of antioxidant enzyme and synthesis of antioxidant compounds (5) synthesis of polyamines, (6) generation of nitric oxide (NO), and (7) hormone modulation. Research advances elucidating these mechanisms are discussed below.

Ion Homeostasis and Salt Tolerance

Maintaining ion homeostasis by ion uptake and compartmentalization is not only crucial for normal plant growth but is also an essential process for growth during salt stress (Hasegawa, 2013).

Irrespective of their nature, both glycophytes and halophytes cannot tolerate high salt concentration in their cytoplasm. Hence, the excess salt is either transported to the vacuole or sequestered in older tissues which eventually are sacrificed, thereby protecting the plant from salinity stress (Zhu, 2003).

Major form of salt present in the soil is NaCl, so the main focus of research is the study about the transport mechanism of Na^+ ion and its compartmentalization. The Na^+ ion that enters the cytoplasm is then transported to the vacuole via Na^+/H^+ antiporter. Two types of H^+ pumps are present in the vacuolar membrane: vacuolar type H^+-ATPase (V-ATPase) and the vacuolar pyrophosphatase (V-PPase) (Dietz *et al.,* 2001; Oliveira *et al.,* 2001). Of these, V-ATPase is the most dominant H^+ pump present within the plant cell. During nonstress conditions it plays an important role in maintaining solute homeostasis, energizing secondary transport and facilitating vesicle fusion. Under stressed condition the survivability of the plant depends upon the activity of V-ATPase.

Increasing evidence demonstrates the roles of a Salt Overly Sensitive (SOS) stress signalling pathway in ion homeostasis and salt tolerance (Hasegawa *et al.,* 2000). The SOS signalling pathway consists of three major proteins, SOS1, SOS2 and SOS3. *SOS1* which encodes a plasma membrane Na^+/H^+ antiporter, is essential in regulating Na^+ efflux at cellular level. It also facilitates long distance transport of Na^+ from root to shoot. Overexpression of this protein confers salt tolerance in plants (Shi *et al.,* 2002). *SOS2* gene, which encodes a serine/threonine kinase, is activated by salt stress elicited Ca^+ signals. This protein consists of a well-developed N-terminal catalytic domain and a C-terminal regulatory domain (Liu *et al.,* 2000). The third type of protein involved in the SOS stress signalling pathway is the SOS3 protein which is a myristoylated Ca^+ binding protein and contains a myristoylation site at its N-terminus. This site plays an essential role in conferring salt tolerance (Ishitani *et al.,* 2000).

Many plants have developed an efficient method to keep the ion concentration in the cytoplasm in a low level. Membranes along with their associated components play an integral role in maintaining ion concentration within the cytosol during the period of stress by regulating ion uptake and transport (Sairam and Tyagi, 2004). The transport phenomenon is carried out by different carrier proteins, channel proteins, antiporters and symporters. Maintaining cellular Na^+/K^+ homeostasis is pivotal for plant survival in saline environments.

A large number of genes and proteins, such as HKT and NHX, encoding K^+ transporters and channels have been identified and cloned in various plant species. During salt stress expression of some low abundance transcripts is enhanced which are found to be involved in K^+ uptake.

Osmoregulation/Osmotic Adjustment

The process of maintenance of cell turgor by a sufficient increase in cell solutes to compensate for the external osmotic stress is called osmoregulation, osmotic compensation or osmotic adjustment (Morgan, 1977). Plants tolerant to salinity must be capable of adjusting to osmotic stress by osmoregulation. Plants can often adjust to decreased potential without losing turgor except when salinisation is abrupt. Rates of osmotic adjustment of about 1 bar/day have been observed (Bernstein, 1963) but both the rate and extent depend upon plant species.

The relative sensitivity of crops to osmotic stress may vary with external salt concentrations. This may indicate that certain crops may benefit from selection pressures, which improve their capacity to adjust osmotically or maintain more favourable water relations under salt stress (Tal and Gardi, 1976; Shannon *et al.,* 1987). Generally poor osmotic adjustment leads to turgor loss and stomatal closure, which is soon followed by reduced gas exchange and photosynthesis. Turgor loss, in turn, can also have detrimental effects on cell division and elongation.

Osmotic adjustment involves both uptake and accumulation of ions and synthesis of organic solutes. Although these two mechanisms have been used as the basis for classifying halophytes and glycophytes, they often operate together. The relationship between salt tolerance and osmotic adjustment is, however, not clear and various workers have shown the way in which the osmotic adjustment is achieved which will determine at least partially the salt tolerance of a species or genotype (Richardson and McCree, 1985). Marschner (1986) indicated that the osmotic adjustment based on the synthesis of organic solutes requires an important energy expenditure, while Flowers *et al.* (1977) argued that the accumulation of ions in the vacuole provides energetically cheap solutes for osmotic adjustment. Since the different contribution of inorganic and organic solutes to osmoregulation has important implications for energy balance (Wyn Jones, 1981), it is necessary to take into account not only the degree of osmotic adjustment but also the type of solutes contributing to the osmotic adjustment in order to achieve a better understanding of the physiological differences among plants with different salt tolerance.

Differences in Ionic Concentration in Halophytes and Glycophytes

Plants which tolerate high external concentration of salt invariability do so with high internal concentrations particularly in the leaves (Table 9.10). The glycophytes can thus be distinguished as plants of non-saline environments, normally having high K/Na ratio in their leaves, from halophytes, the plants of saline environments, in which K/Na ratio tends to be lower and the overall ion concentration higher. Here, it is important to distinguish between the situations in which a plant is able to tolerate high external salt concentration with high salt and that in which a high concentration of salts in the plants is indicative of the death or imminent death of the plant.

Table 9.10. Sodium and potassium levels in plant foliage

Plant types	Sodium	Potassium	Reference
Glycophytes in 5 mM K^+ (g kg^{-1} dry weight) 1 mM Na^+	0.2-20	15-50	Hewitt and Smith (1975)
Halophytes in 5-8mM K^+ and 295-340 mM Na^+ (g kg^{-1} dry weight)	25-154	10-33	Flowers (1975)

Plant Requirement for Sodium and Potassium

Potassium has been shown to be an essential macronutrient and appears to be required for all types of plants. The K content of plant is relatively high and for glycophytes ranges from 1.5 to 5% of the dry weight. In general, root K contents are half those of the shoot, so that the root contains only some 16% of total plant K (Pitman, 1975). Na is not an essential macronutrient, even for the extreme halophytes although there is evidence that it is necessary in low concentrations for the growth of C4 and CAM plants (Brownell, 1979). Many plant species show positive growth responses to Na but have not been shown to have an absolute requirement for this element. Halophytes respond positively to concentrations of salts of even 500 mM and beyond, although the specificity of this growth response is very variable between different species (Flowers *et al.,* 1977). However, for glycophytes such high Na concentrations are lethal. Moreover, Na, although present in the phloem of higher plant, is found to be much less mobile than is K (Raven, 1977). Although the various roles of K in plant metabolism have been fairly established, the role of Na and its substitution for K is less clear.

Selectivity of Ion Transport

Plants have developed a remarkable capability to acquire essential nutrients from a saline solution in which the concentration of a nonessential (or toxic) ions is disproportionately greater than that of essential elements. For example, Na concentrations in saline soil solutions may exceed that of K by many folds and yet, the Na/K ratio in plants growing on these soils may be near one or even less. This high selectivity for K uptake is present in a wide range of plants. Fig. 9.7 shows the rate of absorption of potassium by barley roots as a function of the increasing Na concentrations where potassium being present in the experimental solution at 50 mM. At the highest sodium concentration, the rate of potassium remains fairly constant and equal to the maximal rate attainable by the low concentration mechanism (Rains and Epstein, 1967). Similar type of preferential potassium absorption has been found in case of halophytes such as *Avicennia marina* and *Suaeda monoica* and several glycophytes.

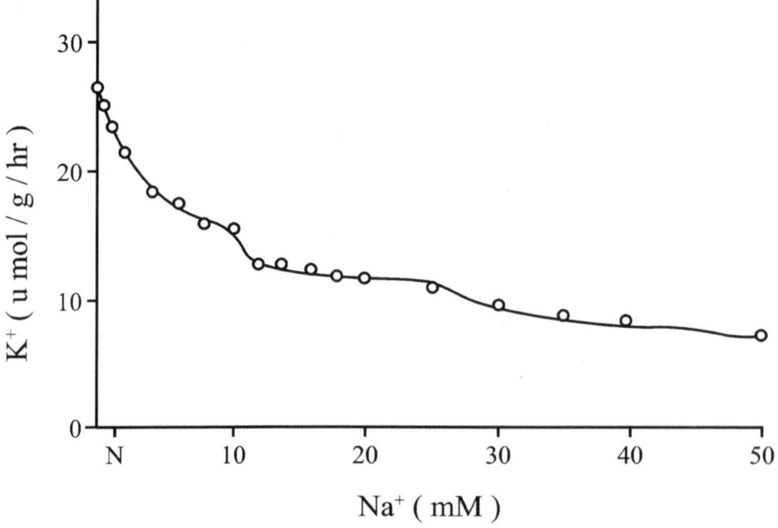

Fig. 9.7. The rate of potassium absorption by excised barley roots as a function of the external concentration of sodium. The potassium concentration was 50 mM

Plants that limit uptake of toxic ions and maintain normal ranges of nutrient ion could be more salt tolerant than those that do not restrict ion accumulation and lose nutrient balance. Selective ion uptake mechanisms capable of discrimination between chemically similar ions such as Na^+ and K^+ could have adaptive value. The mechanisms responsible for ion discrimination probably are located in the membranes of tissues and various organelles throughout the plant (Bliss *et al.,* 1984; Kuiper, 1968). Breeding for efficient nutrient uptake or low ion accumulation under salt stress may be among the simplest ways to improve salt tolerance in sensitive varieties of some species. This also may be accomplished by finding tolerance to the toxicity of a specific ion associated with salt stress.

Munns *et al.* (1988) concluded that high salt concentrations in the phloem of the salt-sensitive *Lupinus albus* are not directly related to either growth reduction or leaf injury but are more important, a symptom of disrupted regulation of ion transport properties in the root. Among plant species, mangroves undoubtedly have the most efficient system of restricting salt uptake through the development of a passive root membrane filtration system.

Localisation of Potassium and Sodium in Plants

In higher plants K is present in various subcellular compartments (Table 9.11). Analyses of mesophyll cells of *Zea mays* demonstrated that the concentration of K in the nucleus and the chloroplasts was rather similar and somewhat higher than in the remainder of the cytoplasm. Cytoplasmic K concentrations appeared to be some 2.7 times those in the vacuole. Like chloroplasts, nuclei contain substantial concentrations of K, but there is little information on the other organellis.

Table 9.11. Potassium distribution and content of various subcellular compartment of plant cells

Plant	Units	Cytoplasm	Nucleus	Chloroplasts	Vacuole	Reference
Barley	mM	100	–	–	20	Jeschke and Steller (1976)
Low salt Roots	mM	71-119	–	–	9-12	Pitman *et al.*, (1981)
Maize (Mesophyll cells)	Count rate	1335	1765	1620	500	Pallaghy (1973)

Na in contrast to the K is not a major component of most plant cells. In glycophytes, vacuolar concentration ranged from 24 to 50 mM and in halophytes the cytoplasmic to vacuolar contents in the leaf cells, where an average of 80% of the ions that are accumulated range from 500 to 700 mM and the evidence to date suggests that the bulk of the Na ions is sequestered in extra cytoplasmic compartments. Nearly all data showing that sodium concentrations are lower in the cytoplasm than the vacuolar have been obtained with roots exposed to low external sodium (1-10 mM). However, there are no estimates of Na^+ or Cl^- in the cytoplasm of vascular non-halophytes at salinities which seriously reduce their growth.

OSMOREGULATION WITH INORGANIC SOLUTES

In most plants, especially the obligate halophytes osmoregulation in response to severe salt stress is achieved mainly by accumulation of inorganic ions from the external medium, particularly in the leaf cells (Flowers *et al.*, 1977; Wyn Jones, 1981). The accumulation appears to be primarily in the vacuoles, the cytoplasmic concentration being 1/2 to 1/3 of the vacuolar. In *Suaeda maritima* grown at 340 mM NaCl, Na^+ was about 150 mM in the cytoplasm and 600 mM in the vacuoles of leaf slices suspended in 600 mM NaCl.

Restriction of ions into roots or shoots is one of the most frequently reported differences between salt-tolerant and sensitive varieties. It is well known that halophytes take up substantially high concentrations of ions as an adaptation to saline environments (Flowers *et al.*, 1977); however, some can sequester toxic ions not only in vacuoles but also in specialized organs such as salt glands and bladders (Levitt 1980; Schirmer and Breckle, 1982). The accumulation of salt in the plant or its excretion onto leaf surfaces is believed to reduce the requirements for increased wall extensibility that might otherwise be required to maintain positive growth and turgor at low soil water potentials. The wild tomato species (*Lycopersicon cheesmanii*) is considered to be more salt tolerant than the cultivated species due to its capacity to accumulate ions (Rush and Epstein, 1981), and the salt tolerant "Edkawy" tomato also accumulates higher concentrations of Na^+ in leaf tissues than does more sensitive cultivars of *L. esculentum* (Hashim *et al.*, 1986). As with salt restriction, salt accumulation within tissues is believed to be well regulated and generally sequestered away from cytosolic compartments containing the salt-sensitive metabolic machinery of the cell. In both glycophytes and halophytes, salt may accumulate preferentially in vacuoles, interstitial compartments stems or older leaves. The physical and genetic factors that influence ion compartmentation and distribution within plants are mostly unknown.

The salt concentration inside the cell wall nevertheless, may be slightly lower than that outside, due to the presence of organic solutes in the cell. The osmotic value of the leaves of salt bush (*Atriplex*), for instance varied directly with that of the solution in which the plants were growing, showing a constant absolute value of about 12 bar greater than that of the solutions throughout the range. As a result, their osmotic pressures ranged from thirty times that of the nutrient solutions for control plants, to only two times for plants growing in solutions having 600 mM NaCl. This relationship was also true in nature. In soils with chloride contents varying from 19 to 9250 ppm, the leaves of the respective *Atriplex* plants had chloride contents ranging only from 9.6 to 13.6% of the dry weight (Beadle *et al.,* 1957).

The relative contribution of various ions to osmotic adjustment depends upon the mechanisms regulating membrane transport : (a) membrane permeability, (b) transport kinetics, energetics and selectivity and (c) negative feed back controls. Rates of uptake differ among ions, thereby affecting ionic balance. Chloride contributes much more to osmotic adjustment than sulphate because it is much more readily absorbed. When salinity consists predominantly of monovalent cations and divalent anions (eg. Na_2SO_4) cation uptake rates exceed those of anion. Ionic balance is achieved by the synthesis and accumulation of organic acids. The imbalance in inorganic cations and anions is further accentuated by the metabolic assimilation of nitrate and sulphate.

Among the halophytes, one class euhalophytes adjusts to saline conditions by accumulating large amounts of NaCl (Waisel, 1972). These plants have adapted mechanisms to withstand the high concentrations of salt either by increasing succulence, by compartmentation, by secreting salts from the tissues or various combinations of these mechanisms. However, in several other halophytes osmoregulation is due to active K^+ uptake. In general the Gramineae, Luncaceae and Cyperaceae take up less Na^+ than any other group of higher plants, becasue they prefer K^+. A more salt tolerant tomato, however, showed opposite changes including accumulation of Na^+ (Rush and Epstein, 1976) while cultivated species *L. esculentum* could not tolerate higher Na^+ concentration and died (Fig. 9.8). However, osmoregulation with inorganic solutes is not uncommon in glycophytes. In case of tomato one variety exhibited higher salt tolerance due to accumulation of Na^+ and Cl^- in their leaves (Alfocea *et al.,* 1993), while the other variety excluded these ions. Similarly, in chickpea lines both types of salt resistance mechanisms were operative (Ashraf and Waheed, 1993). Some glycophytes may accumulate Na in the shoots (Na accumulators) and like euhalophytes tolerate higher concentrations of NaCl than those which exclude Na^+ (Na excluders). In general, roots of all plants accumulate NaCl accounting for a significant part of osmotic adjustment of this tissue even for those species, like *Phaseolus* sppl. in which Na^+ is excluded from the tops.

Several reports in recent years suggest that the characteristics of K^+ and Na^+ transports are determinant of the NaCl tolerance. In suspension cells of *Brassica napus,* increased tolerance to NaCl arises by alteration of the K^+ uptake system (Lefebvre, 1989) and tobacco cell cultures show enhanced K^+ uptake capacity when adapted to NaCl (Watad *et al.,* 1991). It is also clearly established that salt tolerant genotypes of wheat translocate less Na^+ from roots to shoots than salt sensitive genotypes (Davis, 1984; Schachtman *et al.,* 1989). It has been found that low capacity of discrimination of the K^+ uptake system of bean plants in presence of Na^+ can account for the NaCl sensitivity of bean plants (Benlloch *et al.,* 1994).

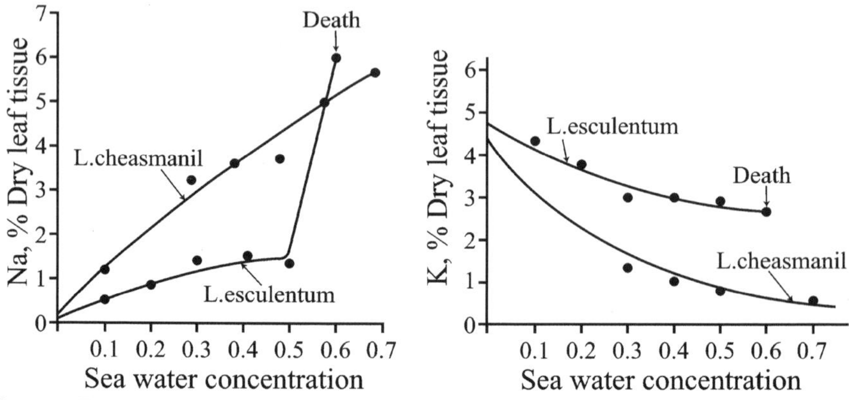

Fig. 9.8. Percentage of sodium and potassium in dry leaf tissue as a function of the salinization of the nutrient solution with seawater salt mix.

OSMOREGULATION WITH ORGANIC SOLUTES (COMPATIBLE SOLUTES)

Compatible solutes, also known as compatible osmolyes, are a group of chemically diverse organic compounds that are uncharged, polar, and soluble in nature and do not interfere with the cellular metabolism even at high concentration. They mainly include proline (Houque *et al.,* 2007), glycine betaine (Wang and Nil, 2000), sugar (Kerepesi and Galiba, 2000), and polyols (Ashraf and Foolad, 2007). Organic osmolytes are synthesised and accumulated in varying amounts amongst different plant species. For example, quaternary ammonium compound beta alanine betaine's accumulation is restricted among few members of Plumbaginaceae, whereas accumulation of amino acid proline occurs in taxonomically diverse sets of plants (Saxena *et al.,* 2013). The concentration of compatible solutes within the cell is maintained either by irreversible synthesis of the compounds or by a combination of synthesis and degradation. The biochemical pathways and genes involved in these processes have been thoroughly studied. As their accumulation is proportional to the external osmolarity, the major functions of these osmolytes are to protect the structure and to maintain osmotic balance within the cell via continuous water influx (Hasegawa *et al.,* 2000).

One response which is universal to changes in external osmotic potential is the accumulation of metabolites that act as "compatible" solutes i.e. they do not inhibit normal metabolic reactions (Brown *et al.,* 1972; Yancy *et al.,* 1982; Ford, 1984). Compatible solutes have also been defined by Galinski (1983) as organic osmolytes responsible for osmotic balance and at the same time compatible with the cells metabolism. Furthermore, such compatibility is a general property of protein/solute interactions, irrespective of species and nature of stress. Frequently observed metabolites are sugars (mainly sucrose and fructose), sugar alcohols (glycerol, methylated inositols), and complex sugars (trehalose, raffinose, fructans). In addition, ions (K^+) or charged metabolites (glycine betaine, dimethyl sulfonium propionate (DMSP), proline and ectoine are encountered. The accumulation of these osmolytes helps in `osmotic adjustment' by which the internal osmotic potential is lowered and may then contribute to tolerance (McCue and Hanson, 1990; Delauney *et al.,* 1993; Louis and Galinski, 1997). Compatible solutes are typically hydrophillic, so they replace water at the surface of proteins, protein complexes, or membranes, thus acting as osmoprotectants and non enzymatically as low-molecular weight chaperones. Compatible solutes at high concentrations can reduce the inhibitory effects of ions on enzyme activity (Brown, 1990; Solomon *et al.,* 1994) to increase thermal stability of enzymes (Galinski, 1993) and to prevent dissociation of enzyme complexes, e.g. the oxygen evolving complex of photosystem II (Papageorgiou and Murata, 1995).

In vitro studies have shown that enzymes and protein synthesizing systems in most of the halophytes are as sensitive or even more sensitive to high salt concentrations as those of glycophytes (Flowers *et al.,* 1977; Greenway and Munns, 1980). This implies that the high content of the salts in the halophytes must be chiefly in the vacuoles and that equivalent osmotic potential in the cytoplasm must be to a large extent be due to non-toxic or compatible organic solutes. The term ``compatible solutes" was introduced by Brown and his coworkers to describe non-inhibitory substances accumulated in the cytoplasm of cells at low external water potentials. Compatibility requires that the organic solutes are non-toxic or less toxic to metabolic functions than the inorganic ions. Solutes which have been recognised to accumulate at high salinity in many species include glycinebetaine (Sotrey and Wyn Jones, 1977; Wyn Jones and Storey,

1981), proline (Stewart and Lee, 1974), sucrose (Gauch and Eaton, 1942; Bernstein and Ayers, 1953; Rathert, 1984) and organic acids such as malate and oxalate (Waisel, 1972; Osmond, 1976). In higher plants as in the salt tolerant organisms other than the extreme halophytic bacteria the compatible solutes fall chemically into these classes: amino acids and the N-methyl-substituted derivatives referred to by the name betaines, sugar alcohols or polyols and organic acids (Table 9.12).

Table 9.12. Taxonomic distribution of possible compatible solutes (Gorham *et al.*, 1985)

Solute	Structure	Distribution
D-sorbitol	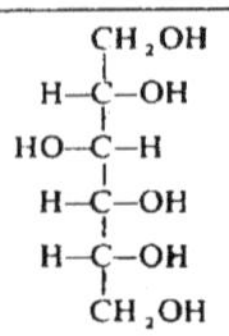 CH_2OH–(H–C–OH)–(HO–C–H)–(H–C–OH)–(H–C–OH)–CH_2OH	Plantaginaceae Rosaceae
D-mannitol	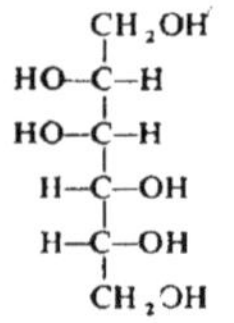 CH_2OH–(HO–C–H)–(HO–C–H)–(H–C–OH)–(H–C–OH)–CH_2OH	Combretaceae Myrsinaceae Rubiaceae
D-pinitol	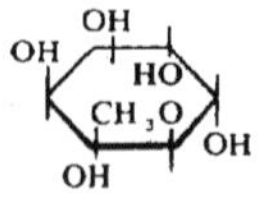 [ring structure: OH, OH, HO, CH_3O, OH, OH]	Leguminoseae Rhizophoraceae Caryophyllaceae
L-quebrachitol	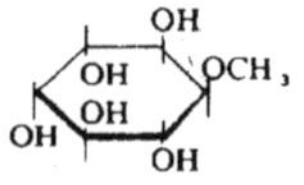 [ring structure: OH, OCH_3, OH, OH, OH, OH]	Euphorbiaceae
Glycinebetaine	$(CH_3)_3N^+CH_2COO^-$	Chenopodiaceae Amaranthaceae Asteraceae Solanaceae Gramineae Avicenniaceae
B-alaninebetaine	$(CH_3)_3N^+CH_2CH_2COO^-$	Plumbaginaceae
Proline	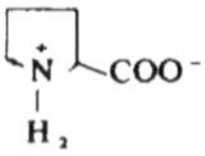 [pyrrolidine ring: $\overset{+}{N}H_2$, COO^-]	Juncaginaceae Asteraceae Gramineae
Prolinebetaine (Stachydrine)	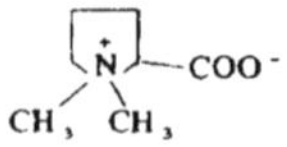 [pyrrolidine ring: $\overset{+}{N}(CH_3)_2$, COO^-]	Labiateae Capparidaceae Leguminoseae
3-dimethylsulfonio -propionate	$(CH_3)_2S^+CH_2CH_2COO^-$	Asteraceae Gramineae

Solutes that are compatible with the cytoplasm are accumulated by all groups of organisms. Plant accumulates these solutes in response to low water availability, low temperature, and salinity (Bohnert *et al.,* 1995). Both glycophytes and halophytes have been found to accumulate a number of different solutes in cytoplasm when they experience saline conditions. Accumulation of cytoplasmically compatible solutes cannot be the only method of osmotic adjustment for plants in long-term hypersaline conditions because much of the total plant photosynthate would be required for osmotic adjustment (Wyn-Jones *et al.,* 1984). Therefore, both inorganic ions (Na^+, K^+, Ca^{2+}) and organic solutes are involved in osmotic adjustment of halophytes. Accumulation of organic solutes in hypersaline conditions occurs in response to the osmotic effects of the soil solution and is not a response to specific ions. Also in any one species, several different organic solutes may accumulate in responses to salinity. For example, reducing sugars, amino acids, proline, and quaternary ammonium compounds all accumulate when *Halimione protulacoides* is exposed to hypersaline conditions. In general, under saline conditions, low soil osmotic potential is associated with osmotic adjustment in tissues by an increase in organic solute concentration in glycophytes, and an increase in both inorganic ions and organic solutes in halophytes.

Greenway and Munns (1980) suggested that high concentrations of organic solutes in the cytoplasm can have the following roles : (a) a contribution to the osmotic balance when *electrolytes* are lower in the cytoplasm than in the vacuole, (b) a protective effect on enzymes in the presence of high electrolytes in the cytoplasm. Protective effects of a large range of solutes have been established under extreme conditions. Sucrose protected isolated chloroplasts against injury during desiccation and proline increased the solubility of proteins. Glycinebetaine (500 mM) alleviated the inhibitory effects of 200 mM NaCl on malic enzyme isolated from barley (Pollard and Wyn Jones, 1979).

Likewise proline has been shown to prevent or diminish the glutamine synthetase precipitating ability of PEG in barley plants (Paleg *et al.,* 1985). However, evidence has yet to be produced in each case to show that accumulation is of adaptive value and not a reflection of impaired metabolism.

Plants engineered to synthesize and moderately accumulate a number of osmolytes showed marginally improved performance under abiotic stress conditions. The effects seen with modest increases in mannitol, fructans, trehalose and with strong increases in proline amount indicate that the purely osmotic contribution of these metabolites to stress tolerance may not describe their function completely. Transgenic plants have been generated to probe the effect of reactive oxygen species (ROS) scavenging on salinity stress tolerance based on observations of gene expression changes in stressed plants. ROS scavenging as an important component of abiotic stress responses is documented by mutant analysis.

The high concentrations of solutes needed for a protective effect on enzymes mean that this role would be experimentally indistinguishable from an osmotic role. In some cases the accumulation at high salinity is very large e.g. proline increased by 300 μ mol g^{-1} FW in shoots of *Triglochin maritima* (Stewart and Lee, 1974) and sucrose by 90 μ mol g^{-1} FW in carrot roots (Bernstein and Ayers, 1953). Similarly many fold increases in free proline levels with increasing salinity stress have been reported in a number of arid zone crops (Lahiri *et al.,* 1987; Garg *et al.,* 1990, 1993). However, one can only speculate whether the primary role is one of storage (of reduced carbon or nitrogen) or in the osmotic balance of the cell as a whole. In case of glycinebetaine its compatibility with *in vitro* protein synthesis, polysome stability, chloroplast CO_2 fixation and coupled mitochondrial oxidative phosphorylation has been established (Wyn Jones and Storey, 1981). Of the nitrogenous compounds glycinebetaine and proline have been most widely studied in higher plants. Therefore, their adaptive roles as solutes localized in cytoplasm are discussed below.

GLYCINE BETAINE

Glycine betaine is an amphoteric quaternary ammonium compound ubiquitously found in microorganisms, higher plants and animals, and is electrically neutral over a wide range of pH. It is highly soluble in water but also contains nonpolar moiety constituting 3-methyl groups. Because of its unique structural features it interacts both with hydrophobic and hydrophilic domains of the macromolecules, such as enzymes and protein complexes. Glycine betaine is a non-toxic cellular osmolyte that raises the osmolarity of the cell during stress period; thus it plays an important function in stress mitigation. Glycine betaine also protects the cell by osmotic adjustment (Gadallah, 1999), stabilizes proteins, and protects the photosynthetic apparatus from stress damages and reduction of ROS (Ashraf and Foolad, 2007; Saxena *et al.,* 2013).

Analysis of vegetative tissues from a large number of species have shown glycinebetaine to be present in substantial concentrations (10 to 100 m moles g^{-1} FW) in many, but by no means all halophytes. It is universally found in members of the chenopodiaceae, a family typified by a halophytic distribution, but is rarely detectable in members of other families except Gramineae. Both glycinebetaine and alanineb-etaine are normally found in higher concentrations in the shoot than in root tissues of mature plants (Storey and Wyn Jones, 1977). Most of the salt tolerant species increase in glycinebetaine when grown at high NaCl. However, little or no betaine has been found in salt sensitive species. Glycinebetaine is also concentrated in partially dehydrated tissues of unstressed cereals, seed embryo, aleurone cells, anthers and paleas as well as in pollen (Gorham *et al.,* 1985).

Glycine-betaine can protect enzyme from heat damage (Nash, 1982), which indicates its capability to proect enzymes from stress damage (Gorham, 1992). Increasing the concentration of glycine-betaine in cyanobacteria stabilizes ribulose-1,5-bisphosphate carboxy-lase/oxygenase to inactivation by NaCl (Nomura *et al.,* 1998). In the Plumbaginaceae, various species accumulate glycine-betaine and other QAMs under salinity stress and the QAMs serve as an osmoprotectant (Hanson *et al.,* 1994). Because glycine-betaine accumulates in many members of the Poaceae but does not accumulate in rice or tobacco, there has been a lot of effort to engineer the production of glycine-betaine in plants that do not normally produce it.

Evidences for progressive accumulation of glycinebetaine in plants exposed to increasing external NaCl concentrations are available for a wide range of species (Storey and Wyn Jones, 1977, 1979; Wyn Jones, 1981). Recently, Varshney and Sangeeta (1992) reported accumulation of choline and betaine in chickpea varieties with increasing salinity due to different salts, such as NaCl, Na_2CO_3, $NaHCO_3$ and $CaCl_2$.

Glycinebetaine accumulation can be brought about by both salt and water stress, although the former appears to be more effective. However, in marked contrast to proline accumulation, glycinebetaine accumulation can not be regarded as a general response to water stress, as many species, for example tomato and pea do not respond in this manner.

Direct evidence for the location of glycinebetaine in the cytoplasm comes from histochemical studies of *Suaeda maritima* grown at high NaCl concentrations and this was detected in the cytoplasm and not in the vacuole. No staining could be detected in cells of plants grown without NaCl (Hall *et al.,* 1978). Furthermore, subcellular fractiona-tion of red beet roots indicated that betaine is lower in the vacuole than in the rest of the cell. Pollen is a tissue which is highly cytopl-asmic and which might be expected to have high concentration of compatible protective solutes. The pollen samples for both the grass *Leymus giganteus* and the chenopod *Spinacia olerecea* grown in nutrient solution have considerably higher glycine betaine contents than the leaves of these species (Table 9.13). Gorham *et al.* (1985) reported that glycinebetaine accumulation is particularly high in apical tissues during salt stress and such tissues are highly cytoplasmic in nature.

Table 9.13. Glycinebetaine contents of leaves and pollen (Gorham *et al.*, 1985)

Species	Glycinebetaine (mmol kg^{-1} Fresh wt.)
Leymus giganteus	
Leaf	30
Pollen	142
Spinacia oleracea	
Leaf	2
Pollen	85

Table 9.14. Content of glycinebetaine, choline and proline in leaves of spinach grown in the absence or presence of 200 mM NaCl (Robinson and Jones, 1986)

Parameter	Control	Salt present
Leaf osmotic potential (MPa)	– 0.91 ± 0.09	–2.09 ± 0.10
Water content (0 fresh wt)	93.0 ± 0.2	92.3 ±0.3
Fresh weight (mg cm^{-2})	28.7 ± 0.7	51.8 ± 0.9
Leaf chlorophyll (mg g fresh wt^{-1})	1.16 ± 0.04	0.79 ± 0.03
Leaf photosynthesis (μmol m^{-2} s^{-1})	28.3 ± 1.5	29.3 ± 1.5
Leaf photosynthesis (μmol $mgChl^{-1}$ h^{-1})	290 ± 12	289 ± 23
Glycinebetaine (μmol g fresh wt^{-1})	2.34 ± 0.13	13.58 ± 0.38
Choline (μmol g fresh wt^{-1})	0.92 ± 0.21	0.24 ± 0.01
Proline (μmol g fresh wt^{-1})	0.85 ± 0.16	0.83 ± 0.8

Robinson and Jones (1986) determined glycinebetaine in leaves and isolated chloroplasts of spinach by nuclear magnetic resonance spectroscopy. The leaf osmotic potential of salt stressed plants was -2.09 MPa compared to -0.91 MPa in non stressed controls (Table 9.14). This was accompanied by a six fold increase in glycinebetaine content in the leaf but the levels of choline and proline were not increased. In chloroplasts from salt stressed plants, glycinebetaine was 20 times the concentration in the leaf as a whole. This glycinebetaine concentration in chloroplasts was equivalent to an osmotic potential of the cell and 64% of the decrease in osmotic potential induced by salt stress (Table 9.15). The study demonstrated that glycinebetaine accumulates in chloroplasts to provide osmotic adjustment during salt stress and this provides support for the hypothesis that glycinebetaine is a compatible cytoplasmic solute which may be preferentially located in the cytoplasm of cells. Exogenous applications of glycinebetaine have been reported to ameliorate salt stress effects in maize and barley seedlings and to partially protect some enzymes against salt inhibition (Polland and Wyn Jones, 1979).

Table 9.15. Glycinebetaine levels in chloroplasts isolated from leaves of spinach grown in the absence or presence of 200 mM NaCl (Robinson and Jones, 1986)

Parameter	Control	Salt present
Chloroplasts intact (%)	66 ± 2	75 ± 3
Chloroplast volume (mm Chl^{-1})	25.4 ± 0.3	23.3 ± 1.0
Glycinebetaine in chloroplasts (mmol mg Chl^{-1})	0.67 ± 0.02	6.65 ± 0.54
Glycinebetaine in chloroplasts (mM)	26 ± 0.9	289 ± 32
Glycinebetaine in leaf (mmol mg Chl^{-1})	2.02 ± 0.06	17.19 ± 1.10
Glycinebetaine in leaf (mM in tissue water)	2.52 ± 0.14	14.57 ± 0.36
Ratio (Chloroplast concen. + leaf concen)	10.5 ± 0.8	19.8±2.0
Glycinebetaine in chloroplasts (%)	33 ± 1	39 ±2

PROLINE

Proline accumulation is a well-known measure adopted for alleviation of salinity stress (Matysik *et al.*, 2002; Saxena *et al.*, 2013). Intracellular proline which is accumulated during salinity stress not only provides tolerance towards stress but also serves as an organic nitrogen reserve during stress recovery. Proline is synthesised either from glutamate or ornithine. In osmotically stressed cell glutamate functions as the primary precursor. The biosynthetic pathway comprises two major enzymes, pyrroline carboxylic acid synthetase and pyrroline carboxylic acid reductase. Both these regulatory steps are used to over produce proline in plants (Sairam and Tyagi, 2004). It functions as an O_2 quencher thereby revealing its antioxidant capability. This was observed in a study carried out by Matysik *et al.* (2002). Ben Ahmed *et al.* (2010) observed that proline supplements enhanced salt tolerance in olive (*Olea europaea*) by amelioration of some antioxidative enzyme activities, photosynthetic activity, and plant growth and the preservation of a suitable plant water status under salinity conditions. It has been reported that proline improves salt tolerance in *Nicotiana tabacum* by increasing the activity of enzymes involved in antioxidant defense system (Hoque *et al.*, 2008).

The physiological significance of changes in nitrogen compounds, in particular, the accumulation of free proline ranges from the view that it is a symptom of injury (Hanson *et al.,* 1977) to the suggestion that it is of adaptive significance and plays role in resistance to drought (Singh *et al.,* 1973) and salinity. Positive roles suggested for proline include a source of carbon and/or nitrogen and reductant for growth in post stress period, an ammonia detoxification product, a dessication protectant and an osmoticum (Heuer, 1994; Chauhan *et al.,* 1995). According to Paleg *et al.* (1985) proline protects enzymes and membrane proteins against dehydration and conformational changes and thus maintains their activity. Proline has also been found to reduce the inhibition of malic dehydrogenase activity caused by 300 mM NaCl. Alia and Mohanty (1993) proposed that proline accumulation lowers the generation of free radicals and thus reduces the lipid peroxidation linked membrane deterioration under salt stress in *Brassica* seedlings.

Proline is accumulated by a number of different plants (e.g. *Arabidopsis,* alfalfa, bean, halophytes, tobacco, spinach, potato, soybean, barley, rice) in response to both water stress and salinity (McCue and Hanson, 1990, Delauney and Verma, 1993). Such accumulation is not unique to plants and has been observed in eubacteria, marine invertebrates, protozoa, and algae (Yoshiba *et al.,* 1997). Therefore, this trait most likely appeared early in the evolution of plants (Bohnert *et al.,* 1995).

Is the production of proline an adaptive response to osmotic stress, or is it a consequence of protein degradation during stress? In order to answer this question, it must be shown that increasing proline concentration by itself will increase osmotic protection. The amount of accumulation by proline and glycine-betaine is enough to account for all the osmotic adjustment of *Spartina alternifora* during high salinity (Cavalieri, 1983). A mutation in *E. coli* (proB74) results in the over production of proline and endows increased resistance to osmotic stress (Csonka *et al.,* 1988). Also, the over production of proline in tobacco leads to an increase in tolerance of osmotic stress (Kishor *et al.,* 1995). All of these studies suggest that the accumulation of proline in tissues confers some salinity tolerance.

Evidences suggest that adaptive role of proline is related to survival rather than to maintenance of growth. Nearly all salt sensitive and salt tolerant species contain substantial concentrations only when growth is severely reduced (Chu *et al.,* 1976; Lahiri *et al.,* 1987; Garg *et al.,* 1990, 1993). Furthermore, there is not evidence that proline accumulates more in salt tolerant than in salt sensitive species. In fact, the available comparisons between closely related genotypes indicate that the reverse is true (Table 9.16). Although some studies indicate more accumulation of free proline in salt tolerant than salt sensitive genotypes (Kumar, 1984) but there are many reports which show higher proline accumulation in the sensitive varieties under salt stress (Garg *et al.*, 1990, 1993). Poljakoff-Mayber *et al.* (1987) did not get a significant correlation between proline content and RWC in many naturally occurring vegetation.

Table 9.16. Accumulation of free proline in leaves of salt-tolerant (T) and salt sensitive (S) genotypes of certain crops under salt stress (10 dSm^{-1})

Crops	Genotypes	Free proline			References
		Control	Salt stressed	Per cent increase	
Wheat	Kharchia-65 (T)	0.62	0.84	35.5	Garg *et al.* (1990)
	HD-4502 (S)	0.84	1.62	113.2	
Clusterbean	Malosen (T)	0.96	1.09	13.5	Lahiri *et al.* (1996)
	FS-277 (S)	0.95	1.74	83.2	
Indian mustard	DIRA-337 (T)	2.56	3.06	19.5	Garg *et al.* (1999)
	RYS-80 (S)	2.56	4.20	67.2	

Accumulation of proline is widespread in plants subjected to low water potentials and not unique to plants growing in saline environments. Results from short term studies in which shoots, tiller or intact plants and seedlings were grown at different salinities, indicate that the amount of free proline is dependent upon the degree of osmotic stress (Stewart and Lee, 1974; Lahiri *et al.*, 1987). Most of the reports agree on the point that proline accumulation increases with increasing level of stress. Proline accumulation was suggested to start only after a threshold of total monovalent cations had been reached. It was found to be approximately 200 m mole g^{-1} FW for sorghum leaves (Weimberg *et al.*, 1982). It required 115 mM NaCl in the culture medium for wheat and 213 mM NaCl for barley as threshold concentration for proline accumulation. Lahiri *et al.* (1996) found a highly significant and positive correlation between free proline accumulation and sodium and chloride contents in the shoot of clusterbean genotypes differing in salt tolerance. Qadar (1994) found that proline was best correlated with sodium content of laminae and leaf sheaths of rice plants under salinity stress.

Proline in marked contrast to glycinebetaine is highly labile and rapidly disappears after the withdrawal of stress and thus can also act as a source of nitrogen and energy store. Furthermore, the rate of degradation of glycinebetaine in plant tissue is slow in contrast to proline. The relative intertness of glycinebetaine led Hanson and Nelson (1978) to suggest that it might be employed as accumulation index of water stress in barley. The difference in the rapidly of accumulation and metabolic liability of proline and glycinebetaine suggests that proline accumulation could be major response to a transient stress or shock and glycinebetaine to a longer possible stress, usually involving salt accumulation (Fig. 9.9). Thus, unlike proline, glycinebetaine is unlikely to represent a potential store of energy and nitrogen.

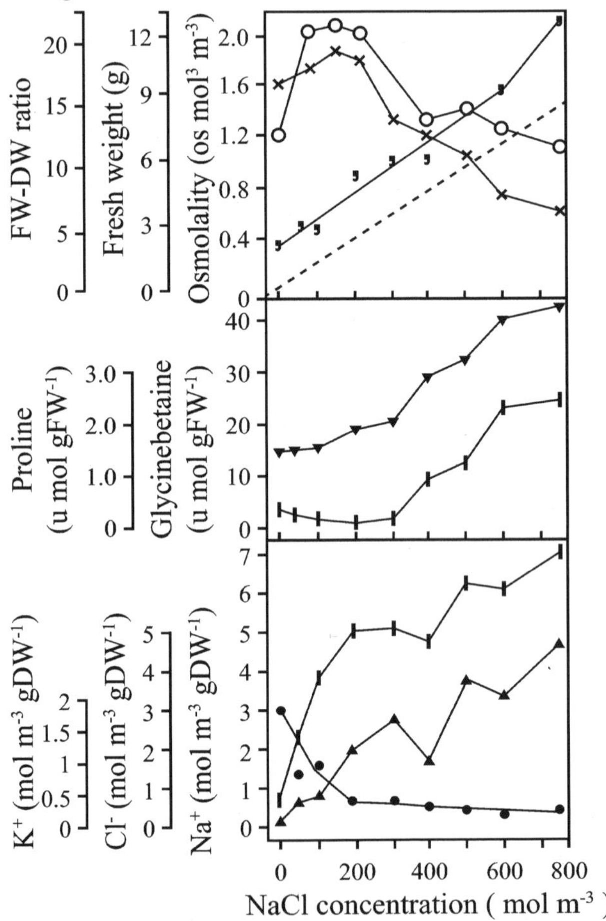

Figure 9.9. Influence of increasing NaCl salinity on growth and various other parameters of *Atriplex spongiosa* (Storey and Wyn Jones, 1979).
X = Fresh weight yield, O = Fresh weight/dry weight ratio.
, = Shoot sap osmolarity, - = Osmolarity of the growth medium.
= Glycinebetaine, I = Proline. = K^+, I = Na^+, = Cl^-

ORGANIC ACIDS, SUGARS AND POLYOLS

Organic acids are important as organic osmoregulatory solutes in cells of some glycophytes and many halophytes, either as free acids or more frequently as organic ions balancing excess of cations (Hellebust, 1976). In certain halophytes organic acid content decreases in response to salinity and inorganic salts, mostly NaCl taken up from the saline medium, become the major osmotica. Although the role of organic acids in adaptation to salinity is restricted in this way, they still are at least partially important as osmotica and organic acids metabolism remains essential. This applied especially to the Chenopodiaceae where oxalate levels were actually more in response to salinity. Sharma and Gupta (1986) reported that MI-48 (a salt sensitive rice variety) suffered a serious set back in organic acid metabolism, whereas CSR-1 (a salt tolerant variety) and *Eleusine* showed slight effects on total content of organic acids as well as individual organic acids under salt stress. The large concentrations of organic acids found in many plant cells are largely confined within vacuoles. The malic acid stored in CAM vacuoles in large amounts constitutes a major cellular osmoticum and the large day and night changes in malic acid levels are associated with marked changes in leaf water relations.

Malate, the organic acid anion that most frequently balances excess cation uptake is involved in several osmotic and turgor responses in plants. Dhindsa *et al.* (1975) found that K^+ and malate levels correlated with the growth rate of cotton fibers and they suggested that these solutes serve as the major osmotica for the turgor-driven cell extension. Among halophytes oxalate is the major organic anion balancing excess cation uptake (Waisel, 1972).

Sucrose accumulates in response to salinity in certain plants. When the amount of reserve carbohydrates has been measured (both soluble and insoluble compounds) it has been found with one exception (Downton, 1977) that a salt treatment which reduces growth causes either no change or an increase in the total concentration of reserve carbohydrates. This is true for shoots as well as roots (Gauch and Eaton, 1942; Rathert, 1984; Aslam *et al.,* 1986; Lahiri *et al.,* 1987). In case of barley it has been reported (Munns and Termaat, 1986) that the carbohydrate status of the youngest leaf was higher after 5 days of NaCl treatment than in controls, especially in most rapidly growing region (Table 9.17), while that of the older leaves was slightly lower. Several workers have suggested that accumulated sugars contribute to osmotic adjustment but in certain studies accumulation of both starch and soluble carbohydrates under salinity stress has been attributed to impaired carbohydrate utilization (Greenway and

Munns, 1980; Munns and Termaat, 1986). Rathert (1984) suggested that the extent of osmotic adjustment via sucrose accumulation depends on the salt tolerance of the crop.

Table 9.17. Effect of NaCl on the carbohydrate concentration of growing and expanded parts of the growing leaf (leaf 3) of barley plants (Munns *et al.*, 1982)

Sugar	Carbohydrate concn. (g $100g^{-1}$ dry wt.) of			
	Growing zone (0-20 mm)		Expanded zone (>40 mm)	
	Control*	NaCl**	Control	NaCl
		Afternoon harvest		
Soluble	32	53	17	19
Starch	12	12	14	15
Total	44	65	31	34
		Dawn harvest		
Soluble	14	29	4	6
Starch	10	11	5	7
Total	24	40	9	13

* Control 0.5 mM NaCl ** NaCl 180 mM for 5 days

Sorbitol, a polyhydric alcohol accumulates to high levels in *Plantago maritima* under salts conditions (Ahmed *et al.,* 1979). This compound is known to be compatible with enzyme and organelle activity. Other polyols, such as glycerol and mannitol are also known to serve as osmotic solutes (Wyn Jones, 1981). Polyols have a role in osmoregulation which is independent of ion transport. Enzymes isolated from *Dunaliella* - a salt resistant alga, are salt sensitive despite the halotolerance of the organism. It was found later on that these enzymes are resistant to glycerol or require glycerol for full enzyme activity.

OTHER METHYLATED ORGANIC OSMOLYTES

Several other groups of methylated compounds increase during osmotic stress, which also may have osmoprotective properties. For example polyamines such as spermine and putracine, and tertiary sulfonium compounds such as dimethylsulfoniopropionate, can increase during osmotic stress (McCue and Hanson, 1990, Bohnert and Jensen, 1996). It is interesting to note that many of the compounds that increase during osmotic or water stress are methylated (e.g. metylated inositols, QAMs, tertiary sulfonium species, and polyamines). Central to the methyltransferases involved in the production of these compounds is the cosubstrate S-adenosylhomocysteins (SAM). During osmotic stress, the mRNA transcripts for SAM synthase increase in the tomato (Espartero *et al.,* 1994), suggesting an importance of methylation for salinity stress tolerance.It is possible that the source of the methyl groups required for the SAM-mediated pathway is photorespiration (Bohnert and Jensen, 1996).

Polyols are compounds with multiple hydroxyl functional groups available for organic reactions. Sugar alcohols are a class of polyols functioning as compatible solutes, as low molecular weight chaperones, and as ROS scavenging compounds (Ashraf and Foolad, 2007). They can be classified into two major types, cyclic (e.g., pinitol) and acyclic (e.g., mannitol). Mannitol synthesis is induced in plants during stressed period via action of NADPH dependent mannose-6-phosphate reductase. These compatible solutes function as a protector or stabilizer of enzymes or membrane structures that are sensitive to dehydration or ionically induced damage.

Antioxidant Regulation of Salinity Tolerance

Abiotic and biotic stress in living organisms, including plants, can cause overflow, deregulation, or even disruption of electron transport chains (ETC) in chloroplasts and mitochondria. Under these conditions molecular oxygen (O_2) acts as an electron acceptor, giving rise to the accumulation of ROS. Singlet oxygen ($^1O_2$), the hydroxyl radical (OH^-), the superoxide radical (O^-_2), and hydrogen peroxide (H_2O_2) are all strongly oxidizing compounds and therefore, potentially harmful for cell integrity (Grob *et al.,* 2013). Antioxidant metabolism, including antioxidant enzymes and nonenzymatic compounds, play critical parts in detoxifying ROS induced by salinity stress. Salinity tolerance is positively correlated with the activity of antioxidant enzymes, such as superoxide dismutase (SOD), catalase (CAT), glutathione peroxidise (GPX), ascorbate peroxidase (APX), and glutathione reducase (GR) and with the accumulation of nonenzymatic antioxidant compounds.

Accumulations of carbohydrates such as sugars (e.g., glucose, fructose, fructans, and trehalose) and starch occur under salt stress. The major role played by these carbohydrates in stress mitigation involves osmoprotection, carbon storage, and scavenging of reactive oxygen species. It was observed that salt stress increases the level of reducing sugars (sucrose and fructans) within the cell in a number of plants belonging to different species (Kareperi and Galifa, 2000). Besides being a carbohydrate reserve, trehalose accumulation protects organisms against-several physical and chemical stresses including salinity stress. They play an osmoprotective role in physiological responses (Ahmad *et al.,* 2013). Roy *et al.* (2014) in their recent review have argued that there are three main traits in plants, which help them in their adaptation to salinity stress: ion exclusion, tissue tolerance, and salinity tolerance. It seems that antioxidants have some role in tissue and salinity tolerance mechanism.

Polyamines (PA) are small, low molecular weight, ubiquitous, polycationic aliphatic molecules widely distributed throughout the plant kingdom. Polyamines play a variety of roles in normal growth and development such as regulation of cell proliferation, somatic embryogenesis, differentiation and morphogenesis, dormancy breaking of tubers and seed germination, development of flowers and fruit, and senescence (Gupta *et al.,* 2013; Groppa *et al.,* 2013). It also plays a crucial role in abiotic stress tolerance including salinity and increases in the level of polyamines are correlated with stress tolerance in plants (Groppa and Benavides, 2008).

Nitric Oxide

Nitric oxide (NO) is a small volatile gaseous molecule, which is involved in the regulation of various plant growth and developmental processes, such as root growth, respiration, stomata closure, flowering, cell death, seed germination and stress responses, as well as a stress signalling molecule (Lamattina *et al.,* 2003; Benson-Bard *et al.,* 2008). NO directly or indirectly triggers expression of many redox-regulated genes. NO reacts with lipid radicals thus preventing lipid oxidation, exerting a protective effect by scavening superoxide radical and formation of peroxynitrite that can be neutralised by other cellular processes. It also helps in the activation of antioxidant enzymes (SOD, CAT, GPX, APX and GR) (Baigu, 2014).

The positive effects of NO on salinity tolerance or stress mitigation have been attributed to antioxidant activities and modulation of ROS detoxification system. Improved plant growth under salinity stress by exogenous applications of NO was associated with increases in antioxidant enzymes such as SOD, CAT, GPX, APX, and GR (Zhao *et al.,* 2004), and suppression of malondialdehyde (MDA) production or lipid peroxidation. Effect of NO on salinity tolerance are also related to its regulation of plasma membrane H^+- ATPase (H^+-PPase), thereby producing a H^+ gradient and offering the force for Na^+/H^+ ratio (Nalousi *et al.,* 2012). NO stimulates H^+-ATPase (H^+-PPase), thereby producing a H^+ gradient and offering the force for Na^+/H^+ exchange. Such an increase of Na^+/H^+ exchange may contribute to K^+ and Na^+ homeostasis.

Importance of Osmolytes

The accumulation of organic osmolytes, such as proline, glycine betaine, sugar alcohol polyamines and proteins from the late embryogenesis abundant (LEA) superfamily, play key role in maintaining the low intracellular osmotic potential of plants and in prevention the harmful effects of salinity stress (Deinlein *et al.*, 2014). The metabolic rearrangement regulatory networks controlling osmolyte levels are therefore, pivotal to understanding salinity tolerance. Molecular analyses have shown that salt stress stimulates proline synthesis whereas its catabolism is enhanced during recovery (Szekely *et al.*, 2008). During this recovery phase, proline may function as an essential signaling molecule and has been proposed to regulate cell proliferation, cell death and expression of stress-recovery gene. In *Arabidopsis*, knockout of the *P5CS1* gene, which encodes a Δ-1-pyrroline-5-carboxylate synthetase, impairs proline synthesis resulting in salt hypersensitivity. For many years, it was presumed that proline plays a crucial role in osmotic adjustment; however, alternative suggestions are that it acts as a reactive oxygen scavenger, redox buffer, or molecular chaperone, stabilizing proteins and membrane structures under stress conditions (Ashraf and Foolad, 2007).

INTRACELLULAR COMPARTMENTATION

In most of the halophytes and in some tolerant glycophytes, salt uptake is the mechanism mainly responsible for the generation and maintenance of low osmotic potentials and hence osmotic adjustment. The use of inorganic ions to establish a favourable osmotic gradient seems to apply to a large fraction of mature plant cells (about 90%) occupied by the vacuole. Osmotic dessication of the cytoplasm is prevented by the synthesis of metabolically nontoxic organic solutes. This is metabolically expensive but applies to only a small fraction of the total cell volume. The cell membranes plasmalemma and tonoplast delimiting the principal cell compartments play key roles in efficient subcellular compartmentation of the ions.

It has generally been thought that halophytes are able to accumulate salts in their apoplastic space and vacuole, thereby keeping the symplast relatively free of salt accumulation. There is very good evidence that Na^+ and K^+ are preferentially accumulated into the vacuole, and that this is one way of moderating salt accumulation into the cytoplasm. When salinity reaches the saturation point for the Na^+/H^+ antiporters, salt compartmentation in the vacuole will stop, and salt will rapidly rise in the cytoplasm. Several studies have shown that there is an equilibrium that develops for salt across the tonoplast membrane that cannot account for salinity tolerance. In maize, vacuolar salt content increases above that in the cytoplasm only when intolerable salinity is reached. In most halophytes, salinity increased faster in older tissues than in young tissues (A notable exception is the ice plant). Salt increase in older tissues may simply be due to longer duration of transpiration in these tissues, or it may be due to directed tanslocation.

A fast growth rate can be an effective salt tolerance mechanisms because the bulk of the salt can be abscised in older tissues that are going senescent.

If the central vacuole is effective at accumulating salinity, there must be an adjustment of cytoplasmic osmotic potential to match that of the vacuole. There is considerable evidence on the accumulation of betaine and proline in response to salinity and drought (McNeil *et al.,* 1999) as well as a number of data sets showing the role of various carbohydrates and polyols in osmoregulation. What is critical is the identification of genes regulating the response, both in terms of overproduction and the effect of these sbustances on the physiology and biochemistry of plants exposed to stress environments (Zhu *et al.,* 1997). The concentration and identity of the osmolytes produced by plants when exposed to salinity stress and the potential role these molecules play in regulating osmotic potential will provide important information to researchers and managers working on salinity.

Gorham *et al.* (1985) have proposed the intracellular model of solute compartmentation of a plant cell. According to this model the large quantities of salts that are absorbed into the leaves and contribute substantially to osmotic adjustment are accumulated mainly in the vacuoles when tissue concentration exceed about 200 mM. In contrast the concentration of inorganic ions in the cytoplasm is held in the range of 100 to 200 mm and the cytoplasm shows a strong selectivity for potassium over sodium, magnesium over calcium and phosphate over chloride or nitrate. When cell osmotic pressures are greater than -0.9 MPa, the maintenance of osmotic equilibrium across the tonoplast requires organic solutes which are commonly called as ``compatible solutes (Borowitzka, 1980). While there is much circumstantial evidence supporting these concepts, obtaining precise quantitative evidence for the model presents many technical and interpretational problems. Although this model is consistent with the biochemical data and allows the interpretation of much of the physiological and anatomical evidence, it is also apparent that little is known at the cellular level about the detailed operation of mechanisms for solute compartmentation.

Salt concentrations in individual leaves of non-halophytes usually increase with time and they do not show any signs of regulation of the salt concentration in their leaves (Flowers and Yeo, 1986). Thus, for non-halophytes it is therefore inevitable that ion concentrations will eventually build up in the older leaves and they will die. The concentration at which this occurs will depend upon the ability of the species to compartment the salts in the vacuole, and the time taken for this to happen will depend mainly on the salinity level, the ability of the roots to exclude the salt and the ambient conditions affecting the rate of transpiration. Build-up in the cytoplasm will interfere with metabolism, and build-up in the cell wall will cause loss of turgor and then excessive loss of water. In case of barley the later condition has been reported to kill the cells (Munns and Termaat, 1986). It thus appears that the transport of salt to the leaf may occur at rates faster than their uptake by the mesophyll cells.

The response of intracellular compartments to external stresses varies amongst species and genotypes. Plant cells avoid the toxic effects of salts and dehydration by partitioning ions and compatible solutes within cells and tissues. This process is regulated by transport processes (Blumwald, 2000; Hasegawa *et al.,* 2000; Zhang *et al.,* 2001). Commonly the vacuole serves as a recipient and is responsible for the sequestering of Na ions, which, if accumulated in the cytoplasm, could result in a toxic response. The cytoplasm, in response to the ion accumulation in the vacuole, produces organic osmolytes to assure the maintenance of appropriate osmoic environment for the essential physiological and biochemical reactions occurring in that compartment (Yeo, 1998). Zhang *et al.* (2001) have produced salt-tolerant transgenic *Brassica* plants via increased vacuolar sodium accumulation.

COST OF OSMOREGULATION

When a plant is exposed to high levels of salt, extra metabolic energy is likely to be consumed in processes related to osmotic adjustment within the cells. Without this adjustment, the plant would lose water to the surrounding saline environment, dehydrate and die. Consuming energy to adjust osmotically to salt stress could reduce the energy available for growth and thus influence productivity. Most crop plants are glycophytes and can tolerate only modest amounts of salt. Under chronic stress, they often can adjust osmotically and maintain turgor, but leaf area production, photosynthesis and yield are often considerably reduced in spite of this adjustment (McCree, 1986).

Tolerance of stress involves regulation of internal cellular environments. This requires concentrating osmoic solutes, either inorganic or organic. Regulation of inorganic material requires transport of ions across membranes from the external environment to intracellular spaces as well as across intracellular membranes into cellular compartments. Concentrating these materials against an electrochemical potential gradient involves the expenditure of metabolic energy. Also regulation of organic osmotica requires metabolism of carbon constituents and utilization of energy to produce the osmotic gradient necessary to maintain hydration of the cellular environment. Therefore, the development of salt tolerant plants involves the development of plants capable of not only tolerating the stress, but also capable of being productive. Plants exposed to stress frequently respond by diverting energy from growth to the maintenance of physiological processes and repair of damage caused by the stress. This reduces the amount of energy used for growth, thus reducing productivity. A plant tolerant of salinity must also use energy to produce yield. It is unlikely for plants to compensate for the effect of salinity by increasing fixation of carbon. What is required are plant types that can use the available energy more efficiently (Stavarek and Rains, 1985).

A question that has often been debated in the literature is how great is the metabolic cost of osmotic adjustment? In the case of salt stress, there surely must be an additional cost of compartmentalizing any salt that is taken up. It appears that the additional requirement for transport of NaCl and storage within the vacuole is significant for plants at high salt, but is not excessive. Greenway and Munns (1980) calculated that, for the extreme care of hypothetical (root) cell growing rapidly in the presence of 500 mM NaCl as in its absence, ATP production would have to be increased by 50-70%. Yeo (1983) estimated that for two halophytes, *Salicornia fruticosa* and *Suaeda maritima* growing rapidly at high salinity, ATP production for ion regulation would have been about 10-20% of their total ATP production.

Synthesis of organic molecules such as proline or glycine betaine should also represent an additional metabolic load on the plant. When sugars are used for osmotic adjustment, this must make them unavailable for biomass synthesis. Osmond (1980) calculated that the total amount of sucrose used for osmotic adjustment in water stressed sorghum leaves could be accumulated in 2 h of photosynthesis in the greenhouse. He also speculated that perhaps 10 times of this quantity might be needed by the whole plant.

Schwarz and Gale (1981, 1984) and Shone and Gale (1983) determined the respiratory costs associated with salt stress in several different species of plants with different degrees of salt tolerance (bean : *Phaseolus vulgaris,* Cocklebur : *Xanthium strumarium*; Maize :

Zea mays; salt bush : *Atriplex halimus* and lucerne : *Medicago sativa*). The whole plant daily carbon balance approach was used and the growth and maintenance components of respiration were separated by varying the light level. They pointed out that while the growth component always decreases in response to stress (being proportional to the growth rate), the maintenance component should increase if there are indeed additional metabolic costs. From their data they calculated the efficiency of conversion of photosynthate to biomass in the growth processes i.e. the yield of growth (Yg). The general pattern in all species was that Yg, was unaffected by salt (NaCl) while the respiratory rate at zero growth (the maintenance component or respiration, per unit of dry weight) increased with increasing salt concentration, upto about the concentration at which salt damage was not observed. At higher concentration, the maintenance rate began to decrease. However, conflicting data were obtained with maize, found to be very sensitive to salt. Schwartz and Gale (1981) estimated that, in the cocklebur plant, about one fifth of the 30% reduction in net 24 h carbon gain at an osmotic potential of -0.5 MPa was attributable to increased maintenance costs, the other four fifths being ascribed to reduced photosynthesis.

According to Gale and Zeroni (1985) if a mechanism of adaptation indeed requires the expenditure of assimilates then the amount available for normal growth will be reduced. It follows that the retardation of growth by such a stress should be reduced if assimilate production is increased. This has been demonstrated by Schwartz and Gale (1984) in case of bean, cocklebur, maize and salt bush plants where CO_2 supplementation increased the relative growth rate more in the salinized than in the control plants. Similar effects were found in sonia roses where at low to moderate concentrations of salt relative response of salt stressed plants was greater with CO_2 supplementation.

Regulatory Molecules for Salt Tolerance

There are two classes of genetic determinants of salt tolerance in higher plants. These are (1) Plant signaling genes as determinants of salt stress; (2) Plant signaling factors that enhance salt tolerance.

A brief discussion of these regulatory moleculues follows:

Plant Signaling Genes

Determinants of plant stress tolerance have been identified by functional complementation of osmotic sensitive yeast mutants. A putative MAP (Mitogen activated protein kinase) (MAPK) has been identified from *Pisum sativum* (PsMAPK), with 47% primary sequence identity to Hog 1p, which is the MAPK in the yeast osmoregulatory pathway that controls glycerol accumulation. PsMAPK functionally suppressed salt-induced cell growth inhibition of *hog* 1 (Popping *et al.,* 1996). AtDBF2 kinase of *Arabidopsis* mediated functional sufficiency of yeast cells for LiCl tolerance (Lee *et al.,* 1999). Over expression of this serine/threonine kinase enhanced tolerance of plant cells to numerous stresses, including salt and drought.

SalT is a cDNA clone containing an open reading frame coding a 145 amino acid protein whose mRNA accumulates rapidly showing organ-specific expression (sheaths and roots of mature plants and seedlings) in response to a variety of salts, osmotic and dessication stresses and the exogenous application of ABA (Claes *et al.,* 1990). The organ specific expression first appeared co relatable with the pattern of sodium accumulation but more recent results showed that organ specificity was also determined by other factors, since salT was expressed only in leaf sheaths of rice even though the salt concentrations in the leaf lamina were equivalent (Garcia *et al.,* 1997).

Several genes induced by salt in salt tolerant alfalfa callus were cloned, although these were not induced by salt-stress in the parent salt sensitive cell-line. Two transcripts were similarly salt-induced in callus, but in plants tissue-specific expression can be seen (Winicov and Shirzadegan, 1997). Fusion of the *E. coli* gene betB coding betaine aldehyde dehydrogenase to the promoter of ars/a encoding the small subunit of Rubisco in *Arabidopsis thaliana* including the N-terminal transit peptide for chloroplast targetting resulting in betB being directed to the chloroplast in transgenic *Nicotiana tabacum* (Holmstrom *et al.,* 1994).

Plant Signaling Factors

Numerous signals or signal like molecules have now been identified and presumed to function in plants as mediators of osmotic adaptation. Some of these affect osmotic tolerance in genetically modified plants. Transcriptional modulation has laways been predicted to play a major role in control of plant responses to salt stress (Shinozaki *et al.,* 1996; Zhu *et al.,* 1997). Transcription factors have been identified based on inter-action with promoters of osmotic/salt stress responsive genes. These factors participate in the activation of stress-inducible genes and lead to osmotic adaptation (Sheen, 1996; Abe *et al.,* 1997; Liu *et al.,* 1998; Winicov and Bastola, 1999). Since the promoters that are controlled by these transcription factors are responsive to several environmental signals, it is not clear which transcription factors, if any, function only in salt stress responses, or if salt specific transcriptional regulation alone is a requisite component of salt tolerance in plants. ABA deficient and insensitive mutants have been used to delineate transcription factors as components of osmotic stress signal transduction pathways that either involve or are independent of the growth regulator. Promoters of ABA dependent osmotic stress-responsive genes include regulatory elements that interact with basic leucine-zipper motif (bZIP), MYB or MYC domains in DNA building proteins (Shen and Ho, 1995). The bZIP transcription factors interact with the ABA responsible element (ABRE). The promoter of the salt stress responsive rd 22 gene contains signature recognition motiffs for MYC and MYB transcription factors. DNA binding proteins rd 22 BP1 (MYC) and Atmyb2 (MYB), interact with their corresponding *cis* - elements and *trans*-activate *rd22.* However, that *rd22* promoter does not contain ABRE motifs, which indicates that ABA dependency is mediated through another process. At least some ABA dependent transcriptional activation appears to involve Ca^{2+} dependent protein kinases (CDPKs) (Sheen, 1996). Transcription factors which function in salt stress gene induction, independent of ABA, include dehydration response element (DRE) binding proteins (Stockinger *et al.,* 1997).

It is clear from physiological studies that regulation of ion uptake and compartmentation within organelles, cells and tissues forms the basis of salt tolerance of higher plants. Glycophytes and halophytes have in common the ability to exclude most of the NaCl in the external solution from entering the plant while admitting water. However, most salt sensitive plants have difficulty controlling Na entry into stete and they lack the ability to efficiently compartmentatlize NaCl in leaf cell vacuoles limiting its role in osmotic adjustment. They may rely mainly on K and organic compounds for osmotic adjustment, at considerable expenditure of metabolic energy.

Studies on Na/H antiports and H/ATPase are just beginning to reveal differences between glycophytes and halophytes in the activity levels and control mechanisms of ion translocation enzymes. However, halophytes and glycophytes differ quantitatively rather than qualitatively and the extreme example really lay on continuum. Their compartmentation is certainly better, their ion selectivity in probably better, their water use is lower and WUE positively higher. Some key metabolic steps seem to be more tolerant. Stress responses appear to show quantitative inheritance, therefore, attempts should be made to engineer quantitative traits to enhance the prospects of intervening in the molecular biology of salt tolerance of higher plants.

PHYSIOLOGICAL MECHANISMS OF SALT TOLERANCE

Salinity stress involves changes in various physiological and metabolic processes depending on severity and duration of the stress and ultimately inhibits crop production. Initially soil salinity is known to repress plant growth in the form of osmotic stress which is then followed by ion toxicity. During the initial phase of salinity stress, water absorption capacity of root system decreases and water loss from leaves in accelerated due to osmotic stress of high salt concentrations. In the second phase ion toxicity leads to changes in leaf metabolic functions, ion homeostatis and damage to photosynthetic machinery (Gupta and Huang, 2014).

Recently genomics, transcriptomics, proteomics and metabolomics have been successfully applied and given exciting outcome to unsaved different components of salt tolerance in plants. These components include:

1. Various genes (SOS-signaling-network)
2. Transcription factors (HLH, MYB, bZ IP, APZ, WRKY etc.)
3. Protein and metabolites (Osmolytes, Phytohormones, lipids and others).

These may be used to engineer plants for their increased salt tolerance. The multidisciplinary approach will be highly helpful in increasing plant as well as crop productivity to meet the increasing demand of food for every increasing population.

CONCLUSIONS AND FUTURE PROSPECTS

Growth under saline conditions could be improved by following ways :

(1) The rate of leaf expansion be increased.

(2) The maximum ion concentration tolerated by the leaf be increased.

(3) The rate of leaf ion uptake is decreased.

(4) The photosynthetic efficiency of the leaf is increased.

The rate of leaf area expansion is a short-term response which is set within a day of exposure to salinity. It is governed by the response of the root to the low water potential in the soil and this response is communicated to the shoot via growth regulators. Understanding these processes might enable us to increase leaf expansion in saline soils.

By increasing the cellular tolerance to NaCl the leaf could function at higher internal NaCl concentrations. This mechanism would be a rapid rate of salt uptake at the plasmalemma and tonoplast i.e. rapid removal of salt from the apaplast, coupled with an intracellular distribution which placed the toxic salts in the vacuole and less toxic salts or organic solutes in the cytoplasm. There is at present no evidence for a change in the properties of enzymes or organelles involved in intermediary metabolism which might make them more tolerant of salt.

A decrease in the rate of salt accumulation in the shoot would improve salt tolerance. The extent of ion accumulation in the leaves of a set of closely related genotypes usually correlates inversely with their salt tolerance. The rate of accumulation of Na^+ and Cl^- in the leaves depends mainly on the ability of the roots to exclude salt from the transpiration stream and the volume flux of the transpiration stream.

A higher rate of photosynthesis per unit leaf area might be adaptive in the long term when leaf area is becoming critical and carbohydrate reserves are diminishing. However, the conditions which change photosynthesis usually also change water use efficiency and therefore ion concentrations in leaves.

Although various mechanisms for salt tolerance have been proposed the one most likely to explain genotypic differences in salt tolerance of crop plants is better regulation of salt transport preventing excessive accumulation of salt in mature leaves. Processes like osmotic adjustment and ionic compartmentation and gene mapping and transfer of genes from bacteria to higher plants are rapidly emerging as new concepts in salt tolerance.

REFERENCES

Abe, H, Yamaguchi-Shinozaki, K, Urao T, Iwasaki T, Hosokawa, D, Shinozaki, K. 1997. Role of *Arabidopsis* MYC and MYB homologs in drought - and abscisic acid-regulated gene expression. Plant Cell, 9 : 1859-58.

Abel, G.H. 1969. Inheritance of the capacity for chloride inclusion and chloride exclusion by soybeans. Crop Sci. 9 : 697-98.

Abel, G.H. and Mackenzie, A.J. 1964. Salt tolerance of soybean varieties (*Glycine max* L. Merill) during germination and later growth. Crop Sci. 4 : 157-161.

Ahmad, P. and Umar, S. 2011. Oxidative Stress: Role of Antioxidants in Plants, Studium Press, New Delhi, India.

Ahmad, P., Jaleel, C.A., Salem, M.A., Nabi, G. and Sharma, S. 2010. Roles of enzymatic and nonencymatic antioxidants in plants during abiotic stress, Critical Review in Biotechnology. 30(3): 161-175.

Ahmad, R., Lim, C.J. and Kwon, S.Y. 2013. Glycine betaine: a versatile compound with great potential for gene pyramiding to improve crop plant performance against environmental stresses, Plant Biotechnology Reports. 7: 49-57.

Ahmed, I., Larher, F. and Stewart, G.R. 1979. Sorbitol, a compatible solute in *Plantago maritima*. New Phytol. 82 : 671-678.

Alarcon, J.J., Sanchez-Blanc, M.J., Bolarin, M.C. and Tolrecielas, A. 1994. Growth and osmotic adjustment of two tomato cultivars during and after saline stress. Plant Soil 166 : 75-82.

Alfocea, F.P., Estan, M.T., Caro, M. and Bolarin, M.C. 1993. Response of tomato cultivars to salinity. Plant Soil 150 : 203-211.

Alia, P.S. and Mohanty, P. 1993. Proline in relation to free radical production in seedlings of *Brassica juncea* raised under sodium chloride stress. Plant Soil 155/156 : 497-500.

Apel, R. and Hirt, F. 2004. Reactive oxygen species: metabolism, oxidative stress, and signal transduction, Ann. Rev. Plant Biol. 55: 373-399.

Ashraf, M. and Foolad, M.R. 2007. Roles of glycine betaine and proline in improving plant abiotic stress resistance, Environ. Exp. Bot. 59(2): 206-216.

Ashraf, M. and Foolad, M.R. 2013. Crop breeding for salt tolerance in the era of molecular markers and marked assisted selection. Plant Breeding 132:10-20.

Ashraf, M. and Waheed, A. 1993. Responses of some genetically diverse lines of chickpea (*Cicer arietinum* L.) to salt. Plant Soil 154 : 257-266.

Aslam, Z., Jesche, W.D., Barret-Lennard, E.G., Setter T.L., Watkin, E. and Greenway, H. 1986. Effects of NaCl on growth, ion relations and carbohydrate status of leaves of *Atriplex amnicola*. Plant Cell Environ.

Ayoub, A.T. 1975. Sodium and cation accumulation by Senna (*Cassia acutifolia*). J. Exp. Bot. 26 : 891-896.

Bajgu, A. 2014. Nitric oxide: role in plants under abiotic stress. In: *Physiological Mechanisms and Adaptation Strategies in Plants under Changing Environment,* pp. 137-159, Springer.

Beadle, N.C.W., Whalley, R.D.B. and Gibson, J.B. 1957. Studies in halophytes. II. Analytic data on the mineral constituents of three species of *Atriplex* and their accompanying soils in Australia. Ecology 38 : 340-344.

Ben Ahmed, C., Ben Rouina, B., Sensoy, S., Boukhriss, M. and Ben Abdullah, F. 2010. Exogenous proline effect on photosxynthetic performance and antioxidant defense system of young olive tree. J. Agricul. Food Chem. 58(7): 4216-4222.

Benlloch, M., Ojeda, M.A., Ramos, J. and Rodriguez-Navarro. A. 1994. Salt sensitivity and low discrimination between potassium and sodium in bean plants. Plant Soil 166 : 117-123.

Bernstein, L. 1963. Osmotic adjustment of plants to saline media. II. Dynamic phase. Amer. J. Bot. 50 : 360-370.

Bernstein, L. and Ayers, A.D. 1953. Salt tolerance of five varieties of carrots. Proc. Amer. Soc. Hort. Sci. 61 : 360-366.

Besson-Bard, A., Pugin, A. and Wendehenne, D. 2008. New insights into nitric oxide signalling in plants. Ann. Rev. Plant Biol. 59:21-39.

Bliss, R.D., Platt-Aloria, K.A. and Thompson, W.W. 1984. Changes in plasma lemma organization in cowpea radicle during inhibition in water and NaCl solutions. Plant Cell Environ. 7 : 600-606.

Bohnert, H.J., Nelson D.E. and Jensen, R.G. 1995. Adaptation to environmental stresses. Plant Cell. 7 : 1099-11.

Bonhert, H.J. and Jensen, R.G. 1996. Strategies for engineering water stress tolerance in plants. Trends Biotechnol. 14 : 89-97.

Borowitzka, L. 1980. Solute accumulation and regulation of cell water activity. In : The Physiology and Biochemistry of Drought Resistance in Plants (Eds. L.G. Paleg and D. Aspinall). pp. 97-130. Academic Press, London.

Bressan, R.A., Hasegawa, P.M. and Pardo J.M. 1998. Plants use calcium to resolve salt stress. Trends Plant Sci. 3 : 411-412.

Brown, A.D. 1990. Microbial Water Stress Physiology, Principles and Perspectives. New York : Wiley and Sons.

Brown, A.D. and Simpson, J.R. 1972. Water relations of salt tolerant yeasts : the role of intracellular polyols. J. Gen. Microbiol. 72 : 589-91.

Brownell, P.F. 1979. Sodium as an essential micro-nutrient element for plants and its possible role in metabolism. Adv. Bot. Res. 7 : 118-219.

Brugnoli, E. and Lauteri, M. 1991. Effects of salinity on stomatal conductance, photosynthetic capacity and carbon isotope discrimination of salt tolerant *Gossypium hirsutum* L. and salt sensitive *Phaseolus vulgaris* L. C3 non halophytes. Plant Physiol. 95 : 628-635.

Chauhan, R.P.S., Singh, M., Singh, S., Singh, B.B. and Singh, R.K. 1995. Free proline accumulation in plants. In: Plant Physiology and Biochemistry (Eds. B.B. Singh and K. Mengal) pp. 173-208. Panima Publishing Corp. New Delhi.

Cheeseman J.M. 1988. Mechanisms of salinity tolerance in plants. Plant Physiol. 87 : 547-550.

Chu, T.M., Aspinall, D. and Paleg, L.G. 1976. Stress metabolism VII. Salinity and proline accumulation in barley. Aust. J. Plant Physiol. 3 : 219-228.

Claes, B., Dekeyser, R., Villarroel, R, Van den Bulcke M, Bauw, G., Van Montagu M. and Caplan, A. 1990. Characterisation of a rice gene showing organ specific expression in response to salt stress and drought. The Plant Cell 2: 19-27.

De Lourdes, M. Oliveira Otoch, Menezes Sobreira, A.C./, Farias De Aragao, M.E., Orellano, E.G., Da Guia Silva Lima, M. and Fernandes De Melo, D. 2001. Salt modulation of vacuolar H^+-ATPase and H^+-Pyrophosphatase activities in *Vigna unguiculata,* J. Plant Physiol. 158(5): 545-551.

Deinlein, U., Stephan, A.B., Horie, T., Lyo, W., Xu, G. and Schroedel J.I. 2014. Plant salt-tolerance mechanisms. Trends Plant Sci. 19: 371-379.

Delauney, A.J. and Verma, D.P.S. 1993. Proline biosynthesis and osmoregulation in Plants. Plant J. 4 : 215-223.

Dhindsa, R.S., Beasley, C.A. and Ting, I.P. 1975. Osmoregulation in cotton fibre : Accumulation of potassium and malate during growth. Plant Physiol. 56 : 394-395.

Dietz, K.J., Tavakoli, N., Kluge, C. *et al.,* 2001. Significance of the V-type ATPase for the adaptation to stressful growth conditions and its regulation on the molecular and biochemical level. J. Exp. Bot. 52(363): 1969-1980.

Downton, W.J.S. 1977. Photosynthesis in salt stressed grape vines. Aust. J. Plant Physiol. 4 : 183-192.

Flowers, T.J. and Lauchli, A. 1983. Sodium versus potassium : Substitution and compartmentation. In : Encyclopedia of Plant Physiology. Vol. 15B (Eds. A. Lauchli and R.L. Bielski) pp. 651-681. Springer-Verlag, Berlin.

Flowers, T.J. and Yeo, A.R. 1990. Salinity and rice : A physiological approach to breeding for resistance. In : Proc. Inter. Cong. Plant Physiol. (Eds. S.K. Sinha, P.V. Sane, S.C. Bhargava and P.K. Agarwal). pp. 953-959. Society for plant Physiology and Biochemistry, New Delhi.

Flowers, T.J., Troke, P.F. and Yeo, A.R. 1977. The mechanism of salt tolerance in halophytes. Ann. Rev. Plant Physiol. 28 : 89-121.

Ford, C.W. 1984. Accumulation of low molecular weight solutes in water stressed tropical legumes. Phytochemistry 23 : 1007-15.

Gadallah, M.A.A. 1999. Effects of proline and glycinebetaine on *Vicia faba* responses to salt stress, Biol. Plant. 42(20: 249-257.

Gale, J. and Zeroni, M. 1985. The cost of plant of different strategies of adaptation to stress and the alleviation of the stress by increasing assimilation. Plant Soil 89 : 57-67.

Galinski, E.A. 1993. Compatible solutes of halophilic eubacteria, molecular principal water-solute interaction, stress protection. Experientia 49: 487-96.

Galston, A.W., Kaur-Sawhney, R., Altabella, T. and Tiburcio, A.F. 1997. Plant polymaines in reproductive activity and response to abiotic stress. Botanic Acta, 110(3): 197-207.

Garcia, A.B., Engler, J.D., Iyer, S., VanMontagu, M. and Caplan, A.B. 1997. Effects of Osmoprotectants upon NaCl stress in rice. Plant Physiol., 115 : 159-69.

Garg, B.K. 2009. Physiological aspects of salt tolerance in arid legumes. J. Arid. Legumes 6 : 1-10.

Garg, B.K. and Gupta, I.C. 1996. Mechanism of varietal differences in Salt tolerance. Curr. Agric. 20 : 1-18.

Garg, B.K. and Gupta, I.C. 1997. Saline Wastelands Environment and Plant Growth. Scientific Publishers (India), Jodhpur. pp. 283.

Garg, B.K., Kathju, S., Vyas, S.P. and Lahiri, A.N. 1990. Effect of saline water irrigation on tolerant and sensitive wheat varieties under disparate soil fertility conditions. Ann. Arid Zone 29 : 179-189.

Garg, B.K., Vyas, S.P., Kathju, S., Lahiri, A.N., Mali, P.C. and Sharma, P.C. 1993. Salinity-fertility interaction on growth, mineral composition and nitrogen metabolism of Indian mustard. J. Plant Nutr. 16 : 1637-1650.

Gates, C.T., Haydock, K.P. and Claringbold, P.J. 1966. Response to salinity in *glycine* II. Differences in cultivars of *Glycine javanica* in dry weight, nitrogen and water content. Aust. J. Exp. Agr. Anim. Husb. 6 : 374-379.

Gauch, H.G. and Eaton, F.M. 1942. Effect of saline substrate on hourly levels of carbohydrates and inorganic constituents of barley plants. Plant Physiol. 17 : 347-365.

Glenn, E.P. and O'Leary, J.W. 1984. Relationship between salt accumulation and water content of dicotyledonous halophytes. Plant, Cell and Environment 7 : 253-261.

Glenn, E.P., O'Leary, J.W., Watson, M.C., Thompson, T.L. and Kuehl, R.O. 1991. *Salicornia bigelovii* Torr.: an oilseed halophyte for seawater irrigation. Science, 251 : 1065-1067.

Glenn, E.P., Olsen, M.W., Frye, R.J. and Moore, D.W. 1994. How much sodium accumulation is necessary for salt tolerance of subspecies of the halophyte, *Atriplex canescens?* Plant, Cell and Environment 17 : 711-719.

Gorham, J. 1992. Salt tolerance of plants. Sci. Prog. Oxford. 76: 273-285.

Gorham, J., Wyn Jones, R.G. and McDonnel, E. 1985. Some mechanisms of salt tolerance in crop plants. Plant Soil 89 : 15-40.

Greenway, H. 1973. Salinity, plant growth and metabolism. J. Aust. Inst. Agric. Sci. 39 : 24-34.

Greenway, H. and Munns, R. 1980. Mechanisms of salt tolerance in nonhalophytes. Ann. Rev. Plant Physiol. 31 : 149-190.

Grob, F., Durner, J. and Gaupels, F. 2013. Nitric oxide, antioxidants and prooxidants in plant defence responses, Front. Plant Sci. 4: 419.

Groppa, M.D. and Benavides, M.P. 2008. Polyamines and abiotic stress: recent advances. Amino Acids, 34(1): 35-45.

Gupta, K., Dey, A. and Gupta, B. 2013. Plant polyamines in abiotic stress responses, Acta Physiologiae Plant. 35(7): 2015-2036.

Gupta, K., Dey, A. and Gupta, B. 2013. Polyamines and their role in plant osmotic stress tolerance, in Climate Change and Plant Abiotic Stress

Tolerance, N. Tuteja and S.S. Gill (Eds.) pp. 1053-1072. Wiley-VCH, Weinheim, Germany.

Hall, J.L., Harvey, D.M.R. and Flowers, T.J. 1978. Evidence for the cytoplasmic localization of betaine in leaf cells of *Suaeda maritima*. Planta 140 : 59-62.

Hanson, A.D. and Nelson, C.E. 1978. Betaine accumulation and 14 C formate metabolism in water stressed barley leaves. Plant Physiol. 62 : 305-312.

Hasegawa, P.M. 2013. Sodium (Na^+) homeostasis and salt tolerance of plants. Environ. Exp. Bot. 92: 19-31.

Hasegawa, P.M., Bressan, R.A., Zhu, J.K. and Bohnert, H.J. 2000. Plant cellular and molecular response to high salinity, Ann. Rev. Plant Biol. 51: 463-499.

He, T. and Cramer, C.R. 1993. Growth and ion accumulation of two rapid cycling *Brassica* species differing in salt tolerance. Plant Soil 153 : 17-31.

Hellebust, J.A. 1976. Osmoregulation. Ann. Rev. Plant Physiol. 27 : 485-505.

Heuer, B. 1994. Osmoregulatory role of proline in water and salt stressed plants. In : Handbook of Plant and Crop Stress (Ed. M. Pessarakli) pp. 360-381. Marcel Dekker, Inc. New York.

Holmstrom, K.O., Welin, B., Mandal, A., Krisitiandottir, I., Teeri T.H., Lamark, T., Strom, T. and Palva, E.T. 1994. Production of the *Escherichia coli* betaine aldehyde dehydrogenase and enzyme required for the synthesis of the osmoprotectant glycine betaine in transgenic plants. The Plant J. 6 : 749-58.

Hoque, M.A., Banu, M.N.A., Nakamura, Y., Shimoishi, Y. and Murata, Y. 2008. Proline and glycinebetaine enhance antioxidant defense and methylglyoxal detoxification systems and reduce NaCl-induced damage in cultured tobacco cells, J. Plant Physiol. 165(8): 813-824.

Hoque, M.A., Banu, M.N.A., Okuma, E. *et al.,* 2007. Exogenous proline and glycinebetaine increase NaCl- induced ascorbate-glutathione cycle enzyme activities, and proline improves salt tolerance more than glycinebetaine in tobacco Bright Yellow-2 suspension-cultured cells, J. Plant Physiol. 164(11): 1457-1468.

Ishitani, M., Liu, J., Halfter, U., Kim, C.S., Shi, W. and Zhu, J.K. 2000. SOS3 function in plant salt tolerance require N-myristoylation and calcium binding, Plant Cell, 12(9): 1667-1677.

Jacoby, B. 1965. Sodium retention in excised bean stems. Physiol. Plant. 18 : 730-739.

James, R.A., Blake, C., Byrt, C.S. and Munns, R. 2011. Major genes for Na^+ exclusion, Nax1 and Nax2 (wheat HKT1:4 and HKT1:5), decrease Na^+ accumulation in bread wheat leaves under saline and waterlogged conditions, J. Exp. Bot. 62(8): 2939-2947.

Jones, R.R. 1986. High salt tolerance potential in *Lycopersicon* species during germination. Euphytica 35 : 575-582.

Kerepesi, I. and Galiba, G. 2000. Osmotic and salt stress-induced alteration in soluble carbohydrate content in wheat seedlings, Crop Sci. 40(2): 482-487.

Kingsbury, R.W. and Epstein, E. 1986. Salt sensitivity in wheat. A case for specific ion toxicity. Plant Physiol. 80 : 651-654.

Kuiper, P.C. 1984. Functioning of plant cell membranes under saline conditions : membrane lipid composition and ATPase. In Salinity Tolerance in Plants : Strategies for Crop Improvement, edited by R.C. Staples and G.H. Toenniessen, pp. 77-91. New York : John Wiley.

Kumar, D. 1984. The value of certain plant parameters as an index for salt tolerance in Indian mustard (*Brassica juncea* L.). *Plant Soil* 79 : 261-272.

Lahiri, A.N., Garg, B.K., Kathju, S., Vyas, S.P. and Mali, P.C. 1987. Response of clusterbean to soil salinity. Ann. Arid Zone 26 : 33-42.

Lahiri, A.N., Garg, B.K., Kathju, S., Vyas, S.P. and Mali, P.C. 1996. Genotypic differences to soil salinity in clusterbean. Arid Soil Res. and Rehabilit. 10 : 333-345.

Lamattina, L., Garcia-Mata, C., Graziano, M. and Pagnussat, G. 2003. Nitric oxide: the versatility of an extensive signal molecule, Ann. Rev. Plant Biol. 54: 109-136.

Lauchli, A. 1984. Salt exclusion : an adaptation of legumes for crops and pasture under saline conditions. In : Salinity Tolerance in Plants : Strategies for Crop Improvement (Eds. R.C. Staples and G.H. Toennissen) pp. 171-188. John Wiley and Sons, New York.

Lauchli, A. and Wieneke, J. 1979. Studies on growth and distribution of Na^+, K^+ and Cl^- in soybean varieties differing in salt tolerance. Z. Pflanzenernaehr. Bodenkd. 142 : 3-13.

Lauter, D.J., Munns, D.N. and Clarkin, K.L. 1981. Salt response of chickpea as influenced by N Supply. Agron. J. 73 : 961-966.

Leary, J.W. and Glenn, E.P. 1994. Global distribution and potential for halophytes. In Halophytes as a Resource for Livestock and for Rehabilitation of Degraded Land, Task in Vegetation Science Series. edited by V.R. Squiresand A. Ayoub, pp. 7-18. The Netherlands : Kluwer Academic Publishers.

Lee, J.H., Van Montagu, M. and Verbruggen, N. 1999. A highly conserved kinase is an essential component for stress tolerance in yeast and plant cells. Proc. Natl. Acad. Sci. USA 96 : 5873-77.

Lefebvre, D.D. 1989. Increased potassium absorption confers resistance to group 1A cations in rubidium selected suspension cells of *Brassica napus*. Plant Physiol. 91 : 1460-1466.

Leschke, D. and Wolf, O. 1993. Importance of mineral nutrient cycling for salinity tolerance of plants. In Towards the Rational Use of High Salinity Tolerant Plants, edited by H. Lieth and A. AJ Masoom, pp. 265-278.

Liu, J., Ishitani, M., Halfter, U., Kim, C.S. and Zhu, J.K. 2000. The *Arabidopsis thaliana* SOS2 gene encodes a protein kinase that is required for salt tolerance, Proceedings National Acad. Sci. United States America, 97(7): 3730-3734.

Liu, Q., Kasuga, M., Sakuma, Y., Abe, A. and Miura, S. 1998. Two transcription factors, DREB1 and DREB2, with an EREBP/AP2 DNA binding domain separate two cellular signal transduction pathways in drought and low-temperature responsive gene expression, respectively, in *Arabidopsis*. Plant Cell 10 : 1391-1406.

Louis, P. and Galinski, E.A. 1997. Characterization of genes for the biosynthesis of the compatible solute ectoine from *Marinococcus halophilus* and osmoregulated expression in *E. coli*. Microbiol. 143 : 1141-49.

Marschner, H. 1986. Mineral Nutrition in Hihger Plants. Academic Press, London, 674 p.

McCree, K.J. 1986. Whole plant carbon balance during osmotic adjustment to drought and salinity stress. Aust. J. Plant Physiol. 13 : 33-43.

McCue, K.F. and Hanson, A.D. 1990. Drought and salt tolerance towards understanding and application. Biotechnol. 8 : 358-62.

Munns, R. 1990. Causes of varietal differences in salt tolerance. In : Proc. Inter. Cong. Plant Physiol (Eds. S.K. Sinha, P.V. Sane, S.C. Bhargava and P.K. Agarwal), pp. 960-968. Society for plant Physiology and Biochemistry, New Delhi.

Munns, R. and Termaat, A. 1986. Whole plant responses to salinity. Aust. J. Plant Physiol. 13 : 143-160.

Munns, R. and Tester, M. 2008. Mechanisms of salinity tolerance, Ann. Rev. Plant Biol. 59: 651-688.

Munns, R., Tonnet, M.L., Shannon, C. and Gardns P.A. 1988. Effect of high external NaCl concentration on ion transport within the shoot of *Lupinus albus*. Plant Cell Environ. 11: 283-289.

Nalousi, A.M., Ahmadiyan, S., Hatamzadeh, A. and Ghasemnezhad, M. 2012. Protective role of exogenous nitric oxide against oxidative stress induced by salt stress in bell-pepper (*Capsicum annum* L.). American Eurasian Journal of Agricultural and Environmental Science, 12(8): 1085-1090.

Niu, X, Bressan, R.A., Hasegawa, P.M., Pardo, J.M. 1995. Ion homeostasis in NaCl stress environments. Plant Physiol. 109 : 735-42.

Niu, X., Zhu, K., Narasimhan, M., Bressan, R. and Hawegawa, P. 1993. Plasma-membrane H^+ - ATPase gene expression is regulated by NaCl in cells of the halophyte *Atriplex nummularia* L. Planta 190: 433-438.

Orcutt, D.M. and Nilsen, E.T. 2000. Physiology of Plants under Stress-Soil and Biotic Factors. John Wiley and Sons, Inc. New York.

Osmond, C.R. 1976. Ion absorption and carbon metabolism. In : Transport in Plants (Eds. U. Luttge and M.G. Pitman). pp. 367-380.

Paleg, L.G., Stewart, G.R. and Starr, R. 1985. Effect of compatible solutes on proteins. Plant Soil 89 : 83-86.

Pang, Q, Chen, S., Dai, S., Chen, Y., Wang, Y. and Yan, X. 2010. Comparative proteomics of salt tolerance in *Arabidopsis thaliana* and *Thellungiella halophila*, J. Proteome Res., 9(5): 2584-2599.

Papageorgiou, G. and Murata, N. 1995. The unusually strong stabilizing effects of glycine betaine on the structure and function of the oxygen - evolving photosystem II complex. Photosynth. Res. 44 : 243-52.

Pitman, M.G. 1975. Whole plants. In : Ion Transport in Plant Cells and Tissues. (Eds. D.A. Baker and J.L. Hall) pp. 267-308. North Holland, Amsterdam.

Poljakoff-Mayber, A., Symon, D.E., Jones, G.P., Naidu, B.P. and Paleg, L.G. 1987. Nitrogenous compatible solutes in native south Australian plants. Aust. J. Plant Physiol. 14 : 341-350.

Pollard, A. and Wyn Jones, R.G. 1979. Enzyme activities in concentrated solutions of glycinebetaine and other solutes. Planta 144 : 291-298.

Popping, B., Gibbons, T. and Watson, M.D. 1996. The *Pisum sativum* MAP kinase homologue (PsMAPK) rescues the *Saccharomyces cerevisiae hog1* deletion mutant under conditions of high osmotic stress. Plant Mol. Biol. 31 : 355-63.

Qadar, A. 1993. Physiological aspects of crop response to saline/sodic irrigation. I. Accumulation of organic and inorganic solutes. In: Advances in Research on Sustainable Irrigation in Saline Environment (Eds. N.K. Tyagi, S.K. Kamra, P.S. Minhas and N.T. Singh) pp. 76-83, CSSRI, Karnal, India.

Rathert, G. 1984. Sucrose and starch contents of plant parts as a possible indicator in salt tolerance of crops. Aust. J. Plant Physiol. 11 : 491-495.

Raven, J.A. 1977. H^+ and Ca^{2+} in phloem and symplast relation of the relative immobility of the ions to the cytoplasmic nature of the transport paths. New Phytol. 79 : 465-480.

Richardson, S.G. and McCree, K.J. 1985. Carbon balance and water relations of sorghum exposed to salt and water stress. Plant Physiol. 79 : 1015-1020.

Robinson, S.P. and Jones, C.P. 1986. Accumulation of glycinebetaine in chloroplasts provides osmotic adjustment during salt stress. Aust. J. Plant Physiol. 13 : 659-668.

Roy, S.J., Negrao, S. and Tester, M. 2014. Salt resistant crop plants, Curr. Opinion Biotechnol. 26: 115-124.

Rozema, J. 1991. Growth, water and ion relationships of halophytic monocotyledonae and dicotyledonae : a unified concept. Aquatic Botany 39 : 17-33.

Rozema, J. and Flowers, T. 2008. Ecology: Crops for a salinized world, Science, 322(5907): 1478-1480.

Rush, D.W. and Epstein, E. 1976. Genotypic responses to salinity. Differences between salt sensitive and salt tolerant genotypes of the tomato. Plant Physiol. 57 : 162-166.

Rush, D.W. and Epstein, E. 1981. Breeding and selection for salt tolerance by the incorporation of a wild germplasm into a domesticated tomato. J. Amer. Soc. Hort. Sci. 106: 669-670.

Sairam, R.K. and Tyagi, A. 2004. Physiology and molecular biology of salinity stress in plants. Curr. Sci. 86: 407-427.

Sairam, R.K. and Tyagi, A. 2004. Physiology and molecular biology of salinity stress tolerance in plants, Curr. Sci., 86(3): 407-421.

Saxena, S.C., Kaur, H., Verma P. *et al.,* 2013. Osmoprotectants: Potential for Crop Improvement under adverse conditions. in Plant Acclimation to Environmental Stress, pp. 197-232, Springer, New York, NY, USA.

Schachtman, B.P., Bloon, A.J. and Dvorak, J. 1989. Salt tolerant *Triticum X Lophopyrum* derivatives limit the accumulation of sodium and chloride ions under saline stress. Plant Cell Environ. 12 : 47-55.

Scholander, P.F., Hammel, H.T., Hemmingsen, E. and Garey, W. 1962. Salt balance in mangroves. Plant Physiol. 37 : 722-729.

Schwarz, M. and Gale, J. 1981. Maintenance respiration and carbon balance of plants at low levels of sodium chloride salinity. J. Exp. Bot. 32 : 933-941.

Schwarz, M. and Gale, J. 1984. Growth response to salinity at high levels of carbon dioxide. *J. Exp. Bot.* 35 : 193-196.

Serrano, R., Culianz-Macia, A. and Moreno, V. 1999. Genetic engineering of salt and drought tolerance with yeast regulatory genes. Sci. Hortic. 78: 261-269.

Shannon, M.C. and Nobel, C.L. 1995. Crop physiology and metabolism: variations in salt tolerance and ion accumulation among subterranean clover cultivation. Crop Sci. 35 : 789-804.

Sharma, S.K. 1986. Mechanism of tolerance in rice varieties differing in sodicity tolerance. Plant Soil 93 : 141-145.

Sharma, S.K. and Gupta, I.C. 1986. Saline Environment and Plant Growth. Agro Botanical Publishers, Bikaner, 172 p.

Sharma, S.K., Joshi, Y.C. and Bal, A.R. 1984. Osmotic and ionic effects in salt sensitive and resistance wheat varieties. Indian J. Plant Physiol. 27: 153-158.

Sheen, J. 1996. Ca^{2+} dependent protein kinases and stress signal tranduction in plants. Science 274 : 1900-2.

Shen, Q. and Ho, T-HD. 1995. Functional dissection of ABA-inducible gene reveals two independent ABA-responsible complexes each containing a G-box and a novel cis-acting element. Plant Cell 7 : 295-307.

Shi, H., Quintero, F.J., Pardo, J.M. and Zhu, J.K. 2002. The putative plasma membrane Na^+/H^+ antiporter SOS1 controls long distance Na^+ transport in plants, Plant Cell 14(2): 465-472.

Shinozaki, K. and Yamaguchi-Shinozaki, K. 1996. Molecular responses to drought and cold stress. Curr. Opin. Biotechnol. 7 : 161-67.

Shone, M.G.T. and Gale, J. 1983. Effects of sodium chloride stress and nitrogen source on respiration, growth and photosynthesis in lucerne (*Medicago sativa* L.) J. Exp. Bot. 34 : 1117-1125.

Singh, T.N., Aspinall, D., Paleg, L.G. and Boggess, S.F. 1983. Changes in proline concentration in excised plant tissues. Aust. J. Biol. Sci. 26 : 57-63.

Solomon, A., Beer, S., Waisel, Y., Jones, G.P. and Paleg, L.G. 1994. Effects of NaCl on the carboxylating activity of Rubisco from *Tamarix jordanis* in the presence and absence of proline related compatible solutes. Plant Physiol. 90 : 198-204.

Starvarek, S.J. and Rains, D.W. 1984. The development of tolerance to mineral stress. Hort Science 19: 377-382.

Stewart, G.R. and Lee, J.A. 1974. The role of proline accumulation in halophytes. Planta 120 : 279-289.

Stockinger, E.J., Gilmour, S.J. and Thomashow, M.F. 1997. *Arabidopsis thaliana CBF 1* encodes and AP2 domain-containing transcriptional activator that binds to the C-repeat/DRE, a *cis-acting* DNA regulatory element that stimulates transcription in response to low temperature and water deficit. Proc. Natl. Acad. Sci. USA 94 : 1035-40.

Storey, R. and Wyn Jones, R.G. 1977. Quarternary ammonium compounds in plants in relation to salt resistance. Phytochemistry 16 : 447-453.

Storey, R. and Wyn Jones, R.G. 1979. Responses of Atriplex spongiosa and *Suaeda maritima* to salinity. Plant Physiol. 63 : 151-162.

Subbarao, G.V., Johansen, C., Kumar Rao, J.V.D.K. and Janna, M.K. 1990. Salinity tolerance in F1 hybrid of pigeonpea and a tolerant wild relative. Crop Sci. 30 : 785-788.

Sze, H., Li, X. and Palmgren M.G. 1999. Energization of plant cell membranes by H-pumping ATPases : regulation and biosynthesis. Plant Cell 11 : 677-89.

Tal, M. and Gardi, I. 1976. Physiology of polyploid plants : Water balance in autotetraploid and diploid tomato under low and high salinity. Physiol. Plant. 38: 257-261.

Tal, M. and Gavish, V. 1973. Salt tolerance in the wild relatives of the cultivated tomato. Water balance and abscisic acid in *Lycopersicon esculentum* and *L. peruvinum* under low and high salinity. Aust. J. Agric. Res. 24 : 353-367.

Tattini, M. 1984. Ionic relations of aeroponically grown olive genotypes during salt stress. Plant Soil 161 : 251-256.

Torres, B.C. and Bingham, F.T. 1973. Salt tolerance of Mexican wheat I. Effect of NO_3 and NaCl on mineral nutrition, growth and grain production of four wheats. Soil Sci. Soc. Am. Proc. 37 : 711-715.

Varshney, K.A. and Sangeeta 1992. Accumulation of certain organic metabolites in chickpea during salt stress. Indian J. Plant Physiol. 35 : 295-299.

Waisel, Y. 1972. Biology of Halophytes. Academic Press. New York. 395 p.

Watad, A.E.A., Reuveni, M., Bressan, R.A. and Hasegawa, P.M. 1991. Enhanced net K^+ uptake capcity of NaCl adapted cells. Plant Physiol. 95 : 1265-1269.

Weimberg, R., Lerner, H.R. and Poljakoff-Mayber, A. 1982. A relationship between potassium and proline accumulation in cucumber cotyledons. New Phytol. 77 : 593-598.

Winicov, I. and Bastola D.R. 1999. Transgenic overexpression of the transcription factor *Alfin* 1 enhances expression of the endogenous *MsPRP2* gene in alfalfa and improves salinity tolerance of the plants. Plant Physiol. 120 : 473-80.

Winicov, I. and Shirzadegan, M. 1997. Tissue specific modulation of salt inducible gene expression : callus versus whole plant response in salt tolerant alfalfa. Physiol. Plant. 100 : 314-19.

Wyn Jones, R.G. 1981. Salt tolerance. In : Physiological Processes Limiting Plant Productivity (Ed. C.B. Johnson) pp. 272-292. Butterworths, London.

Wyn Jones, R.G. and Storey, R. 1981. Betaines. In : The Physiology and Biochemistry of Drought Resistance in Plants. (Eds. L.G. Paleg and D. Aspinall) pp. 171-204, Academic Press, New York.

Wyn-Jones, R.G., Gorham, J. and McDonnell, E. 1984. Organic and inorganic solute content as selective criteria for salt tolerance in the Triticeae. In Salinity Tolerance in Plants : Strategies for Crop Improvement, edited by R.C. Staples and G.H. Toenniessen, pp. 189-212. New York : John Wiley.

Yancey, P.H., Clark, M.E., Hand S.C., Bowlus R.D. and Somero, G.N. 1982. Living with water stress : evolution of osmolyte system. Science 217: 1214-22.

Yensen, N.P. 1999. Halophyte Data-Base: Salt tolerant plants and their uses. From internet: http//www.ussl.ars.usda.gov/pls/caliche/halophyte.preface.

Yeo, A. 1998. Molecular biology of salt tolerance in the context of whole plant physiology . Journal Experimental Botany 49 (323) : 915-929.

Yeo, A.R. 1983. Salinity resistance, physiologies and prices. Physiol. Plant. 58 : 214-222.

Yeo, A.R. and Flowers, T.J. 1986. Salinity resistance in rice (*Oryza sativa* L.) and a pyramiding approach to breeding varieties for saline soils. Aust. J. Plant Physiol. 13 : 161-173.

Yeo, A.R. and Flowers, T.J. 1986a. Ion transport in *Suaeda maritima* : its relation to growth and implications for the pathway of radial transport of ions across the root. Journal Experimental Botany 37 : 143-159.

Zhang, H.X., J.N. Hodson, J.P. Williams, and E. Blumwald 2001. Engineering salt-tolerant *Brassica* plants : Characterization of yield and seed oil quality in transgenic plants with increased vacuolar sodium accumulation. Proc. Natl. Acad. Sci. *USA* 98 : 12832-12836.

Zhao, L., Zhang, F., Guo, J., Yang, Y., Li, B. and Zhang, L. 2004. Nitric oxide functions as a signal in salt resistance in the calluses from two ecotypes of reed, Plant Physiol. 134 (2): 849-857.

Zhu, J.K. 2003. Regulation of ion homeostasis under salt stress, Curr. Opinion Plant Biol. 6(5): 441-450.

Zhu, J.K., Hasegawa, P.M. and Bressan, R.A. 1997. Molecular aspects of osmotic stress in plants. Crit. Rev. Plant Sci., 16 : 253-255.

■■■■

Chapter - 10

MOLECULAR MECHANISMS OF SALT TOLERANCE

INTRODUCTION

Soil salinization is a major threat to agriculture in arid and semi-arid regions, where water scarcity and inadequate drainage of irrigated lands severely reduce crop yield. Salt accumulation inhibits plant growth and reduces the ability to uptake water and nutrients, leading to osmotic or water-deficit stress. Salt is also causing injury of the young photosynthetic leaves and acceleration of their senescence, as the Na^+ cation is toxic when accumulating in cell cytosol resulting in ionic imbalance and toxicity of transpiring leaves. To cope with salt stress, plants have evolved mainly two types of tolerance mechanisms based on either limiting the entry of salt by the roots, or controlling its concentration and distribution. Understanding the overall control of Na^+ accumulation and functional studies of genes involved in transport processes, will provide a new opportunity to improve the salinity tolerance of plants relevant to food security in arid regions. A better understanding of these tolerance mechanisms can be used to breed crops with improved yield performance under salinity stress (Hanin *et al.,* 2016).

Plant adaptation or tolerance to salinity stress involves complex physiological traits, metabolic pathways, and molecular or gene networks. A comprehensive understanding on how plants respond to salinity stress at different levels and an integrated approach of combining molecular tools with physiological and biochemical techniques are imperative for the development of salt-tolerant varieties of plants in salt-affected areas. Recent research has identified various adaptive responses to salinity stress at molecular, cellular, metabolic, and physiological levels, although mechanisms underlying salinity tolerance are from being completely understood (Gupta and Huang, 2014). This chapter provides a comprehensive review of major research advances on biochemical, and molecular mechanisms regulating plant adaptation and tolerance to salinity stress.

Crop performance is severely affected by high salt concentrations in soils. To engineer more salt-tolerant plants it is crucial to unravel the key components of the plant salt tolerance network. Recent studies have shown that stress sensing and signaling components may play important roles in regulating the plant salinity stress response. (Deinlein *et al.,* 2014). In addition, the progress that has been made toward engineering salt tolerance in crops, including marker assisted selection and gene stacking techniques has been presented in this chapter.

There are four major salinity tolerance mechanisms in plants which can be practically manipulated for fruiting and genetic engineering of salt tolerance in crops. These are:

1. Antioxidative defence
2. Ion-homeostatis
3. Compatible solutions
4. Transcription factors

In this chapter these mechanisms are described from the molecular aspects based on recent research findings.

Sensing and Signaling to Regulate Plant Salt Stress Response

To tolerate salt in soil solution, plants deploy a variety of traits to control cell function and development that relies on signal perception, signal integration, and processing. The development of high-throughput sequencing technologies during the last few years has generated huge quantity of data and led to the discovery of several signaling molecules. Among them, are those involved in the activation of signaling pathways that can enhance plant ability to tolerate salt stress. ROS, despite their potential toxicity, have the advantage of being versatile signaling molecules with regard to their properties and mobility within cells. The nitogen-activated protein kinase (MAPK) can trigger plant response to biotic and abiotic stresses by activating the antioxidant enzymes. Many different MAPKs cascades are activated upon ROS accumulation. These include the ROS-responsive MAPKKK, MEKK1, MPK4, and MPK6 (Xing *et al.,* 2008; Jammes *et al.,* 2008). ROS signaling is tightly linked to cellular homeostasis and highly integrated with hormonal signaling networks, allowing plants to regulate development processes, as well as adaptive responses to environmental constraints (Miller *et al.,* 2010). Increased generation and accumulation of ROS such as superoxide (O_2^-), hydrogen peroxide (H_2O_2), and nitric oxide (NO) cause oxidative damages in the apoplastic compartment and lipid peroxidation of cellular membranes, and have an extensive impact on ion homeostasis by interfering with ion fluxes (Baier *et al.,* 2005). Excess of ROS levels are particularly scavenged by antioxidants metabolites such as ascorbate, peroxidase (APX), and CAT. Moreover, increased ROS levels can cause salicyclic acid accumulation contributing to plant defense, cell death, and induced stomatal closure (Khokon *et al.,* 2011).

To cope with salt stress, plants have developed the ability to sense both the hyperosmotic and the ionic Na^+ components of the stress. However, the molecular identities of the plant Na^+ and hyperosmotic sensors remain unknown until now. Nevertheless, plant hyperosmotic sensors are likely to be coupled with Ca^{2++} channels due to the rapid rise in cytosolic Ca^{2+} levels within seconds of exposure to NaCl or mannitol. The Ca^{2+} signal occurs in roots and in several cell types (Maiti *et al.,* 2013). This finding indicates that hyperosmotic stress may be sensed by a mechanically gated Ca^{2+} channel (Kurusu *et al.,* 2013). In a recent study, Choi *et al.* (2014) suggested that Ca^{2+} - dependent signaling plays a role in the process of systemic signaling in response to salt stress. Thus, local salt stress of the root tip leads to the spreading of a Ca^{2++} wave that propagates preferentially through cortical and endodermal cells to distal shoot tissues at speeds of up to 400 μm/s. Salt stress induced long-distance Ca^{2+} wave is dependent on the activity of the ion channel protein Two Pore Channel 1 (TPC1), which appears to contribute to whole-plant stress tolerance. Under salinity the increase in cytoplasmic Na^+ and reduction of K^+ result in changes of membrane potential, osmotic pressure, turgor pressure, turgor pressure, calcium signaling, ROS signaling, etc. Recent results on ion fluxes in glycophyte *Arabidopsis thaliana* and the halophytic relative *Thellungiella halophila* showed lower Na^+ fluxes and higher K^+/Na^+ selectivity of ion currents in the roots and root protoplasts of the halophyte under salt treatment (Volkov and Amtmann, 2006; Amtmann, 2009). Maintenance of K^+ homeostasis is essential for enzyme activities, ionic and pH homeostasis, and cytosolic K^+ is considered to be an attribute of plant adaptive responses to a broad range of environmental constraints (Shabala and Pottosin, 2014). In addition, a strong correlation between the K^+ retention ability and plant salinity stress tolerance was reported for several species including wheat (Cuin *et al.,* 2012), barley (Wu *et al.,* 2015), and *Brassica* species (Chakraborty *et al.,* 2016).

On another hand, it is well established that electrolyte leakage which is considered as a hallmark of plant cell response to abiotic (including salinity) and biotic stresses, is based mainly on K^+ efflux. This stress induced K^+ leakage is often accompanied by ROS generation and leads to cell death (Demidchik *et al.,* 2014). In addition, there are K^+ outwardly rectifying channels that are induced by ROS and especially hydroxyl radicals (Demidchik *et al.,* 2010). Therefore, under stress K^+ leakage, ROS and plant cell death (PCD) seem to be intimately connected. Such hypothesis was elegantly strengthen by Demidchik *et al.* (2010) who showed through pharmacological and genetic approaches that blocking K^+ channel and the lack of functional GORK both inhibit the stress-induced activation proteases and endonucleases that led to PCD.

Transport of Sodium and Detoxification Pathways

Na^+ influx pathways into roots occur via different channels and transporters. Na^+ may cross the plasma membrane through nutrient channels and calcium permeable nonselective cation channels (NSGCs), including the cyclic nucleotide-gated channel (CNGC) and the glutamate-like receptor (GLR), which represent a likely entry point of Na^+ into the cell (Tyerman and Skerrett, 1999; Deinlein *et al.,* 2014). The role of NSCG is not restricted to mediate Na^+ influx, which triggers K^+ efflux via KORG, but they can also contribute to K^+ efflux directly. Under salt stress and for efficient storage of Na^+ to the vacuole, NSCG may play important role by preventing Na^+ leak to the cytosol and without perturbing K^+ release (Pottosin and Dobrovinskaya, 2014).

The several KHT-type transporters were shown to be involved in Na^+/K^+ symport in different plant species including *Arabidopsis,* rice and wheat. Based on sequence and transport analyses, HKT transporters can be classified into two distinct subgroups class I and II, with the first being more Na^+ -selective transport and the second as Na^+-K^+ co-transporter (Munns and Tester, 2008; Deinlein *et al.,* 2014). The rice Na^+ transporter OsHKT2; 1 (previously named OsHKT1) has been shown to mediate Na^+ influx into roots under K^+ starvation (Horie *et al.,* 2007). Other members of the HKT gene subfamily 1, are thought to mediate Na^+ influx in root cells and to regulate the Na^+ distribution between roots and shoots (Horie *et al.,* 2009).

To balance the toxic effect of Na^+ accumulation and to control ion homeostasis during salinity stress, plants require the maintenance of stable K^+ acquisition and distribution. The tonoplast-localized Na^+/H^+ exchangers (NHX1 and NHX2) and the plasma membrane-localized

Na^+/H^+ antiporter (SOS1), are thought to play important roles in osmoregulation and to maintain low cytoplasmic Na^+ concentration in plant cells (Shi *et al.,* 2002; Brini *et al.,* 2007; Olias *et al.,* 2009). Most NHX proteins are essential for Na^+ detoxification through sequenstration into the vacuole, whereas, SOS signaling pathways were responsible to export Na^+ outside the cell. The concept that vacuolar NHX proteins were capable of exchanging Na^+ and H^+ across the tonoplast as claimed by Apse *et al.* (1999) and Blumwald (2000).

Adaptive Mechanisms of Salt Tolerance at the Cell and Organ Level

Research carried out on salinity tolerance mechanisms was mainly performed on model plants (i.e., *A. thaliana*) and only few cases were reported on crop plants. In glycophytes, salinity tolerance is mainly achieved through more than one strategy operating either simultaneously or separately, depending on the duration and intensity of the stress. According to Chakraborty *et al.* (2016), the overall superior salinity tolerance observed in *Brassica napus* was achieved by the high osmotolerance matched by the moderate tissue tolerance and superior K^+ retention ability in the leaf mesophyll. Moreover, Chakraborty *et al.* (2016) provide strong evidence that higher salt tolerance in *B. napus* is conferred by at least three complementary physiological mechanisms : (i) higher Na^+ extrusion ability from roots correlated with increased expression and activity of plasma membrane SOS1-like Na^+/H^+ exchanges; (ii) better root K^+ retention ability resulting from stress-inducible activation of H^+. ATPase and the ability to maintain a more negative membrane potential under saline conditions; and (iii) reduced sensitivity of *B. napus* root K^+ permeable channels to ROS. Shabala *et al.* (2015) has argued that changes in SOS1 activity strongly correlated with changes in net K^+ and H^+ fluxes, making the involvement of the H^+-ATPase/GORK tanden system as a potential sensor a plausible hypothesis. Therefore, it was suggested that Na^+ exclusion from uptake plays an important but not a crucial role as a determinant of genetic variability in salinity stress tolerance in *Brassica.*

Transcriptional Regulation and Gene Expression of Salinity Tolerance

Regulation of gene expression in salinity stress includes a wide array of mechanisms that are used by plants to upregulate or downregulate (increase or decrease) the production of specific gene products (protein or RNA). Various mechanisms of gene regulation have been identified during the central dogma, from transcriptional initiation, to RNA processing, and to the post-translational modification of a protein.

Transcriptomic analysis provides detailed knowledge about the gene expression at mRNA level, which is widely used to screen candidate genes involved in stress responses. Genomic approaches play a significant role in encoding, cloning, and characterization of important genes. A huge number of salt-responsive transcription factors and genes which are either upregulated or downregulated in response to salinity stress have been identified and characterized using transcriptomic and genomic approaches.

A large number of genes and transcription factors are upregulated in response to salinity in different plant species, which serve diverse functions (Gupta and Huang, 2014). Examples of salt responsive genes are listed in the Table 10.1, and these genes are mainly classified into the following functional categories: ion transport or homeostasis (e.g., SOS genes, *AtNHX1,* and *H^+-ATPase*), senescence-associated genes (e.g., *SAG*), molecular chaperones (e.g., *HSP* genes), and dehydration-related transcription factors (e.g., *DREB*). Among stress-responsive genes, the *SOS* gene family is believed to play a very intriguing role in ion homeostasis, thereby conferring salt tolerance. Some ROS-scavenging and osmotic-regulating genes are also upregulated by salinity in some plant species (Shi *et al.,* 2002). For example, a continuous exposure of rice plants to salinity for about 24 hours resulted in upregulation of glutathione-S-transferase and ascorbate peroxidase, both of which were known to play an active role in ROS scavenging, and with the increase in duration of exposure to salinity stress, upregulation of metallothionein and water channel proteins was also observed (Kawasaki *et al.,* 2001). Halophyte plant species *Spartina alterniflora* when subjected to salt stress exhibits upregulation of 10 genes associated with osmotic regulation (Kawasaki *et al.,* 2001).

Recently, Schmidt *et al.* (2013) identified *SALT-RESPONSIVE ERF1 (SERF1),* a rice (*Oryza sativa*) transcription factor gene that showed a root-specific induction upon salt and H_2O_2 treatment. Loss of *SERF1* impaired the salt-inducible expression of genes encoding members of a nitogen-mediating TFs. Furthermore, they showed that SERF1-dependent genes are H_2O_2 responsive and demonstrated that SERF1 binds to the promotors of *MAPK KNASE KNIASE6 (MAP3K6), MAPK5, DEHYDRATION-RESPONSIVE ELEMENT BINDING2A (DREB2A),* and *ZINC FINGER PROTEIN179 (ZFP179) in vitro* and *in vivo,* SERF1 also directly induces its own gene expression. In addition, it was observed that SERF1 is a phosphorylation target of

MAPK5, resulting in enhanced transcriptional activity of SERF1 toward its direct target genes. Finally, they demonstrated that the plants deficient for SERF1 are more sensitive to salt stress compared with the wild type, while constitutive overexpression of SERF1 improves salinity tolerance.

Table 10.1. Examples of upregulated genes in response to salinity stress (Adapted from Gupta and Huang, 2014)

Species	NaCl concentration (mM)	Gene name	Gene functions
Brassica juncea and *Brassica campestris*	25 and 50	*SOS1* *SOS2* *SOS3* *AtNHX1*	(i) Plasma membrane Na^+/K^+ antiporter. (ii) Protein kinase (iii) Calcium-binding protein (iv) Vacuolar Na^+/K^+ antiporter
Oryza sativa	50	*PRP* *SAG* *HSPG025*	(i) Proline rich proteins and cell-wall protection. (ii) Senescence associated genes, regulatory processes, and cellular signal transduction (iii) Heat-shock proteins, protein stabilizing
Oryza sativa	100	*OsHSP23.7* *OsHSP71.1* *OsHSP80.2*	Heat-shock proteins, molecular chaperones, folding, assembling, and transporting proteins
Arabidopsis thaliana	150	*AtSKIP*	Transcription factor, transcriptional pre-initiation, splicing, and polyadenylation
Oryza sativa	200	*OsHsp17.0* *OsHsp23.7*	Heat-shock proteins, molecular chapterones, and folding, assembling, and transporting proteins
Carrot	300	*DcHsp17.7*	Cell viability and membrane stability under heat stress
Arabidopsis thaliana	300	*JcDREB*	Transcription factor.

There are some transcription factors which are regulated by different kinases and have been found to be significant players of plant adaptation to salinity stress. Serra *et al.* (2013) showed that *OsRMC* encodes a receptor-like kinase described as a negative regulator of salt stress responses in rice. Two transcription factors,

OsEREBP1 and OsEREBP2, belonging to the AP2/ERF family were shown to bind to the same GCC-like DNA motif in *OsRMC* promoter and to negatively regulate its gene expression. Serra *et al.* (2013) further revealed that *OsEREBP1* transcript level is not significantly affected by salt, ABA, or severe cold (5°C) and is only slightly regulated by drought and moderate cold. On the other hand, the *OsEREBP2* transcript level increased after cold, ABA, drought, and high salinity treatments, indicating that *OsEREBP2* may play a central role mediating the response to different abiotic stresses. Gene expression analysis in rice varieties with contrasting salt tolerance further suggests that *OsEREBP2* is involved in salt stress response in rice, A bZIP class of ABRE binding transcription factor, known as OSBZ8, has also been identified from rice and has been shown to be highly expressed in salt tolerant cultivars than in salt sensitive one.

Downregulated genes are emerging now as essential components of the response to salinity. For example downregulation of β-carotene hydroxylase increase β-carotene and total carotenoids enhancing salt stress tolerance in transgenic cultured cells of sweet potato. It seems that mutual regulation mechanisms exists between different genes and proteins and signals underlying different processes of plant adaptation to abiotic stress (Gupta and Huang, 2014).

In addition to protein coding genes, recently discovered microRNAs (miRNAs) and endogenous small interfering RNAs (siRNAs) have emerged as important players in plant stress responses. Initial clues suggesting that small RNAs are involved in plant stress responses stem from studies showing stress regulation of miRNAs and endogenous siRNAs, as well as from target predictions for some miRNAs.

Proteomic and Metabolic Responses to Salinity Stress

Genomics technologies have helped to address the multigenicity of the plant abiotic stress responses. Analysis of genome sequences, and specific transcript collections and their dynamic changes, has provided a more global picture of stress-dependent responses at the cell, tissue, and whole plant level and moved the field from a single-gene approach toward an understanding of interactions between multiple components in cells, facilitating the dissection of abiotic stress circuits and coexpression hubs (Barkla *et al.,* 2013). However, directly focusing on genes may not accurately portray conditions in the cell at a particular state and time during stress due to regulation at the RNA and protein level, including post-translational regulation. Proteomics, and in particular quantitative proteomics, is emerging as a powerful technique to be applied to the field of crop abiotic stress tolerance research; it has the potential to allow rapid identification and quantification of novel stress and tolerance-related proteins. Understanding the dynamics of expression and postranslational modifications of these proteins, and gaining direct insight into their function and interactions, can provide essential information that can be applied to engineer stress-tolerant crops with novel traits through biomarker selection and transgenic strategies (Barkla *et al.*, 2013; Gupta and Huang, 2014). Available data suggest that several common stress responsive proteins are expressed in response to various abiotic stresses in different plant species. About 2171 proteins from 34 plant species have been identified and characterized as salt-responsive proteins, which are either upregulated or downregulated by salinity stress (Zhang *et al.*, 2012). Based on gene ontology; BLAST alignment and literature information, salt-responsive proteins can be grouped into 14 functional categories. Specifically, there appears to be a general regulation of proteins involved in carbohydrate, nitrogen and energy metabolisms, with particular emphasis on glycolytic and Krebs cycle enzymes.

Another significant research approach in plant system biology is the metabolomics which involves the study of metabolome. Higher plants have a remarkable ability to synthesize a vast array of metabolites that differ in chemical complexity and biological functions playing an indispensible role in stress alleviation (Saito and Matsuda, 2010). Examples of plant metabolites that are involved in salinity tolerance include polyols such as mannitol and sorbitol, dimethylsulfonium compounds, glycine betaine, sugars such as sucrose, trehalose and fructans, or amino acids such a s proline that serve as an osmolyte or osmoprotectant. Plants when subjected to salinity stress show an increase in the concentration of these osmolytes thus playing significant role in stress mitigation. Results indicate that the metabolic processes are highly specific for given tissues, species, and plant-environment interactions. The clusters of identified compounds not only serve as base in the quest of novel defense compounds but also as markers for the characterization of the plants' defensive state. The latter is especially useful in agronomic applications where meaningful markers are essential for crop protection (Gupta and Huang, 2014).

Bioengineering for Improving Salinity Tolerance

Conventional plant breeding approaches through which beneficial traits can be introgressed into elite varieties have been adopted since a long time to generate stress tolerant varieties (Varshney *et al.*,

2011). Hence, traditional breeding allowed the development of new salt-tolerant rice and wheat varieties (Munns *et al.,* 2006; Singh *et al.,* 2010). However, as salinity tolerance is a multigenic trait, such approaches have only limited success, which can explain the absence of commercially available salt-tolerant crops. Moreover, as a procedure, plant breeding is time consuming and labor intensive, relies on well characterized germplasms and can result in introducing undesirable traits along with the selected one. Therefore, biotechnological approaches including molecular breeding and genetic engineering seem to be more attractive alternatives.

In the last 30 years, tremendous progress has been made toward the isolation and molecular characterization of genes involved in plant salt stress responses. These candidate genes can be classified into two main groups: effectors and regulatory genes. The first group includes mainly genes encoding ion transporters, channels, enzymes involved in osmolyte biosynthesis, antioxidant systems and protective proteins such as heat shock proteins and late embryogenesis abundant (LEA) proteins. The second group is composed of genes involved in transcriptional and post-transcriptional regulation as well as in signaling pathways. Within this group we find essentially transcription factors, protein kinases, phosphatases, and proteases. Numerous reports described how the overexpression of these genes can improve plant tolerance to various abiotic stresses including salinity (Ashraf, 2009; Turkan and Demiral, 2009; Cominelli *et al.,* 2013; Roy *et al.,* 2014) (Table 10.2).

Genetic engineering as a solution to abiotic stress tolerance is based on introgression and expressions of genes involving signaling and controlling pathways. There may also be genes encoding proteins producing stress tolerance or enzymes related to metabolic pathways resulting in the functional and structural metabolite synthesis (Roy *et al.,* 2014). The candidate genes involved in introducing salinity tolerance by genetic engineering are listed below.

S.No.	Genes involved in salinity tolerance by Genetic engineering
1	Antioxidant Defence System
2	Ion Homeostatis and Ion Transport
3	Compatible Solutes
4	Transcription Factors

These mechanisms have been dealt in great details by Bahmani *et al.* (2015) in a recent review where all probable molecular aspects have been considered. It has been suggested that when traditional methods are not applicable and feasible plant breeders should use marker assisted selection and genetic engineering to develop salinity tolerant crops. However, to make a better use of new technologies like engineering of genes, a complete and deeper understanding of molecular and cellular basis of salt tolerance mechanisms is necessary.

Table 10.2. Example of genes leading to improvement of salt stress tolerance of crop plants through genetic engineering (Adapted from Hanin *et al.*, 2016)

Transgene	Function	Donor	Transgenic plant	Description	Reference
Na^+/H^+ antiporter (*AtNHX1)*	Vacuolar sequestration of Na^+ and K^+?	*Arabidopsis*	Tomato	Enhanced salt tolerance with higher accumulation in leaves but not in fruits	Znag and Blumwald, 2001
			Brassica napus	Maintenance of seed yield and seed oil quality under high salinity	Zhang *et al.*, 2011
			Wheat	Improved grain yield in saline soils	Zue *et al.*, 2013
			Cotton	Increased fiber yield under salt stress	He *et al.*, 2005
H^+-pyrophosphatase (AVP1) *AtNHX* + *AVP1*	Vacuolar membrane-bound proton pump	*Arabidopsis*	Cotton	Increased fiber yield under salt stress in field conditions	Pasapula *et al.*, 2011
			Barley	Higher biomass production and grain yield in saline field	Schilling *et al.*, 2013
			Cotton	Further enhancement of salt tolerance compared to single- gene overexpressing plants	Shen *et al.*, 2014

Transgene	Function	Donor	Transgenic plant	Description	Reference
			Tomato	Increased salt stress tolerance compared to single gene over expression	Goulab and Khoudi, 2015
Ncl	Homologous to NHX gene family	Soybean	Soybean	Improved salt tolerance	Do *et al.*, 2016
Mannitol-1-phosphate dehydrogenase *mt1D*	Mannitol biosynthesis	*E. coli*	Tobacco	Increased salt tolerance	Tarczynski *et al.*, 1992
Choline synthase *CodA*	Betaine biosynthesis	*E. coli*	Rice	Enhanced tolerance to salinity and cold	Sakamoto *et al.*, 1998
delta 1-pyrr-oline-5- carbo-xylate synt-hase (*P5CS*)	Proline biosynthesis	*Arabidopsis*	Tobacco	Increased tolerance to drought and salt stress	Kishor *et al.*, 1995
Ascorbate peroxidase (*AtAPX*)	ROS-scavenging	*Arabidopsis*	Tobacco	Enhanced tolerance to salt and osmotic stress	Badawi *et al.*, 2004
Late embryo-genesis abu-ndant protein (*HVA7*)	Osmo-protection	Barley	Rice	Enhanced tolerance to salt and osmotic stress	Xu *et al.*, 1996
Transcription factor *DREB1B/ CBF1*	Transcription regulation	*Arabidopsis*	Rice	Enhanced tolerance to salinity and drought	Oh *et al.*, 2005
Transcription factor *DREB1AS/ CBF3*	Transcription regulation	*Arabidopsis*	Rice	Increase in salinity tolerance	Hu *et al.*, 2006
SNAC1	Transcription regulation	Rice	Rice	Increase in salinity tolerance	Hu *et al.*, 2006

Transgene	Function	Donor	Transgenic plant	Description	Reference
Transcription factor *SIAREB1*	Transcription regulation	Tomato	Tomato	Enhanced tolerance to salinity and water stress	Orellana *et al.*, 2010
Transcription factor *ZFP179*	Transcription regulation	Rice	Rice	Increase in salinity tolerance	Sun *et al.*, 2010
Calcium-dependent protein kinase *OsCDPK21*	Calcium signaling	Rice	Rice	Increase in salinity tolerance	Asano *et al.*, 2011
MAP kinase *GhMPK2*	MAPK signaling	Cotton	Tobacco	Enhanced tolerance to salinity and drought	Zhang *et al.*, 2011

Limitations of Bioengineering

Despite the major advances in our understanding of the mechanisms of salt tolerance there are several factors in the way of engineering commercial crops more tolerant to salinity. Firstly the laboratory conditions are quite different from field conditions. There is a need to consider field tests because of the reproducibility factors and to measure and monitor environmental conditions that may affect crop yields.

Secondly, stress tolerance of transgenic crop grown in a greenhouse or a growth chamber is assessed by measuring survival or rate of recovery. However, for the farmers the main trait for stress tolerant is the crop yield. Therefore, the need to conduct field trials under a given set of environmental conditions is necessary condition to ensure the development of sustainable genetically improved salt-tolerant crops.

Ion Transporters

Ion transporters are obvious candidates among effectors since they participate in salt detoxification and ion homeostasis. Overexpression of genes involved in Na^+ transport was therefore, largely investigated. The *Arabidopsis* vacuolar Na^+/H^+ antiporter (AtNHX1) was among the first candidate genes which when overexpressed lead to enhanced salinity tolerance (Apse *et al.,* 1999). These transgenic plants were reported to exhibit higher ability for vacuolar sequestration of Na^+ to avoid its toxic accumulation into the cytoplasm. Later on, overexpression of AtNHX1 and related NHX proteins from various sources have been shown to increase salt tolerance in other plant species including tomato (Zhang and Blumwald, 2001), *B. napus,* wheat and cotton (Munns and Tester, 2008). Interestingly, the tomato AtNHX1 overexpressing plants are more salt stress tolerant, with an accumulation of NaCl in leaves but not in fruits. However, as was indicated above, its role in salt stress tolerance may also result from its capacity to facilitate K^+ uptake at the tonoplast (Leidi *et al.,* 2010). Nevertheless, engineering of vacuolar cation/H^+ antiporter coupled with its H^+-translocating pyrophosphatase (H^+-PPase) often leads to increased salt stress tolerance. Therefore, overexpression of the wheat TNHX1 and TVP1 improves salt stress tolerance in *Arabidopsis* (Brini *et al.,* 2007), toabcco and tomato (Gouiaa and Khoudi, 2015). Again the *Arabidopsis* H^+-PPase (AtAVP1) was also demonstrated to increase stress tolerance in a crop plant. Indeed, Schilling *et al.* (2013) reported that barley transgenic plants overexpressing AtAVP1 are more tolerant to salinity under greenhouse conditions but they also showed increased shoot biomass production and grain yield in a saline field.

On another hand, the HKT family and the SOS pathway play also a relevant role in controlling Na^+ transport within the plant and have been considered in transgenic approaches to increase salt stress tolerance in crops. However, knowing their role in the removal of Na^+ from the xylem sap into the surrounding xylem parenchyma cells, improving salinity tolerance using HKTs can be successful only when its expression is targeted to the stele or driven by salt-inducible promoter (Moller *et al.,* 2009; Roy *et al.,* 2014). Recently, the overexpression of *Ncl* gene (homologous to the Na^+/H^+ antiporter gene family) into a Japanese soybean salt-sensitive cultivar Kariyutaka, resulted in improved salt tolerance in transgenic soybean. A close association was observed between the high expression of the salt tolerance gene *Ncl* in the root, the lower accumulation Na^+, K^+, and Cl^- in the shoot under salt stress and salt tolerance in the transgenic lines (Do *et al.,* 2016).

Osmolyte Accumulation

Under salt stress, along with Na^+ exclusion from the cytoplasm, plant cell accumulates a wide range of compatible solutes or osmolytes to balance the osmotic pressure of ions in vacuoles. Sucrose, proline, and glycine betaine are among the most studied osmolytes accumulating upon salt stress in some plant species including halophytes. Therefore, plant engineering for higher accumulation of these compounds was considered as a possible way in improving crop tolerance to salinity. The bacterial *mt1D* gene encoding the mannitol-1- phosphate dehydrogenase, enzyme involved in mannitol biosynthesis was, early on, successful in enhancing salt tolerance after its ectopic expression in tobacco (Tarczynski *et al.,* 1993). Similarly, transgenic rice overexpressing choline oxidase showed increased levels of glycine betaine and enhanced tolerance to salinity and cold (Sakamoto *et al.,* 1998). Also, plants such as tobacco transformed with delta 1-pyrroline-5-carboxylate synthase (P5CS) gene exhibited higher proline production, correlated with increased tolerance to drought and salt stress (Kishor *et al.,* 1995). Similarly, transgenic rice plants expressing the mothbean P5CS gene under constitutive or stress inducible promoter showed significant tolerance to high levels of NaCl (Su and Wu, 2004).

Antioxidant Systems and Protective Proteins

Despite their importance as signaling molecules regulating cellular responses to various stresses (Apel and Hirt, 2004), ROS can also damage plant tissues during salinity stress by perturbing enzyme, cell wall and membrane function. Plants detoxify then ROS generated by salt stress by upregulating antioxidative enzymes such as SOD, CAT, APX and glutathione peroxidase. Therefore, overexpressing ROS-scavenging enzymes were shown to promote tolerance of plants to various stresses including salinity (Miller *et al.,* 2008). For example, the overexpression of ascorbate peroxidase in tobacco chloroplasts enhances the tolerance to salt stress and water deficit (Badawi *et al.,* 2004).

Other proteins like osmotin and LEA proteins contribute in alleviating salt stress by protecting macromolecules from damages caused by ion toxicity and/or water deficit. HVA7, a LEA from barley, when transferred to rice, confers water and salt stress tolerance (Xu *et al.,* 1996). It is worth to note that, among the different groups, the group 2 of LEA proteins known as dehydrins are particularly interesting and were shown to enhance plant tolerance to various stresses (Hanin *et al.,* 2011). Indeed Brini *et al.* (2007), showed that the expression of the wheat dehydrin DHN-5 in *A. thaliana* led to an increase in salt and osmotic stress tolerance, but also evidence is provided for the involvement of DHN-5 in other abiotic stress responses (Brini *et al.,* 2011; Drira *et al.,* 2015).

Transcription Factors and Signaling Proteins

Plant response to salinity is complex and involves multiple genes involved in distinct or overlapping regulatory pathways. Therefore, the engineering of a single downstream effector gene might have limited success when one considers multiple stress combination as occurring in the field. In contrast, regulatory proteins such as transcription factors and signaling proteins will gain increasing interest as they are expected to modulate the expression of numerous downstream genes involved in stress responses. It is well documented that transcription factors belonging to the families of DREB, NAC, MYB, MYC, Cys2/His2 zinc finger, bZIP, AP2/ERF, and WRKY are relevant in salt stress tolerance (Golldack *et al.,* 2011, 2014).

In this regard, several transcription factors such as DREBs, MYCs, AP2/ERFs, and NACs were tested in model plant species and few crops. In some cases, the overexpression of these transcription factors was successful to enhance salinity tolerance in crops.

The expression of DREB1B/CBF1 or DREB1A/CBF3 under the control of the cauliflower mosaic virus 35S promoter in *Arabidopsis* plants increases significantly tolerance to freezing, drought, and high salinity stresses (Kasuga *et al.,* 1999). Similarly, transgenic rice plants constitutively expressing DREB1A/CBF3 were reported to be more tolerant to drought and salinity (Oh *et al.,* 2005).

SIAREB1 is a bZIP transcription factor from tomato (*Solanum lycopersicum*), member of the ABA responsive element binding protein (AREB)/ABA-responsive element binding factor (ABF) subfamily. Its overexpression in tomato was reported to improve tolerance to water and salt stresses (Orellana *et al.,* 2010). Likewise, Sun *et al.* (2010) have shown that the overexpression of ZFP179, a salt responsive gene encoding a Cys2/His2 zinc finger protein enhanced salt tolerance in rice.

Moreover, effects of the overexpression of genes encoding signaling proteins such as kinases and phosphatases on salt tolerance were reported. As conserved signaling proteins at the crossroads of several cascades, MAPKs play pivotal role in plant responses to various stresses.

In addition, CDPKs, which are involved in salt stress response, were found efficient in transgenesis approaches, as was reported for the overexpression of *OsCDPK21* that resulted in increased salinity tolerance of transgenic rice (Asano *et al.,* 2011).

CONCLUSIONS AND FUTURE PERSPECTIVES

Salinity tolerance involves a complex of responses at molecular, cellular, physiological and whole plant levels. Research during last two decades has revealed that among various salinity responses, mechanisms and strategies controlling ion uptake, transport and balance, osmotic regulation, hormone metabolism, antioxidant metabolism and stress signaling play critical role in plant adaptation to salinity stress.

Furthermore, research conducted on *Arabidopsis* and several crops over the last 20 years have added valuable information on the molecular mechanisms controlling the salt tolerance in plants. Available tools including molecular breeding and advanced biotechnology methods when combined may speed up the release of salt-tolerant crop varieties. To achieve our goals, we need to have diversity of germplasm resources, high-throughout phenotyping platforms, genome sequencing of crops and their relatives. Many traits related to salt stress tolerance have been identified and shown effective for engineering stress tolerance in model plants. Finally, the targeted genome editing using CRISPR-CaS 9 technology has emerged as an alternative to classical plant breeding and transgenic methods (Zhang *et al.,* 2014). However, in the coming years an indepth study of genome editing to engineer mechanisms of salt stress tolerance needs to be undertaken.

Presently, plant biologists are focusing on the development of a complete profile of genes, proteins and metabolites responsible for different mechanisms of salt tolerance in different species. This has become possible due to the latest developments in the field of genomic, transcriptomic, proteomic and metabolmic techniques. However, a combined approach is essential for the determination of the key pathways and processes controlling salinity tolerance.

Therefore, future focus should be on the study of intercellular and intracellular molecular interactions involved in salinity stress response. In future years, genetic engineering will become more powerful as more candidate genes associated with salt tolerance will be identified and utilized widely.

REFERENCES

Amtmann, A. 2009. Learning from evolution. *Thellungiella* generates new knowledge on essential and critical components of abiotic stress tolerance in plants. *Mol. Plant* 2: 3-12.

Apel, K. and Hirt, H. 2004. Reactive oxygen species metabolism, oxidative stress and signal transudction. *Annu. Rev. Plant Biol.* 55: 373-399.

Apse, M.P., Aharon, G.S., Snedden, W.A. and Blumwald, E. 1999. Salt tolerance conferred by overexpression of a vacuolar Na^+/H^+ antiport in *Arabidopsis. Science* 285: 1256-1258.

Asano, T., Hakata, M., Nakamura, H., Aoki, N., Komatsu, S., Ichikawa, H., *et al.,* 2011. Functional characterisation of OSCPK21, a calcium-dependent protein kinase that confers salt tolerance in rice. *Plant Mol. Biol.* 75: 179-191.

Ashraf, M. 2009. Biotechnological approach of improving plant salt tolerance using antioxidants as markers. *Biotechnol. Adv.* 27: 84-93.

Ashraf, M. and Foold, M.R. 2013. Crop breeding for salt tolerance in the era of molecular markers and marker-assisted selection. *Plant Breed.* 132: 10-20.

Badawi, G.H., Kawani, N. and Yamauchi, Y. 2004. Over-expression of ascorbate peroxidase in tobacco chloroplasts enhances the tolerance to salt stress and water deficit. *Physiol. Plant.* 121: 231-238.

Baier, M., Kandlbinder, A., Golldack, D. and Dietz, K.J. 2005. Oxidative stress and ozone perception, signalling and response. *Plant Cell Environ.* 28: 1012-1020.

Barkla, B.J., Vera-Estrella, R. and Pantoja, O. 2013. Progress and challenges in a biotic stress proteomics of crop plants. *Proteomics* 13: 125-140.

Blumwald, E. 2000. Sodium transport and salt tolerance in plants. *Curr. Opin. Cell. Biol.* 12: 431-434.

Brini, F., Hanin, M., Mezghani, I., Berkowitz, G.A. and Masmoudi, K. 2007. Overexpression of wheat Na^+/H^+ antiporter TNHX1 and H^+ - pyrophosphatase TVP1 improve salt-and drought stress tolerance in *Arabidopsis thaliana* plants. *J. Exp. Bot.* 58: 301-308.

Brini, F., Yazmamoto, A., Jlaiel, L., Takeda, S., Hobo, T., Dinh, H.Q., *et al.,* 2011. Pleiotropic effects of the wheat dehydrin DHN-5 on stress responses in *Arabidopsis. Plant Cell Physiol.* 52: 676-688.

Chakraborty, K., Bose, J., Shabala, L. and Shabala, S. 2016. Difference in root K^+ retention ability and reduced sensitivity of K^+ permeable channels to reactive oxygen species confer differntial salt tolerance in three *Brassica* species. *J. Exp. Bot.* 67: 4611-4625.

Choi, W.G., Toyota, M., Kim, S.H., Hilleary, R. and Gilroy, S. 2014. Salt stress-incuded Ca^{2+} waves are associated with rapid, long-distance root-to-shoot signaling in plants. *Proc. Natl. Acad. Sci. U.S.A.* 111: 6497-6502.

Cominelli, E., Conti, L., Tonelli, C. and Galbiati, M. 2013. Challenges and perspectives to improve crop drought and salinity tolerance. *Nat. Biotechnol.* 30: 355-361.

Cuin, T.A., Zhou, M., Parsons, D. and Shabala, S. 2012. Genetic behaviour of physiological traits conferring K^+/Na^+ homeostasis in wheat. *Plant Biol.* 14: 438-446.

Deinlein, U., Stephan, A.B., Horie, T., Luo, W., Xu, G. and Schroeder, J.I. 2014. Plant salt-tolerance mechanisms. *Trends Plant Sci.* 19: 371-379.

Deinidchik, V., Cuin, T.A. and Svistunenko, D. 2010. *Arabidopsis* root K^+ efflux conductance activated by hydroxyl radicals single channel properties, genetic basis and involvement in stress-induced cell death. *J. Cell. Sci.* 123: 1468-1479.

Demidchik, V., Straltsova, D., Medvedev, S.S., Pozhvanov, G.A., Sokolik, A. and Yurin, V. 2014. Stress-induced electrolyte leakage: the role of K^+ permeable channels and involvement in promed cell death and metabolic adjustment. *J. Exp. Bot.* 65: 1259-1270.

Do, T.D., Chen, H., Vu, H.T.T., Hamwieh, A., Yamada, T., Sato, T., *et al.*, 2016. Net synchronously regulates Na^+, K^+ and Cl^- in soybean greatly increases the grain yield in saline field conditions. *Sci. Rep.* 6: 19147.

Drira, M., Saibi, W., Amara, I., Masmoudi, K., Hanin, M. and Brini, F. 2015. Wheat dehydrin K-segments ensure bacterial stress tolerance, antiaggregation and antimicrobial effects. *Appl. Biochem. Biotechnol.* 175: 3310-3321.

Golldack, D., Li, C., Mohan, H. and Probst, N. 2014. Tolerance to drought and salt stress in plants unraveling the signaling networks. *Front. Plant Sci.* 5: 151.

Golldack, D., Lucking, I. and Yang, O. 2011. Plant tolerance to drought and salinity: stress regulating transcription factors and their functional significance in the cellular transcriptional network. *Plant Cell Rep.* 30: 1383-1391.

Gouiaa, S. and Khoudi, H. 2015. Co-expression of vacuolar Na^+/H^+ antiporter and H^+-pyrophosphatase with an IRES-mediated dicistronic vector improves salinity tolerance and enhances potassium biofortification of tomato. *Phytochemistry* 117: 537-546.

Gupta, B. and Huang, B. 2014. Mechanisms of salinity tolerance in plants: physiological biochemical and molecular characterization. *International J. Genomics.*

Hanin, M., Brini, F., Ebel, C., Toda, Y., Takeda, S. and Masmoudi, K. 2011. Plant dehydrins and stress tolerance versatile proteins for complex mechanisms. *Plant Signal. Behav.* 6: 1503-1509.

Hanin, M., Ebel, C., Ngom, M., Laplazi, L. and Masmoudi, K. 2016. New insights on plant salt tolerance mechanisms and their potential use for breeding. *Frontiers Plant Sci.* 7: 1787-1803.

Horie, T., Hauser, F. and Schroeder, J.I. 2009. HKT transporter-mediated salinity resistance mechanisms in *Arabidopsis* and monocot crop plants. *Trends Plant Sci.* 14: 668.

Jammes, F., Song, C., Shin, D., Munemasa, S., Takeda, K., Gu, D., *et al.*, 2009. MAP kinases MPK9 and MPK12 are preferentially expressed in guard cells and positively regulate ROS-mediated ABA signaling. *Proc. Natl. Acad. Sci. U.S.A.,* 106: 20520-20525.

Kasuga, M., Liu, Q., Miura, S., Yamaguchi-Schinozaki, K. and Shinozaki, K. 1999. Improving plant drought, salt, and freezing tolerance by gene transfer of a single stress-inducible transcription factor. *Nat. Biotechnol.* 17: 287-291.

Kawasaki, S., Borehert, C., Deyholos, M. and Wang, M. 2001. Gene expression profiles during the initial phase of salt stress in rice. *Plant Cell* 13: 889-905.

Khokon, M.A.R., Okuma, E., Hossain, M.A., Munemasa, S., Uraji, M., Nakamura, Y., *et al.,* 2011. Involvement of extracellular oxidative burst in salicyclic acid induced stomatal closure in *Arabidopsis. Plant Cell Environ.* 34: 434-443.

Kishor, P.B.K., Hong, Z., Miao, G.H., Hu, C.A.A., and Verma, D.P.S. 1995. Over expression of Δ1-pyrroline-5-carboxilase synthetase increase proline, production and confers osmotolerance in transgenic plants. *Plant Physiol.* 108: 1387-1394.

Kurusu, T., Kuchitsu, K., Nakano, M. Nakayama, Y. and Lida, H. 2013. Plant mechanosensing and Ca^{2+} transport. *Trends Plant Sci.* 18: 227-233.

Leidi, E.O., Barragan, V., Rubio, L., El-Hamdaoui, A., Ruiz, M.T. and Cubero, B. 2010. The AtNHX1 exchanger mediates potassium compartmentation in vacuoles of transgenic tomato. *Plant J.* 61: 495-506.

Marti, M.C., Stancombe, M.A. and Webb, A.A. 2013. Cell- and stimulus type-specific intracellular free Ca^{2+} signals in *Arabidopsis. Plant Physiol.* 163: 625-634.

Miller, G., Shulaev, V. and Mittler, R. 2008. Reactive oxygen signaling and abiotic stress. *Physiol. Plant.* 133: 481-489.

Miller, G., Suzuki, N., Ciftci-Yilmaz, S. and Mittler, R. 2010. Reactive oxygen species homeostasis and signalling during drought and salinity stresses. *Plant Cell Environ.* 34: 453-467.

Moller, I.S., Gilliham, M., Jha, D., Mayo, G.M., Roy, S.J., Coates, J.C., *et al.,* 2009. Shoot Na^+ exclusion and increased salinity tolerance engineered by cell type-specific alteration of Na^+ transport in *Arabidopsis. Plant Cell* 21: 2163-2178.

Munns, R., James, R.A. and Lauchli, A. 2006. Approaches to increasing the salt tolerance of wheat and other cereals. *J. Exp. Bot.* 57: 1025-1043.

Munns, R. and Tester, M. 2008. Mechanisms of salinity tolerance. *Ann. Rev. Plant Biol.* 59: 651-681.

Oh, S.J., Song, S.I., Kim, Y.S. Jang, H.J., Kim, S.Y., Kim, M., *et al.,* 2005. *Arabidopsis* CBF3/DREB1A and ABF3 in transgenic rice increased tolerance to abiotic stress without stunting growth. *Plant Physiol.* 138: 351-351.

Olias, R., Eljakaoui, Z., Li, J., De Morales, P.A., Marin-Manzano, M.C., Pardo, J.M., *et al.,* 2009. The plasma membrane Na^+/H^+ antiporter SOS1 is essential for salt tolerance in tomato and affects the partitioning of Na^+ between plant organis. *Plant Cell Environ.* 32: 904-916.

Orellana, S., Yanez, M., Espinoza, A., Verdugo, I., Gonzalez, E., Ruiz-Lara, S., *et al.,* 2010. The transcription factor SIAREB1 confers drought salt stress tolerance and regulates biotic and abiotic stress related genes in tomato. *Plant Cell Environ.* 33: 2191-2208.

Pottosin, I. and Dobrovinskaya, O. 2014. Non-selective cation channels in plasma and vacuolar membranes and their contribution to K^+ transport. *J. Plant Physiol.* 171: 73-742.

Roy, S.J., Negrao, S. and Tester, M. 2014. Salt resistant crop plants. *Curr. Opin. Biotechnol.* 26: 115-124.

Saito, K. and Matsuda, F. 2010. Metabolomics for functional genomics, systems biology and biotechnology. *Annu. Rev. Plant Biol.* 61: 463-489.

Sakamoto, A., Murata, A. and Murata, N. 1998. Metabolic engineering of rice leading to biosynthesis of glycine betaine and tolerance to salt and cold. *Plant Mol. Biol.* 38: 1011-1019.

Schilling, R.K., Marschner, P., Shavrukov, Y., Berger, B., Tester, M., Roy, S.J., *et al.,* 2013. Expression of the *Arabidopsis* vacuolar H^+ pyrophosphoatase gene (AVP1) improves the shoot biomass of transgenic barley and increases grain yield in a saline field. *Plant Biotechnol. J.* 12: 378-386.

Shabala, S. and Pottosin, I. 2014. Regulation of potassium transport in plants under hostile conditions: implications for abiotic and biotic stress tolerance. *Physiol. Plant.* 151: 257-279.

Shabala, S., Wu, H. and Bose, J. 2015. Salt stress sensing and early signalling events in plants roots: current knowledge and hypothesis. *Plant Sci.* 241: 109-119.

Shi, H., Quintero, F.J., Pardo, J.M. and Zhu, J.K. 2002. The putative plasma membrane Na^+/H^+ antiporter SOS1 controls long-distance Na^+ transport in plants. *Plant Cell* 14: 465-477.

Singh, R.K., Redona, E. and Refuezo, L. 2010. Varietal improvement for abiotic stress tolerance in crop plants: special reference to salinity in rice, in *Abiotic Stress Adaptation in Plant: Physiological, Molecular and Genomic Foundation,* eds. A. Pareek, S.K. Sopory and B.H. Govindjee (New York, NY: Springer), 387-415.

Su, J. and Wu, R. 2004. Stress-inducible synthesis of proline in transgenic rice confers faster growth under stress conditions than that with consititutve synthesis. *Plant Sci.* 166: 941-948.

Sun, S.J., Guo, S.Q., Yang, X., Bao, Y.M., Tang, H.J., Sun, H., *et al.,* 2010. Functional analysis of a novel Cys2/His2 type zinc finger protein involved in salt tolerance in rice. *J. Exp. Bot.* 61: 2807-2818.

Tarczynski, M.C., Jensen, R.G. and Bohnert, H.J. 1993. Stres protection of transgenic tobacco by production of the osmolyte mannitol. *Science* 259: 508-510.

Turkan, L. and Demiral, T. 009. Recent developments in understanding salinity tolerance. *Environ. Exp. Bot.* 67: 2-9.

Varshney, R.K., Bansal, K.C., Aggarwal, P.K., Datta, S.K. and Graufurd, P.Q. 2011. Agricultural biotechnology for crop improvement in a variable climate : hope or hype. *Trends Plant Sci.* 16: 363-371.

Volkov, V. and Amtmann, A. 2006. *Thellungiella* halophila, a salt-tolerant relative of *Arabidopsis thaliana,* has specific root ion-channel features supporting K^+/Na^+ homeostasis under salinity stress. *Plant J.* 48: 343-353.

Wu, H., Zhu, M., Shabala, L., Zhou, M. and Shabala, S. 2015. K^+ retention in leaf mesophyll, an overlooked component of salinity tolerance mechanism: a case study for barley. *J. Integr. Plant Biol.* 57: 171-185.

Xing, Y., Jia, W. and Zhang, J. 2008. AtMKK1 mediates ABA-induced CAT1 expression and H_2O_2 production via AtMPK6-coupled signaling in *Arabidopsis. Plant J.* 54: 440-451.

Xu, D., Duan, X., Wang, B., Hong, B., Ho, T.H.D. and Wu, R. 1996. Expression of a late embryogenesis abundant protein gene, HVA1 from barley confers tolerance to water deficit and salt stress in transgenic rice. *Plant Physiol.* 110: 249-257.

Zhang, H.X. and Blumwald, E. 2001. Transgenic salt tolerant tomato plants accumulate salt in the foliage but not in the fruits. *Nat. Biotechnol.* 19: 765-768.

Zhang, H.X., Tan, X, Wei, L.X., Wang, Y.L., Zhang C.P. and Niu, X.M. 2012. Morphology regulatory metabolites from *Arthrobotrys oligospora. J. Nat. Proc.* 75: 1419-1423.

■■■■

Chapter - 11

STRATEGIES FOR INCREASING SALT RESISTANCE IN PLANTS

INTRODUCTION

Salt affected soils and saline irrigation waters are wide spread in arid and semi-arid zones of the world and present a serious problem of great practical importance for crop production in view of their inhospitable characteristics for plant growth despite inherent potentialities. Because of the food shortage besetting mankind in various parts of the world, the need to produce more food is continuously pushing agriculture further on to marginal lands often characterized by soils and waters with a high degree of natural salinity. Salinity tolerance would therefore, be a highly desirable character to introduce into crop plants if yield potential could also be maintained.

The various strategies proposed to overcome the problem of salinity stress can be deciphered from a number of comprehensive reviews published on plant salt tolerance (Epstein *et al.,* 1980, Flowers, 2004, Chinnusamy *et al.,* 2005, Yamaguchi and Blumwald, 2005, Munns *et al.,* 2006, Ashraf *et al.,* 2008, Ashraf and Akram, 2009, Rai *et al.,* 2011). However, there are two major approaches/strategies to minimize the deleterious effects of high soil or water salinity and both must be applied simultaneously to achieve sustainable crop production in the presence of excessive salts (Epstein *et al.,* 1980). One is a technological approach, that is, implementing large engineering schemes for reclamation, drainage and irrigation with high-quality water. No doubt, these practices have had continuing success in some areas, the associated costs are high and often provide only a temporary solution to the problem. The second approach, that is a complement to technological approach, entails biological strategies focused upon the exploitation or development of plants capable of tolerating excessive levels of salts. This approach includes (i) diversifying cropping systems to include crops that are known to be salt tolerant, (ii) exploiting wild species that are adapted to saline environments (e.g. by domestication); and/or (iii) genetically modifying domesticated crops by breeding and selection to develop cultivars with enhanced ST. Breeding for salt-tolerant genotypes that can grow more efficiently than the conventional varieties under high salinity stress is a fundamental approach which is considered economically feasible (Blum, 1988, Ashraf *et al.,* 2008).

Both the approaches for increasing crop production on saline soils are feasible. Technological approach changing environment to improve conditions for the plant i.e. better soils for the crops we have; or biological approach changing the plant genetically to better tolerate the saline environment i.e. better crops for the soils we have. While efforts to adjust edaphic environment to suit crop production are continuing on a massive scale, there is growing realization now that a more effective, less costly, non-polluting and longer lasting contingency lies with genetic modification of the plant itself (Epstein and Norlyn, 1977; Epstein *et al.,* 1980; Downton, 1984; Epstein and Rains, 1987; Subbarao and Johansen, 1994). Several reports available on this aspect in recent years have emphasized the need to develop improved crop varieties best suited to specific adverse soil conditions (Staples and Tonniessen, 1984; Tal, 1985; Rana, 1986; Pessarakli, 1994; Garg and Gupta, 1997). Cultivar tolerance for salinity is a substitute for amendments on moderately saline soils and a supplement to amendments in highly saline soils. It must, however, be admitted that the two approaches are complimentary to one another. The first approach has been used successfully in the past, but, with the need to utilize marginal lands and poor quality saline waters, attention is also to be paid to the second approach. The rising cost of energy, inputs and scarcity of nonsaline waters has put new constraints on the economical use of the technological approach.

The adaptation of plants to saline soils is an exciting prospect but present plant scientists face with some formidable challenges. However, this task may be best achieved by the application of classical genetics (conventional plant breeding) in combination with modern techniques of cell and tissue culture, somatic hybridization, molecular biology and genetic engineering (Rains *et al.,* 1980; Shannon, 1985, 1997; Tal, 1985; Epstein and Rain, 1987; Subbarao and Johansen, 1994). An attempt has been made here to discuss these biological approaches based on genetical aspects currently followed for increasing salt tolerance in crop plants.

Richards (1992) argued that there was no need to breed for salt resistant crops because of the patchiness of salt affected soils. He advocated that it is preferable to concentrate on yield potential; plant the highest yielding cultivar, take the yield from the areas with lower salt concentrations and accept the losses on the most salt affected patches, rather than accept any losses in potential yield that might be associated with breeding for salt resistance. This strategy may work in some situations but not in many others. For most of the salt sensitive crops, which unfortunately are many, even the patches with the lowest salinity may be too much. A few relatively more salt tolerant crops, such as sugar beet, barley, cotton, asparagus, and sugarcane, can substitute for the more sensitive species. Such substitution is still practised in some saline areas. However, shifting to other crops may not be feasible in most of the cases due to social, food habit or lack of alternative compulsions.

The options for the developed and the developing countries may be different, with stakes being very high for the developing countries. For the countries of Asia, Africa, and Latin America, future expansions of population, subsistence agriculture along with increased dependence on irrigation will result in making agricultural salinization a life-threatening problem. Often there are no options to tolerate reduced yields either from bad fields or bad years. In such situations, yield stability may be more important than yield potential.

One school of thought is that despite lack of knowledge of a mechanism for tolerance, improved salinity tolerance has been achieved in several crops (McNeily, 1990). Most of the efforts on this front owe their failure to the repeated tendency to treat salinity as a single, simple character ignoring the complex nature of this character suggested by a burgconing literature in recent years. Use of yield, a complex character involving physiological, biochemical, and morphological characteristics whose genetics is largely unknown and whose exact relationship with plant productivity is mostly not clear, as a selection criterion for other complex traits such as drought and salinity has a low efficiency. However, recent demonstrations that salt tolerance can be greatly raised by intro-gression of a single gene making for increased Na^+ sequestration in the vacuole, have created a breakthrough in genetic engineering of salinity tolerance in crops (Zhang and Blumwald, 2001; Zhang *et al.*, 2001).

There appears to be another prevailing opinion which supports the view that progress towards achieving salinity tolerance in crops can be made only when the physiological mechanisms of tolerance are understood and advocates the use of relevant traits and breeding strategies (Shannon, 1985; Shannon and Noble, 1990; Flowers and Yeo, 1995). Blum (1989) suggested that the use of physiological parameters should simplify the genetics and breeding procedures.

GENETIC IMPROVEMENT FOR INCREASED SALT TOLERANCE

Genetic improvement strategies aim at identifying and/or developing crop cultivars with higher yield potential, higher yield stability, and greater production efficiency under salt affected conditions. Crop cultivars with higher yield potential have been the key to increased productivity. Major advances have occurred in the past in increasing the yield potential of crop varieties and further improvements may continue. Conventional hybridization and selection procedures are the time-tested strategies for selecting the crop cultivars with higher yield potential and have two phases. The first phase involves the creation of variability and in the second phase desirable individuals are selected. Selected genotypes are evaluated for yield and the superior ones are released as varieties. It has been estimated that on an average about 1% increase has occurred per year in the yield potential of major crops. The yield potential of wheats developed at CIMMYT increased by 0.83% per year over the last 30 years (Fischer, 1994). The yield of cotton lint in the US increased from about 600 kg ha^{-1} in 1950 to about 900 kg ha^{-1} in 1980 showing an increase of 50% over a 30-year period (Meredith and Bridge, 1984). The yields of most of the crops where enough investment has been made have shown a continuous improvement and there is every reason that this can be achieved on salt affected soils also.

With increased plant resistance to salinity, smaller fractions of applied water need to be leached through the soil to maintain salinity within acceptable limits (Hoffman, 1990); the smaller the quantity of salts to be leached, the lower the demands for drainage. Therefore, crops having increased resistance to salinity and/or having lower water requirement will be more suitable for cultivation and maintaining/sustaining these environments (particularly, arid and semiarid).

The genetic improvement of salt tolerance in crops can be achieved by :

(1) Selection among agronomic cultivars.

(2) Hybridization and ploidy changes.

(3) Use of wild germplasm to develop new cultivated species.

(4) Employing modern techniques of cell and tissue culture, somaclonal variation, somatic hybridization and recombinant DNA technology.

Promising studies have been undertaken with barley, wheat, wheat grass, rice, soybean and tomato. Barley lines which survive and provide average growth yield of 1.1 t/ha under irrigation with undiluted sea water have been developed (Ramage, 1980; Epstein and Rains, 1987). According to Evans *et al.* (1983) the application and development of cell culture techniques are opening the door to a second green revolution. Borlaug agrees that the potential of the modern techniques is high but because of the complexity of the plant genome, many years will elapse before these techniques will be used successfully for breeding better crop varieties.

According to Epstein (1983) the prerequisites for improving salt tolerance are as under :

(1) Vigorous plant growth must be possible on saline substrates. This is demonstrated by the existence of halophytic species which include approximately 550 species belonging to 220 genera in 75 families.

(2) Suitable genetic variability should be available in the cultivated species or their wild relatives. Genotypic variation for salt tolerance in crop species has been reviewed by many workers. Such variations seem relatively limited probably because most cultivated species were selected in non-saline environments where salt tolerance traits would have been gradually lost from their gene pools. However, salt tolerant mutants selected in cell cultures or genes from wild salt tolerant plants can be used to enrich these gene pools.

(3) Suitable methods for screening large number of genotypes should be available.

(4) The fourth and final requirement is diagnostic criteria by which salt tolerant genotypes can be easily identified in large segregating populations. This is a major limiting factor in breeding for salt tolerance (Greenway and Munns, 1980; Ponnamperuma, 1983; Shannan, 1985, 1997; Tal, 1985).

The study of the genetics and physiology of salt tolerance has proceeded slowly. The reasons for this are the following :

(1) The specific physiological effects of salinity have not been determined.

(2) Inadequate means of detecting and measuring salinity.

(3) Varietal differences have only recently been detected.

(4) Salt tolerance changes with plant development.

(5) Poor information on the effects of salt alone and in interaction with other environmental factors such as soil water, temperature, light intensity and relative humidity.

(6) Lack of reliable selection criteria.

GENETIC VARIATIONS FOR SALT TOLERANCE

One requirement for breeding for salt tolerance is that genetic variation exists for the character in the gene pool. Such variation may be between individuals, varieties, or even species that have some degree of sexual compatibility so that genes may be transferred from one individual to another. Another necessity is that salt tolerance can be identified in segregating generations or that specific information exists concerning its genetic control in terms of numbers of genes involved and their heritability. Although considerable research has been devoted to quantifying the salt tolerance of various crop species (Francois and Maas, 1978, 1985; Maas and Hoffman, 1977; Maas, 1987), data for many species are usually based on comparisons among only a few varieties. In studies that have examined a range of varieties, some species exhibit wide intraspecific variation of salt tolerance, whereas others have limited variation. Usually, only a relatively small portion of the existing germplasm base has been screened. Many wild progenitors of cultivated species have not been tested (Shannon, 1997).

Differences in salt tolerance occur within every genus and species examined. Lot of published literature is available on the relative salt tolerance of crops. Maas and Hoffman (1977), Maas (1984) and Francois (1994) and Sharma (1997) reviewed the available salt tolera-nce data and classified crops on the basis of relative salt tolerance. Several perennial species within the tribe Triticeae (intergeneric variation) have been shown to be significantly more salt tolerant than wheat (Wyn Jones and Gorham, 1990). Green alga *Dunaliella viridis* thrives in the highly saline dead sea at a salinity level of above 35 dSm^{-1} and is being tried in Israel for extraction of glycerol, occurring in this species in high concentrations. Mangrove swamps of highly productive large trees growing in sea water are examples of higher plants adapted to perform remarkably well in saline environment.

Inter-specific Differences

A survey of reported data on responses of crop plants to increasing salinity reveals about eight fold differences among them regarding tolerance to salinity stress (Rana, 1986). Based on the work done at CSSRI, Karnal (India), an interesting point has emerged that some allopolyploid crop species turned out to be remarkably more tolerant to both alkali and saline soils as compared to their putative progrenitors and relatives of lower ploidy levels. Bread wheat (2 n = 6 x = 42) was found to be far more tolerant compared to the durums (2n = 4 x = 28) and einkorn wheat (2n = 2 x = 14) (Table 11.1). Similarly,

Brassica juncea (2n = 4 x = 36) showed its superiority to salt tolerance in comparison with its diploid progenitors *B. compestris* (2n = 2x = 20) and *B. nigra* (2n = 2x = 16) (Kumar, 1995). Alloploid species *Medicago sativa, Sesbania grandiflora, Oryza coarciata, Prosopis juliflora* and *Zizyphus rotundifolia* were also characterized by a high level of salinity/alkalinity stress tolerance. However, in cotton there was a reverse situation as diploid species *arboreum* and *herbaceum* were found to be more tolerant to the new world tetraploid species *hirsutum* and *barbadense.* Aswathappa *et al.* (1986) recorded that some *Acacia* species grew in NaCl concentrations of 1000 mM NaCl, while others stopped growing at 200 mM NaCl.

Table 11.1. Grain yield of hexaploid, tetraploid and diploid wheat as influenced by salinity and alkalinity

Soil grades		Grain yield (g pot^{-1})		
pH	ESP	Hexaploid *T. aestivum* (2n = 6x = 42)	Tetraploid *T. durum* (2n = 4x = 28)	Diploid *T. monococcum* (2n = 2 x = 14)
8.2	7	19.2	17.3	4.0
8.8	28	17.9	9.6	0.0
9.1	41	13.8	4.1	0.0
9.4	52	9.4	0.0	0.0

Intra-specific Differences

Early reports indicated absence of appreciable varietal differences within a crop species regarding salt tolerance. However, numerous workers have subsequently reported/reviewed notably large differences in this respect among genotypes of the same crop species paving the way for plant breeding efforts to combine high yield potential with superior salt tolerance (Norlyn and Epstein, 1984, Rana, 1986). Many recent studies have shown significant varietal differences in a number of crop species such as wheat, barley, rice, pearl millet, pigeon pea, chickpea, clusterbean, olive, Indian mustard and tomato (Greenway and Munns, 1980; Garg *et al.,* 1990; Subbarao *et al.,* 1990; Garg and Gupta, 1998; Kumar, 1995; Lahiri *et al.,* 1996).

Shannon (1997) has reviewed the genetic variability in large number of cereals, field crops, oil seed crops, forages and grasses, vegetable and fruit crops and listed number of tolerant and sensitive varieties within different species. He has advocated the need to identify genetic variability within the germplasm to improve salt tolerance of various crops. This also includes varieties of closely related species among which genes can be transferred.

Table 11.2. Classification of some wheat varieties with respect to tolerance under sodic soils

Tolerant	Medium tolerant	Medium sensitive	Sensitive
Kharchia	HD 1981	WG 377	HD 4502
KRL II-10	HD 1982	WG 357	HD 4530
KRL III-2	WH 957	HD 2009	Raj 911
KRL III-4	HD 916	HD 2177	LSW 34
KRL III-7	Sonalika	HD 2204	Malavraj
KRL III-9	HD 2286	WL 711	Jairaj
KRL III-23	HD 2329	C 306	Moti
	HD 2285	HD 2270	Hira

At CSSRI, Karnal nearly 1000 wheat genotypes were evaluated for their performance in saline-sodic soils, under comparable conditions of field, micro-plot, pot culture etc. and varieties were classified in four categories viz. tolerant, medium tolerant, medium sensitive and sensitive. Some known wheat varieties in aforesaid classes have been listed (Table 11.2). Remarkably, salt tolerant wheat comprising red grained selection of locally adapted *Kharchia* wheats of Rajasthan (Pali district) and *Rata* wheats of Bhal tract of Gujarat has generated valuable materials for study of genetic mechanisms governing salt tolerance and also for developing varieties best suited to specific salt affected areas. Kharchia-65 maintained relatively lower Na levels, higher K levels, lower Na/K ratio and high NR activity and these features enabled them to avoid the dangers of saline or sodic stress to a large extent (Sharma and Gupta, 1986; Garg *et al.,* 1990). Like wise some promising salt resistance selections from locally adapted indigenous rice cultivars are presented in Table 11.3.

Table 11.3. Some salt tolerant rice cultivars

State	Cultivars
West Bengal	Damodar, Getu, Dasal, Nonabokhra, Nonasail, Rupsail, Matla, Hembilton
Kerala	Pokkali, Vytila-1, Oorpandy, Oormundakan, Cheruvirippu, Chotupokkali
Maharashtra	Bhurarata, Kalarata
Tamil Nadu	PVR-1
Andhra Pradesh	MCM-1, MCM-2
Orissa	SR 22B

It has been observed that, in general, the varieties of a crop which were tolerant to adverse soil conditions had a somewhat lower yield potential under relatively nonstress environments.

CRITERIA FOR SALT TOLERANCE

Criteria that have been used to evaluate salt tolerance in species include germination, emergence, shoot dry weight, root weight, shoot number, seedling survival, leaf damage, flowering and seed or fruit yield. The degree of salt tolerance between and within species can vary according to the criteria used for evaluation. Besides the above mentioned growth indices, several physiological responses such as ion accumulation, ion exclusion, Na/K ratio, compatible solute production, osmotic adjustment, late maturation, pollen sterility and chlorophyll flourescence have been considered as selection criteria. Some of these major potential selection criteria are summarized in Table 11.4.

Table 11.4. Potential selection criteria for salt resistance (Tal, 1985)

Growth	
Germination percentage	Useful only for crops that are equally resistant during germination and later stages
Emergence percentage	
Seedling survival	
Rooting	
Ions	
Accumulation	Indicates tolerance in salt includers
Exclusion	Indicates tolerance in salt excluders
Membranes	
Leakage of electrolytes abnormal plasmolysis	Indicates sensitively
Vital staining	May indicate disturbed or undisturbed metabolism
Chlorophyll fluorescence	Changes in its parameters may be used for salt tolerance

Selection Methods

Several screening and selection schemes have been proposed for salt tolerance (Epstein *et al.,* 1980; Sykes, 1985, Shannon, 1997). Because of the many interactions of salinity with environmental variables and specific ions, a target environment is important to a screening program for salt tolerance. Initial considerations should be made concerning the balance of major and minor in so that could prove toxic or have detrimental effects on nutrients. Recent advances in soil physics and chemistry allow the prediction of soil water compositions in the plant root zone based on applied water quality and soil chemistry.

Many examples have been cited to support the generalization that salt tolerance varies with ontogeny and development. Salt tolerance measured at one growth stage usually does not correlate well with salt tolerance at other growth stages. Intervarietal variation in salt tolerance during germination has been found, in a number of cases, to differ with tolerance at other growth stages. Germination rates as measured in saline filter paper or gels have very little relationship to emergence potential. Many seeds that germinate on salinized filter papers or in gels may be too weak to break a soil crust and establish as viable plants. Emergence rate might be a more practical screening criterion than germination rate, and seedling vigor may also be an additional useful screening factor for conditions in which saline soils form hard crusts.

Germination or emergence trials under saline conditions should always be conducted using healthy and viable seed lots. Hard or damaged seed should be discarded; however, too much selection may also be a problem. Large seed size has been shown to have greater salt tolerance than small seed selected from the same lot in wheat and sorghum (Grieve and Francois, 1992).

Available information on the salt tolerance of the crop should be collected. Such information should include relative salt tolerance range, potential variability among cultivated varieties and closely related species, and sensitivities to specific ions and environmental interactions.

The process of developing salt-tolerant varieties should begin with the identification of the precise growth stage that is limiting to productivity. It should be determined if economical management techniques can be used to overcome the limitation. For example, if stand establishment is a limiting factor, this limitation could be overcome by better bed preparation to move dissolved salts away from the seed, by applying a timely irrigation of high or medium quality water; or by more dense seeding or plant spacing. Seed priming has also been proposed as a method to enhance seed germination and emergence under conditions of salinity and low temperatures.

Determine whether other locally adapted varieties are more tolerant at this growth stage. Develop a screening procedure for the sensitive growth stage. The procedure must be based on information concerning average salt concentration and composition of the soil water during sensitive growth periods and the environmental conditions during the period of salt damage in the field. A selection criterion needs to be one that is related to mean yield response in the field.

Field experiments should be conducted at an early stage to prove the relationship of the criteria selected to the desired field characteristics.

Evaluate a range of varieties and introductions to determine genetic variance for the desired characteristics. Proper controls must be included to separate genetic from environmental effects under both nonsaline and saline conditions. Information from the collected data can be used to determine if intravarietal selection will be effective. If genetic variance is low or if a greater degree of tolerance is required. Wild relatives and lines developed from hybridizations can be evaluated.

A point of interest is that a newly selected rice variety, Giza 160, was found to be sensitive to two new blast races prior to its release, and germplasm derived from the salt-tolerant Kharchi wheat line was found to be susceptible to both rusts and powdery mildew (Rana, 1985). Research has not been conducted to determine whether there is a mechanistic link between salt tolerance and susceptibility to rusts and mildews. It has been speculated that a weaker cell wall structure could favor easier expansion growth under saline conditions. Weaker cell walls could also decrease resistance to rusts and mildews.

Field Screening Techniques

Selection for tolerance under field conditions is inaccurate because of the difficulty in assessing the physical and temporal variability in salinity in soil solutions proximal to the root zone. The ability of plants to extract preferentially water from the most nonsaline portions of their root zones further complicates this assessment. It may be possible to compensate for these problems to some extent by using very large populations, making a judicious selection of breeding strategy and experimental design, and precisely controlling irrigation in artificially salinized plots. Nevertheless, a low selection efficiency can be expected even under ideal conditions and the efforts will be very expensive. The advantages of field screening are that space is not a major limiting factor to replication and population sizes, and that the environmental conditions are more nearly what will be encountered under real field conditions.

PHYSIOLOGICAL FACTORS RELATED TO SALT TOLERANCE

Ion Restriction

Plants that limit uptake of toxic ions and maintain normal ranges of nutrient ions could be more salt tolerant than those that do not. Selective ion uptake mechanism that can differentiate between chemically similar ions, such as Na^+ and K^+ could have adaptive value. These mechanisms probably operate in the membranes of tissue and various organelles throughout the plant. Breeding for efficient nutrient uptake or low ion accumulation under salt stress may be one of the simplest ways to improve salt tolerance in sensitive varieties of some species.

Genetic differences result in variability in selective ion transport between varieties. Gene differences may affect salt tolerance, as well as nutrient supply. Salt sensitivities in barley, soybean, citrus and grape relate to ability of plant to regulate the transport of Cl^- adequately from the root to the shoot (Greenway and Munns, 1980). In some plants, such as soybean a single may be responsible. In others such as citrus, multiple genes are involved and the response is complex. Distinguishing whether toxic effects are due to Na^+ or Cl^- sometimes is difficult as the concentrations of these two ions often increase concomitantly. Woody species seem sensitive to Cl^- while varieties of bean, corn and wheat that regulate accumulation of Na^+ more effectively in the shoot can tolerate salinity better than those which cannot. Each species should therefore be examined individually to determine ion toxicity thresholds, ion restriction and the capacity for ion selectivity.

Ion Accumulation

Although ion exclusion or restriction to roots or shoots is one of the frequently reported differences between salt tolerant and sensitive varieties, halophytes take up very high concentrations of ions to cope with saline environments (Flowers *et al.,* 1977). The accumulation of salt is thought to reduce the requirements for increased wall extensibility, leaf thickness to maintain growth and turgor at low soil water potentials. The wild tomato species, *Lycopersicon cheesmani* is thought to be more salt tolerant than the cultivated species as a result of its capacity to accumulate ions (Rush and Epstein, 1976). However, salt accumulation within tissues is well regulated and sequestered from cytosolic compartments containing the salt sensitive metabolic machinery of the cell. The physical and genetic factors that influence ion compartmentation and distribution within plants are mostly unknown.

Osmotic Adjustment

The decrease in plant osmotic potential through an increase in solute content in response to a decrease in external water potential is known as osmotic adjustment. The accumulation of organic solutes, such as sugars, proline and glycinebetaine, improves salt tolerance by contributing to osmotic balance and preserving the enzyme activity in the presence of toxic ions (Greenway and Munns, 1980; Garg and Gupta, 1997). Rathert (1984) has suggested that leaf sucrose and starch contents could be used as a selective index in screening for improved salt tolerance. High betaine genotypes of barley maintained lower solute potentials than near isoline, low betaine genotypes grown at the same salinities. However, these characters alone may prove to be inadequate criteria if other salt tolerant characteristics are not maintained.

Other mechanisms that can prevent turgor loss through better water use efficiency include increased leaf resistance i.e. fewer stomata, increased mesophyll resistance, increased cuticle thickness, a higher root: shoot ratio and osmotic adaptation through ion accumulation and the production of compatible solutes. However, most of these strategies are associated with some aspect detrimental to the maximum production of crops. The energy costs of the various physiological mechanisms of salt tolerance need to be investigated to select the most efficient strategies for a particular crop.

Use of Physiological Traits in Breeding

Given the importance of the major crops to world agriculture, we want to return to the question of generating salt -resistant cultivars. If secondary salinisation continues, at the present rate, then there must inevitably be a time when overall productivity of some regions decline. New cultivars may not yet have been developed but they will be needed in the future.

The efficiency of conventional approaches to breeding for complex traits such as resistance to drought and salinity, in which yield is the main selection criterion is low and hence the use of physiological parameters should simplify the genetics and the breeding procedures (Blum, 1989). A consideration of the use of physiological characters as a basis of selection in plant breeding need not apply simply to attempts to enhance resistance to salinity : it can be applied to the basic problem of yield (Tal, 1985). As far as salt is concerned, traits associated with resistance have been identified and recently breeding procedures have been advocated (Yeo and Flowers, 1984; Yeo *et al.*, 1990; Noble and Shannon, 1990; Noble and Rogers, 1992).

The approach relies upon some knowledge of the physiology of plants that tolerate salinity and of the effects of the salinity on more sensitive species. The work published in the literature indicates that there is no single physiological reason why a plant is resistant or sensitive : resistance arises from the interplay of a number (as yet undefined) of separate (physiological) process, whose linkage in genetic terms is generally unknown. Comparisons made between sensitive and resistant lines of a range of species commonly indicate differences in net ion accumulation, with the more resistant line having the lower leaf ion concentration. This itself is a complex trait, involving transport across the root, movement to the shoot and hence transpiration and its control. An ability to maintain leaf potassium concentrations in the presence of high external sodium has also been seen as important. Other processes that may contribute to resistance include the ability to sequester ions in roots and stems, the ability to compartmentalise ions in vacuoles and to synthesise organic compatible solutes for the cytoplasm : the list is long and touches on all aspects of the physiology of the plant. Nevertheless in identification of those processes dominating in the resistance/sensitivity differential within a species can be used to enhance resistance, given reasonable levels of heritability of a given character and natural variation of that character within the species.

Flowers and Yeo (2002) found that selection for low sodium transport in rice did not discriminate against the agronomically desirable dwarf/intermediate type. Thus low sodium transport can be selected independently of plant type making it possible to select for low sodium plants with desirable agronomic characteristics. The most important is that early selection for low sodium transport preserved variation for desirable plant types, while early selection in agronomic characters strongly depleted the population in desirable low sodium transporting genotypes.

Traits conferring salt tolerance must be identifiable in segregating generations so that information is available about the numbers of genes involved and their inheritance. The breeding procedure depends on the inheritance pattern (qualititative or quantitative), the number of genes with major effects, and nature of gene action. The heritability and inheritance of salt tolerance in a species determines the selection intensity and the number of selection cycles required.

The lack of genetic variation and a poor understanding of physiology and genetics are generally the major barriers in improving salt tolerance of plants (Shannon and Noble, 1990).

The interactions between salinity and other environmental stresses complicate assessments that use yield or growth as an index of tolerance. Hence, other indices have been proposed. Survival after massive doses of salinity has been used as a criterion, but with limited success. No such protocol has been widely adopted or applied.

Development of a breeding programme for improved ST requires (i) efficient screening techniques for the selection and evaluation of specified characters, (ii) identification of genetic variability, (iii) knowledge of the inheritance of tolerance trait(s) at specific developmental stages, (iv) knowledge of biological mechanisms underlying tolerance, (v) reliable direct or indirect selection criteria, and (vi) designing the most appropriate breeding methodologies/strategies to transfer the tolerance trait(s) into improved genetic backgrounds. Among these, the identification of genetic variability, attainment of genetic knowledge of ST-related traits and development of reliable selection criteria are most crucial for the establishment of a successful breeding programme. Accumulating evidence in different crop species has indicated that ST is a developmentally regulated, stage-specific phenomenon. Tolerance at one stage of plant development may be poorly correlated with tolerance at other developmental stages (Foolad, 1999, Zhang and Blumwald, 2001, Ashraf and Akram, 2009, Uddin *et al.,* 2011). Specific stages throughout the ontogeny of the plant such as germination and emergence, seedling survival and growth, and vegetative and reproductive growth should be evaluated separately for the assessment of ST and identification of contributing genetic components.

Physiological Marker for Salt Tolerance (ST)

One suggested approach to improve the efficiency of the breeding programmes is adoption of new selection criteria based on the knowledge of the physiological processes limiting crop production under conditions of salt stress (Ashraf, 2004) and of their genetic control (Tal, 1985). Some of the physiological responses to high salinity, which may be used as indirect selection criteria for improving crop ST, are tissue water potential, tissue ion content, K^+/Na^+ ratio, succulence, water-use efficiency, chlorophyll fluorescence, content of chlorophyll, contents of proline, phenolics and sugars, and activities and levels of enzymatic and non-enzymatic antioxidants (Ashraf, 2004).

Wide-hybridization has been successful in the quest for characters such as insect and disease resistance, where single genes may be introgressed, and selection to maintain that character is relatively straight forward in the backcross generations that are needed to dilute the rest of the genome from the resulting hybrid. The situation for salt tolerance is, however, different. Most studies on plant-environment interactions have shown that the response of crop plants to abiotic stresses (including salinity) involves several gene alterations at the same time, and is multi-or polygenic in nature. Molecular studies further support this notion. In response to salinity stress, it has been shown that a large number of transcripts are affected at the same time. O'Toole (1989) discussed the use of molecular marker technology to assist breeding for such characters, and concluded that genetic engineering of such traits may be years into the future. The situation does not seem to be promising now (Prasad *et al.,* 2000; Jeong *et al.,* 2001; Zhang and Blumwald, 2001; Zhang *et al.,* 2001).

Though the multigenic, quantitative nature of salt tolerance imposes several limitations on breeding, the genetic basis of the variability, with a significant additive genetic component, suggests that further improvements in tolerance and plant performance in saline conditions can be achieved through further cycles of selection and breeding. It may be more practical and feasible to look for more modest resistance closer to the cultivated species, rather than going for the extremes. Research on the use of alien genetic material for salt resistance is still some distance away from practical plant breeding.

PYRAMIDING TRAITS FOR SALT TOLERANCE

Salinity tolerance is a multivariable whole plant phenomenon, each aspect of which needs to be separately optimized and combined, necessitating a building block approach to a tolerant variety. Yeo *et al.* (1990) advocated an approach to enhance salt tolerance using selection of parents on the basis of selection of physiological criteria and the subsequent combining of these traits in a process termed ``pyramiding." This led to the successful release of two cultivars of rice with enhanced salt tolerance (IRRI, 1997).

Pyramiding involves some change from normal plant breeding practices: (i) simple screening is not an effective way of selecting parents, and (ii) characters without visual expression can be used in selecting parents and handling breeding populations. The process, however, remains slow and labor intensive and is not widely used because of the extent of physiological work that has to be undertaken. On the otherhand, recent work from Blumwald's laboratory has successfully demonstrated single-gene transfer resulting in salt tolerance and hence explicit rejection of the notion that ``pyramiding'' is necessarily obligatory (Zhang and Blumwald, 2001; Zhang *et al.,* 2001).

To improve salt tolerance of plants, we need to transfer into plant genes determining limiting traits. Significant advances have been made in the identification and transfer of the relevant genes by genetic engineering. Further work should be aimed at gene expression analysis during early events following exposure of plants to high salinity and understanding signalling cascades (Flowers *et al.,* 2000; Bohnert *et al.,* 2001; Borsani *et al.,* 2001).

BREEDING APPROACHES

Any breeding approach for producing salt tolerance crops has certain limitations such as limited sources of genes for salinity tolerance, non-availability of reliable markers for such resistance and lack of rapid, precise screening methods.

The most important and pivotal step in breeding for resistance to edphic stress is a reliable and efficient method capable of isolating the resistant genotypes. As response of plant to salinity differs at different growth stages, hence, it becomes important for breeders to determine at which growth stage a species is most susceptible to salinity, in its particular cropping system. Some tolerant plants accumulate ions such as sodium and chloride in their leaves and other tissues, while others have ways of preventing ions from entering the plants. The two mechanisms may have important effects on salt tolerance. Breeders are hopeful that simple evaluations for the presence or absence of high or low levels of ions or other osmotically important chemical compounds in plants could be developed that would indicate salt tolerance.

The screening methods advocated by Shannon (1979, 1985), Kingsbury and Epstein (1984), Bona *et al.* (1991) and Timm *et al.* (1991) could be effectively used for isolating salt tolerant genotypes/ segregating materials. Grain yield of these selections can then be tested over a range of salinities and compared with that of other lines. For screening of germplasm at CSSRI for tolerance ratings, criteria of seedlings emergence index and early vigour were used for reducing the number of entries. However, selections for advancement of generation are based on spike weight which is strongly correlated with yielding ability in problem soils. There are several physiological tests such as measurement of chlorophyll disintegration, ratio of plasmolyzed to non plasmolyzed cells in 0.1 mM NaCl, plasma salt tolerance, stomatal opening in salt solutions, degree of salt damage and salt level at which 50% yield reduction occurs which can be effectively used for measurement of relative salt tolerance of genotypes. Besides this, growth and yield components and metabolic parameters are also helpful in selection of tolerant genotypes (Lahiri *et al.,* 1996).

Theoretical aspects of selection for yield under stress and non-stress conditions have been discussed by Rosielle and Hmablin (1981) who suggested that selection should be made for increased mean productivity i.e. for average yield under stress and non-stress environments rather than solely for tolerance to stress. However, Redhu and Chowdhury (1991) concluded that selection under favourable condition was equally good to that of saline environment as some times it would not be possible to develop and maintain the desired level of salinity. Under high stress conditions, selection may not be reliable and it would be advisable to select tolerant types under low stress in early generations.

CONVENTIONAL BREEDING METHODS

Breeding approach include backcrossing, progeny testing and pure line selection. The main factor that determines the method of plant breeding, apart from the induction of useful variation, is the breeding system. Two-thirds of the world's food supply is provided by inbred species. Two of the major cereals, rice and wheat and most of important oil and protein crops are inbred species. Outbreeding crops include most of the forages, sugarcane and many tree fruits.

Although standard and specific breeding methodologies for developing crop varieties best suited to saline/sodic soils are not available, but conventional plant breeding methods comprising mainly introduction, selection and hybridisation have greatly contributed towards increased productivity in problem soils in several crops. Large number of indigenous locally adapted materials of rice, wheat, barley and some other crops have been successfully exploited by Indian breeders for obtaining salt resistant selections which have been used for direct cultivation as well as for recombination breeding programmes as donors for salt resistance. This fact emphasises that there is an urgent need to collect more indigenous cultivars and their wild relatives from areas where salt stress is recurrent and then screening at different stress levels for combining high yield potential along with high stress tolerance.

(A) Genes for Salt Tolerance

Salt sensitivities in barley, soybean citrus and grape have been found to be related to the inability of the plant to adequately regulate Cl^- transport from the root to the shoot (Shannon, 1997). In some cases (e.g. soybean) a single gene may be responsible (Abel and Mackenzie, 1964) but in other cases multiple genes are involved, making the response quantitative (e.g. citrus). Woody species seem to be sensitive to Cl^- whereas bean and corn varieties that better regulate accumulation of Na^+ in the shoot can tolerant low salinity better than those that cannot. It is important that each species be examined without bias for variability in ion toxicity thresholds, ion restriction ability and the capacity for ion selectivity (Shannon, 1997).

Dvorak and co-workers (Dvorak and Ross, 1986; Dvorak *et al.,* 1985) have been successful in transferring salt tolerance from *E. pontica* and *E. elongata* to wheat by hybridizing the two grasses to wheat. Some of the derivatives, which varied in chromosome composition, had improved salt tolerance compared to wheat; furthermore, amphidiploids showed tolerance to a wide range of ions (Na^+, K^+, Mg^{2+}, Cl^-, and $SO^{2-}{}_4$) as well as to seawater. Genes that control K/Na discrimination in wheat have been located on the long arm of chromosome 4D through the use of conventional genetic manipulation of chromosomes and chromosome fragments (Gorham *et al.,* 1987). These investigations made use of disomic D-substitution lines of tetraploid wheat previously developed from crosses between tetraploid and hexaploid wheats with unbalanced A, B and D genome sets.

(B) Heritability

Breeding procedure depends on the pattern of inheritance (qualitative or quantitative), the number of genes with major effects, and the nature of gene action. Knowledge of the heritability and inheritance of salt tolerance in a species will determine the selection intensity and number of selection cycles necessary to effectively improve tolerance. Breeding for improved salt tolerance has been discussed for several decades and has been reviewed on numerous occasions (Norlyn, 1980; Ramage, 1980; Shanno, 1982, 1984, 1985; Noble, 1983; Blum, 1988; Rana, 1985; Tal, 1985; Epstein and Rains, 1987; Jones, 1987).

Salt tolerance in *Agropyron intermedium* was found to be a heritable characteristic that could be selected based on growth in mature plants. Norlyn (1980) found that salt tolerance in barley was heritable, but the genetic control was complex.

The lack of genetic variation and a poor understanding of genetic control are most often the greatest barriers to improvement of salt tolerance. The nature of salt tolerance as a multigene, quantitative character imposes several constraints on the type of breeding approach that might be successful. Ramage (1980) emphasized that a favourable background genotype is needed on which to improve the salt tolerance characteristics. He suggested the use of recurrent selection for both cross-pollinated species and those self-pollinated species that have male sterility systems that can be genetically manipulated. Use of male sterility systems would reduce the work involved in hybrid production during the recombination phase of the recurrent selection procedure with self-pollinators.

(C) Screening and Breeding

Salt resistance must be sought and obtained shortly through screening. There may exist variability within extant genotypes - those with greater or lesser resistance to salinity and, in order to discover these, all that is required is a suitable screening procedure (e.g., Flowers and Yeo, 1981). Developing the procedure may itself prove problematical, due to the complexity of environment by genotype interactions and ontogenetic drift in the response to salt, but the procedure is conceptually simply : expose a range of genotypes to salinity and choose the ones that perform best under the conditions (Norlyn, 1980). Once a salt resistant genotype or plant is discovered, a plant breeder may then wish to improve its agronomic characteristics and this will be the extent of the breeding programme. Any crosses that are made are designed to improve the yield of the resistant line. For inbreeding species the conventional procedures would be pedigree selection, where one parent is chosen for resistance and the other for desirable agronomic traits (Shannon, 1985).

Ramage (1980) suggested that the generation of variation could be attempted by the use of recurrent selection for outbreeders and species, where male sterile lines are available. The most resistant plants from a population with variation for salt resistance are selected and selfed : all possible crosses are made amongst these progeny, seed bulked and the progeny grown for at least another generation to allow genetic recombination. Similar cycles are repeated until there is no further improvement in resistance (Ramage, 1980). The procedure is demanding and, as far as we are aware, has yet to be applied to increasing salt resistance in any crop.

Overall, while there is variation between genotypes within a species, such variation is limited : whether this is due to loss of variation during domestication (Tal, 1985) would require a comparison of the range of resistance in a number of related wild and domesticated species. Certainly crops do not cover the range of resistance seen between wild species. Various alternative ways of generating variation have been advocated : these include the use of mutagens, tissue culture and the import of foreign genes.

Based on the results of biometrical studies in several crops, reciprocal recurrent selection or diallel selective matting with some modifications are suggested for genetic improvement of grain yield for inducing salt tolerance. The strategy of varietal diversification has been advocated to meet the challenges posed by abiotic stresses as it would reduce the damage of genetic uniformity. Sokolova *et al.* (1991) laid emphasis on breeding for hybrid varieties in self pollinated crops as in their study based on *Gossypium barbadense* most F1 hybrid surpassed their parents for salinity tolerance. Joshi (1992) considered multiple cross approach as the most effective for identification of salinity tolerant wheat materials.

The choice of breeding procedure will ultimately depend upon knowledge of the traits, pattern of inheritance, number of genes with major effects and the nature of gene action involved. In most cases resistance to specific ion effects show a qualitative type of inheritance governed by single or few genes with major effects. However, resistance to osmotic effects usually follows a quantitative pattern of inheritance with polygenic control. This complicates the task of the scientists and poses challenge to the development of appropriate breeding methodology with a view to combine superior resistance with high yield potential.

GENETIC MARKERS AND QTLS IN SALT TOLERANCE

A contemporary approach to improving the efficiency of selection and breeding for complex traits such as ST is to discover genetic markers that are associated with the trait(s) of interest and which could be used as indirect selection criteria for the evaluation of large breeding populations. A common method for the identification of associated genetic markers is the detection of linkage between genetic markers already mapped on the chromosomes and the loci or gene complexes (e.g. QTLs) that control quantitative variation and are inherited as major hereditary factors.

QTLs Associated with Salt Tolerance

It has been known that salt stress imposes two primary effects on plants, osmotic and ionic, and both reduce plant growth and final yield to varying extents in different crop species. In addition, salt stress imposes secondary effects such as nutrient imbalances, generation of oxidative stresses and hormonal imbalances (Ashraf *et al.,* 2008). To counteract such salt-induced adverse effects plants respond in different ways including ion exclusion/accumulation to provide ion homeostasis, accumulation of organic osmolytes to establish osmoregulation, production of antioxidants to counteract salt-induced generation of reactive oxygen species (ROS), and changes in uptake and accumulation of essential mineral nutrients (Mittler, 2002, Byrt *et al.,* 2007, Ashraf *et al.,* 2008). During the past two decades, numerous studies have been undertaken in different crop species to identify QTLs that directly or indirectly affect the various plant responses to salt stress at different development stages. These studies on wheat and rice are summarized in Table 11.5, and the major findings have been briefly outline here.

Table 11.5. Identification of quantitative trait loci (QTLs) for salt tolerance (ST) in wheat and rice.

Crop plants	Molecular markers	Locus	Traits governed
Wheat (*Triticum aestivum* L.)	ESTs	*TmHKT7-A2*	Reduces Na^+ concentration in leaf blades by retaining Na^+ in the sheaths
	SSR	*Knal*	Controls the selection of Na^+ and K^+ transport from root to shoot and maintains high K^+/Na^+ ratio
	SSR	*Naxl*	Both are involved in decreasing Na^+ uptake and enhancing K^+ loading into the xylem
	EST	*Nax2*	Increase biomass, shoot length, root length, chlorophyll and proline contents at both the germination and seedling stages under saline conditions.
	Not mentioned in the article	*Xglk-683-Xcdo-460* and *Xjbb168-Xbcd147*	
Rice (*Oryza sativa* L.)	SSRs	*qRL-7, qDWRO-9a* and *qDWRO-9b qBI-1a* and *qBI-1b*	Play important roles in root length and root dry weight at seedling stage under saline conditions.
	ESTs	*qSNC-7* and *qSKC-1*	After shoot Na^+ and K^+ concentrations, respectively.
	SSRs	*QNa, QNa:K, SKC1/OsHKT8*	Regulate K^+/Na^+ homoeostasis

Crop plants	Molecular markers	Locus	Traits governed
	AFLPs	*QNa, QK1, QK2* and *qSTR-6*	Improve Na^+, K^+ and discrimination of Na^+ or K^+ uptake
	SSRs	*qST1* and *qST3*	Improve Na^+/K^+ ratio under saline conditions
	RFLPs, SSRs, AFLPs and isozymes	*qST1* and *qST3*	Enhance ST in shoots
	SSRs	*qNAK-2* and *qNAK-6*	Improve Na^+/K^+ ratio
	SSRs	*Saltol*	Controls shoot Na^+/K^+ homoeostasis
	SSRs	*Saltol* an non-*Saltol*	Control shoot Na^+/K^+ ratio homoestasis
	SSRs	*QKrl.2*	Controls K^+ content in root

QTLs for Ion Uptake and Accumulation

The key transport systems involved in ion homoeostasis in plants grown under saline conditions are regulated by the salt overly sensitive (SOS) signal pathway (Hasegawa *et al.,* 2000, Zhu, 2000). The SOS_2, a serine/threonine protein kinase, regulates SOS1 (Plasma membrane Na^+/H^+ antiporter)-mediated Na^+ efflux from the cytosol, as well as regulating vacuolar Na^+/H^+ antiporter-mediated Na^+ sequestration into the vacuole (Shi *et al.,* 2000, Qiu *et al.,* 2003). In addition to this mechanism of ion homoeostasis, the high-affinity K^+ trans-porters (HKT), which mediate Na^+-specific transport of Na^+/K^+ co-transport, play an important role in maintaining Na^+ homoeostasis in plants under saline conditions.

An important locus (*Knal*) has been reported to control the selectivity of Na^+ and K^+ transport from root to shoot, thereby maintaining a high K^+/Na^+ ratio in leaves (Gorham *et al.,* 1987, 1990, Dubcovsky *et al.,* 1996, Luo *et al.,* 1996). However, Na^+ exclusion mechanism in durum wheat (*Triticum turgidum* L. ssp. *durum* Desf.) was found to be linked to *Naxl* (Na^+ exclusion 1) and *Nax2* loci, which most probably relate to the Na^+ transports HKT1:4 (HKT) and HKT1:5 (HKT8), respectively (Huang *et al.,* 2006, 2008, Byrt *et al.,* 2007). The *Nax1* and *Nax2* loci have been reported to effectively decrease Na^+ transport from root to shoot, thereby to effectively decrease Na^+ transport from root to shoot, thereby maintaining reasonably low Na^+ content as well as high level of K^+ in the leaf blades of durum wheat plants by excluding Na^+ from, and loading K^+ into, the xylem (James *et al.,* 2006). It is apparent that many studies have identified QTLs for ion uptake and/or accumulation in different crop species grown under salt stress. However, the utility of these QTLs information for improving plant ST via Marked Assisted Selection (MAS) has not been vertified. Additional work is necessary to determine the utility of such QTLs for use in crop breeding for improved ST.

QTLs for Organic Osmolytes and Osmoprotectants

When subjected to stress conditions, including salt stress, most plants accumulate a variety of organic osmolytes (osmoprotectants), including sugar alcohols, proline and quaternary ammonium compounds such as glycine betaine. Such organic solutes not only contribute to osmoregulation in stressed plants, but also provide protection to many enzymes that are active in the cytosol (Bohnert *et al.,* 1995, Ashraf and Foolad, 2007). There are great variations within and among plant species in the accumulation of organic osmolytes in response to saline conditions. In general, plants accumulating higher levels of these osmolytes are more salt tolerant than those accumulating lower amounts (Ashraf and Harris, 2004). However, limited research has been conducted to identify putative QTLs underlying various osmoprotectants, besides a few reports related to proline. For example, two putative proline QTLs were identified in barley on chromosomes 2 and 4, and recently, Siahsar and Narouei (2010) identified 29 QTLs for a number of physiological traits including proline content under saline conditions in barley. Unlike the QTL approach, transgenic approaches have been used frequently to engineer plants with overproduction of different types of osmoprotectants, including glycine betaine, proline and trehalose (Hussain *et al.,* 2011). Ziaf *et al.* (2011) transferred an early responsive-to-dehydration gene (*SpERD15*) from a drought- and salt-tolerant wild tomato *S. pennellii* to tobacco (*Nicotiana tobacum* L.). The resultant transgenic tobacco line showed improved drought and ST because of high accumulation of proline and soluble sugars, which occurred because of the overexpression of their respective genes, *P5CS* and *sucrose synthase.*

Because different osmolytes play important roles in conferring ST, it is important to conduct research to discern the genetic control of osmolyte production in different plant species. Such knowledge may facilitate the development of plants with improved ST via breeding for osmolyte accumulation.

QTLs for Oxidative Defence System

Under saline stress, similar to that under a variety of other stress conditions, reduction of oxygen (O_2) frequently takes place in plants, which leads to the production of different types of ROS, including superoxide (O^-_2), hydrogen peroxide (H_2O_2) and hydroxyl radical (OH). ROS are known to adversely interact with a number of metabolites, which may result in the impairment of normal functioning of the cell (Mittler, 2002, Ashraf, 2009, Mittler *et al.*, 2011). However, plants have the ability to generate a variety of enzymatic and non-enzymatic compounds that act as antioxidants to detoxify the ROS. The type of antioxidants depends on plant species. Some key enzymatic antioxidants include superoxide dismutase, catalase ascorbate peroxidase (APX), dehydroascorbate reducatase (DHAR), monodehydroascorbate reducatse (MDHAR) and glutathione reducatse (GR; Mittler, 2002, Ashraf, 2009). Important non-enzymatic antioxidants include glutathione (GSH), ascorbate (AsA), carotenoids, tocopherols, phenolics and flavonoids (Mittler, 2002, Mateo *et al.*, 2004, Gupta *et al.*, 2005, Ashraf, 2009). Overexpression of these antioxidants has been reported to be associated with ST in many plant species (Ashraf and Foolad, 2013). Up-regulation of the genes for different antioxidants has been reported in plants under stress conditions (Zhu *et al.*, 2005). Thus, attempts have been made to identify QTLs underlying the expression of various antioxidants in different plant species. For example, Frary *et al.* (2010) reported that *Solanum pennellii* accession LA716 (a salt-tolerant wild accession of tomato) had considerably more accumulation of antioxidants, such as total phenolics, flavonoids and some other key antioxidant enzymes, than the salt-sensitive cultivated species, *Solanum lycoperscium*. In this research, a total of 125 QTLs for antioxidants were identified under saline and non-saline conditions.

QTLs for Growth-related Traits under Salt Stress

Quantitative trait loci have been identified for growth-related traits as well as physiological and biochemical attributes related to ST during different plant development stages (Table 10.5). For example, QTLs for ST during the germination stage have been identified in different plant species, including tomato (Foolad and Jones, 1993, Foolad *et al.*, 1997, 1998), rice (Prasad *et al.*, 2000, Cheng *et al.*, 2008), barley (Mano and Takeda, 1997) and wheat (Ma *et al.*, 2007). Similarly, QTLs for ST at the seedling stage (i.e. vegetative growth under salt stress) have been detected in a number of plant species, including rice, Arabidopsis, barley, soybean, white clover and wheat

(Ashraf and Foolad, 2013). The results in wheat and rice are summarized in Table 10.5.

USE OF SALT TOLERANT WILD SPECIES

Three approaches are being used by researchers in different parts of the world to develop salt tolerant crops. In the first approach, searches have been initiated within traditional crops for salt tolerant cultivars. This effort has met with some success, especially with rice, barley and wheat. A second approach that has proven successful has been to cross commercial species low in salt tolerance with wild salt tolerant relatives. Rush and Epstein (1976) and Sacher *et al.,* (1982) have pursued this avenue of research, producing salt tolerant tomato plants with moderate sized fruit. The third approach has been to begin with wild species that have evolved high salt tolerance in their native habitats. These species are first evaluated for their potential agronomic value and desirable cultivars are then developed using both traditional and modern biotechnological methods (Somers, 1979; Gallagher, 1985).

There is a wide range of species in which salt tolerance (as defined above) has already evolved. These halophytes, over 1560 species are included in Aronson's (1989) database, have great potential for use on severely salt-affected land as well as in coastal areas, where the use of irrigation with sea water is possible. In the past two years there have been over 100 papers published in which the word halophyte appears either in the title or keywords. Already some species are being used for fodder and fuel wood (Aronson, 1989) and there is potential for their use as oil seed crops among others (O'Leary, 1994). There is still an argument to be evaluated that it may be cheaper to domesticate a wild salt-tolerant species than modify an existing crop.

On a limited scale wild plant adapted to saline substrate including sea water are being used and many have been evaluated and selected for development as crops (Gallagher, 1985; Pasternak *et al.,* 1985; Carlson, 1994). These plants can be grown using land and water unsuitable for conventional crops and can provide food, fuel, fodder, fibre, resins, essential oils and pharmaceutical feedstocks (Table 11.6).

Wild salt-tolerant but economically worthless relatives of crop species may be used as sources of germplasm to transfer salt tolerance into economic species. Those salt tolerant species and genera related to a crop plant which exhibit high levels of chromosome pairing with it could contribute useful sources of tolerance to the crop species through interspecific and intergenetic hybridization (Nobel, 1983). Rush and Epstein (1976) compared two species of tomato : the domesticated

Lycopersicon esculentum (salt sensitive) and *L. cheesmanii,* a wild species whose seeds survived in solution culture salinized to full sea water salinity. These two species were crossed and the progeny of the cross could grow in upto 70% sea water. Subbarao *et al.,* (1990) have found that wild relative *Atylosia albicans* of pigeonpea (*Cajanus cajan*) and its F1 hybrids are highly salt tolerant as compared to its salt sensitive cultivated species.

Table 11.6. Potential uses of salt-tolerant plants as food, fuel, fodder, fibre and other products

Species	Common name	Potential uses
Food		
Zostera marina	Eel grass or sea grass	Seeds-Flour/Bread
Distichlis palmeri	Palmer salt grass	Seeds-Flour/Bread
Kosteletzkva virgincia	Sea shore mellow	Seeds-Proteins/Oils
Argania sphiosa	Argan	Seeds-Edible oil Browse
Fuel		
Prosopis juliflora, P. articulata	Mesquite	Fuelwood
Casuarina equisetifolia	–	Fuelwood/Charcoal
Rhizophora mangle	Mangroves	Fuelwood/Charcoal
Tamarix aphylla	–	Fuelwood/Windbreak
Fodder		
Leptochloa fusca	Kallar grass	Fodder/Biogas production
Distichlis spicata	Salt grass	Cattle forage
Sporobolus virginicus, S. airoids	Alkali saction	Cattle forage
Atriplex spp.	Salt bush	Forage source in arid zones
Fibre and other products		
Sesbania bispinosa	Dhaincha	Fibre/Fuel/Gums
Phragmite australis	Common Reed	Roofing/Basket making/ Fuel/Fencing
Simmondsia chinensis	Jojoba	Oil-Cosmetics/Lubricants/ Waxes
Pandanus fascicularis	Kewda	Perfume/Flavouring ingredients

Intergenetic and interspecific hybridisation within the Triticeae is a promising technique for the improvement of salt tolerance in wheat.

Attempts are already underway for exploiting the outstanding superior salt tolerance of the wheat grass genus *Elytrigia* (Agropyron) for enhancing adaptation of wheat cultivars to salt affected soil (Dvorak *et al.,* 1985). Because use of exotic salt-tolerant germplasm can enlarge the potential genetic variability available, the approach of using wild crosses holds great promise of success (Epstein and Rains, 1987). The conservation of wild germplasm is, therefore, very essential for the development of salt tolerant in crop species.

MARKER ASSISTED SELECTION (MAS)

Improving yield performance of staple crops is an ultimate objective of plant breeding programs (Schroder, 2013; Ashraf and Foolad, 2013). As obtaining salt-tolerant crops using conventional breeding methods takes a long time, alternative approaches are being considered in parallel. Traditionally, crops were outcrossed to genetically diverse germplasms and were then selected based on their phenotype as evaluated in the field. This process is now streamlined using QTL analyses coupled with marker-assisted selection (MAS). MAS is an approach that requires the linkage of a quantitative trait with a genetic marker that is polymorphic between parental lines. Hence the essential basis for successful breeding with MAS is an in-depth knowledge of genetic traits and variability within the desired plant species. One example is *Saltol,* a favourable QTL identified in rice that is responsible for the bulk of genetic variation in ion uptake under saline conditions (Ashraf and Foolad, 2013).

MAS clearly has the potential of contributing to the development of more salt-tolerant crops. A major bottleneck remains the selection of appropriate markers (i.e. key players in the context of salinity tolerance). Many studies are attempting to improve plant salt tolerance by genetic manipulation of certain genes; however, some of these genes might not have sufficient impact to improve crop viability significantly in highly saline environments. It has been proposed that enhancing plant stress tolerance is practicable by manipulating only one or a few main components of the regulatory gene network instead of engineering several molecular mechanisms (Goeldack *et al.,* 2011). Besides conventional breeding methods, stacking of traits (also known as pyramiding) is a promising approach based on the introduction of several beneficial genes to improve plant performance. However, this approach is limited by independent segregation of traits, which complicates breeding strategies. An emerging method to this issue involves the use of trait landing pads, whereby engineering sequence-specific nucleases, such as zinc-finger nucleases, are used to target multiple transgenes to the same locus (Ainley *et al.,* 2013). While so far only Zinc-finger nucleases have been used in practice for generating trait landing pods, it is expected that the rapidly emerging CRSPR/Cas method will further facilitate targeted insertion of promoters and genes of interest.

In the case of ST, however, a few studies can be cited from the literature where gene or QTL information has been used to develop lines or cultivars with improved ST through marker-assisted breeding (Table 11.7). One example is the recent development of a highly salt-tolerant cultivar of durum wheat by R. Munns and her research team at CSIRO, Australia (Munns *et al.,* 2012). Durum wheat is generally known for its more sensitivity to salt stress compared with the common bread wheat, and this has been attributed to its inherent lower ability of excluding Na^+ from the leaf blade. In an earlier study, Munns *et al.* (2000) had identified a durum line (known as Line 149) with Na^+ exclusion ability similar to that of bread wheat. In a genetic analysis of a cross between Line 149 and a durum wheat accession (Tamaroi) with normal Na^+ exclusion ability, two genes *Naxl* and *Nax2* were identified for controlling Na^+ exclusion (Munns *et al.,* 2003). *Nax1* was mapped to the distal region of chromosome 2AL (Lindsay *et al.,* 2004), and *Nax2* was mapped to chromosome 5L (Byrt *et al.,* 2007). In subsequent studies, these two genes were introduced into various durum wheat lines through marker-assisted breeding. The newly developed durum wheat lines were tested under natural saline fields in northern New South Wales, Australia, and the lines possessing particularly *Nax2* were able to produce approximately 25% more yield than the control durum wheat lines under such saline conditions (Munns *et al.,* 2012). The findings of this research are encouraging and provide confidence to the use of MAS for breeding crops for improved ST (Table 10.7).

In a similar study in rice, scientists at the International Rice Research Institute (IRRI) identified a single major QTL (*Saltol*) on the short arm of chromosome 1, which explained much of the variation for ST in a segregating rice population (Bonilia *et al.,* 2002). In subsequent studies conducted at the Bangladesh Rice Research Institute (BRRI), attempts have been made to transfer *Saltol* to two high yielding popular commercial rice varieties, BR11 and BR28. In this study, various markers closely linked to *Saltol* were identified and used to transfer the QTL to the two commercial varieties via marker-assisted backcrossing (Table 10.7). In tomato, MAS was employed to develop salt-tolerant breeding lines using an F_2 population of a cross between the cultivated tomato (salt-sensitive) and a salt-tolerant accession of the tomato wild species *Solamum pimpinellifolium* (Monforte *et al.,* 1996). In this study, a combination of phenotypic selection and MAS was used to develop lines with improved ST.

Table 11.7. Improving salinity tolerance in different crop lines/ varieties using marker-assisted selection

Crop	QTL used	QTL donor line/ cultivar	Recipient line/ cultivar	Line/ cultivar developed	Traits improved	References
Durum wheat (*Triticum turgidum* L.)	*Nax1* and *Nax2*	Line 149	Tamaroi	Durum wheat	Improved 25% yield by enhancing Na^+ exclusion from the leaf blade	Munns *et al.* (2000)
Rice (*Oryza sativa* L.)	*Saltol*	FL378	BR11 or BR28	BR11	Higher yield under saline conditions	Rahman *et al.* (2008)
Tomato (*Solanum lycopersicum* L.)	*TG24*	*Solanum pimpinellifolium* (wild tomato)	*S. lycopersicum* (cultivated tomato)	F_3 population of *S. lycopersicum*	Improved total fruit weight under salinity	Monforte *et al.* (1996)

In summary, despite the extensive efforts made to identify QTLs contributing to ST in different crop species, limited attempts have been made to use such QTL information for marker-assisted breeding for improved ST. However, the limited use of markers for improving crop ST is attributable to various reasons, including (i) limited efforts that have been made by breeders to develop plants with improved ST, compared to efforts devoted to other economically important traits such as disease resistance and improved quality and quantity of yield, (ii) limited familiarity of many plant breeders with the marker technology, (iii) insufficient reliability of the identified QTLs or lack of QTL confirmation and (iv) population specificity of the reported QTLs and their associated markers. However, with the new advancement in marker technology and trait phenotyping and the greater need in crop plants with improved ST, it is expected that more progress will be made in developing new cultivars with improved ST by adding marker- assisted breeding approaches.

Recent advances in the field of biotechnology have led to the development of more efficient selection tools to replace phenotype-based selection systems. Molecular markers associated with genes or quantitative trait loci (QTLs) affecting important traits have been identified which could be used as indirect selection criteria to improve breeding efficiency via marker-assisted selection (MAS).

Despite the fact that the use of MAS for manipulating simple units has been stream lined in several plant breeding programmes, MAS for improving more complex traits appears to be at infancy stage. Notwithstanding the identification of numerous QTLs involved in salt tolerance, only few commercial cultivars or breeding lines with improved salt tolerance have been developed via MAS. However, the utility of MAS for breeding for salt tolerance will gain momentum in the coming years.

BIOTECHNOLOGICAL APPROACHES

A number of *in vitro* techniques have been developed during the past 2-3 decades which can be used to supplement the conventional plant breeding methods. These modern techniques of cell and tissue culture, somatic hybridization, molecular biology and genetic engineering have a great potential for the future. Some of these approaches are briefly outlined here.

Cell and Tissue Culture

The use of undifferentiated cells in tissue culture to improve salt tolerance has been explored in many crops (Nabors *et al.,* 1975; Croughan *et al.,* 1978; Ben-Hayyim and Kochba, 1983; Pandey and Ganapathy, 1984; Winicov, 1991). Large populations of cells in cultures can be maintained under precisely controlled environmental conditions and easily manipulated. Mutagenesis, haploid production, somatic hybridization, and transformation are relatively easy in tissue cultures. The major limitation of tissue cultures with respect to salt tolerance is that the selected character cannot be maintained during the regeneration process. In addition, there is still insufficient understanding of the relationship between cellular and whole plant response. An additional limitation is the possibility that epigenetic change, or physiological adaptation, may occur. Tolerance mechanisms that depend on the integrated function of differentiated tissues cannot be readily identified in cell cultures, but it is hoped that some intracellular tolerance characteristics can be improved. Potential characteristics that may be selected from cell cultures include the ability to regulate or partition ions more efficiently, the production of organic solutes to improve osmoregulation under salt stress, and improved metabolic efficiency (Hasegawa *et al.,* 1980).

Tissue culture has been advocated as means to generate variation as well as a selection procedure. The success, or otherwise, of such procedures with respect to salt resistance remains controversial (Dracup, 1993; Hasegawa *et al.,* 1994). In reviewing the papers quoted by Dracup and Hasegawa *et al.* (1994) concluded that there was no published evidence of improvements being expressed in the field (Flowers and Yeo, 1997; Rowland *et al.,* 1989) but improvements could be demonstrated under controlled environment conditions (Winicov, 1991). Selection in tissue culture is not a process that can be used simply to generate salt-resistant plants, in spite of the fact that salt tolerant cell lines can readily be selected. Rather, as Hasegawa *et al.* (1994) suggested the value of *in vitro* experiments to further the intensity of cellular basis of salt tolerance in plants (Flowers and Yeo, 2002).

Somaclonal Variation

Somaclonal variation is the variation among plants derived from cell culture. Regeneration of plants from plant cells is not a guarantee that the regenerated plant will be an exact copy or a clone of the parential line used to culture the cell lines. These phenotypic variants provide a potential source of genetic variation and are very important to selection systems. Even though salt selection *in vitro* has produced regenerated plants with increased tolerance (Nabors *et al.,* 1980), selection *in vitro* for salt tolerance will not detect all potentially salt tolerant plants. This reaffirms the importance of screening somaclonal variants at whole plant level. Several salt tolerant genotypes derived from somaclones have been reported in crop plants such as tobacco and pepper (Hess, 1990). Somaclones haploids from anthers and pollen are being largely screened for resistance to salinity in *Brassica* (Jagannathan, 1991). Jain *et al.* (1991) developed three somaclones of *Brassica juncea* L. cv. Prakash and reported that SR3 (somaclone 3) derived line had about three time higher inherent free proline as well as high total free amino acids as compared to parent and other somaclones.

In vitro culture systems have been exploited to obtain salt-resistant lines followed by utilizing somaclonal variation to obtain salt-tolerant rice plants (Winicov, 1996, Lutts *et al.,* 1999). Somaclonal variation can be induced as a positive shift to generate new properties. Recently Aditya and Baker (2006) selected salt tolerant somaclones from four Bangaldeshi indica rice genotypes enough *in vitro* and *ex vitro* NaCl stress applied through step wise and non-step wise methods. Satisfactory seedlings survival was observed in all the genotypes upto 150 mM NaCl stress. Advance flowering was obtained mostly for all the genotypes when stress was applied at the seedlings stage compared with combined stress at both seedling at booting stages.

Somatic Hybridization

Protoplast fusion with formation of somatic hybrids provides an opportunity to transfer genetic information from one species to another. Protoplast fusion is currently a reasonable approach for salt tolerant genotype production only if (a) genetic variability for salt tolerance within a protoplast, culturable genes has been identified, and (b) this variability must be inaccessible to the cultivated plant by sexual hybridization *in vivo* or *in vitro*. At present very few important agronomic crops have produced protoplasts, whether single or fused which have been successfully regenerated. Good success has been met with tobacco, potato, carrots, petunia and others but the major food crops are only now being incorporated into the line of plant species amenable to somatic manipulation (Epstein and Rains, 1987).

Recombinant DNA Technology

The concept of genetic engineering of plants has commonly focussed on the use of recombinant DNA technology. Two gaps in our knowledge, namely, how to control introduced gene expression and what genes to introduce make speculation on use of the actual genetic engineering of salt tolerance difficult. Efforts to identify genes responsible for salt tolerance are essential not only for future engineering attempts but also to help us discover how plants respond to stress.

Recombinant DNA techniques have been effectively utilized in bacteria for the transfer of OSM genes govering production of osmoprotective molecules e.g. glycinebetaine, prolinebetaine and proline etc. These molecules accumulate in plant cells during osmotic stress and prevent damage from cellular dehydration by balancing the osmotic strength of the cytoplasm with that of the environment. Using the recombinant DNA technique a nitrogen fixing strain of *Rhizobium* normally highly sensitive to osmotic stress was converted into a tolerant type with the help of broad host range plasmid by transferring a DNA segment of about 10,000 base pairs from *Escherichia coli* carrying the OSM gene (s) which led to proline over production and consequent osmotic tolerance (Valentine, 1984). The discovery of osmoprotective molecules and new classes of OSM genes may lead to applications such as genetic enhancement of salinity tolerance in crop plants. However, for attaining maximum gains from this emerging technology, an intergrated approach with standard plant breeding procedures is to be closely followed as suggested by Rana (1986).

According to Epstein and Rains (1987) a general programme for applying genetic engineering techniques to improve stress tolerance and WUE of plants might involve the following steps :

(1) Molecular cloning of stress tolerance genes from beneficial microorganisms.

(2) Recombinant DNA analyses of the structure of genes regulating tolerance to salt or water stress.

(3) Development or plant gene vectors for genetic engineering of stress tolerant plants.

(4) Application or recombinant DNA technology for molecular cloning of stress tolerant genes.

(5) Use of cell culture techniques for the selection of stress tolerant mutant plants.

(6) Transfer of these genetic traits for stress tolerance into a range of crop plants for agriculture.

Recent advances in molecular biology have broadened the possibilities for gene manipulation at the level of cell cultures and higher units of organization. The development of new or improved technologies for monoclonal antibodies, endonuclease digestion, cDNA libraries, purified protein isolation, and rapid amino acid sequencing now make gene identification, isolation, and transformation realities. The major problems that prohibit the use of these techniques to develop salt tolerant crops are that salt tolerance is a complex, multigenic trait and is often a composite response of the integrated biological system. At its current level of technology, molecular biology is not able to engineer traits that involve whole plant responses. However, the power of molecular biology goes beyond the ability to manipulate genes: It is also useful as a method to monitor environmental effects at the level of the genome.

Although the applications of these techniques have great potential but our understanding of mechanisms related to stress tolerance coupled with the very preliminary status of our ability to apply genetic engineering techniques to higher plants will limit the usefulness of this approach. Success will not be realised without substantial investments in time and resources devoted to the basic studies of plant processes and plant genetic systems as well as to transferring genetic engineering technology from the prokaryotes to the eukaryotes.

Success of Breeding for Salt Tolerance

Efforts have been made in many countries to develop salt-tolerant plant materials and varieties. This is reflected in the registration of such materials with national and international authorities, publication in journals, patents and release of improved varieties at regional and national levels (Sharma and Goyal, 2003). A search of literature reveals a number of varieties/cultivars of different crops developed and released for improved salt tolerance during the last 25 years. The list of such successful attempts from different countries has been presented in Table 11.8.

Table 11.8. List of salt-tolerant varieties/cultivars of various crops (Shannon and Noble, 1990)

Species	Variety developed and country	Process of development	Features
Avocado	Arsola 1-18	Cross between salt tolerant west indian and cold hardy Mexican varieties	Cold hardy and highly salt tolerant
Wheat grass	Nebraska 10	Selection	Salt tolerant strain
Rice	Giza 159	Cross between salt tolerant Agam-1 and local var. Giza 14	Highly salt tolerant
Rice	Giza 160	Cross between Giza 159 and Salt tolerant var. Reiho	Highly salt tolerant
Tomato	Edkway	Selection by Farmers of Egypt	Highly salt- tolerant in northern coast of Egypt
Alfalfa	A2-Zerm Salt-1 A2-Zerm Salt-2	Selection	High germination rates under salinity
Red fescue	Saltol	Naturally selected ecotype	Natural tidal marsh lands
Muskmelon	BG 84-3	Selection for locally adapted varieties from Israel	More productive and salt tolerant
Maize	Arizona 8601	Natural selection prog.	Suitable for saline irrigation water 2000-4000 ppm

In India, Scientsts at CSSRI, Karnal have successfully led to the development and release of varieties of rice, wheat, chickpea and Indian mustard suitable for cultivation in alkali and saline soils. Some of these varieties are listed in Table 11.9. Thus a good beginning has been made and many success stories will emerge in future. However, many of these superior cultivars have yet to prove their worth in actual stress situations. Therefore, efforts in this direction have to be more focussed and multidisciplinary in approach, in view of the enormity of the situation and immense challenges involved (Sharma and Goyal, 2003).

Table 11.9. Salt tolerant varieties of rice, wheat, chickpea and Indian mustard developed in India (Sharma and Goyal, 2003)

Crop	Genotype developed	Process of development	Suitability for cultivation
Rice	CSR 10 (1989)	Mutation breeding from a cross between M40-431-24-114 and Jaya	Alkali and inland saline soils pH upto 10.0, EC_e upto 11.0 dS/m
Rice	CSR 13	Three way cross of CSR 1/ Basmati 370/ CSR 5	Alkali and inland saline soils (pH 10.0, EC_e 9.0 dS/m)
Rice	CSR 27 (1999)	Derivative of a cross Nina Bokara and IR 5657-33-2	Alkali and Coastal saline soils (pH 9.7, EC_e 10.0 dS/m)
Wheat	KRL 1-4 (1990)	Cross of Kharchia and WL 711	Alkali soils (pH upto 9.3 and saline soils EC_e upto 7.0 dS/m)
Wheat	KRL 19 (1999)	Safe replacement of KRL 1-4	Better tolerance to rusts and other diseases (pH 9.3, EC_e 7.3 dS/m)
Chickpea	CSG 8962 (Karnal Chana-1)	Selection for moderate Salt stress	pH upto 9.0, EC_e 4-6 dSm
Indian Mustard	CS 52 (1997)	Selection	Saline and Sodic soils pH 9.5, EC_e 10.0 dS/m

CONCLUSIONS

The progress in the breeding and selection of salt tolerant plants has been relatively limited so far. The efficiency of breeding programmes can be increased only by adoption of new selection criteria based on knowledge of the physiological processes limiting crop production under stress conditions and their genetic control. The main prerequisites for efficient breeding programmes for increasing salt tolerance are the availability of sufficient genetic variability and of reliable selection variability is incomplete and must therefore be supplemented with more research. Except in few cases, the value of the various attributes proposed as selection criteria requires experimental testing. The ultimate success will depend upon the adoption of proper selection criteria. The development of selection criteria could also include detailed studies of growth as well as yield components and metabolic parameters. However, simultaneous selection in at least two directions i.e. yield and tolerance is advocated.

In view of the complexity, diversity and integration of the mechanisms rendering a plant salt tolerant, it will take considerable time before these modern, biotechnological techniques can make their mark. And their utility, once it is realized, will lie not in their replacing plant breeding, but instead as additional tools to create genetic diversity, screen it and incorporate selected traits into crops.

Salinity is an ever-present threat to agriculture, especially in area where secondary salinisation has developed through irrigation or deforestation. Attemps to improve the salt tolerance of crops have met with limited success due to the complexity of the trait, both genetically and physiologically. Tolerance shows all the characteristics of a multigenic trait, with quantitative trait loci (QTLs) identified in barley, citrus, rice, and tomato (Flowers and Flowers, 2005).

Attempts to produce salt tolerant crops have involved both the domistication of halophytes and the manipulation of existing crop species through conventional breeding programmes. The use of *in vitro* selection, pooling physiological traits, interspecific hybridisation, the use of marker aided selection and the development of transgenic plants are promising approaches. After 15 years of research, the value of using transgenic plants to alter salt tolerance has yet to be tested in the field. The use of physiological traits in breeding programmes and the domestication of halophytes currently offer viable alternative to the development of tolerance through the use of transgenic technologies.

REFERENCES

Abel, G.H. and Mackenzie, A.J. 1964. Salt tolerance of soybean during germination and later growth. Crop Sci. 4: 157-160.

Aditya, T.L. and Baker, D.A. 2006. Selection of salt tolerant somaclones from indica rice through continuous *in vitro* and *ex vitro* sodium chloride stress. Indian J. Plant Physiol. 11: 349-357.

Aronson, J.A. 1989. Haloph A database of salt tolerant plants of the world. Office of Arid Land Studies. Tucson, pp. 77.

Ashraf, M. 2004. Some important physiological selection criteria for salt tolerance in plants. Flora 199: 361-376.

Ashraf, M. 2009. Biotechnological approach of improving plant salt tolerance using antioxidants as markers. Biotechnol. Adv. 27: 84-93.

Ashraf, M. and Akram, N.A. 2009. Improving salinity tolerance of plants through conventional breeding and genetic engineering: an analytical comparison. Biotechnol. Adv. 27: 744-752.

Ashraf, M. and Foolad, M.R. 2007. Roles of glycine betaine and proline in improving plant abiotic stress resistance. Environ. Exp. Bot. 59: 206-216.

Ashraf, M. and Foolad, M.R. 2013. Crop breeding for salt tolerance in the era of molecular markers and marked - assisted selection. Plant Breeding 132: 10-20.

Ashraf, M. and Harris, P.J.C. 2004. Potential biochemical indicators of salinity tolerance in plants. Plant Sci. 166: 3-16.

Ashraf, M., Athar, H.R., Harris, P.J.C. and Kwon, T.R. 2008. Some prospective strategies for improving crop salt tolerance. Adv. Agron. 97: 45-110.

Aswathappa, N., Marcar, N.E. and Thomson, L.A.J. 1986. Salt tolerance of Australian tropical and subtropical *Acacias*. In : Proc. Inter. Workshop Aust. Acacias in Developing Countries (Ed. J.W. Tumbull) pp. 70-73. Gympie, Australia.

Ben-Hayyim, G. and Kochba, J. 1983. Aspects of salt tolerance in a NaCl-selected stable cell lines of *Citrus sinenis*. Plant Physiol. 72; 685-690.

Binzel, M.L., Hasegawa, P.M., Bressan, R.A., Handa, S. and Handa, A.K. 1983. Adaptive responses of cultured tolerance cells to NaCl. Plant Physiol. 72: 5136.

Blum, A. 1988. Plant Breeding for Stress Environments. CRC Press, Boca Raton, El. 223 p.

Blum, A. 1989. Breeding methods for drought resistance. In: Plants under Stress (Eds. Jones, H.G., Flowers, T.J. and Jones, M.B.) Cambridge University Press, Cambridge. pp. 197-215.

Bohnert, H.J., Ayubi, P., Borchert, C., Bressan, R.A. and Burnap, R.L. 2001. A genomics approach towards salt stress tolerance. Plant Physiol. Biochem. 39: 295-311.

Bohnert, H.J., Nelson, D.E. and Jensen, R.G. 1995. Adaptation to environmental stresses. Plant Cell 7: 1099-1111.

Bona, L., Wright, R.J. and Baligar, V.C. 1991. A rapid method of screening cereals for acid soil tolerance. Cereal Res. Commun. 19 : 465-468.

Borsani, O., J. Cuartero, J.A. Fernandez, V. Valpuesta and M.A. Botella. 2001. Identification of two loci in tomato reveals distinct mechanisms for salt tolerance. The Plant Cell. 13: 873-887.

Byrt, C.S., Platten, J.D., Spielmeyer, W., James, R.A., Lagudah, E.S., Dennis, E.S., Tester, M. and Munns, R. 2007. HKT1;5-like cation transporters linked to Na^+ exclusion loci in wheat, Nax2 and Kna1. Plant Physiol. 143: 1918-1928.

Carlson, R. 1994. Potential plants and crops for cultivation under moderately saline conditions. In : Handbook of Plant and Crop Stress (Ed. M. Pessarakli) pp. 531-541. Marcel Dekker, Inc. New York.

Chembarisova, E.J. and Azimov, R.A. 1991. Obtaining salt tolerant cell cultures of cotton cultivars. Golden Jubilee Symp. Genet. Res. and Edu. Current Trends and the Next Fifty Years. 3 : 890.

Cheng, H.T., Jiang, H., Xue, D.W., Guo, L.B., Zeng, D.L., Zhang, G.H. and Qian, Q. 2008. Mapping of QTL underlying tolerance to alkali at germination and early seedling stages in rice. Acta Agron. Sin. 34: 1719-1727.

Chinnusamy, V., Jagendorf, A. and Zhu, J.K. 2005. Understanding and improving salt tolerance in plants. Crop Sci. 45: 437-448.

Croughan, T.P., Stavarek, S.J. and Raini, D.W. 1978. Selection of a NaCl toleant line of cultured alfalfa cells. Crop Sci. 18: 959-963.

Downton, W.J.S. 1984. Salt tolerance of food crops: Prospective for improvements. CRC Critical Rev. Plant Sci. 1 : 183-201.

Dracup, M. 1993. Why does *in vitro* cell selection not improve the salt tolerance of plants? In: Genetic Aspects of Plant Mineral Nutrition. (Eds. Randall, P.J. Dalhaze, E.D., Richards).

Dubcovsky, J., Maria, G.S., Epstein, E., Luo, M.C. and Dvorak, J. 1996. Mapping of the K^+/Na^+ discrimination locus Knal in wheat. Theor. Appl. Genet. 92: 448-454.

Dvorak, J. and Ross, K. 1986. Expression of tolerance of Na^+, K^+, Mg^{2+}, Cl^-, and SO^{-2} ions and sea water in the amphliploid of *Triticum aestivum* x *Elytigia elongata*. Crop Sci. 26: 658-660.

Dvorak, J., Ross, K. and Mendlinger, S. 1985. Transfer of salt tolerance from *Elytrigia pontica* (Podp) Holub to wheat by the addition of an incomplete *Elytritia* genome. Crop Sci. 25 : 306-309.

Epstein, E. 1983. Crops tolerant of salinity and other mineral stresses. In: Better Crops for Food. Ciba Foundation Symp. 97: 61-82, Pitman Books, London.

Epstein, E. and Noryln, J.D. 1977. Sea water based crop production: A feasibility study. Science 197 : 249-251.

Epstein, E. and Rains, D.W. 1987. Advances in salt tolerance. Plant Soil 99 : 17-29.

Epstein, E., Noryln, J.D., Rush, D.W., Kingsbury, R.W., Kelley, D.B., Cunningham. G.A. and Wrona, A.F. 1980. Saline culture of crops: A genetic approach. Science 210 : 399-404.

Evans, D.A., Sharp, W.R., Ammirato, P.V. and Yamata, Y. 1983. Handbook of Plant Cell culture Vol. 1. Techniques for Propagation and Breeding. Macmillan, New York.

Fischer, R.A. 1994. Key issues in wheat yield improvement. In: Breaking the Yield Barrier. ed. K.G. Kassman. Proc. of a workshop on rice yield potential in favourable environments. International Rice Research Institute, Manila. The Phillippines.

Flowers, T.J. 2004. Improving crop salt tolerance. J. Exp. Bot. 55: 307-319.

Flowers, T.J. and Flowers, S.A. 2005. Why does salinity pose such a difficult problem for plant breeders. Agric. Water Manage. 78: 15-24.

Flowers, T.J. and Yeo, A.R. 1981. Variability of sodium chloride resistance within rice (*Oryza sativa* L.) varieties. New Phytol. 88: 363-373.

Flowers, T.J. and Yeo, A.R. 1986. Ion relations of plants under drought and salinity. Aust. J. Plant Physiol. 13: 75-91.

Flowers, T.J. and Yeo, A.R. 1995. Breeding for salinity resistance in crop plants - where next? Aust. J. Plant Physiol. 22: 1057-1063.

Flowers, T.J. and Yeo, A.R. 1997. Breeding for salt resistance in plants. In: Strategies for Improving Salt Tolerance in Higher Plants (Eds. P.K. Jaiwal, R.P.Singh and A. Gulati) pp. 243-264. Oxford/IBH, New Delhi.

Flowers, T.J., M. Koyama and A.R. Yeo, 2000. Breeding for salt tolerance in crop plants in the age of molecular biology. In: Salinity Management in Agriculture. Eds. S.K. Gupta, S.K. Sharma and N.K. Tyagi. Proc. National Conference on Salinity Management in Agriculture. PP. 303-313.

Foolad, M.R. 1999. Comparison of salt tolerance during seed germination and vegetative growth in tomato by QTL mapping. Genome, 42: 727-734.

Foolad, M.R. and Jones, R.A. 1993. Mapping salt-tolerance genes in tomato (*Lycopersicon esculentum*) using trait-based market analysis. Theor. Appl. Genet. 87: 184-192.

Gallagher, J.L. 1985. Halophytic crops for cultivation at seawater salinity. Plant Soil 89 : 323-326.

Garg, B.K. and Gupta, I.C. 1997. Saline Wasteland Environment and Plant Growth. Scientific Publishers (India), Jodhpur. 287 p.

Garg, B.K., Kathju, S., Vyas, S.P. and Lahiri, A.N. 1990. Effect of saline water irrigation on tolerant and sensitive wheat varieties under disparate soil fertility conditions. Ann. Arid Zone. 29 : 179-187.

Gorham, J. 1992. Salt tolerance of plants. Sci. Prog. Oxford. 76: 273-285.

Gorham, J., Hardy, C., Jones, R.G.W., Joppa, L.R. and Law, C.N. 1987. Chromosomal location of a K/Na discrimination character in D-genome of wheat. Theor. Appl. Genet. 74: 584-588.

Greenway, H. and Munns, R. 1980. Mechanisms of salt tolerance in non-halophytes. Ann. Rev. Plant Physiol. 31 : 149-190.

Grieve, C.M. and Francois, L.E. 1992. The importance of initial seed size in wheat plant response to salinity. Plant Soil 147: 197-205.

Gupta, K.J., Stoimenova, M. and Kaiser, W.M. 2005. In higher plants, only root mitochondria, but not leaf mitochondria reduce nitrite to NO, *in vitro* and *in situ*. J. Exp. Bot. 56: 2601-2609.

Hasegawa, P.M., Bressan, R.A. and Handa, A.K. 1980. Growth characteristics of NaCl- selected and non-selected cells of *Nicotiana tabacum*. Plant Cell Physiol. 21: 1347-1355.

Hasegawa, P.M., Bressan, R.A., Nelson, D.E., Samaras, Y. and Rhodes, D. 1994. Tissue culture in the improvement of salt tolerance in plants. In: Soil Minerals Stresses: Approaches to Crop Improvement (Eds. Yeo, A.R. and Flowers, T.J.) Springer-Verlag. Berlin. pp. 83-125.

Hasegawa, P.M., Bressan, R.A., Zhu, J.K. and Bohnert, H.J. 2000. Plant cellular and molecular responses to high salinity. Annu. Rev. Plant Physiol. Plant Mol. Biol. 51: 463-499.

Hess, D. 1990. Gene Technology and Plant Breeding. Proc. Inter. Cong. Plant Physiol. 1 : 34-59.

Huang, S.B., Spielmeyer, W., Lagudah, E.S. and Munns, R. 2008. Comparative mapping of HKT genes in wheat, barley, and rice, key determinants of Na^+ transport, and salt tolerance. J. Exp. Bot. 59: 927-937.

Huang, S.B., Spielmeyer, W., Lagudah, E.S., James, R.A., Platteen, J.D., Dennis, E.S. and Munns, R. 2006. A sodium transport (HKT7) is a candidate for Naxl, a gene for salt tolerance in durum wheat. Plant Physiol. 142: 1718-1727.

IRRI 1997. Rice varieties boost yield and improve saline soils. Partners making a difference, IRRI, Los Banos, 37 p.

Jagannathan, V. 1991. Biotechnological approaches to biotic and abiotic stress tolerance in *Brassicae*. Golden Jubilee Symp. Genet. Res. and Edu. Current Trends and the Next Fifty years 1 : 192-193.

Jain, S., Nainawatee, H.S., Jain, R.K. and Chowdhury, J.B. 1991. Solute accumuation and osmoregulation in *in vitro* salt tolerant *Brassica juncea* L. plants. Golden Jubilee Symp. Genet. Res. and Edu. Current Trends and the Next Years. 1 : 273.

Jeong, M.J., Park, S.C. and Byun, N.O. 2001. Improvement of salt tolerance in intransgenic potato plants by glyceraldehyde-3 phosphate dehydrogenase gene transfer. Molecules and Cells 12: 185-189.

Jones, R.A. 1987. Genetic advances in salt tolerance. In: Tomato Biotechnology (Eds. D.J. Evans and R.A. Jones) pp. 125-137. A.R. Liss New York.

Joshi, P. 1992. Genetic variability in *Kharchia* wheat (*Triticum aestivum*) for salt affected areas in Rajasthan. Ann. Arid Zone 31 : 103-106.

Kingsbury, R.W. and Epstein, E. 1984. Selection of salt resistant spring wheat. Crop Sci. 24 : 310-315.

Kumar, D. 1995. Salt tolerance in oil seed Brassicas- present status and future prospects. Plant Breeding Abstr. 65 : 1439-1447.

Lahiri, A.N., Garg, B.K., Kathju, S., Vyas, S.P. and Mali, P.C. 1996. Genotypic differences to soil salinity in clusterbean. Arid Soil Res. Rehabilit. 10 : 333-345.

Lutts, S., Bouharmont, J. and Kinet, J.M. 1999. Physiological charcterization of salt-resistant rice (*Oryza sativa*). somaclones. Aust. J. Bot. 47: 835-849.

Ma, L.Q., Zhou, E.F., Huo, N.X., Zhou, R.H., Want, G.Y. and Jia, J.Z. 2007. Genetic analysis of salt tolerance in a recombinant inbred population of wheat (*Triticum aestivum* L.). Euphytica 153: 109-117.

Maas, E.V. 1984. Crop tolerance. California Agri. 38 : 20-21.

Maas, E.V. and Hoffman, G. 1977. Crop salt tolerance-current assessment. J. Irrig. Drain. Div. Am. Soc. Civil Engg. 103 : 115-134.

Mano, Y. and Takeda, K. 1997. Mapping quantitative trait loci for salt tolerance at germination and the seedling stage in barley (*Hordeum vulgare* L.) Euphytica 94: 263-272.

Mateo, A., Muhlenbock, P., Rusterucci, C., Chang, C.C.C., Miszalski, Z, Karpinska, B., Parker, J.E., Mullineaux, P.M. and Karpinski, S. 2004. Lesion simulating disease 1 - Is required for acclimation to conditions that promote excess excitation energy. Plant Physiol. 136: 2818-2830.

McNeilly, T. 1990. Selection and breeding for salinity tolerance in crop species: A case for optimism? Acta Oecologica 11: 595-610.

Mittler, R. 2002. Oxidative stress, antioxidants and stress tolerance. Trends Plant Sci. 7: 405-410.

Mittler, R., Vanderauwera, S., Suzuki, N., Miller, G., Tognetti, V.B., Vandepoele, K., Gollery, G., Shulaev, V. and Van Breusegem, F. 2011. ROS signaling: the new wave? Trends Plant Sci. 16: 300-309.

Munns, R., James, R.A. and Lauchli, A. 2006. Approaches to increasing the salt tolerance of wheat and other cereals. J. Exp. Bot. 57: 1025-1043.

Munns, R., James, R.A., Xu, B., Athman, A., Conn, S.J., Jordans, C., Byrt, C.S., Hare, R.A., Tyreman, S.D., Tester, M., Plett, D. and Gilliham, M. 2012. Wheat grain yield on saline soils is improved by an ancestral transporter gene. Net Biotechnol. 30: 360-364.

Munns, R., Rebetzke, G.J., Husain, S., James, R.A. and Hare, R.A. 2003. Genetic control of sodium exclusion in durum wheat. Aust. J. Agric. Res. 54: 627-635.

Nabors, M.W. 1983. Increasing the salt and drought tolerance of crop plants. In : Current Topics in Plant Biochemistry and Physiology (Ed. D.D. Randall) Vol. 2. Univ. of Misouri, Columbia.

Nabros, M.L., Daniel, A., Nadolny, L. and Owns, C. 1975. Sodium Chlorine tolerant lines of tobacco cells. Plant Sci. Lett. 4: 155-159.

Naqvi, S.M.A. and Tandon, J.P. 1991. Screening of wheat germplasm for tolerance to salt affected soils. Golden Jubilee Symp. Genet. Res. and Edu. Current Trends and the Next Fifty Years 1 : 210-211.

Nobel, C.L. 1983. The potential for breeding salt tolerant plants. Proc. R. Soc. Viet. 95 : 133-138.

Noble, C.L. and Rogers, M.E. 1992. Arguments for the use of physiological criteria for improving the salt tolerance of crops. Plant Soil. 146: 99-107.

Noble, C.L. and Shannon, M.C. 1990. Salt tolerance selection of forage legumes using physiological criteria. In: Proceedings of the International

Congress of Plant Physiology (Eds. Sinha, S.K., Sane, P.V., Bhargava, S.C. and Aggarwal, P.K.) Society for Plant Physiology and Biochemistry, New Delhi pp. 989-994.

Norlyn, J.D. 1980. Breeding salt tolerant crop plants. In: Genetic Engineering of Osmoregulation : Impact on Plant Productivity for Food, Chemicals and Energy. (Eds. Rains, D.W., Valentine, R.C. and Hollaender, A.). Plenum Press, New York and London, pp. 293-309.

Norlyn, J.D. and Epstein, E. 1984. Variability in salt tolerance of four triticale lines at germination and emergence. Crop Sci. 24 : 1090-1092.

O'Leary, J.W. 1994. The agricultural use of native plants on probelm soils. In: Soil Mineral Stresses: Approaches to Crop Improvement. (Eds. Yeo, A.R. and Flowers, T.J.) Springer, pp. 127-143.

Orion, T.J. 1980. Comparison of salt tolerance between *Hordeum vulgare* and *H. jubatum* in whole plants and callus cultures. *Z.* Pflanzenphysiol. 98 : 105-118.

Pandey, R. and Ganapathy, P.S. 1984. Effects of sodium chloride stress on callus cultures of *Cicer arietinum* L. Cv. BG-203. Growth and ion accumulation. J. Exp. Bot. 35: 1194-1999.

Pasternak, D., Danon, A. and Aronson, J.A. 1985. Developing the seawater agriculture concept. Plant Soil 89 : 337-348.

Pessarakli, M. 1994. Handbook of Plant and Crop Stress. Marcel Dekker. Inc. New York. 697 p.

Ponnamperuma, F.N. 1984. Role of cultivar tolerance in increasing rice production in saline lands. In : Salinity Tolerance in Plants. Strategies for Crop Improvement (Eds. R.C. Staples and G.A. Toenniessen) pp. 255-271, Willey, New York.

Prasad, K.V.S.K., Sharmila, P., Kumar, P.A. and Saradhi, P.P. 2000. Transformation of *Brassica juncea* (L.) Czern with bacterial codA gene enhances its tolerance to salt stress. Molecular Breeding. 6: 489-499.

Prasad, S.R., Gagali, P.G., Hittalmani, S. and Shashidhar, H.E. 2000. Molecular mapping of quantitative trait loci associated with seedling tolerance to salt stress in rice (*Oryza sativa* L.). Curr. Sci. 78: 162-164.

Qiu, Q.S., Barkla, B.J., Vera-Estrella, R., Zhu, J.K. and Schumaker, K.S. 2003. Na(+)/H(+) exchange activity in the plasma membrane of Arabidopsis. Plant Physiol. 132: 1041-1052.

Rai, M.K., Kalia, R.K., Singh, R., Gangola, M.P. and Dhawan, A.K. 2011. Developing stress tolerant plants through *in vitro* selection - an overview of the recent progress. Environ. Exp. Bot. 71: 89-98.

Rains, D.W., Valentine, R.C. and Hollaender, A. (Eds.) 1980. Genetic enginerring of osmoregulation: Impact on plant productivity for food, chemicals and energy. Plenum Press, New York.

Ramage, R.T. 1980. Genetic methods to breed salt tolerance in plants. In: Genetic Engineering of Osmoregulation. Impact on plant productivity for food chemicals and energy (Eds. D.W. Rains, R.C. Valentine and A. Hollanender). pp. 311-318. Pleunm, New York.

Rana, R.S. 1986. Breeding crop varieties for salt affected soils. In: Approaches for Incorporating Drought and Salinity Resistance in Crop Plants (Eds. V.L. Chopra and R.S. Paroda). pp. 24-55. Oxford and IBH Publishing Co. Pvt. Ltd. New Delhi.

Redhu, A.S. and Chowdhury, R.K. 1991. Genetic component analysis for yield and related traits in bread wheat under salinity. Golden Jubilee Symp. Genet. Res. and Edu. Current Trends and the Next Fifty Years 1 : 208.

Richards, R.A. 1992. Increasing salinity tolerance of grain crops: is it worthwhile? Plant Soil. 32: 89-98.

Richards, R.A. 1995. Improving crop production on salt affected soils: by breeding or management? Expt. Agric. 31: 395-408.

Rosielle, A.A. and Hamblin, J. 1981. Theoretical aspects of selection for yield in stress and nonstress environments. Crop Sci. 21 : 943-946.

Rowland, G.G., McHughen, A. and McOnie, C. 1989. Field performance at saline-affected sites of a somaclonal variant of McGregor flax selected for salt tolerance *in vitro*. Can. J. Plant Sci. 69: 49-60.

Rush, D.W. and Epstein, E. 1976. Genotypic responses to salinity : Differences between salt tolerant and sensitive genotypes of the tomato. Plant Physiol. 57 : 162-166.

Shannon, M.C. 1979. In quest of rapid screening techniques for plant salt tolerance. Hort. Sci. 14 : 587-589.

Shannon, M.C. 1982. Genetics of salt tolerance: New Challenges. In: Biosaline Research: A look into the future. (Eds. A. Sanpieter) pp. 271-273. Plenum New York.

Shannon, M.C. 1985. Principles and strategies in breeding for higher salt tolerance. Plant Soil 89 : 227-241.

Shannon, M.C. 1997. Adoptations of plants to salinity. Advances in Agronomy. 60: 75-120.

Shannon, M.C. and Noble, C.L. 1990. Genetic approaches for developing economic salt tolerant crops. In Agricultural Salinity Assessment and Management (Ed. Tanji, K.) Scientific Publishers (India), Jodhpur pp. 161-185.

Sharma, S.K. and Goyal, S.S. 2003. Progress in plant salinity resistance research: Need for an integration paradigm. In: Crop Production in Saline Environments; Global and Integrade Perspectievs (Eds. S.S. Goyal, S.K. Sharma and D.W. Rains) pp. 387-407. Food Products Press. The Howarth Press Inc.

Shi, H.Z., Ishitani, M., Kim, C.S. and Zhu, J.K. 2000. The *Arabidopsis thaliana* salt tolerance gene SOS1 encodes a putative Na^+/H^+ antiporter. Proc. Nat. Acad. Sci., U.S.A., 97: 6896-6901.

Siahsar, B.A. and Narouei, M. 2010. Mapping QTLs of physiological traits associated with salt tolerance in `Steptoe' x `Morex' doubled haploid lines of barley at seedling stage. J. Food Agric. Environ. 8: 751-759.

Smith, M.K. and McComb, J.A. 1981. Effect of NaCl on the growth of whole plants and their corresponding callus cultures. Aust. J. Plant Physiol. 8 : 267-275.

Sokolova, T.B., Tomakov, A.A. and Shlyakhov, Y.V. 1991. Breeding value in *Gossypium barbadense* L. cultivars with respect to salinity tolerance. Golden Jubilee Symp. Genet. Res. and Edu. Current Trends and the Next Fifty Years 1 : 235.

Somers, G.F. 1979. Natural halophytes as a potential source for new salt tolerant crops : Some progress and prospects. In : The Biosaline Concept : An approach to the utilization of underexploited resources (Ed. A. Hollaender *et al.*) pp. 101-115, Plenum Press, New York.

Subbarao, G.V. and Johansen, C. 1994. Strategies and scope for improving salinity tolerance in crop plants. In : Handbook of Plant and Crop Stress. (Ed. M. Pessarakli) pp. 559-579. Marcel Dekker, Inc. New York.

Subbarao, G.V., Johansen, C., Kumar Rao, J.V.D.K. and Jana, M.K. 1990. Salinity tolerance in F1 hybrids of pigeonpea and a tolerant wild relative. Crop Sci. 30 : 785-788.

Sykes, S.R. 1985. A glass house salinity procedure in identifying citrus by fruits which redirect chloride accumulation in shoot tissues. Aust. J. Agric. Res. 36: 779-789.

Tal, M. 1985. Genetics of salt tolerance in higher plants. : Theoretical and practical considerations. Plant Soil 89 : 199-226.

Timm, D.A., Waskom, R.M. Miller, D.R. and Nabors, M.W. 1991. Greenhouse evaluation of regenerated spring wheat for enhanced salt tolerance. Cereal Res. Commun. 19 : 451-457.

Uddin, K.M., Juraimi, A.S., Ismail, M.R., Othman, R. and Rahim, A.A. 2011. Relative salinity tolerance of warm season turfgrass species. J. Environ. Biol. 32: 309-312.

Valentine, R.C. 1984. Genetic engineering. California Agri. 38 : 36-37.

Winicov, I. 1991. Characterisation of salt tolerant alfalfa (*Medicago sativa* L.) plants regenerated from salt tolerant cell lines. *Plant Cell. Rep.* 10: 561-564.

Winicov, I. 1996. Characterization of rice (*Oryza sativa* L.) plant regenerated from salt tolerant cell lines. *Plant Sci.* 105-111.

Wyn Jones, R.G. and Gorham, J. 1990. Physiological aspects of salinity-scope for genetic improvement. Proc. Inter. Cong. Plant Physiol. 2 : 943-952.

Yamaguchi, T. and Blumwald, E. 2005. Developing salt-tolerant crop plants: challenges and opportunities. Trends Plant Sci. 10: 615-620.

Yeo, A.R. and Flowers, T.J. 1984. Mechanisms of salinity resistance in rice and their role as physiological criteria in plant breeding. In Salinity Tolerance in Plants: Strategies for Crop Improvement (Eds. Staples, R.C. and Toenniessan, G.A.) New York, pp. 151-170.

Yeo, A.R. and Flowers, T.J. 1989. Selection for physiological characters - examples from breeding for salt tolerance. In: Plant Culture Stress

Biochemistry. (Eds. H.G. Jones, T.J. Flower and T.J. Jones) pp. 217-234. Cand. Univ. Press. Cambridge.

Yeo, A.R., Yeo, M.E., Flowers, S.A. and Flowers, T.J. 1990. Screening of rice (*Oryza sativa* L.) genotypes for physiological characters contributing to salinity resistance and their relationship to overall performance. Theor. Appl. Genet. 79: 377-384.

Zhang, H.X. and Blumwald, E. 2001. Transgenic salt-tolerant tomato plants accumulate salt in foliage but not in fruit. Nat. Biotechnol. 19: 765-768.

Zhang, H.X., Hodson, J.N. Williams, J.P. and Blumwald, E. 2001. Engineering salt-tolerant *Brassica* plants : Characterization of yield and seed oil quality in transgenic plants with increased vacuolar sodium accumulation. Proc. Natl. Acad. Sci. USA 98: 12832-12836.

Zhu, J.K. 2000. Genetic analysis of plant salt tolerance using arabidopsis. Plant Physiol. 124: 941-948.

Ziaf, K., Loukehaich, R., Gong, P.J., Liu, H., Han, Q.Q., Wang, T.T., Li, H.X. and Ye, Z.B. 2011. A multiple stress-responsive gene ERD15 from *Solanum pennellii* confers stress tolerance in tobacco. Plant Cell Physiol. 52: 1055-1067.

■■■■

Chapter - 12

AFFORESTATION AND AGROFORESTRY IN SALT AFFECTED SOILS

INTRODUCTION

Salinity is reported to adversely affect about 1 billion hectares (Szabolcs, 1992) mostly located in arid and semi-arid regions of the world. The problem which extends to more than 100 countries, is encountered in all types of climate and is the consequence of both natural processes as well as human interference. In India, several agencies have given divergent estimates (ranging for 1.5 to 9.8 million hectare) of the extent of salt-affected soils. The data from various source were critically evaluated at Central Soil Salinity Research Institute (CSSRI) Karnal and figure has been modified to 7.4 million ha (Sharma, 1996). The status of salt affected soils in India in different states has been presented in Table 12.1 which indicates that Gujarat, Rajasthan, Uttar Pradesh and West Bengal are most affected by salinity problems.

Table 12.1. Status of salt affected soils in India in different states (Sharma, 1996)

State/UT	Area (.000 ha)
Andhra Pradesh	394
Bihar	85
Goa	17
Gujarat	1649
Haryana	555
Punjab	480
J & K	80
Karnataka	179
Kerala	45
Madhya Pradesh	242
Maharashtra	127
Orissa	135

Rajasthan	1183
Tamil Nadu	470
Uttar Pradesh	958
West Bengal	820
Andaman & Nicobar	1
Delhi	0.6
Pondicherry	0.3
Total	7420.9

Reclamation of salt affected soils for crop production requires additional input such as amendments, fertilizer and infrastrucure for leaching and drainage. Despite the available technical know-how for the reclamation and management of salty soils and waters, the pace of agriculture production is rather slow. This is primarily due to poor resources of marginal farmers and because of the fact that the salt affected lands belong to village community or the governmental agencies. However, saline soils can be successfully put to alternative land use through afforestation and agroforestry programmes. Planting trees is now a viable option for those situations where salt affected soils cannot be reclaimed economically through conventional techniques. This has become possible due to research efforts in the recent past and a better understanding of the biology and management of forestry plantation and agroforestry systems on saline soils. Besides providing fuel, fodder and timber, afforestation and biosaline agroforestry would help in bio-amelioration of salt-affected soils. Therefore, an attempt has been made to compile and synthesize the available information on this subject in the following sections of this chapter.

AFFORESTATION OF SALINE SOILS

Both alkali and the saline soils differ in their physical and chemical characteristics and thus in the nature of constraints to plant growth. Therefore, the diagnosis and magnitude of salt problem is a prerequisite for undertaking afforestation ventures, as the practices for creating favorable root environment for proper tree establishment and suitability of tree species will vary accordingly. Keeping this in view, various practices have been discussed separately for the alkali and saline soils.

Saline Soils

These soils, with white salt encrustation on the surface, have predominantly chlorides and sulphate of Na, Ca and Mg. The soils with neutral soluble salts have saturation paste pH <8.2. The electrical conductivity of saturation extract, ECe of saline soils, is generally more than 4 dSm^{-1} at 25°C. Such soils invariably have sodium adsorption ratio (SAR) of the soil solution >15. In saline soils, the excess of neutral salts restricts the normal plant growth.

Alkali Soils

These soil contain excess of salts capable of alkaline hydrolysis such as sodium carbonate, sodium bicarbonate and sodium silicate; and sufficient exchangeable sodium to impart poor physical conditions to soils and thus affecting growth of most plants. These soils have saturated paste pH >8.2; exchangeable sodium percentage (ESP) >15 and different levels of ECe. The pH of the calcareous alkali soils is highly related to their ESP.

Saline-Sodic Soils

In saline soils when chlorides and sulphates of Ca and Mg are the predominant salts, the SAR usually remains less than 15. However, predominance of Na invariably results in soil solution SAR > 15. Such soils are termed saline-sodic. Many saline-sodic soils contain, in addition to soluble carbonates, the excess of neutral salts. Such soils manifest alkaline properties. From management point of view, the saline-sodic-soils that do not contain soluble carbonates are grouped with saline soils and the others with alkali soils.

Trees and Salinity

A range of salt-tolerant trees and shrubs can be used to revegetate saline sites, reduce erosion and improve biodiversity. Tree planting play a significant role in control of erosion, improvement of biodiversity and reduction of water logging at saline discharge sites. Trees can also be grown for productive purposes e.g. for firewood sawlogs of carbon storage, although farm-forestry is test suited for high-quality non saline sites.

In general, tree planting on discharge areas is unlikely to significantly reduce the salinity problem by lowering ground water levels. Planting of larger saline areas can be effective in this respect but this depends on local and regional hydrogeology. Dense planting directly on discharge this can potentially increase soil salinity over time because salt accumulate in the root zone as a result of tree water use. However, salt accumulation depends strongly on site characteristics, rainfall and evaporation as well as transpiration by trees and other plants.

Tree Planting Technique in Alkali Soils

Tomar and Singh (1998) have discussed various planting methods and concluded that "Auger Hole Technique" developed at CSSRI, Karnal was the best and an improvement over pit planting technique.

In this technique, 100-140 cm deep and 20-25 cm diameter auger holes are dug with a tractor-operated auger and saplings are planted after suitably amending the dug-out soil. The performance of trees planted with this method has been highly satisfactory in field trials than the routine pit or trench methods (Table 12.2). This method has picked up very well with the foresters because of reduced manual labour costs and speedy operations. In addition to piercing of hard kankar layer, the advantage of this technique includes encouraging and training of deeper rooting. Thus the trees are able to probe deeper soil layers for water and nutrients to sustain their growth.

Table 12.2. Comparative performance of trees planted with pit, trench and augerhole techniques in highly alkali soils (Gill and Abrol, 1991; Singh *et al.*, 1993)

Dimension (cm)		*Eucalyptus tereticonrnis* (8 y)		*Acacia nilotica* (8 y)		*Prosopis juliflora* (6y)	
Depth	Width	Height (m)	DSH (cm)	Height (m)	DSH (cm)	Height (m)	DSH (cm)
Pit							
90	90	9.2	17.3	7.5	19.6		
30	30					7.0	8.4
Trench							
30	30					6.9	7.9
Auger hole							
10	120	8.2	14.2	6.7	14.6		
10	120	8.9	16.2	7.7	16.2		
15	90					7.7	9.6
15	180	8.4	14.8	6.8	16.6		
15	180	9.1	16.6	7.4	17.6		

Plantation Establishment Technique in Saline Soils

The method showing the best results to establish tree plantations on salt affected soils is the furrow method. This method shows the highest sapling survival in different field studies (Tomar, 1997) and is recommended by the handbook of Qureshi and Barret-Lennard (1998). It is an example of how to control soil salinity in the root zones (salinity management). The furrow method works as follows : furrows about 0.6 m wide and 0.2 to 0.3 m deep are created with a tractor mounted furrow maker. The furrows are used for irrigating the saplings for at least two years. The seedlings are planted at a uniform distance. The place where they are planted, depends on the degree of internal drainage of the soil. In soils with good internal drainage the seedlings should be planted at the bottom of the furrow. In soils with poor drainage the saplings are planted on the shoulder of the furrows so that they are not affected by water logging after irrigation. In waterlogged soils the trees should be planted on bunds or mounds to avoid water logging. The trench-ridge technique is widely used in Pakistan and India on waterlogged soils. As soils are heavier, more salt-affected and waterlogged mounds should be taller and wider, and have a distinct trough. Such double ridge mounds have proven very effective by providing (1) site drainage, (2) elevation of the seedling above the water table (thereby increasing oxygen availability) and (3) salt leaching from the root zone. These mounds are particularly useful where there is sufficient rain and/or irrigation water before planting time to leach salts from below the trough at this critical time. It is best to construct mounds several months before planting. Single-ridge and flat-topped mounds are not as effective, but are better than no mounds. Several techniques are available for mechanically forming mounds. Equipment includes : press wheels attached to the mounds of the plough, modified direct seeding machinery and twin discs (Marcar and Crawford, 1996; Lambert and Turner, 2000).

At CSSRI, Karnal several long-term field experiments have been conducted for developing suitable techniques for growing salt tolerant trees. For this purpose about three dozen tree species were evaluated for use in the saline waterlogged soils of semi-arid regions (Tomer *et al.,* 1998). The soils of the study area were saline sandy loams with a preponderance of chloride and sulfates of Na^+, Ca^{2+}, and Mg^{2+}. The soil's initial electrolytic conductivity (ECe) was 36.4 dSm^{-1} in the upper 30 cm. The water table was shallow, fluctuating between 1.5 m depth to the surface in different seasons of the year, and the water was brackish (average EC was 29.8 dS m^{-1}). Three methods of planting, namely, ridge-trench, subsurface, and furrow, were compared. The furrow technique provided favorable niches for plant survival and growth and was also the most economical for such soils. *Prosopis juliflora, Tamarix* sp., *Casuarina glauca, Acacia farnesina, A. nilotica, A. torti-lis* and *Parkinsonia aculeata* were found to be the most promising species for these saline soils. *Casuarina glauca* and *Salvadora oleoides* survived even prolonged stagnation of floodwaters for 9 months.

Comparing the three planting techniques Tomar *et al.* (1998) concluded that the furrow planting technique was superior to ridge and subsurface methods of planting for waterlogged saline soils. The establishment and growth of the various tree species markedly improved with furrow planting and it was possible to keep salt concentrations lower in the rooting zones of trees, by adopting this technique of planting and thus able to escape the adverse effects of salinity.

Irrigation and Spacings

Results of the study by Minhas *et al.* (1997) indicate that irrigation should at least be applied in the first two years after planting. In most cases the trees have developed their rooting system after two years and are able to reach the groundwater. Irrigation should be managed for trees on saline soils in such a way that it avoids water logging and aeration problems, and the accumulation of salt in the root zone. These objectives are met with the furrow irrigation method (Tomar, 1997). With the furrow irrigation method, saline waters of EC up to 15 dS/m in sandy loam and on loamy sand soils can be used. In finer textured soils, waters of EC up to 8-10 dS/m can be used for irrigating saplings. The recommended irrigation interval for brackish water is the same as for normal water. In India they have the following scheme : a monthly irrigation during October-March and fortnightly, irrigation during April-June (Tomar, 1997).

The distance between the saplings in a row and between rows is dependent on the species, but also on whether the trees are to be planted in blocks, mixed with shrub species, or used in alley cropping. In case of the presence of a dense soil layer or hard pan the furrow planting method is combined with the auger hole method. Auger holes (0.2 m in diameter and 1.2 m deep) are dug at the sill of the furrows spaced at the desired intervals. These holes may be refilled with a mixture of original soil, farmyard manure, super-phosphate, zinc sulphate and iron sulphate to give the saplings a good start.

Management of Natural Salt Affected Soils

These soils have high concentration of salt distributed through out the profile and sodium chloride is the dominant salt. For management of these salt affected soils species of shrubs, grasses and trees tolerant to salinity have been identified and techniques for their establishment developed. Jain (1984) found *Prosopis juliflora* and *Tamarix articulata* to be the most tolerant to salinity followed by *Acacia tortilis* and *Eucalyptis camaldulensis*. Similar observations have been reported by Shakarnaraynan and Kolarkar (1988). Jain

(1986) further reported the tree and grass species which can tolerate soil salinity in the range of EC 4-8, 8-16 and more than 16 dSm^{-1} (Table 12.3). They have also described the ridge and furrow technique for tree plantation.

Table 12.3. Tolerance of different plant species to salinity (Jain, 1986)

Vegetation	Salinity tolerance mmhos/cm			
	Very high (>16)	High (16-8)	Medium (8-4)	Low (<4)
Trees	*Prosopis juliflora, Tamarix articulata, Acacia tortilis, Chenopodium* sp.	*Eucalyptus camaldelensis, Eucalyptus hybrid, Leucaena leucocephala, Dichrostachys glomerata*	*Azadirachta indica, Acacia aneura, Colophospermus mopane*	*Casia siamea, Albizzia lebbek*
Grass/shrub species	*Sporobolus marginatus, Eleusine compressa, Dactyloctenium sindicum*	*Eremopogon faveolatus, Sehima nervosum, Cenchrus setigerus*	*Heteropogon tortus, Zizyphus nummularia*	*Eragrostis ciliaris, Dichanthium annulatum*

Tomar and Gupta (1985) reported the salinity tolerance limits of a number of forest tree species (Table 12.4). According to this data, *Tamaris troupi, T. articulata* and *Prosopis juliflora* were highly salt tolerant trees and could tolerate above 25 dSm^{-1}.

Table 12.4. Salinity tolerance limits of some forest species (Tomar and Gupta, 1985)

Tolerant (EC_e 25-35 dSm^{-1})	*Tamarix troupii* (Pharans), *T. articuleata* (Pharans), *Prosopisi juliflora* (Pahari Kikar), *Pithecellobium dulce* (Jangle jalebi), *Parkinsonia aculeata* (Parkinsonia), *Acacia farnesiana* (Pissi Babul)
Moderately tolerant (EC_e 15-25 dSm^{-1})	*Prosopis juliflora* (Pahari Kikar), *Pithecellobium dulce* (Jangle jalebi), *Parkinsonia aculeata* (Parkinsonia), *Acacia farnesiana* (Pissi Babul), *Callistemon lancheolatus* (Bottle brush), *Acacia nilotica* (Kikar), *A. pennatula* (Babul), *A. tortilis* (Israeli kikar), *Casuarina glauca* 13155 (Casuarina), *C. glauca* 13987 (Casuarina), *C. obesa* 27, *C. glauca* (FRI), *C. equisetifolia* (FRI) (Casuarina), *Eucalyptus camaldulensis* (Safeda), *Leucaena leucocephala* (Subabool), *Erescentia alata* (Ereseentia)
Moderately sensitive (EC_e 10-15 dSm^{-1})	*Casuarina crominghamiana* (FRI) (Casuarina), *C. cuminghamina* (Aust) (Casuarina), *Eucalyptus tereticornis* (Safeda), *Acacia auriculkiformis* (Australian kikar),

	Terminalia arjuna (Anjar), *Pongamia pinnata* (Papri), *Gausuma ulmifloia, Leucaena shennioni* (Subabool), *Samemea saman* (Rain tree), *Albizia cartibea* (Siris), *Senna atomaria* (Senna)
Sensitive (EC_e 7-10 dSm^{-1})	*Syzgium cuminii* (Jamun), *S. fruitocsum* (Jamoa), *Tamarindus indica* (Imli), *Salix* spp. (Willow), *Acacia deami* (Acacia), *Albizia guachepela* (Siris), *Alelia herbertismithi* (Alelia), *Caesalpinia eriostachyyas* (Caesalpinia), *C. velutina* (Caesalpinia), *Haemotoxylon brasiletto* (Haemataxylon)

Table 12.5. Relative tolerance of forest tree species to soil pH/sodicity

Highly tolerant pH_2 >10.0	Tolerant pH_2 9.5-10.0	Moderately tolerant pH_2 9.0-9.5	Sensitive pH_2 <9.0
Pahari kikar *(Prosopis juliflora)*, Kikar *(Acacia nilotica)*, Pharans *(Tamarix articulata),* Jor-tor *(Casuarina* equistifolia)*	Jangle jalebi *(Pithecellobium dulce)*, Safeda *(Eucalyptus tereticornis)*, Arjan tree *(Terminalia arjuna),* Imly *(Tamarindus indica)*, Papri *(Pongamia pinnata)*, Kharia *(Acaca catechu),* Dhak *(Butea monosperma)*, Lasura *(Cordia rothii)*, Siris *(Albizia labbek),* Parkinsonia *(Parkinsonia aculeata)*	Kingelia *(Kingelia pinnata)*, Shisham *(Dalbergia sissoo)*, Siamea *(Cassia siamea)*, Neem *(Azadirachta indica)*, Silver oak *(Grevilla robusta)*, Dhaincha *(Sesabnia sesban),* Bakain *(Melia azadarch)*, Subabul *(Leucaena leucocephala),* Saunjana *(Moringa olifera),* Bamboo *(Bambusa* spp.), Schoot *(Aforus alba)*, Siris *(Albizia procera)*, *Acacia auriculiformis, Acacia senegal*	Poplar *(Populus deltoides)*, Teak *(Tectonia grandis), Ficus virnes, Bombax ceiba, Hardwickia binnata, Thespesia populnea, Bauhinia variegata, Lagerstroemia speciosa, Cadrella toona*

*Not suitable for deep ground water area, Source: Singh *et al.* (1993)

Chaturvedi (1984) prepared the list of firewood species that are suited to saline water irrigation: *Prosopis juliflora, Acacia nilotica, Terminalia arjuna, Syzygium cumini, Albizzia lebbek, Pongamia pinnata, Cassia auriculata, Adhatoda vasica, Cassia siamea.* Plantation of these species could yield about 230 tonnes of firewood in five years. In terms of decreasing tolerance to saline water, trees can be placed in the following order: *prosopis juliflora, Tamarix articuleata, Acadia tortilis, Eucalyptus camaldulensis, Azadirachta indica, Albizzia lebbek* (Jain *et al.,* 1983). Tomar and Yadav (1980) reported that waters having EC up to 10 dSm^{-1} could be used in *Acacia nilotica, Pongamia pinnata* and *Prosopis juliflora* plantations whereas waters of EC upto 5 dSm^{-1} could be used for *Acacia tortilis, Albizzia lebbek, Azadirachta indica, Lawsonia glauca* and *Parkinsonia aculeata.* Eucalyptus hybrid did not tolerate EC_{iw} greater than 2 dSm^{-1} The alkali tolerance limits of useful tree species have not been worked out under similar conditions. However, approximate limits of tolerance to soil pH for some trees are listed in Table 12.5.

Tree species like *Casuarina equisetifolia, Tamarix articulata* and *Prosopis juliflora* can be planted under shallow and saline water table conditions. Whereas, *Acaca auriculiformis, Acacia nilotica* and *Terminalia arjuna* can also be planted but on ridges (Tomar and Gupta, 1985). *Salvadora oleoides* (Jal) and *Salvadora persica* (Kharajal) are suitable for saline wastelands having medium to heavy soil texture. Seeds of *Salvadora* spp. yield oil which has very good export value. Plantation of this tree along with *Sporobolus marginatus,* a highly salt tolerant grass is being tried on an extensive scale in saline wastelands in the western Rajasthan (Mathur, 1986).

Afforestation of Salt Affected Soils in Rajasthan

About 0.40 m ha of area in south-eastern part of arid Rajasthan suffers from primary salinity. The main problem of these areas is salinity of the substrata and the ground water. The major areas of salt affected lands have mean ECe from 18 to 65 dSm^{-1} with predominance of sodium chloride and sodium sulphate salts. High cost of their amelioration and cultivation technology often limits crop production on these soils. Trees, shrubs and grasses being more hardy and salt tolerant could be not only grown economically but may have ameliorative influence on such soils in the long term. Particularly in developing and under developed countries where shortage of fuel, fodder and timber is acute, bringing salt affected lands under tree and shrub cover holds tremendous scope.

A number of studies conducted by Forest Research Institute, Central Soil Salinity Research Institute and Central Arid Zone Research Institute have made it possible to undertake afforestation in salt affected soils and with saline water irrigation (Gupta *et al.,* 1995; Minhas *et al.,* 1997; Tomar and Minhas, 1998, Tomar *et al.,* 1998, Singh and Dagar, 1998, Dagar *et al.,* 2001, Mishra *et al.,* 2003). Based on these investigations following are the list of recommended tree species for reforestation of salt affected soils (Table 12.6).

Table 12.6. Recommended tree species for alkali soils and saline and waterlogged soils (Gupta *et al.*, 1995)

Average pH_2 (0-120 cm)	**Soil alkalinity**
	Tree species
More than 10.0	*Prosopis juliflora, Acacia nilotica, Casuarina equisetifolia*
9.0 to 10.0	*Tararix articulata, Terminalia arjuna, Eucalyptus tereticornis, Albizzia lebbek, Pongamia pinnata, Sesbania sesban, Embica offinalis*
8.2 to 9.0	*Dalbergia sisoo, Morus alba, Grevilla robusta, Azadirachta indica, Tectona grandis, Populus deltoides*
EC_e (dSm^{-1})	**Saline and waterlogged soils (Salinity below 0.3 m)**
20-30	*Acacia famesiana, Prosopis juliflora, Parkinsonia aculeata*
14-20	*Acacia nilotica, Acacia tortilis, Callistemon lanceolatus, Casuarina graica, C. obesa, C. equisetifolia, Eucalyptus camaldulensis*
10-14	*Casuarina cunninghamiana, Eucalyptus tereticornis, Terminalia arjuna*
5-10	*Albizzia caribea, Dalbergia sissoo, Guazuma ulmifolia, Pongamia pinnata, Samanea saman*
<5	*Acacia auriculiformis, A. demil, A. catechu, Syzglum cuminii, Salix* spp. *Tamarindus indica*

Studies conducted at CAZRI, Jodhpur have indicated that salt affected lands can be afforested with salt tolerant species. Different tree and grass species have been identified by Jain (1984) according to their relative tolerance under arid conditions. Most of these species are sensitive at germination or early stages of establishment. However, *Prosopis juliflora* and *Colophospermum mopane* were quite tolerant at germination stage even to 9.0 dSm^{-1} salinity water (Jain, 1984). Furthermore, survival and performance of different tree species to saline water irrigation have shown that survival of *Prosopis juliflora* and *Tamarix articulata* was almost 100 per cent even with 9 dSm^{-1} saline water. Studies conducted at Pali (Jain, 1986) showed that pits of the size of 0.6 m^3 to 1m^3 with manure @ 25 kg/pit and gypsum 3.6 kg/pit in case of sodic soils with auger holes of 150 cm depth and 15 cm diameter filled with original soil and 2 kg gypsum +7 kg manure helped in improving soil environment and establishment of tree seedlings of *Acacia nilotica, Prosopis juliflora* and *Casuarina equisetifolia*. Seedlings of *Eucalyptus hybrid* and *C. mopane* were found sensitive to high salinity waters. Growth of *Acacia aneura,*

Tamarix articulata and *D. glomerata* decreased with high salinity irrigation water (Jain, 1986).

Many studies have indicated that some of the tree species found to occur naturally on highly salt affected soils are *Prosopis juliflora, Capparis decidua, Ziziphus* spp., *Tamarix articulata* and *Butea monosperma* depending upon the degree of salinity (Gupta and Prem Lal, 1993). The species suitability for saline and sodic soils is briefly dealt below.

Saline soils : Gupta *et al.* (1986) reported that *Acacia nilotica* and *E. camaldilensis* could be grown with less than 50% growth reduction upto 5 dSm^{-1} in clay soil and upto 10 dSm^{-1} in sandy soil. *Casuarina equisetifolia* showed moderate tolerance whereas *Ceiba pentandra* and *Acacia acriculiormis* were highly sensitive and did not survive beyond 2.5 dSm^{-1}. A comparative evaluation of some forest species made on sandy loam soil (in Pali district of arid zone) by Jain *et al.* (1985) indicated that *Prosopis juliflora* and *Tamarix articulta* were most tolerant with 87% survival and no decrease in height even at the highest salinity levels (Table 12.7). *Acacia tortilis* and *E. camaldulensis* exhibited medium tolerance while *A. indica* and *Albizia lebbeck* were most sensitive.

Table 12.7. Effect of soil salinity levels on survival and height growth of tree species, five years after planting in field (Jain, 1986)

Species	Soil salinity levels EC (dSm^{-1})					
	Medium		High		Highest	
	Survival (%)	Height (cm)	Survival (%)	Height (cm)	Survival (%)	Height (cm)
Prosopis juliflora	100	320	94	327	87	320
Tamarix articulata	100	204	100	199	87	200
Acacia tortilis	100	363	100	249	44	157
Eucalyptus camaldulensis	100	588	75	440	0	–
Azadirachta indica	81	147	94	104	0	–
Albizia lebbeck	12	350	19	90	0	–

Sodic soils: Yadav and Prakash (1990) studied the survival, growth and nutrition of seedlings of seven forest species under different levels of soil salinity and sodicity to find out the critical limits. *Terminalia arjuna, Albizia procera* and *Dalbergia sissoo* were the most tolerant to soil sodicity as they survived upto pH 10.0 (ESP 80). *Eucalyptus hybrid* was next which survived upto pH 9.7 (ESP 60) while *Leuceaena leucocephala* and *Halophragma adenophyllum* had only marginal tolerance as they survived only upto pH 8.5 (ESP 20), Gupta *et al.* (1988, 1991) found *Acacia nilotica* and *Eucalyptus camaldulensis* as the alkaline resistant tree species. They reported that *A. nilotica* could be grown upto pH 9.7 on sandy soil, 9.2 on clay soil and 8.8 on loam soil, while for *E. camaldulensis,* such pH value was upto 11.0 on sand, 9.2 on clay and 8.6 on loam soil. More recently Dagar *et al.* (2001) observed that out of the 30 species planted on Alkali-sodic soils only *Prosopis juliflora, Tamarix articulata* and *Acacia nilotica* were found suitable on such soils. Similar results have been reported by other investigators (Singh and Dagar, 1998; Mishra *et al.,* 2003, 2004).

BENEFICIAL EFFECT OF PLANTING TREES ON SALINE LANDS

Economical Benefits

Recent researches have shown that potentially tree plantations are at least as economic as some of the cropping alternatives on salt lands. The afforestation of a highly alkali soil with appropriate technologies leads to biomass production of 197, 162, 141 and 124 Mg ha^{-1} from 8-10 years growth of *P. juliflora, A. nilotica, C. equisetifolia* and *E. tereticornis* respectively (Singh *et al.,* 1993). Field investigations have further shown a great potential of *Sesbania* species in yielding high amounts (120-150 Mg ha^{-1}) of biomass even in relatively shorter period (2.5 - 3 years) (Rao *et al.,* 1986). Abrol and Joshi (1984) worked out the annual cost and returns from raising *A. nilotica* and *E. tereticornis* on a highly alkaline soil at Karnal (Table 12.8). The returns from *E. tereticornis* were taken as only one time return at the end of 7 years, whereas loppings from *A. nilotica* in the third and fifth year provided for additional income. Although the cost of raising *E. tereticornis* worked out to be lower, the returns were considerably higher from *A. nilotica* (C : B ratio 1.63). Computation by Singh *et al.* (1993) have shown that the tentative net income per hectare per annum was Rs. 8,175 and 3,587 for alkali and saline soils, respectively. Similar computations by Iyengar show the total cost of cultivation of *Salvadora persica* per hectare for 6 years to be Rs. 26,650 and gross income to be Rs. 50,000, thereby giving a net benefit of Rs. 23,350. The cost and returns for *Atriplex* for 10-15 years worked out to be Rs. 18,750 and 52,000 respectively. Thus, in addition to providing the additional fuel and fodder, afforestation can lead to generation of employment opportunities for the rural masses.

Table 12.8. Annual cost and returns (Rs) from raising trees on alkali soil

Year	*A. nilotica*		*E. tereticornis*	
	Cost	Return	Cost	Return
1	6,416		6,416	
2	445		445	
3	459	1,500	135	
4	876		871	
5	1,363	3,000	986	
6	986		685	
7	7,735	40,000	6,485	3,200
Total	18.276	44,500	16,321	32,000
Benefit : cost ratio		1.63		1.22
Annuity value		1580		521

Soil Amelioration

The tree-soil interactions can lead to amelioration of alkali soils through physical and biological processes. The most important benefit of growing trees on salt lands is enrichment of barren soils through litter fall, which will be governed by the growth habit of planted trees, their age, density of plantation and N-fixing capacity of trees etc. Gill *et al.* (1987) observed that litter production by *A. nilotica* and *E. tereticornis* plantations on a highly alkali soil was 2,537 - 5,746 and 1,027 - 1,125 kg ha^{-1}, respectively. The accumulation of N, P, K and S was substantially more in litter of *Acacia* whereas that of Ca, Mg, Na, Fe and Mn was more in *Eucalyptus*. The annual litter yield of *P. juliflora* and *C. equisetifolia* was 7,483-9,967 and 6,957-8,769 kg ha^{-1} respectively and the litter from the latter was more lignified and hence decomposed at a slower rate. In addition to recycling of nutrients, the decomposition of litter leads to evolution of CO_2, which helps to mobilise the inherent Ca. The released Ca can hasten the reclamation by replacing the exchangeable Na from the soil, thus reducing the soil's sodicity and pH levels.

Plant growth at 20 years of *P. juliflora, A. nilotica, E. tereticornis, A. lebbeck* and *T. arjuna,* on a highly alkali soil (pH 10.2-10.5) reduced the pH of surface 1.2 m soil to 8.1, 9.03, 9.1, 8.67 and 8.15, respectively (Singh and Gill, 1990). The organic carbon of the profile increased several-folds, being highest under *P. juliflora* and lowest under *E. tereticornis* (Table 12.9). The improvement in organic carbon and reduced pH of alkali soils in turn lead to improvement in their infiltration characteristic and water retention properties (Gill *et al.*, 1987; Singh *et al.*, 1992).

Table 12.9. Ameliorative effect in upper 15 cm sodic soil under plantations of 20 years (Singh and Gill, 1990).

Species	pH_2	EC (dSm^{-1})	Org. C (%)	Av. P (kg. ha^{-1})	Av. K (kg ha^{-1})
Acacia nilotica	8.4	0.25	0.85	59.0	498.6
Eucalyptus tereticornis	8.5	0.44	0.66	33.3	358.9
Prosopis juliflora	7.5	0.51	0.93	110.5	701.5
Terminalia arjuna	7.9	0.32	0.86	67.8	409.8
Albizia lebbek	7.9	0.32	0.62	42.6	387.0
Original soil	10.2	1.18	0.22	28.0	278.0

Biodrainage

One of the most-quoted benefits of tree plantations in relation to salinity management is their ability to draw down the water-tables through seepage reduction and increased water use. Reforestation strategies of selected sites for evaluating the potential of trees for salt concentration and checking water-table rise were taken on a large scale in late 1970s in Australia (Schofield, 1992). Some plantations of *Eucalyptus* species have been estimated to transpire 3-4 times (220 cm) the rainfall per annum though the water use by the tree was quite sensitive to local conditions of soil water and groundwater salinity. Trees can be used to lower these risen water tables. They are relatively deep rooting and can therefore, reduce water leakage, they may use water all year around whenever it is available, and intercept a significant fraction of rainfall before it reaches the soil. In this way, strategic tree planting being part of agro-forestry and farm forestry systems, can contribute to lowering risen (salt) water tables.

Similarly, the potential of tree species to transpire large amount of better-quality water fed from the bottom of lysimeters has been observed by Chhabra and Thakur (1996) and their potential to check seepage (60-84%) from canal was reported under the black soil conditions of Gangawati by Patil *et al.* (1994). Although there is controversy on the role of trees as "biological pump", or a "biological wick", for taking out water from water-table and in the process protecting the surrounding agricultural areas from salinity problems, studies are still awaited in India to determine the processes involved and predict tree-water table interaction involving some indigenous tree species native to saline areas. A fall in water table level of 1.5 m after 10 years of reforestation was observed in south-west Australia, whereas, the water level rose by 1.8 m under pasture (Schofield, 1992).

Trees are usually more tolerant to water logging and salts than many arable crops. Many of such plants are also capable of evapotranspiring at a very high rate. Thus, cultivation of such plants can help to biodrain the surface stagnated or the ground water in water logged normal/saline lands (Dagar, 2011). The depth and distance to which the plant would be effectively tapping the water would depend upon soil and plant characteristics.

Since water gets evapotranspired, salts remain in the soil profile resulting in a group to question the claims of such a system for land reclamation. Moreover, questions have also been raised on the diversification of land from arable cropping to growing of trees. Apparently, this strategy is suitable for areas where abundant wasteland is available and can be utilized to grow plantations. Nonetheless, many new innovations are being made to adopt bio-drainage as a strategy to arrest seepage from canal irrigation network and to cultivate plants on killa lines to arrest water logging. It may also provide a long-term income resource to the farmers (Gupta, 2005, Gupta, 2014).

BIOSALINE (AGRO) FORESTRY

About two thirds of world's land area is arid and semi-arid, where crop prodution is dependent upon rains. Practising agriculture in such areas is always risky. However, integration of woody perennials with annual agricultural crops reduces the risk of crop failure and holds sustainability of such arid lands. Trees of *Prosopis, Acacia, Zizyphus, Parkinsonia* are promising for agroforestry in arid and semi-arid areas for food, forage and fuel-wood production. Owing to their low water requirement these trees do not compete with the companion crops. References are available where yields of grasses and other herbaceous plants increased when grown in association with *Prosopis* and *Acacia* trees in water-scarcity situations. The most promising agroforestry systems for arid and semi-arid areas are : *Prosopis - Opuntia, Prosopis cineraria - Cenchrus cilliaris, Prosopis cineraria- millets, Acacia albida- millets, Acacia tortilis- millets.* For salt-affected arid and semi-arid lands *Prosopis juliflora - Opuntia, Ficus indica* could be excellent combinations because both are salt tolerant, have low water requir-ement and are good biomass producers. Recently fruit, vegetable and forage clones of edible cactus have been introduced, multiplied and successfully cultivated at CSSRI, Karnal (Singh and Felker, 1995).

Agroforestry Systems

Many of the species that may be suitable for the establishment of tree plantations on salt affected soils are used in agroforestry systems, especially in arid regions. In agroforestry systems the trees are used as sources of fodder during the dry season when there is no grass, serve as windbreaks and hedges and provide shade. If the trees are used as fodder or as shade trees, they can be planted scattered over the fields that are used as pastures, but can also be planted in blocks or as timber belts. The trees can be directly browsed by cattle or the leaves and pods can be harvested to feed cattle.

A well known species that is favoured in agroforestry systems is *Prosopis cineraria :* it fixes large amounts of nitrogen and does not affect growth of plants under the canopy (Qureshi and Barrett-Lennard, 1998).

Agroforestry (sometimes called : alley cropping) combines trees and shrubs with pastures and agricultural crops to maximise productivity and water use. Planting layout and density is flexible; 100 to 450 trees/ha in multiple rows up to 50 m apart will provide a satisfactory ratio of trees to pasture/crops. Tree belts in alley cropping systems have the potential to use saline groundwater. The key issue generally is to maximise the distance between tree belts in order to reduce production losses due to competition between trees and pasture or crops and to balance the local recharge of groundwater with groundwater use by the trees in the belts. Qureshi and Barrett-Lennard (1998) indicate that it is better to use a few wide belts of trees than using a lot of narrow belts of trees. Using a few wide belts of trees reduces the area at which an edge effect is experienced by the pasture or the crop. In this way, the area at which loss of crop growth caused by competition with invading tree roots and by shading takes place is minimized.

The extension of tree roots depends on tree species and tree size. The roots of most trees extend 1.5 tree height from the base of the tree. Competition between tree roots and crops occurs most often within 1.5 heights of the base of the tree. However, the roots of some species such as *Acacia* spp. and *Prosopis juliflora* may extend three to five times the average tree heights.

Promising Tree species for Biosaline (Agro) Forestry

A large number of species are reported in literature as being salt tolerant. However, the level of salt tolerance differs between species , but also within species and between provenances. For instance different provenances of *Eucalyptus camaldulencis* (Dehnh.) show different levels of salt tolerance. Promising salt tolerant trees species, subdivided into three different salinity classes are summarized in Table 11.8. A species is marked with a `W' if it is tolerant to water logging. A `I' indicates that a species is suitable for irrigation with saline/brackish water (Table 12.10).

In a number of studies field trials have been performed to determine the level of salt tolerance of different tree species and their provenances. The field trial with 31 tree species of Tomar *et al.* (2003) recommends the following species to be planted in arid regions with moderate levels of saline irrigation (EC 8.5-10.0 dS/m) with the furrow planting method: *Tamarix articulata, Acacia nilotica, Prosopis juliflora, Eucalyptus tereticornis, Acacia tortilis, Acacia tortilis* (hybrid) and *Casuarina glauca* (Table 12.10). These species not only produce economic yields (>20 tons/ha) but also improved soil conditions, i.e. in terms of organic matter.

Table 12.10. Salt tolerant tree species.

Very tolerant (25-35 dS/m)			
Acacia	*famesiana*		
Melaleuca	*halmaturorun* (F. Muell)		
Parkinsonia	*aculeata* L.		
Prosopis	*juliflora* (Swartz) D.C.		1
Tamarix	*aphylla* (L.) Kartsen / articulata (Vahl.)	W	
Tamarix	*Troupii*		
Tolerant (15-25 dS/m)			
Acacia	*ampliceps* (Maslin)		
Acacia	*nilotica* (L.) Del.		1
Acacia	*stenophylla* (A. Cunn. ex Benth.)*		
Acacia	*tortilis* (Forsk.)		1
Casuarina	*equisetifolia* (L.)		
Casuarina	*glauca* (Caloundra)*	W	
Casuarina	*obesa* (Miq.)	W	
Eucalyptus	*camaldulensis* (Dehnh.)*	W	1
Eucalyptus	*campaspe* (S. Moore)		
Eucalyptus	*kondininensis* (Maiden & Blakely)		
Eucalyptus	*occidentalis* (Endl.)*		
Eucalyptus	*platypus* (Hook) var. *heterophylla* Blakely		
Eucalyptus	*sargentii* (Maiden)	W	
Eucalyptus	*spathulata* (Hook)	W	
Leucaena	*leucocephala* (Lam.) de Wit		
Melaleuca	*lanceolata* (Otto)		

Melaleuca	*lateriflora*		
Pithecellobium	*dulce* (Roxb) Benth		
Prosopis	*cineraria* (L.) Druce		1
Moderately tolerant (10-15 dS/m)			
Cassia	*siamea*		1
Casuarina	*cunninghamiana*		
Dalbergia	*sissoo*		
Eucalyptus	*tereticornis* (Sm.)	W	1
Melia	*azedarach* (L.)		
Samanea	*saman*		
Sesbania	*bispinosa* (Jacq.) W.F. Wight	W	
Sesbania	*sesbana* (L.) Merr.	W	

* Species known for their difference in the level of salt tolerance between different provenances.

Fagg and Stewart (1994) give a number of *Prosopis* and *Acacia* species that are salt tolerant and are well suited for afforestration of salt-affected soils in arid regions with the furrow planting method. For *Prosopis* these species are : *Prosopis juliflora, P. glandulosa, P. tamarugo, P. cineraria, P. pallida* and *P. farcta.* For *Acacia* spp. these are: *Acacia nilotica, A. tortilis, A. ampliceps, A. maconochieana* L. *Pedley, A. stenophylla A. Cunn.* ex Benth. and *A. silicina.*

Experiments by Van der Moezel *et al.* (1991) in Australia on seedlings of *Eucalyptus* and *Melaleuca* showed that the most tolerant species under salt and water logging conditions were : *E. occidentalis, E. sargentii, E. spathulata, E. striaticalyx, E. tereticornis, Melaleuca lateriflora, M.* sp. aff. *lanceolata* and *M. thyoides.* In this experiement salt concentration increased every week by 7 dS/m until a concentration of 35 dS/m was reached. Under these well-drained salt conditions, survival rate of all species remained high, but a significant growth reduction took place. The species which grew best under these adverse conditions were : *E. yilgarnensis, E. micranthera, E. myriadena, Melaleuca adnata, M. eleuterostachya, M. lateriflora, M.* sp. aff. *lanceolata* and *M. thyoides.*

Diverse Agroforestry Systems

Singh and Dagar (1998) have described different agroforestry systems depending upon diverse topographic, climatic and water resources situations. These systems are suitable for agroforestry on community lands, commercial farms, reclaimed alkali soils or rainfed saline lands. A brief summary is presented in Table 12.11 about these various systems of agroforestry.

Table 12.11. Most promising agroforestry systems for various situations

Agroforestry systems	Major components	Important species/ compatible combinations
(A) Community Lands (Alkali soils)		
(i) Silvipastoral model	Trees and grasses for fire-wood and forage produc-tion and soil ameliotation	*Prosopis juliflora, Acacia nilotica, Casuarina equise-lifolia, Leptochloea fusca* (Kallar grass) *Chloris gay-ana* (Rhodes grass) *Brach-iaria mutica* (Para grass)
(ii) Silvi-agricultural model	Trees and crops for fuel food and soil improvement	*P. juliflora, A. nilotica, Albizia lebbek,* Wheat and oat crops.
(iii) Silvi-horti-pasture model	Fruit crops, grasses and trees for fuel, fodder and fruits.	Guava + Eucalyptus + Sababul, Tomato, Cabbage and Bottle guard as inter crops.
(B) Commercial Farms/ Alkali Soils	Trees and inter-crops with suitable amendments adopting auger-hole technique	Poplar (*Populus deltoides*) Eucalypts (*E. tereticurnis)* Babul (*A. nilotica*) Rice-wheat, Rice-Berseen, Sorg-hum-mustard rotations.
(C) Reclaimed Alkali Soils		
(i) Agri-Silvicultural model	Trees and inter-crops	*Eucalyptus,* Poplar, Teak, *Dalbergia sissoo,* Rice-Berseem, Rice-wheat
(ii) Sequential Agroforestry	Trees and arable crops in sequence	Short duration, N_2-fixing trees/crops, *Sesbania Laucaena* and *Cajanus*
(D) Rainfed Saline Lands	Trees and grasses in arid and semi-arid areas	*Prosopis opuntia, P. cineraria, C. ciliaris, P. cineraria, millets, A. albida-* Millets.

Productivity of Plantations

The ultimate aim of any agroforesry programme is to get maximum biomass per unit time. Estimates on biomass production for plantations of 8 to 10 years in alkali soils followed the order:

Prosopis juliflora — Acacia nilotica — Casuarina equselifolia — Eucalyptus tereticarus (Singh and Singh, 1993). Some *sesbania* species can yield 120-150 Mg ha^{-1} of biomass in a relatively small span of 3 years when planted on amended alkali soils.

In silvipastoral model, *P. juliflora* and Kallar grass has been found most promising for firewood and forage production as well as for soil amelioration. Kallar grass in association with mesquite produced 46.5 Mg green fodder ha^{-1} in 50 cuttings. Mesquite produced 160 Mg ha^{-1} air dried firewood in 6 years, when planted at 2 x 2 m spacing. Singh and Dagar (1998) have also described the benefits of various agroforestry systems in terms of fuel, fodder and grain production, as well as on soil properties and economic gains.

Considering the vast scope that salt-affected soils offer for agroforestry, a number of tree species have been evaluated for their tolerance to sodic and saline conditions at CSSRI, Karnal (Singh *et al.*, 1993). Among these, the most tolerant species for alkali soils identified are: *Prosopis juliflora, Acacia nilotica, Casurina equisetifolia, Terminalia arjuna, Pongamia pinnata* and *Tamarix articulata.* The moderately tolerant species are: *Albizzia lebbek, Eucalyptus tereticornis, Sesbania sesban, Sesbania aegyptica* and *Dalbergia sissoo.* While in alkali soils, adaptation is aided by the mode of site preparation; under saline conditions, tolerance is governed by the ability of trees to withstand salt stress and or waterlogging. In areas where soil salinity is not accompanied by a shallow water table, species like *Acacia auriculiformis, Terminalia arjuna* and *Leucaena leucocephala* showed good promise. When high salinity and shallow water table co-existed *Casuarina equisetifolia, Prosopis juliflora* and *Tamarix articulata* performed better than the former group.

Tree-Crop Interactions and Trees Compatibility

When plants are grown as mixed communities, their companion components interact with each other for sharing of resources of growth medium. Such interactions can be complementary, supplementary or antagonistic. Kumar *et al.* (1990–91) studied the effect of 12 tree species on selected intercrops of berseem, sorghum, pearl millet and maize in a reclaimed alkali soil and found great variations in yields under different canopies (Table 12.12). Under all tree canopies berseem yield was least affected, followed by that of maize. Yields of almost all the intercrops were less affected under *Casuarina* and *Eucalyptus* canopies compared with *Acacia* and *Sesbania* canopies, probably owing to less canopy spread and straight posture. Singh *et al.* (1989) studied the effect of different spacings and pruning schedules on growth performance of *Prosopis juliflora* and on inter-planted Karnal grass (*Leptochloa fusca*). The maximum tree biomass (161 Mg ha^{-1}) was obtained when planted in 2 m x 2 m spacing and minimum (61 Mg ha^{-1}) in 4 m x 4 m spacing. However, maximum grass yield (81 Mg ha^{-1}) was harvested under 4 m x 4 m tree spacing, as probably more light was available to the grass in this treatment (Table 12.13). Pruning of side branches of *Prosopis* during dormant period helps the tree grow better and also improves the forage yield of inter-planted grass. When large canopy-forming trees like *Prosopis* and *Acacia* are pruned of side branches, light penetrates through the cavities formed owing to their removal and increases productivity of under-storey vegetation.

Table 12.12. Fodder yield of intercrops under tree canopies in a reclaimed alkali soil (Kumar *et al.,* 1990)

Tree species	Berseem	Sorghum (Mg ha^{-1})	Pearl millet	Mean
Acacia salicina	95.3	22.4	10.7	30.1
Acacia stenophylla	64.9	7.6	5.4	22.3
Acacia machanochiana	81.4	16.6	13.3	18.6
Acacia saligna	94.4	22.4	10.7	23.1
Acacia ampliceps	57.5	8.1	8.9	29.1
Acacia vicotoriae	54.0	9.3	9.5	24.0
Acacia hemignosta	58.9	12.9	8.2	27.5
Casuarina glauca	92.5	19.3	14.0	21.1
Casuarina obesa	108.6	24.0	27.1	24.0
Eucalyptus microtheca	113.0	26.7	22.9	24.7
Eucalyptus camaldulensis	108.9	23.3	22.9	30.3
Sesbania formosa	69.7	11.4	9.6	13.4
Without tree canopy	127.8	32.6	27.8	34.7

Table 12.13. Performance of *Prosopis* with or without *Leptochloa* grass 6 years after planting (Singh, 1995)

Treatment	Height (m)	GBH (cm)	Biomass (Mg ha^{-1})	
			Lopped	Harvested
Prosopis alone	7.1	23.3	4.1	33.6
Prosopis + Leptochloa grass	6.5	20.3	2.9	28.0

The ultimate aim of any agroforestry programme is to get maximum biomass per unit time. Estimates on biomass production for plantations of 8 to 10 years in alkali soils followed the order : *Prosopis juliflora* > *Acacia nilotica* > *Casuarina equisetifolia* > *Eucalyptus tereticornis* (Singh and Singh, 1993). Some *Sesbania* species can yield 120-150 Mg ha^{-1} of biomass in a relatively short span of 3 years when planted on amended alkali soils (Rao *et al.,* 1990).

Hoever, most biosaline trees are not true halophytes and they produce reasonably well in the salinity range of 15-25 dS/m. (the salinity of seawater is about 40 dS/m). To prevent salt accumulation in the root zone of trees, effective salinity management is needed, which include irrigation and drainage. This, of course, is a capital intensive management system which needs to be very productive in order to cover the costs. Hoek *et al.* (2004) suggest that a maximum production level of 15 t/ha is considered realistic for salinity levels in the range 10-20 dS/m (i.e. brackish water) and that even higher potential production levels can be obtained with tree improvements (selection and breeding) and by appropriate salinity management.

Experimental evidences accumulated so far have identified agroforestry as an economically viable land use system for the utilisation and improvement of salt-affected soils, especially the alkali soils of the Indo-Gangetic Plains. In many salt-affected soils, where good quality irrigation water is not available, afforestration with hardy species is the only practical solution for their proper utilisation. Moreover, agroforetry is an alterative to check wide spread water-logging and salinity in canal command areas. Agro-forestry should not only be viewed as a source of wood/fuel, fodder and food production system, but also as an alternative form of biological reclamation of salt affected soils (Singh and Dagar, 1998).

Growing Fruit Trees

Since most of the fruit plants were considered sensitive to salt stress, little efforts were made in the past to standardize techniques for raising fruits in salt lands. Singh *et al.* (1993) screened large number of fruit trees in pots for sodicity tolerance and observed sapota (*Achras japota*) to be quite tolerant, followed by *Karaunda* (*Cassia carandas*). The chemical composition of plant parts of sapota indicated that it was able to maintain its K uptake even at pH_2 10.0 which was not observed with other species. Singh *et al.* (1993) successfully grew fruits especially amla by either changing sodic soil of the pit (1 m^3) with good normal soil or treating the pit soil with amendments. However, making large size pits manually in sodic soils having $CaCO_3$

pans in the subsurface layers is quite laborious, costly and time consuming and hence limit its wider applicability. Recently Singh *et al.* (1997) compared pit method (90 cm^3 pits) with pit auger-hole method (45 cm^3 pits with auge-bores of 20-25 cm diameter and 160-180 cm depth) for growing 10 fruit species in an alkali soil of pH_2 10.5. The amendment dose for pit auger-hole technique was 50 per cent of the dose used in pits. Growth observations recorded 26 months after planting showed that survival, height and girth growth of all the species remained unaffected owing to planting methods and amendments used. Irrespective of site preparation techniques, *jamun, guava, ber* and *imli* gave best performance. Date palm and baelpather performed poorly. Sapota, which was growing satisfactorily, was found highly sensitive to frost. Similarly, pomegranate, which was performing exceedingly well, was found very sensitive to prolonged water stagnation during monsoon season.

BIODRAINAGE

A primary requirement for sustainable agriculture is the maintenance of a satisfactory balance of water, air and salt in the root zone favorable for plant growth. This balance can be achieved by adequate drainage. Drainage can be either natural or artificial. When natural drainage is inadequate, artificial drainage is required to increase the drainage capacity. Artificial drainage is essential to sustain irrigated agriculture. Often subsurface drainage is needed in irrigation schemes to control the rising water table and avoid water logging and salinization. Conventional subsurface drainage systems are of two types, vertical (tubewells) and horizontal (drainpipe) systems. When properly designed, installed and maintained, these systems are efficient in lowering the water table and preventing salinization of irrigated lands. However, they have two drawbacks, namely, they are costly and they generate drainage effluents, which will have to be either carefully reused or safely disposed of. An option is biodrainage, which can be less costly and more environmentally friendly. It is a combined drainage-cum disposal system. Biodrainage relies on vegetation, rather than mechanical means, to remove excess water. The driving force behind the biodrainage concept is the consumptive water use of plants. Biodrainage is economically attrac-tive because it requires only an initial investment for planting the vegetation, and when established, the system could produce economic returns by means of fodder, wood or fibre harvested.

Biodrainage may be defined as the vertical drainage of water through evapo-transpiration of strategically planted vegetation particularly deep rooted trees with the intention of lowering the water table.

Biodrainage is an ecologically attractive concept which has the merits of low cost and environmental friendliness.

However, the limitations of biodraiange are a requirement of land for tree plantations, slow and uncontrolled lowering of water table, limited evacuation of salts from the system and vulnerability of trees to highly saline conditions.

Th areas where biodrainage plantations are raised to offset the recharge water are known as discharge areas. On average about 10 per cent of land in a water logged agricultural landscape is to be marked as a discharge area.

Biodrainage certainly depress the water table underneath plantations, but in an agroforestry system the objective is to lower the water table to a safer depth well below the crop root zone in the cultivated area that surrounds the plantation.

Akram *et al.* (2009) study showed that the major constraint of the biodrainage is salt accumulation in tree plantation strips in arid and semi-arid region. Maximum soil water salinity which can be controlled by biodrainage is around 3 dSm^{-1} and sustainability may only be achieved where a salt removal mechanism is considered.

Scientific Basis of Biodrainage

In natural environments the component of the hydrological system, i.e. rainfall, evapotranspiration, change in soil-water storage and drainage, are (on average) in balance. Periods of high rainfall might temporarily result in increased drainage flows, a rise in the groundwater table and/or soil moisture storage, then over a period of about 5-10 years equilibrium is established. Vegetation plays a vital role in the evapotranspiration and soil-water storage components of this balance. When natural vegetation is cleared and replaced by crops or tree plantations, the seepage losses to the groundwater table under the new land use system are either higher or lower than under the pre-clearing situation. The increased seepage scenario prevails and the development of land for either rainfed or irrigated agriculture generally increases groundwater recharge rates; the new landscape systems ``leak"! Where this increased recharge results in shallow water tables, evaporation at the soil surface causes groundwater to move upwards through the soil, and consequently salts accumulate in the root zone, limiting plant growth.

Plantations of fast growing tree species such as Eucalyptus, when grown on previously cleared land, could also result in strongly reduced accessions to the groundwater and in the drying-up of wells and springs.

The driving force behind the biodrainage concept is the consumptive water use of plants. One of the major factors determining the sustainability of plant productivity (and thus plant water use) processes is salt balance. If the salts moving into the root zone are not either (1) taken up by the vegetation and harvested or (2) removed from the root zone by leaching, the vegetation is doomed to succumb to salinity. The deep-rooting characteristics of trees make them extremely efficient users of water. While shallow-rooted grasses and crops have limited access to underlying water tables, deep-rooted trees can access water tables down to several metres.

Principles of Planning and Design

The aim of biodrainage is to remove excess groundwater through the process of transpiration by vegetation. This is achieved by enhancing the transpiration capacity of the landscape by introducing high-water use vegetation types in large enough areas to balance recharge/discharge processes to maintain groundwater balance below the rootzone of the agriculture crops. The following issues should be considered in the development of biodrainage systems :

i. *Water balance* : Biodrainage plantations should be able to extract groundwater volumes equal to the net recharge. The water balance is to be maintained such that the water table is kept below the rootzone.

ii. *Plantation area :* The biodrainage plantation area should be kept as small as possible. Agriculture (particularly irrigated agriculture) is practised primarily to produce high-value crops. Conversion of high-value cropping land to relatively low-return forestry may be difficult. Often good quality water is in short supply while land is not a limited resource. Particularly in arid and semi-arid regions, dryland areas surrounded by irrigated land could be earmarked for tree plantations without loss of productive resources.

iii. *Salt tolerance* : Biodrainage crops need to be salt tolerant. Groundwater qualities can vary greatly spatially, normally they have a higher salinity than irrigation supplies. The water use capacity of trees and other crops decreases with increase in water salinity. In the case of Eucalyptus species, it reduces to about one-half of potential when the water salinity increases to about 8 dS/m.

iv. *Drawdown of water table :* Crops, including trees, act as biopumps; they depress the water table directly underneath plantation areas and consequently lower table in the surrounding area. The drawdown effect under trees/crops depends on the tree/crops water use, the rate of recharge in the surrounding area, the hydraulic conductivity of substrata and the depth to deeper barrier layers. Biodrainage plantings should be established in blocks or strips and spaced to keep water table levels in the irrigated farmland in between the plantings below the rootzone. The harvesting of the biodrainage plantations would need to be planned in such a manner that the ``drainage'' function is not lost.

v. *Salt balance* : The introduction of irrigation always upsets the salt balance. Although irrigation supplies often have relatively low salinities, the large volumes of water that are introduced in the landscape increase salt imports significantly. Drainage of effluent to export these salts is therefore generally considered a necessity. To achieve salt balance without conventional drainage, the irrigated crops, along with interspersed biodrainage plantings, would have to accumulate the salts introduced by irrigation, and would subsequently have to be harvested and removed from the region. This is only (potentially) achievable in situations where very low-salinity water is available to the plants.

vi. *Economic aspects :* The growing of biodrainage trees and crops requires a different operational management approach than the growing of agricultural crops. Up-front costs associated with planting and maintenance precedes the income from harvesting by many years. Some form of contract growing, based on annual payments might have to be considered to make the system acceptable to landholders.

vii. *Social acceptance* : The introduction of new crops such as tree plantations affects rural social societies. New markets might have to be developed, security arrangements differ from those for normal crops (illegal pruning or cutting for fuelwood) and fires could destroy the results of many years of labour in a single day. Active participation of local communities in the development of tree plantation-based biodrainage systems is extremely important to overcome problems and ascertain that the benefits of the biodrainage systems are reaped to the maximum extent.

Suitable Plant, Tree and Shrub Species

Selection of species for biodrainage purposes will depend on the environmental conditions for which they are planned. Salt tolerance will be an important criterion for (potentially) saline discharge environments, water use considerations will prevail in recharge control situations where salinity is of no concern and in channel seepage scenarios with low-salinity water supply.

Crop Selection

Literature on salt tolerance for agricultural crops is commonly based on Maas and Hoffman (1977). For non-agricultural tree and bush species, reliable information is more difficult to obtain. Marcar *et al.* (1995) provide detailed information on the use of 30 tree species for use on salt affected land and less detailed summary descriptions for an additional 30 species. Schulz (1994) provides comparisons for five saltbush species grown on a range of saline irrigation regimes and other authors have investigated water use of different tree species under a range of saline conditions (Slavich *et al.,* 1999; Cramer *et al.,* 1999; Morris and Collopy, 1999; Benyon *et al.,* 1999).

Eucalyptus species are generally considered to be effective for biodrainage purposes. *Eucalyptus camaldulensis* is a hardy tree that grows under a wide range of climatic conditions and soil types. Some provenances of the species tolerate saline conditions quite well. They grow fast when good quality water is available. In a study in IGNP, Rajasthan, India, five-year old subirrigated plantations (channel seepage) produced dry biomass of 185 tonnes/ha. The utilizable biomass production was 29 tonnes/ha/year.

Acacia nilotica, Dalbergia sissoo, Tecomella undulata and *Ziziphus manuritiana* are other species that have performed quite well in plantations along leaking canals in arid conditions. Species suitable for non-irrigated conditions are *Acacia tortilis, Prosopis cineraria, Prosopis juliflora* and *Parkinsonia.* Poplar and tamarix trees are also reported to perform well for biodrainage (Bhutta and Chaudhry, 2000).

Tree Cropping Systems Management

Selection of species for biodrainage purposes will have to be based on currently prevailing and expected future environmental and economic conditions, rather than natural conditions as they existed before agricultural clearing and development. For example, in regions where water tables were deep before agricultural development, clearing and the introduction of irrigation have often resulted in the formation of shallow water tables. Under this scenario tree species selected for biodrainage should be able to survive and grow in shallow-water table environments, rather than in the pre-clearing deep-water table environment.

The planting of crops between widely spaced trees rows can be considered, especially at an early stage before competition for light becomes an issue. Crops such as wheat, mustard, lentil, berseem and gram or pasture species can be grown. However, the main objective of biodrainage plantations is the prevention of waterlogging and salinity management. The use of plantation plots for the growing of inter-crops may divert landholders attention from the trees in favour of the crops and for this reason inter-cropping is not recommended until the water table and salinity conditions are fully and finally stabilized. In general it is recommended to assign a certain function to a certain area to avoid conflict of interests negatively affecting the biodrainage function.

Tree Water Use

Often fairly dependable data are available on the rate of evapo-transpiration from standard crops (ET_0) and the rate of evaporation from a free water surface in a pan of standard size such as a Class A pan which is generally 1.15 to 1.20 times of Et_0 (Allen *et al.,* 1998).

Tree plantations often use water at higher rates than shorter vegetation types. This is for three reasons : (1) the high aerodynamic roughness of forests leads to greatly enhanced evaporation rates, which on an annual basis can be as much as twice that for grass : (2) this effect may be even more pronounced because of the so-called clothesline effect prevailing in rows of trees, substituting for a conventional drain pipe; (3) deep root system of trees with access to good-quality groundwater leads to high annual transpiration rates. The water use by tree plantations is not less than 1.5 times that of agriculture crops or about 1.25 times of Class A pan. A conservative tree water use figure of 1.0 Class A pan could therefore be assumed to estimate the potential biodraining capacity of tree plantations under conditions of good water quality. For design purposes this must be corrected for future estimated salinity levels.

In the Indira Gandhi Nahar Project, Rajasthan, India, a perched water table along a seeping irrigation canal resulted in the development of pools in the borrow pits. Plantations of (mainly) *Eucalyptus camaldulensis* and *Acacia nilotica*, established around the waterlogged areas, transpired enough water to lower the water table by 15 m over a period of 6-7 years. Open pits excavated in the plantations down to a depth of 10 m, showed that tree roots were extending at least to that depth.

CONCLUSIONS

Although the term biodrainage is relatively recent, the concept is not new. The drainage of waterlogged land through the introduction or modification of vegetation has been practised historically in different parts of the world. The increasing concern about the environmental impact of intensive agricultural development and associated monocultures has resulted in increased interest in the use of tree and salt-tolerant crops as an integrated part of the landscape and farming on saline land.

Waterlogging: Biodrainage can effectively contribute to strongly reducing the problems as experienced from waterlogging in irrigated agriculture and non-irrigated agriculture.

Salinization : The problems associated with a rise in salinity in the root zone can be effectively delayed using biodrainage systems in semi-arid and arid areas.

Phytoremediation of Salinity by Hyperaccumulator Plants

In recent years, considerable diverse research efforts have been made in the use of plants to remove inorganic, organic and heavy metal contaminants from saline soils (Kumar, 1996; Khan *et al.,* 2000, Suresh and Ravishankar, 2004). Therefore, several studies have been undertaken on phytoremediation by various investigators with a view to develop a sustainable technology for desalinization of saline areas by the use of various salt hyperaccumulator plants. The available information which is scanty, so far, has been summarized here.

Phytoremediation is envisaged as a benign and cost effective plant based green technology that depends on the remarkable ability of some plants to extract and accumulate the toxic metals or ions in the above ground shoots of certain plants or remove or neutralize various other organic chemicals from soil or aqueous environment (Salt *et al.,* 1998, Prasad, 2004). Phytoremediation technique is less costly, non-invasive and is an acceptable way to redress the removal of environmental contaminants (Schwitzguebel, 2004). According to Salt *et al.* (1998), phytoextraction is a phytoremediation technique that exploits the unique ability of some plants to accumulate unusually high amount of chemicals/toxic metals including salt ions in their harvestable tissues. Although more emphasis is on phytoextraction which is one of the sub-sets of phytoremediation. However, there are few research studies with respect to phytoextraction of salts/heavy metals (Elizabeth, 2005).

Ash content is reported to increase in most of the salt hyper accumulators with enhancement of soil salinity. Joshi (1992) advocated that ash content in *Salicornia branchiata* mainly consisted of Na^+ and Cl^- and it had direct correlation with increased salt accumulation. Likewise, *Suaeda nudiflora* and *Salicornia brachiata* have high ash content as a result of excessive accumulation of salts. Miyamoto *et al.* (1996) also reported higher ash content with salinity in *Atriplex nummularia, Distichlis palnaeri* and *Suaeda esteroa.*

Increase in salinity is known to enhance the Na^+ and Cl^- content in both shoots and roots of many halophytes with decrease in Ca^{2+}, K^+ and Mg^{2+} contents (Khan *et al.,* 2000). Maggio *et al.* (2000) revealed that most of the Na^+ concentration was centered in the leaves of *Salvodora persica* under salinity and it was 40 fold greater than the non-salinized controls with significant reduction in K^+ and Ca^{2+} concentrations. Devi *et al.* (2008) demonstrated similar results in ten hyperaccumulators under Cl^- dominated salinity (Table 12.14).

Table 12.14. Effect of 8 and 16 dSm^{-1} of Cl-dominated salinity on phytoaccumulation of salt ions and decrease in soil ECe per year in 10 salt hyperaccumulator plants (Devi *et al.,* 2008).

Species	Phytoaccumulation of salt ions in above groun biomass of plants (g m^{-2})		Mean	Total salt ion content left in soil after harvesting of salt hypera-ccumulator plants (g kg^{-1} soil)		Mean	Decrease in soil ECe per year (dSm^{-1}) in a microplot		Mean
	8 dSm^{-1}	16 dSm^{-1}		8 dSm^{-1}	16 dSm^{-1}		8 dSm^{-1}	16 dSm^{-1}	
Arundo donax	693.55	2181.58	1437.57	6.38	15.76	11.07	2.24	2.40	2.32
Atriplex amnicola	1275.92	3783,26	2519.74	6.23	15.16	10.70	2.35	2.44	2.39
Atriplex lentiformis	1372.12	4045.03	2708.57	6.48	15.53	11.00	2.64	2.52	2.58

Atriplex nummularia	881.74	1753.33	1317.54	6.48	15.65	11.06	2.73	2.65	2.69
Haloxylon recurvum	1207.70	4140.08	2473.89	5.37	15.57	10.47	3.13	2.31	2.72
Heliotropium eichwaldi	1119.73	3228.10	2173.92	6.52	15.87	11.20	2.23	2.31	2.27
Portulaca oleracea	1936.45	5963.21	3949.84	6.46	15.54	11.00	2.50	2.65	2.57
Salsola baryosma	1519.78	4451.53	2985.66	6.35	15.26	10.81	2.19	2.35	2.27
Suaeda fruticosa	2181.55	4984.03	3976.42	6.18	15.20	10.69	2.55	2.86	2.70
Suaeda nudiflora	2180.11	6759.72	4469.92	6.20	15.32	10.76	2.35	2.35	2.35
Mean	1354.90	4128.99	5483.89	6.29	15.49	10.88	2.49	2.48	2.48
CD at 5% leve S = Species T = Treatment	S = 0.126, T = 0.174 & S x T = 0.210			S = 0.19, T = 0.32 & S x T = 0.52			S = 0.012 T = 0.014 & S x T = 0.026		

The time taken for phytoextraction of salts by these salt hyperaccumulators to bring down soil salinity up to the safe critical limit of 2 dSm^{-1} was computed (Devi *et al.*, 2008). It was least (1.66 to 2.21 years) by plants like *Haloxylon recurvum, Atriplex nummularia, Atriplex lentiformis, Suaeda fruticosa, Portulaca oleracea, Suaeda nudiflora* and *Atriplex amnicola* at 8 dSm^{-1} of salinity. On the other hand at 16 dSm^{-1} of salinity it was phytoremediated to a safe limit in 4.89 to 5.95 years by plants like *Suaeda fruticosa, Portulaca oleracea, Atriplex nummularia, Arundo donax, Atriplex lentiformis, Atriplex amnicola* and *Suaeda nudiflora.* Conversely, more number of years (6.06 years) were taken to remediate soil salinity by *Haloxylon recurvum* and *Heliotropium eichwaldi* (Table 12.15).

Estimates showed that best salt hyperaccumulators were in the order of *Suaeda fruticosa, Suaeda nudiflora, Portulaca oleracea, Atriplex amnicola, Atriplex lentiformis* and *Haloxylon recurvumi* which could phytoextract 32.72, 32.70, 29.04, 20.63, 20.58 and 18.11 kg salts ha^{-1} $year^{-1}$ at 8 dSm^{-1} and *Suaeda nudiflora, Portulaca oleracea, Suaeda fruticosa, Salsola baryosma, Haloxylon recurvum* and *Atriplex lentiformis* could accumulate 101.39, 89.44, 74.76, 66.77, 62.10 and 60.67 kg toxic salts ha^{-1} $year^{-1}$ in their above ground biomass at 16 dSm^{-1} of salinity, respectively. It is deducted that time taken for phytoremediation of soil salinity up to safe level of 2 dSm^{-1}

was 1.66 to 2.37 years for 8 dSm^{-1} and it was 4.89 to 6 years for 16 dSm^{-1} of soil salinity (Devi *et al.*, 2008).

In view of the above phytoremediation of salts by hyperaccumulator plants is an interesting possibility and there is a need to extend such studies to field conditions under saline environments.

Table 12.15. Effect of 8 and 16 dSm^{-1} of Cl-dominated salinity on time taken for phytoremediation of soil salinity and percent remediation of 10 salt hyperaccumulator plants (Devi *et al.*, 2008).

Species	Time taken calculated (years) for phytoremediation of soil salinity up to safe limit of 2 dSm^{-1}		Mean	Percent remediation per year		Mean
	8 dSm^{-1}	16 dSm^{-1}		8 dSm^{-1}	16 dSm^{-1}	
Arundo donax	2.32	5.36	3.84	18.60	15.80	17.20
Atriplex amnicola	2.21	5.73	3.97	16.38	20.33	18.35
Atriplex lentiformis	1.96	5.55	3.75	18.76	20.53	19.64
Atriplex nummularia	1.90	5.28	3.59	20.08	22.39	21.23
Haloxylon recurvum	1.66	6.06	3.86	14.96	17.13	16.04
Heliotropium eichwaldi	2.33	6.06	4.19	13.60	13.78	13.69
Portulaca oleracea	2.08	5.28	3.68	14.72	18.56	16.64
Salsola baryosma	2.37	5.97	4.17	16.77	14.65	15.71
Suaeda fruticosa	2.03	4.89	3.46	15.28	13.26	14.27
Suaeda nudiflora	2.21	5.95	4.08	15.12	17.35	16.23
Mean	2.10	5.61	3.86	14.41	17.54	16.90
CD at 5% leve	S = 0.09, T = 0.08 & S x T = 0.13			S = 0.037, T = 0.082 & S x T = 0.117		

S = Species
T = Treatment

Concluding Remarks and Future Research Needs

Salt affected lands under the possession of government or village communities which are not suitable for agriculture can be put to other uses such as for growing trees, grasses, medicinal and aromatic plants either alone or by adopting agroforestry system. However, this will need creation of favourable conditions for better root proliferation and establishment of tree saplings. Technologies for making auger hole to pierce through hard pan of calcium carbonate in alkali soils alongwith amelioration of auger hole soil have been widely adopted. Likewise, surface furrow planting and irrigation methods help to maintain salt and water balance in waterlogged saline soils and use of saline irrigation water is arid regions. The tree-based agroforestry land use systems can also contribute towards minimisation of the secondary salinisation due to higher evapotranspiration demands. More concentrated efforts are, however, required for choice of species, planting density, site preparation and other practices alongwith the cost: benefit ratio of such systems.

Furthermore, there is an urgent need of strengthing research on developing alternative land use on salt affected soils owned by village communities, forest and other government departments which are still lying barren. Programmes are needed for screening and conservation of germplasm of trees, identification of compatible tree-crop combinations, suitable agro-techniques for growing higher value crops and their economic viability. Long-term studies on the comparison of agricultural, silvicultural, horticultural, silvipastoral and hortipastrol systems of landuse in terms of productivity, economic benefits, soil health and the ability to sequester green house gases have to be taken up on priority. In the agroforestry systems, nitrogen-fixing trees should be preferred for soil enhancement of nitrogen and efforts to be made to transfer fixed N to companion crops.

REFERENCES

Abrol, I.P. and Joshi, P.K. 1984. Economic viability of reclamation of alkali lands with special reference to agriculture and forestry. Proceedings of Seminar on Economics of Wastelands Development, New Delhi, pp. 19-30.

Akram, S., Kashkouli, H.A. and Pazira, E. 2008. Sensitive variables controlling salinity and water table in a bio-drainage system. Irrig. Drainage Syst. 22: 271-285.

Chaturvedi, A.N. 1984. Fire wood crops in area of brakish water. Indian Forester 110: 364-365.

Dagar, J.C. 2011. Bio-drainage for amelioration of water logging. In: Salinity Management for Sustainable Agriculture in Cannal Command. (Dey and Gupta, eds.) CADA, Haryana and CSSRI, Karnal, 54-58.

Dagar, J.C., Singh, G. and Singh, N.T. 2001. Evaluation of forest and fruit trees used for rehabilitation of semi arid Alkali-soils in India. Arid Soil Research and Management 15 : 115-133.

Devi, S., Rani, C., Datta, K.S., Bishnoi, S.K., Mahala, S.C. and Angrish R. 2008. Phytoremediation of soil salinity using salt hyperaccumulator plants. Indian J. Plant Physiol. 347-356.

Elizabeth, P.S. 2005. Phytoremediation. Annu. Rev. Plant Biol. 56 : 15-39.

Fagg, C.W., J.L. Stewart 1994. The value of *Acacia* and *Prosopis* in arid and semi-arid environments, Journal of Arid Environment, 27 : 3-25.

Gill, H.S. and Abrol, I.P. 1991. The utilisation of salt affected soils and water resources for tree based land management systems in India. Proceedings of Symposium on Trees as a Tool for Resource Management in Irrigation Areas, p. 74-82. Institute for Sustainable Agriculture, Tatura, Australia.

Gill, H.S., Abrol, I.P. and Gupta, R.K. 1990. Afforestation of salt affected soils. In : Technologies for Wastelands Development, pp. 355-80. ICAR, New Delhi.

Gill, H.S., Abrol, I.P. and Samra, J.S. 1987. Nutrient recycling through litter production in young plantations of *Acacia nilotica* and *Eucalyptus tereticornis* in a highly alkaline soil. Forest Ecol. Mgmt. 22 : 57-69.

Gupta, G.N. and Premlal 1993. Afforestation of salt affected lands. In : Management of Salt-affected Soils and Waters. (Eds. L.L. Somani and K.L. Totawat) pp. 289-320. Agrotech Publishing Academy, Udaipur.

Gupta, G.N., Prasad, K.G. and Mohan, S. 1991. Effect of salinity, alkalinity and mulch on tree growth in dry region. Ann. Arid Zone 30 : 29-36.

Gupta, G.N., Prasad, K.G., Mohan, S. and Manivachakam, P. 1986. Salt tolerance in some tree species at seedlings stage. Indian Forester 112 : 101-113.

Gupta, Raj. K., Tomar, O.S. and Minhas, P.S. 1995. Managing salt affected soils and waters for afforestation. p. 23. Bull. 7/95, CSSRI, Karnal.

Jain, B.L. 1984. Saline water use in agro-forestry. In : Agro-forestry in Arid and Semi-arid Zones. (Ed. K.A. Shankarnarayan) pp. 232-237, CAZRI, Jodhpur.

Jain, B.L. 1986. Afforestation and pasture development on salt affected lands. In : Proc. Summer Institute on Wastelands in Arid and Semi-arid Zone and Technologies for their Utilization, CAZRI, Jodhpur pp. 277-287.

Jain, B.L., Goyal, R.S. and Muthana, K.D. 1983. Performance of some tree species in relation to irrigation with saline waters. Ann. Arid Zone. 22: 233-238.

Khan, M.A., Ungar, I.A. and Shawalter, A.M. 2000. The effect of salinity on the growth, water status and ion content of a leaf succulent perennial halophyte *Suaeda fruticosa* (L.) Forssk. J. Arid Environ. 45 : 73-84.

Kumar, A. 1996. Use of *Leptochloa fusca* for the improvement of salt affected soils. Expl. Agric. 32 : 143-149.

Lambert, M. and J. Turner 2000. Commercial forest plantations on saline lands, CSIRO, Collingwood, 198 p.

Maggio, A., Reddy, M.P. and Joly, R.J. 2000. Leaf gas exchange and salt accumulation in the halophyte *Salvadora persica* grown at moderate salinity. Environ. Expt. Bot. 44 : 31-38.

Marcar, N.E. and Crawford, D.F. 1996. Tree growing strategies for the productive use of saline land, Australian Journal of Soil and Water Conservation, 9 : 34-40.

Mathur, K.B.L. 1986. Afforestation and pasture development on wastelands in arid and semi-arid zones. CAZRI, Jodhpur.

Minhas, P.S., Singh, Y.P., Tomar, O.S., Gupta, R.K. and Gupta, Raj K. 1997. Saline water irrigation for the establishment of furrow planted trees in north western India. Agroforestry Systems 35 : 177-186.

Mishra, A., Sharma, S.D. and Gupta, M.K. 2003. Soil rehabilitation through afforestation: Evaluation of the performance of *Prosopis juliflora, Dalbergia sissoo* and *Eucalyptus tereticornis* plantations in sodic environment. Arid Soil Research and Management 17 : 257-269.

Mishra, A., Sharma, S.D. and Pandey, R. 2004. Amelioration of degraded sodic soils by aforestation. Arid Soil Research and Management 18 : 13-23.

Miyamoto, S., Glenn, E.P. and Olsen, M.W. 1996. Growth, water use and salt uptake of four halophytes irrigated with highly saline water. Journal of Arid Environ. 32 : 141-159.

Prasad, M.N.V. 2004. Phytoremediation of metals in the environment for sustainable development. Proc. Indian Natl. Sci. Acad. 370 : 71-98.

Qureshi, R.H., E.G. Barret-Lennard, 1998. Saline agriculture for irrigated land in Pakistan: A handbook, Canbarra: Australian Centre for International Agricultural Research, VI, 142 p.

Rao, D.L.N., Gill, H.S. and Abrol, I.P. 1986. Nitrogen fixation and biomass production by *Sesbania sesban* in alkali soils. In : Amelioration of Soil by Trees, pp. 74-79. Prinsley, R.T. and Swift, M.J. (Eds.) Commonwealth Science Council, London.

Schofeld, N.J. 1992. Tree planting for dryland salinity culture in Australia. Agrofor. Systems. 20: 1-23.

Shankarnarayan, K.A. and Kolarkar, A.S. 1988. Some promising tree species suitable for afforestation in salt affected soils of Indian arid zone. International Symposium on Salt Affected Soils. CSSRI, Karnal 2 : 77-93.

Sharma, R.C. 1996. Comments on the repot on status of land degradation in India. Dept. of Soils and Water Conservation. Govt. of India, New Delhi, 6 p.

Singh, G. and Dagar, J.C. 1993. Agroforestry in salt affected soils. In: Agricultural Salinity Management in India (Eds. N.K. Tyagi and P.S. Minhas) pp. 473-487. CSSRI, Karnal.

Singh, G.B. and Gill, H.S. 1990. Raising trees in alkali soils. Wastelands News 6 : 15-8.

Singh, G.B., Abrol, I.P. and Cheema, S.S. 1989. Agroforestry system on an Alkaline soil. Effects of spacing and lopping on mesquite (*Prosopis juliflora)* - Karnal grass (*Leptochloa fusca*). Exp. Agric. 25 : 401-8.

Singh, G.B., Singh, N.T. and Tomar, O.S. 1993. Agroforestry in salt-affected soils. p. 65, *Res. Bull.* 17 No., CSSRI, Karnal.

Singh, G.G., Singh, N.T. and Tomar, O.S. 1993. Agroforestry in salt affected soils. Bull. No. 17, CSSRI, Karnal, 65 p.

Singh, Gurbachan 1995. An agroforestry practice for the development of salt lands using *P. juliflora* and *Leptochola fusca*. Agrefor. Systems. 29:61-75.

Singh, Gurbachan and Felkar, P. 1995. Edible cactus: A new hope for wasteland development. Wastelands News, 11: 58-61.

Singh, Gurbachan, Abrol, I.P. and Cheema, S.S. 1989. Effect of spacing and lopping on mesqute-karnal grass agroforestry system in a sodic soil. Exp. Agric. 25: 405-408.

Singh, Gurbachan, Dagar, J.C. and Singh, N.T. 1997. Growing fruit trees in a highly alkali soil. A case study. Land Degrad. Dev. 8: 257-268.

Suresh, B. and Ravisankar, G.A. 2004. Phytoremediation - A novel and promising approach for environmental clean up. Critic. Rev. Biotech. 24 : 97-124.

Szabolcs, I. 1992. Salt affected soils as the ecosystem for halophytes, International workshop on halophytes for reclamation of saline wastelands and as resource for livestock, Nairobi, Kenya, Nov. 22-27.

Tomar, O.S. 1997. Technologies of afforestation of salt-affected soils. International Tree Crops Journal 9: 131-158.

Tomar, O.S. and Gupta, R.K. 1985. Performance of some forest tree species in saline soils under shallow and saline water table conditions. Plant Soil. 87: 329-335.

Tomar, O.S. and Minhas, P.S. 1998. Afforestation of salt affected soils. In: Agricultural Salinity Management in India (Eds. N.K. Tyagi and P.S. Minhas) pp. 453-471, CSSRI, Karnal.

Tomar, O.S. and Yadav, J.S.P. 1980. Effect of saline irrigation water of varying EC, SAR and RSC levels on germination and seedling growth of some forest species. Indian J. Forestry 3: 306-314.

Tomar, O.S., Gupta, R.K. and Dagar, J.C. 1998. Afforestation techniques and evaluation of different trees species for waterlogged saline soils in semi arid tropics. Arid Soil Research and Rehabilitation 12 : 301-316.

Tomar, O.S., P.S. Minhas, V.K. Sharma, Y.P. Singh and Raj K. Gupta 2003. Performance of 31 tree species and soil conditions in a plantation established with saline irrigation. Forest Ecology and Management, 177 : 333-346.

Yadav, R.B. and Prakash, O. 1990. Soil sodicity and salinity tolerance in some forest tree species. National Seminar on Technology for Afforestation of Wastelands. FRI, Dehradun.

■■■■

Annexure - 1

SALINITY MEASURES, UNITS AND CLASSES

Most salinity measures use electrical conductivity to estimate salinity of soil and water. These measures are cheap and easy to do, and can even be done (with some care) in the field. To understand the `salinites' of a salinity reading, it makes a big difference to measure soil or water properties, the way in which it is measured (or estimated in most cases), and the units which are used. For example, moderately salty soil could be presented as 600 mS/m (ECe), 125 mS/m (EM-38 hor), or 60 mS/m (EC 1:5 (w/v) loam).

- Salinity measures include electrical conductivity of a solution or soil and water mix, weight of salts in a given amount of water, and the quantity of molecules of salts in a solution. Each of these measures have particular uses. A popular measuring tool is the EM 38 (Electro Magnetic Induction Metre).
- Units includes grains per gallon, milligrams per litre, millimens per meter, deci Siemens per centimetre and many more.
- Classes range from non-saline to extremely saline. In most of the international publications, the salinity classes refer to plant tolerances to salinity.

Salt tolerance of plants is affected too by factors other than the salinity reading : e.g. by waterlogging, soil type, sodicity, depth to water table, salinity of the water table, rainfall and other factors. Waterlogging interacting with salinity is probably the most important influencing factor on most sites.

Soil Salinity

EC 1 : 5

Soil samples can be measured by the 1:5 w/v method : one part by weight (g) air dried soil to five parts by volume (ml) distilled water, which is agitated then allowed to settle. The solution is measured for Electrical Conductivity (EC 1 : 5). This measure is used to allow for soil texture differences. Sand particles will not hold as much salt from the soil water as will clay. Therefore, sand will give apparently lower readings than from clay, even though the salt concentration in the soil water is the same*.

ECe

The Electrical Conductivity of a saturated soil extract (ECe) is the most useful and most reliable measure of salinity for comparing between soil types. The 1:5 reading can be used to estimate the ECe. Conversion factors from EC 1:5 to ECe are given in table below. Note that these are guidelines - the actual conversion figure can vary quite widely for particular soils. Note also, that proper measurement of the ECe is a laboratory technique, and an expensive one.

A technique for converting reported ECe to EC 1:5 can be used in the reverse direction for converting EC 1:5 (less than 350 mS/m) to ECe. The technique is based on a limited data set, and care should be taken for extrapolating the technique.

Conversion factors to estimate ECe from EC 1:5 (w/v) in relation to soil texture

Soil texture	Multiply EC 1 : 5 (w/v) by factor	Example
sand	15	10 mS/m (EC1:5) = 150 mS/m (ECe)
sandy loam	12	10 mS/m (EC1:5) = 120 mS/m (ECe)
loam	10	10 mS/m (EC1:5) = 100 mS/m (ECe)
clay loam	9	10 mS/m (EC1:5) = 90 mS/m (ECe)
light/medium clay	8	10 mS/m (EC1:5) = 80 mS/m (ECe)
heavy clay	6	10 mS/m (EC1:5) = 60 mS/m (ECe)

EM38

Another measure, which is increasingly becoming common, is the EM 38 or EM 31 using Electro-Magnetic induction. EM readings are useful to compare within and between similar sites, but use EM readings with caution unless they are calibrated against soil salinities (ECe for preference) and other influencing factors. The EM 38 is quite

*(Same field surveys estimate an EC 1:5 based on volume : volume (v/v). That is, one scoop of soil to five scoops of water. This can lead to large errors in reporting in comparison with the w/v method and between samples from the same site, especially for clay soils. It is important to `calibrate' any EC 1 : 5 estimated from the v/v technique against the w/v estimated technique. The technique is not recommended for critical measurements.

an expensive unit, but relatively easy to use in the field, and gives estimates of salt concentrations at the root zone.

Water Salinity

Nearly all testing methods use electrical conductivity. The results are reported in various units, although the official standard is *milli-Siemens per meter* (mS/m). Other units, which are still used are : deci-Siemens per meter (dS/m); parts per million' (ppm), which can be thought of as the same as milligrams per litre (mg/L); osmotic potential (OP); Percentage of Total Soluble Salts and grains per gallon (gpg).

Coversion factors

- mS/m x 0.01 = dS/m (deci-Siemens per metre). For soil or water measures.
- mS/m x 5.5 = mg/L = ppm (assuming sodium chloride is the only salt present; (this figure increases with other common salts and may be as high as 6). For water.
- mS/m x 0.42 = gpg (Grains per Gallon). This conversion is rough, and the factor depends on salts present). For water.
- mS/m x 0.0034 = % TSS in soil from a EC 1:5 (w:v) measure (conversion varies based on salts present). For soil.
- mS/m x 0.00055 = % TSS in water from an ECw (conversion varies based on salts present). For water.
- mS/m x 10 = mmol/dm^3 NaCl = mmol/litre NaCl. For water.

All these salinity measures are trying to relate plant growth to a certain salinity level. Some of these measures are more useful in predicting plant growth (for instance EM38), others are cheap and easy (eg EC 1:5 v/v). The ECe figures are widely accepted by research and are `standards', but the groundwater, EM38 and EC1:5 measures can vary between sites for the same apparent salinity.

Salinity classes for plants. Note these are specific to soil or water measures

Class	ECe (dS/m)	ECe (mS/m)	ECgw (mS/m)	EM-38 hor (mS/m)	NaCl (sol. mmol/ l)	EC 1:5 (w/v) loam (mS/m) approx	EC 1:5 (v/v) loam (mS/m) approx	Total Dissolved Solids (TDS %)
	Soil (a)	Soil (b)	Water (c)	`Soil' (d)	Water (e)	Soil (f)	Soil (g)	Water
Non-saline	<2	<200	<500	<50	<20	<20	<40	<0.275%
Slightly	2-4	200-400	500-1000	50-100	20-40	20-40	40-80	0.275-0.55%
Moderately	4-8	400-800	1000-2000	100-150	40-80	40-80	80-160	0.55-1.1%
Very	8-16	800-1600	2000-3000	150-200	80-160	80-160	160-320	1.1-1.65%
Extremely	>16	>1600	>3000	>200	>160	>160	>320	>1.65%

(a) Based on USDA 1954 categories

(b) Units used in Western Australia

(c) Groundwater from within potential rooting distance of plant (bores). Suitability for `tree' growth

(d) From D Bennett and R George, DAWA Bunbury

(e) `Irrigation' water used in pot trials

Soil Texture Effects on EC(w/v) Measures and Salinity Class

Because EC(w/v) measures are very commonly used to estimate salinity from field samples, an additional table is included. These measures are usually done in a laboratory.

Note that these values are estimates of salinity in the plant root zone. When salinity is measured from the surface layer, the estimate of salinity at the root zone may be incorrect. Note also, that the EC1:5 is one part soil by weight to five parts of water by volume. When the EC 1:5 volume : volume technique is used in the field, it is prone to large errors.

Soil salinity measurements using EC1:5 (w/v) in mS/m

Salinity Ratiing	Sand	Sandy loam	Loam	Clay loam	L/Med clay	Heavy Clay
Non-saline	<13	<17	<20	<22	<25	<33
Slightly	13-26	17-33	20-40	22-44	25-50	33-67
Moderately	26-52	33-67	40-80	44-89	50-100	67-133
Very	52-106	67-133	80-160	89-178	100-200	133-267
Extremely	>106	>133	>160	>178	>200	>267

Source : http://agspsrv34.agric.wa.gov.au/environment/salinity/measurement/ salinity levels for revegetation.htm

■■■■

Annexure - 2

PREPARATION OF SALINE-ALKALI SOILS

Soil salinity or/and alkalinity are one of the major constraints of crop production. In India, approx. 10 m ha land is affected by salt. Salt affected soils are mainly classified as :

(i) saline and

(ii) alkali soils.

Saline soils

Such soils contain large quantities of neutral soluble salts of sodium and calcium. The pH of saline soils is always less than 8.5, exchangeable sodium percentage less than 15 and electrical conductivity more than 4 dS m^{-1}. High concentrations of soluble salts in the soil solution lowers the osmotic potential of saline soils inducing osmotic stress to growing plants.

Alkali soils

Alkali soils are rich in carbonate and bicarbonates of sodium and calcium. They possesses large amount of sodium on the soil micelles replacing calcium in dispersed soil structure, low hydraulic conductivity and poor aeration. Alkali soils have pH more 8.5, exchangeable sodium greater than 15 per cent and electrical conductivity less than 4 dSm^{-1}. It also possesses hard Kankar pans of calcium carbonate of varying thickness below the soil surface at various depths which restricts root growth and water movement.

Yet another type of salt affected soil also exists which possesses adverse qualities of saline and alkali soils together. Such soils are difficult to handle for crop production. Physiological experiments under saline-alkaline conditions require uniform level of salinity/ alkalinity in the growing medium, which seldom prevails under field conditions. Thus, there is a need to prepare constant levels of saline/alkali soils artificially by the addition of desired quantities of specific salts to normal soil. Such artificially prepared soil could be filled either in pots or microplots for conducting physiological experiments with uniform levels of salinity and or alkalinity.

(A) Preparation of alkali soil

Alkali soil could be preapred artificially by using following formula and adding $NaHCO_3$ salt in the soil :

$$ESP = \frac{Exchangeable\ Na}{CEC} \times 100$$

Where, ESP is the exchageable sodium percent, CEC is the cation exchange capacity of the soil (me/100g).

For example, if a soil with CEC of 12 me/100 g is to be changed into alkali soil of ESP 50, then,

$$50 = \frac{Exchangeable\ Na}{12} \times 100$$

$$\text{Exchangeable Na} = \frac{50 \times 12}{100} \times 6 me/100\ g$$

Hence, 23 x 6 = 138 mg Na/100 g soil

Or 5.04 g $NaHCO_3$/Kg soil will be required for 6 me/100 g Na

Preparation of Soil of 50 ESP

Procedure

Find out the water saturation percentage of normal soil (variable for different soils). Divide 100 Kg soil into 5 equal parts. Spread one lot of soil on a thick polythene sheet. The thickness of soil layer should be less than 5 cm. Dissolve 504 g $NaHCO_3$ in 30 litres of water (soil saturation being 30 percent). Spray 6 litres of this solution on the first layer of soil. When this solution is fully absorbed by the soil, spread another layer of soil over the first layer and spray next 6 litres of solution. This process is continued till 5th layer of soil. After completion of spraying, leave this soil for 20-25 days for equilibration. At the end mix the soil thoroughly and fill such soil in pots. All these processes should be done in partial shade (Check the level of alkalinity at the end).

Preparation of Saline Soil

Artificial saline soil, like sodic soil, can also be prepared by the addition of chlorides and sulphates of sodium and calcium. Salinity may be created either by single salt or a mixture of these salts in some fixed ratios.

Procedure

The initial salinity level (electrical conductivity) and water saturation percentage of the soil being used is determined before the addition of salt. For example, the initial salinity of the soil may be 0.5 dSm^{-1} or so and saturation percentage as 30%.

The magnitude of salinity present (say 0.5 dSm^{-1}) should be subtracted from the level to be prepared. The balance level of salinity could then be prepared by the addition of desired quantity of salt. The amount of total salt concentration could be calculated by the following equation :

$$\text{Total salt concentrations (TSC)} = EC \times 1^{01.065} \text{ (me/L)}$$

In the above example with a soil having initial salinity of 0.5 dSm^{-1} and saturation percentage as 30, if we have to prepare a salinity level of 10, the amount of salt can be calculated as under :

Initial salinity = 0.5 dSm^{-1}

Desired salinity = 10.0 dS/m^{-1}

Balance salinity = 10-0.5

To be added = 9.5 dS^{-1} m

Now, TSC = $9.5 \times 10^{1.065}$

= 110.34 me/L

Thus, if only NaCl is to be used for the creation of salinity, 110.34 me/L NaCl (MW, 58.5) is to be added i.e.

$$110.34 \times 58.5 = 6454.89 \text{ mg/L} = 6.455 \text{ g/L}$$

Amount of NaCl for 30 litres of water (Soil saturation being 30%) will be 193.65 g (dissolve this amount in demineralized water). Spread 100 Kg soil on a thick polythene sheet and spray NaCl solution so as to saturate it similar to that mentioned for the prepared of alkali soil. After completion of spraying, leave this soil for few days for equilibration. Finally mix the soil and use for filling in pots or trays for experiment. Before filling in the pots check for the salinity level in the soil as per standard procedure.

Salinity can also be created using several salt mixture in order to simulate natural conditions. For example, NaCl, Na_2SO_4, $CaCl_2$, $CaSO_4$, $MgCl_2$, $MgSO_4$ can be mixed in different proportions to create a desired salinity levels.

■■■■

Annexure - 3

GUIDELINES FOR USING SALINE IRRIGATION WATER IN INDIA

A. Saline Water (RSC <2.5 me/L)

Soil texture (% clay)	Crop Tolerance	ECw (dS/m) limit for rainfall region		
		<350	350-550	>550mm
Fine (>30)	Sensitive	1.0	1.0	1.5
	Semi-tolerant	1.5	2.0	3.0
	Tolerant	2.0	3.0	4.5
Moderately fine (20-30)	Sensitive	1.5	2.0	2.5
	Semi-tolerant	2.0	3.0	4.5
	Tolerant	4.0	6.0	8.0
Moderately coarse (10-20)	Sensitive	2.0	2.5	3.0
	Semi-tolerant	4.0	6.0	8.0
	Tolerant	6.0	8.0	10.0
Coarse (<10)	Sensitive	--	3.0	3.0
	Semi-tolerant	6.0	7.5	9.0
	Tolerant	8.0	10.0	12.5

B. Sodic Waters (RSC > 2.5 me/L, EC_{iw} < 4.0 dS/m)

Soil texture (% clay)	SAR $(mmole/l)^{1/2}$	Upper limit of RSC (me/L)	Remarks
Fine (>30)	10	2.5-3.5	Limits pertain to *kharif* fallow/*Rabi* crops rotation when annual rainfall is 350-550 mm. When the waters have Na < 75% (Ca + Mg > 25%) or rainfall is > 550 mm, the upper limit of the RSC range becomes safe. For double cropping neutralization of RSC with gypsum is essential based on quantity of water used during the rabi season. Grow low water requiring crops during kharif. Avoid rice.
Moderately fine (20-30)	10	3.5-5.0	
Moderately coarse (10-20)	15	5.0-7.5	
Coarse (>10)	20	7.5-10.0	

Source : P.S. Minhas and N.K. Tyagi (1998) Guidelines for irrigation with saline and alkali waters. Bull. No. 1/98, C.S.S.R.I., Karnal, India, p. 36.

■■■■

Annexure - 4

INDIA-STATE-WISE DISTRIBUTION OF WASTELANDS

Sl.No.	State Name	TGA	Total WL	Area in Sq km % to TGA
1	Andhra Pradesh	275068	38788.22	14.10
2	Arunachal Pradesh	83743	5743.84	6.86
3	Assam	78438	8778.02	11.19
4	Bihar	94171	6841.09	7.26
5	Chattisgarh	135194	11817.82	8.74
6	Delhi	1483	83.34	5.62
7	Goa	3702	496.27	13.41
8	Gujarat	196024	21350.38	10.89
9	Haryana	44212	2347.05	5.31
10	Himachal Pradesh	55673	22470.05	40.36
11	Jammu & Kashmir	101387	73754.38	72.75
12	Jharkhand	79706	11670.14	14.64
13	Karnataka	191791	14438.12	7.53
14	Kerala	38863	2458.69	6.33
15	Madhya Pradesh	308252	40042.98	12.99
16	Maharashtra	307690	38262.81	12.44
17	Manipur	22327	7027.47	31.48
18	Meghalaya	22429	3865.76	17.24
19	Mizoram	21081	6021.14	28.56
20	Nagaland	16579	4815.18	29.04
21	Orissa	155707	16648.27	10.69
22	Punjab	50362	1019.50	2.02
23	Rajasthan	342239	93689.47	27.38
24	Sikkim	7096	3280.88	46.24

Sl.No.	State Name	TGA	Total WL	Area in Sq km % to TGA
25	Tamilnadu	130058	9125.56	7.02
26	Tripura	10486	1315.17	12.54
27	Uttarakhand	53483	12790.06	23.91
28	Uttar Pradesh	240928	10988.59	4.56
29	West Bengal	88752	1994.41	2.25
30	Union Territory	9490	337.30	3.55
	Total	3166414	472261.95	14.91

* Unsurveyed areas (J & K) : 120849.00

Total geographical area : 3287263.00

Source : 1:50,000 Wasteland Maps - 2005-06 prepared based on IRS-P6, LISS III Three season data

■■■■

Annexure - 5

INDIA - CATAGORY - WISE DISTRIBUTION OF WASTELANDS

Area in Sq. km.

S.N.	Category	Total WL	% to TGA
1	Gullied and/or ravinous land Medium	6999.03	0.22
2	Gullied and/or ravinous land- Deep-very deep ravine	1714.83	0.05
3	Land with dense scrub	93389.55	2.95
4	Land with open scrub	91633.00	2.89
5	Waterlogged and Marshy land-Permanent	2532.46	0.08
6	Waterlogged and Marshy land- Seasonal	2990.84	0.09
7	Land affected by salinity / alkalinity- Moderate	5429.83	0.17
8	Land affected by salinity/ alkalinity - Strong	1737.81	0.05
9	Shifting cultivation area - Current Jhum	5625.07	0.18
10	Shifting cultivation area - Abandoned Jhum	4608.45	0.15
11	Under utilised/ degraded forest - Scrub dominated	85809.54	2.71
12	Agricultural land inside notified forest land	16386.08	0.52
13	Degraded pastures / grazing land	7196.44	0.23
14	Degraded land under plantation crops	316.22	0.01
15	Sands - Riverine	2439.85	0.08
16	Sands - Coastal sand	719.00	0.02
17	Sands - Desert Sand	5280.07	0.17
18	Sands - Semi - stabilized to stabilized (>40 m) dune	11188.21	0.35
19	Sands - Semi- stabilized to stabilized moderately high (15-40 m) dune	15627.63	0.49
20	Mining Wastelands	505.35	0.02

S.N.	Category	Total WL	% to TGA
21	Industrial wastelands	63.99	0.00
22	Barren rocky area	69373.92	2.19
23	Snow cover and/or glacial area	40694.80	1.29
	Total	472261.95	14.91

Source: 1: 50,000 Wasteland Maps - 2005-06 prepared based on IRS-P6, LISS III Three season data.

Note: 1,20,849.00 sq.km. in Jammu & Kashmir is not mapped and hence not considered for calculating the percentage.

■■■■

SUBJECT INDEX

■■■■